APPLIED MYCOLOGY FOR AGRICULTURE AND FOODS

Industrial Applications

APPLIED MYCOLOGY FOR AGRICULTURE AND FOODS

Industrial Applications

Edited by

Sanjay K. Singh, PhD

Deepak Kumar, PhD

Md. Shamim, PhD

Rohit Sharma, PhD

AAP | APPLE ACADEMIC PRESS

First edition published 2024

Apple Academic Press Inc.
1265 Goldenrod Circle, NE,
Palm Bay, FL 32905 USA

760 Laurentian Drive, Unit 19,
Burlington, ON L7N 0A4, CANADA

CRC Press
2385 NW Executive Center Drive,
Suite 320, Boca Raton FL 33431

4 Park Square, Milton Park,
Abingdon, Oxon, OX14 4RN UK

Library and Archives Canada Cataloguing in Publication

Title: Applied mycology for agriculture and foods : industrial applications / edited by Sanjay K. Singh, PhD, Deepak Kumar, PhD, Md. Shamim, PhD, Rohit Sharma, PhD

Names: Singh, Sanjay K. (Mycologist), editor. | Kumar, Deepak (Research scientist), editor. | Shamim, Md., 1985- editor. | Sharma, Rohit (Mycologist), editor.

Description: First edition. | Includes bibliographical references and index.

Identifiers: Canadiana (print) 20230451888 | Canadiana (ebook) 20230451934 | ISBN 9781774913130 (hardcover) | ISBN 9781774913147 (softcover) | ISBN 9781003369868 (ebook)

Subjects: LCSH: Mycology. | LCSH: Fungi in agriculture. | LCSH: Fungi—Industrial applications. | LCSH: Food—Microbiology.

Classification: LCC QK603 .A67 2024 | DDC 579.5—dc23

Library of Congress Cataloging-in-Publication Data

CIP data on file with US Library of Congress

ISBN: 978-1-77491-313-0 (hbk)
ISBN: 978-1-77491-314-7 (pbk)
ISBN: 978-1-00336-986-8 (ebk)

About the Editors

Sanjay K. Singh, PhD
Scientist F and In-Charge, Biodiversity and Palaeobiology Group, MACS' Agharkar Research Institute, Pune, India

Sanjay K. Singh, PhD, is presently working as a Scientist F and In-Charge, Biodiversity and Paleobiology Group, MACS' Agharkar Research Institute (ARI), Pune. After completing his doctorate in 1996, from Gorakhpur University, Gorakhpur, UP, he continued research as a post-doc fellow for about four years. He availed post-doctorate fellowships of CSIR, New Delhi (1996), a young scientist fellowship of DST, New Delhi (1997–1999), and a research associateship of a Planning Commission-sponsored project at Goa University (2000–2001). He received training in the field of fungal biodiversity, conventional and molecular taxonomy, systematics, fungal germplasm conservation, and various other applications. In 2001, he joined MACS' Agharkar Research Institute, an autonomous institute of the Department of Science and Technology (DST), Government of India at Pune. He has a research experience of over 25 years in the field of mycology, plant pathology, *in vitro* culture conservation, and applications of fungi. He has been instrumental in establishing state-of-the-art National Facility for Culture Collection of Fungi at ARI. Under this facility National Fungal Culture Collection of India (NFCCI) was established which is an affiliate member of the World Federation for Culture Collections (WFCC). NFCCI houses over 5,000 indigenous fungal cultures which are conserved and distributed to researchers when needed. He is the curator at NFCCI and Ajrekar Mycological Herbarium (AMH) and is also serving as coordinator of ARI-Sophisticated Analytical Instrumentation Facility (SAIF). This facility (NFCCI) extends nationwide services to more than 700 academic and research institutions and over 100 industries across India under the leadership of Dr. Singh. He was selected for DBT (Government of India) associate program for young scientists, visited Montana State University, USA, and underwent training with Prof. Gary Strobel on biology of endophytes and their applications. He is an elected fellow of the Maharashtra

Academy of Sciences (Life Sciences), Indian Phytopathological Society (IPS), and Society of Applied Biotechnology (SAB). He has been an active member of the Mycological Society of India (MSI). He has delivered several invited lectures at conferences/seminars. He has operated several extramural and industry-sponsored projects, and guided students for MSc and PhD. Dr. Singh supervised an international faculty selected for CV Raman International Fellowship for African Researchers sponsored by DST, Government of India. Dr. Singh has published over 105 research papers in internationally reputed journals and contributed 14 book chapters, and authored two books. He has organized three-international and national symposia/conferences and more than 12 batches capacity building national level workshops and other courses. He has served as a PhD examiner and reviewed research papers for several prestigious journals and is a member of editorial boards. He was selected and attended International Advance Level Training on Genetic Resources and Intellectual Property (GRIP) training program sponsored by Swedish International Development Authority (SIDA), at SLU, Sweden, and in Johannesburg, South Africa.

Deepak Kumar, PhD
Executive Director (R&D), Nextnode Bioscience Pvt. Ltd., Kadi, Gujarat, India

Deepak Kumar, PhD, is presently working as an Executive Director (R&D) at Nextnode Bioscience Pvt. Ltd., Kadi, Gujarat. Previously, he worked as Director (R&D) and research scientist for five years in a DSIR-recognized (DSIR Reg. No.: TU/IV-RD/3945) laboratory of Shri Ram Solvent Extractions Pvt. Ltd., Jaspur, Uttarakhand, India. He has done his early research work on the production of soybean meal-based bioactive peptide fertilizer production through microbial fermentation and optimization of the extraction process for triterpenoids (azadirachtin and nimbin) from neem (*Azadirachta indica*) at the commercial level. He is the author or coauthor of 20 peer-reviewed journal articles, two books, one patent, and 15 book chapters. He has been involved in the quality production of biofertilizers and biopesticides for the last 10 years. He obtained his MSc (Biotechnology) from the C.C.S. University, Meerut, his MBA (Human Resource Management) from Punjab Technical University, Jalandhar, and his PhD (Ag. Biotech) from the N.D. University of Agriculture and Technology, Faizabad, Uttar Pradesh, India. He has got many prestigious awards from different scientific and market research organizations, like the Indo-Global Award for Innovation in Biotech (2021) from the Indian Economic Development and Research Association (IEDRA), New Delhi, Distinguished Industrial Microbiologist Award (2020) from the Microbiologists Society of India (MSI), India Leadership Awards (2019) in the Best Agribiotech Entrepreneur category from Blindwink Bangalore, Science Entrepreneur Award (2018) from Biologix Research and Innovation Center (BRICPL, India) and Young Scientist Award (2018) from Doctor's Krishi Evam Bagwani Vikas Sanstha, Lucknow, India, respectively. He also completed and guided six indigenous and externally-funded (DSIR and DST) research projects. Moreover, he has worked as a technical consultant for more than a dozen companies for customized technical support. He is a member of the Fertilizer Associations of India (FAI), the Society for Technology Management (STEM), the International Competence Center for Organic Agriculture (ICCOA), and the Association of Microbiologists of India (AMI).

Md. Shamim, PhD
Assistant Professor cum-Jr.-Scientist, Department of Molecular Biology and Genetic Engineering at Dr. Kalam Agricultural College, Bihar Agricultural University, Sabour, India

Md. Shamim, PhD, is presently working as an Assistant Professor cum-Jr.-Scientist in the Department of Molecular Biology and Genetic Engineering at Dr. Kalam Agricultural College, Bihar Agricultural University, Sabour, India. He is the author or coauthor of 30 peer-reviewed journal articles, five books, 20 book chapters, and two conference papers. He has three authored books, two edited books, and one practical book to his credit. He has proved himself as an active scientist in the area of biotic stress management in rice, especially in yellow stem borer management by isolating protease inhibitor from jackfruit seeds and sheath blight resistance mechanisms in wild rice, cultivated rice, and other hosts. He is an editorial board member of several national and international journals. Recently, Dr. Shamim received the Young Faculty Award 2016 from Venus International Foundation, Chennai, India. Dr. Shamim acquired his master (Biotechnology) and PhD (Agricultural Biotechnology) degrees from Narendra Deva University of Agriculture and Technology, Kumarganj, Faizabad, India, with specialization in biotic stress management in rice through molecular and proteomic tools. Before joining Bihar Agricultural University, Sabour, Dr. Shamim worked at the Indian Agricultural Research Institute, New Delhi, where he was engaged in heat-responsive gene regulation in wheat. Dr. Shamim also has working experience at the Indian Institute of Pulses Research, Kanpur, India, on molecular and phylogeny analysis of several *Fusarium* fungi and has also done research at the Biochemistry Department of the Dr. Ram Manohar Lohia Institute on plant protease inhibitor isolation and their characterization. He is a member of the soil microbiology core research group at Bihar Agricultural University (BAU), where he helps with providing appropriate direction and assisting with prioritizing the research work on PGPRs.

Rohit Sharma, PhD
Scientist C, National Centre for Microbial Resource (NCMR), National Centre for Cell Science (NCCS), Pune, Maharashtra, India

Rohit Sharma, PhD, completed his doctoral thesis in 2008 from Rani Durgavati University, Jabalpur. During his doctoral thesis, he explored the forests of Madhya Pradesh and Chhattisgarh for the diversity of ectomycorrhizal fungi. He collected several ectomycorrhizal mushrooms belonging to *Russula*, *Lactarius*, *Boletus*, *Sclreroderma*, *Geaster*, *Astraeus*, etc., from forests, establishing fungal plant ectomycorrhizal associations by root morphologies. He joined the National Center for Cell Science (NCCS) under the Microbial Culture Collection project (MCC) in 2009 and is presently working as Scientist C under the National Center for Microbial Resource Project (NCMR) at the National Center for Cell Science (NCCS). At NCCS, as a curator, he manages more than 15,000 fungal cultures preserved and provides identification services. At NCCS, he shifted his interest to micro-fungi and has described more than 10 novel species and two novel genera by using a polyphasic approach. His present research interests involve both basic and applied mycology. Presently, he is focusing on the use of fungal cultures in enzyme production, waste management, antimicrobial properties, and other applications in industries, as well as solving present problems. He is exploring various habitats in India, collecting fungi and screening them for various properties. He has published more than 40 research papers in national and international journals. He is keen on taking lectures for kids in schools and informing them about the beneficial and harmful effects of this amazing group.

Contents

Contributors

Krishnendu Acharya
Molecular and Applied Mycology and Plant Pathology Laboratory, Center of Advanced Study, Department of Botany, University of Calcutta, Kolkata, West Bengal, India

Santosh K. Arya
Department of Microbiology, Shri Guru Ram Rai University, Dehradun, Uttarakhand, India

N. Balasubramani
Coordinator, CFA Program, MANAGE, Rajendranagar, Hyderabad, Telangana, India

Akhilesh Bind
Department of Biochemistry and Biochemical Engineering, JIBB, Sam Higginbottom University of Agriculture Technology and Sciences, Uttar Pradesh, India

Mahesh Borde
Department of Botany, Savitribai Phule Pune University, Pune, Maharashtra, India

Soumi Bose
Molecular and Applied Mycology and Plant Pathology Laboratory, Center of Advanced Study, Department of Botany, University of Calcutta, Kolkata, West Bengal, India

Reena Buade
Department of Microbiology, Barkatullah University, Bhopal, Madhya Pradesh, India

Chandra Bhanu
ICAR–Indian Institute of Farming Systems Research, Modipuram, Meerut, Uttar Pradesh, India

Dipanti Chourasiya
Microbiology Section, ICAR–Indian Institute of Soybean Research, Indore, Madhya Pradesh, India

Lucimara M. C. Cordeiro
Department of Biochemistry and Molecular Biology, Federal University of Paraná, Curitiba, Paraná, Brazil

Himanki Dabral
School of Agriculture Sciences, Shri Guru Ram Rai University, Dehradun, Uttarakhand; Samvet Bharat, Dehradun, Uttarakhand, India

Jayita De
Molecular and Applied Mycology and Plant Pathology Laboratory, Department of Botany, University of Calcutta, Kolkata, West Bengal, India

L. J. Desai
Center for Research on Integrated Farming System, S.D. Agricultural University, Sardarkrushinagar, Banaskantha, Gujarat, India

Loknath Deshmukh
Fungal Biotechnology and Invertebrate Pathology Laboratory, Department of Biological Science, R.D. University, Jabalpur, Madhya Pradesh, India

Khirood Doley
Department of Botany, Savitribai Phule Pune University, Pune, Maharashtra, India

Kakoli Dutt
Department of Biosciences, Banasthali Vidyapith, Rajasthan, India

Rahul Gajghate
Crop Improvement Division, ICAR–Indian Grassland and Fodder Research Institute, Jhansi, Uttar Pradesh, India

Manju M. Gupta
Sri Aurobindo College, Delhi University, Malviya Nagar, New Delhi, India

Raja Husain
Department of Agriculture, Himalayan University, Itanagar, Arunachal Pradesh, India

V. B. Jha
Department of Plant Breeding and Genetics, Dr. Kalam Agricultural College, Kishanganj, (Bihar Agricultural University), Sabour, Bhagalpur, Bihar, India

Somanjana Khatua
Molecular and Applied Mycology and Plant Pathology Laboratory, Center of Advanced Study, Department of Botany, University of Calcutta, Kolkata, West Bengal, India
Department of Botany, Faculty of Science, University of Allahabad, Prayagraj, Uttar Pradesh, India

Arun Kumar
Department of Agronomy, Bihar Agricultural University, Sabour, Bihar, India

Deepak Kumar
R&D Division, Nextnode Bioscience Pvt. Ltd., Kadi, Gujarat, India

Mahesh Kumar
Department of Molecular Biology and Genetic Engineering, Dr. Kalam Agricultural College, Kishanganj (Bihar Agricultural University), Sabour, Bhagalpur, Bihar, India

Manish Kumar
Amity University, Maharajpura, Dang, Gwalior, Madhya Pradesh, India

Pankaj Kumar
Department of Botany, Purnea University, Purnia, Bihar, India

Pradeep Kumar
Department of Forestry, North-Eastern Regional Institute of Science Technology, Nirjuli, Arunachal Pradesh, India

Santosh Kumar
Department of Plant Pathology, Mandan Bharti Agricultural College (Bihar Agricultural University), Sabour, Bhagalpur, Agwanpur Saharsa, Bihar, India

Rima Kumari
Department of Botany, Purnea University, Purnia, Bihar, India

Hemant R. Kushwaha
School of Biotechnology, Jawaharlal Nehru University, New Delhi, India

Ajay C. Lagashetti
National Fungal Culture Collection of India (NFCCI), Biodiversity and Paleobiology Group, Agharkar Research Institute, Pune, Maharashtra, India

Deepak K. Maurya
National Fungal Culture Collection of India (NFCCI), Biodiversity and Paleobiology Group, MACS' Agharkar Research Institute, Pune, Maharashtra, India

Gautam Kumar Meghwanshi
Department of Microbiology, M.G.S. University, Bikaner, Rajasthan, India

Tatiana Minkina
Academy of Biology and Biotechnology, Southern Federal University, Rostov-on-Don, Russia

Arghya Naskar
Molecular and Applied Mycology and Plant Pathology Laboratory, Department of Botany, University of Calcutta, Kolkata, West Bengal, India

Prakash Pradhan
Molecular and Applied Mycology and Plant Pathology Laboratory, Department of Botany, University of Calcutta, Kolkata, West Bengal, India; West Bengal Biodiversity Board, Kolkata, West Bengal, India

Anil Prakash
Department of Microbiology, Barkatullah University, Bhopal, Madhya Pradesh, India

S. Pranank
Department of Environmental and Applied Sciences, Algonquin College, Canada

Vishnu D. Rajput
Academy of Biology and Biotechnology, Southern Federal University, Rostov-on-Don, Russia

Shiwali Rana
National Fungal Culture Collection of India (NFCCI), Biodiversity and Paleobiology Group, MACS' Agharkar Research Institute, Pune, Maharashtra, India

Tushar Ranjan
Department of Molecular Biology and Genetic Engineering, Bihar Agricultural University, Sabour, Bhagalpur, Bihar, India

Sapna Rawat
University of Delhi, Department of Botany, Delhi, India

Kishor Roy
Molecular and Applied Mycology and Plant Pathology Laboratory, Department of Botany, University of Calcutta, Kolkata, West Bengal, India

A. Sadalaxmi
Coordinator, CFA Program, MANAGE, Rajendranagar, Hyderabad, Telangana, India

Sardul Singh Sandhu
Fungal Biotechnology and Invertebrate Pathology Laboratory, Department of Biological Science, R.D. University, Jabalpur, Madhya Pradesh, India

Baishakhi Santra
Molecular and Applied Mycology and Plant Pathology Laboratory, Department of Botany, University of Calcutta, Kolkata, West Bengal, India

Anik Sarkar
Molecular and Applied Mycology and Plant Pathology Laboratory, Department of Botany, University of Calcutta, Kolkata, West Bengal, India

Sanjai Saxena
Department of Biotechnology, Thapar Institute of Engineering and Technology (Deemed to be University), Patiala, Punjab, India; Agpharm Bioinnovations LLP, Incubatee: Science and Technology Entrepreneurs Park (STEP), Thapar Institute of Engineering and Technology, Patiala, Punjab, India

Md. Shamim
Department of Molecular Biology and Genetic Engineering, Dr. Kalam Agricultural College, Kishanganj (Bihar Agricultural University), Sabour, Bhagalpur, Bihar, India

Rahul Sharma
Center for Biodiversity Exploration and Conservation (CBEC), Jabalpur, Madhya Pradesh, India

Mahaveer P. Sharma
Microbiology Section, ICAR–Indian Institute of Soybean Research, Indore, Madhya Pradesh, India

Ragini Sharma
Department of Zoology, Punjab Agricultural University, Ludhiana, Punjab, India

Rohit Sharma
National Center for Microbial Resources, National Center for Cell Science, Pune, Maharashtra, India

Sushil K. Sharma
ICAR–National Institute of Biotic Stress Management, Baronda, Raipur, Chhattisgarh, India

Abhishek Singh
Department of Agricultural Biotechnology, College of Agriculture, Sardar Vallabhbhai Patel University of Agriculture and Technology, Meerut, Uttar Pradesh, India

Anu Singh
Samvet Bharat, Dehradun, Uttarakhand, India; School of Biotechnology, Jawaharlal Nehru University, New Delhi, India

Awani Kumar Singh
Center for Protected Cultivation, ICAR–Indian Agricultural Research Institute, New Delhi, India

K. P. Singh
Vice Chancellor, M.J.P. Rohilkahand University, Bareilly, Uttar Pradesh, India

Omkar Singh
Subject Matter Specialist (Plant Protection), Krishi Vigyan Kendra, P.G. College, Ghazipur, Uttar Pradesh, India

P. N. Singh
National Fungal Culture Collection of India (NFCCI), Biodiversity and Paleobiology Group, Agharkar Research Institute, Pune, Maharashtra, India

S. K. Singh
Dr. Rajendra Prasad Central Agricultural University, Pusa Samastipur, Bihar, India

Sanjay K. Singh
National Fungal Culture Collection of India (NFCCI), Biodiversity and Paleobiology Group, Agharkar Research Institute, Pune, Maharashtra, India

U. B. Singh
ICAR–National Bureau of Agriculturally Important Microorganisms (NBAIM), Maunath Bhanjan, Uttar Pradesh, India

Gurleen Kaur Sodhi
Department of Biotechnology, Thapar Institute of Engineering and Technology, Patiala, Punjab, India

Deepti Srivastava
Integral Institute of Agricultural Science and Technology, Integral University, Lucknow, Uttar Pradesh, India

Manjusha Tyagi
Department of Microbiology, Shri Guru Ram Rai University, Dehradun, Uttarakhand, India

Helga Willer
Research Institute of Organic Agriculture FiBL, Frick, Switzerland

Abbreviations

AICRP	All India Coordinated Research Project
ALT	alanine aminotransferase
AM	alveolar macrophages
AM	arbuscular mycorrhizal
AMF	arbuscular mycorrhizal fungi
AMH	Ajrekar mycological herbarium
ARDB	antibiotic resistance gene database
ARI	Agharkar Research Institute
AST	aminotransferase
ATA	alimentary toxic aleukia
AXOS	arabinoxylan-oligosaccharides
Bas	bile acids
BCA	biocontrol agents
BGL	β-glucosidase
BP	banana peels
BSEP	bile salt export pump
BSG	Brewer's spent grain
CAGR	compound annual growth rate
CARD	comprehensive antibiotic-resistance database
CBD	convention on biological diversity
CBP	consolidated bioprocessing
CC	corn cobs
CD	Crohn's disease
CDC	Centers for Disease Control
CIB & RC	Central Insecticides Board & Registration Committee
CMCase	carboxymethyl cellulase
CP	cyclophosphamide
CRISPR	clustered regularly interspaced short palindromic repeats
CSIRO	Commonwealth Scientific and Industrial Research Organization
CUPP	conserved unique peptide patterns
CWDEs	cell-wall-degrading enzymes
DDH	DNA-DNA hybridization
DPs	dust particles

DSS	decision support systems
DSS	dextran sulfate sodium
DST	Department of Science and Technology
ECM	ectomycorrhizae
eDNA	environmental DNA
ERM	ericoid mycorrhiza
EMP	earth microbiome project
EPA	Environmental Protection Agency
EPF	entomopathogenic fungi
FAME	fatty acid methyl ester
FFA	free fatty acids
FOS	fructo-oligosaccharides
FPase	filter paper cellulase
FYM	farmyard manure
GABA	gamma-aminobutyrate
GAP	good agricultural practices
GFP	green fluorescent protein
GH11	hydrolase family 11
GI	gastrointestinal
GM	gut microbiota
GOS	galacto-oligosaccharide
GPR	G protein-coupled receptor
GR	granules
GSO	GCC Standardization Organization
HACCP	hazard analysis and critical control points
HDL-c	high-density lipoprotein-cholesterol
HFD	high fat diet
HIV	human immunodeficiency virus
HMG-CoA	3-hydroxy-3-methylglutaryl-coenzyme A
HPCs	high-performance computing system
HSV	herpes simplex virus
IAR	International Agency for Research on Cancer
IBD	inflammatory bowel disease
IBS	irritable bowel syndrome
ICRISAT	International Crops Research Institute for the Semi-Arid Tropics
IFFCO	Indian Farmers Fertilizer Cooperative Limited
IFN	interferon
IL	interleukin
IMO	isomalto-oligosaccharides

IMZ	imazalil
IPM	integrated pest management
ITCC	Indian type culture collection
LDL-c	lipoprotein-cholesterol
l-DOPA	l-3,4-dihydroxyphenylalanine
LPMO	lytic polysaccharide monooxygenases
LPS	lipopolysaccharide
LShF	liquid shaking fermentation
LSSF	liquid state surface fermentation
MBCA	microbial biological control agents
MCC	microbial culture collection
MDA	malondialdehyde
MEA	Middle East & Africa
MnPs	manganese peroxidases
MNPs	metal nanoparticles
MOS	mannano-oligosaccharides
MPO	myeloperoxidase
MS	member state
MTCC	microbial type culture collection
NAIMCC	National Agriculturally Important Microbial Culture Collection
NATP	National Agricultural Technology Project
NCCPF	National Culture Collection of Pathogenic Fungi
NCCS	National Center for Cell Science
NCIM	National Collection of Industrial Microorganisms
NF	nuclear factor
NFCCI	National Fungal Culture Collection of India
NOP	National Organic Program
OCSPP	Office of Chemical Safety and Pollution Prevention
OFAT	one factor at a time
OP	orange peel
OPP	Office of Pesticide Programs
PAH	polycyclic aromatic hydrocarbons
PBMC	peripheral blood mononuclear cells
PCBs	polychlorinated biphenyls
PCR	polymerase chain response
PDA	potato dextrose agar
PGA	potato glucose agar
PGIMER	Post Graduate Institute of Medical Education and Research
PKS	polyketide synthase

PPAR-γ	proliferator-activated receptor-γ
ppb	parts per billion
PPNs	plant parasitic nematodes
PUFA	polyunsaturated fatty acids
qPCR	quantitative PCR
RANKL	receptor activator of nuclear factor kappa-B ligand
RFOs	raffinose oligosaccharides
RH	relative humidity
RHG	reaper, hid, and grim
RON	research octane number
rRNA	ribosomal RNA
SAB	Sabouraud agar
SC	suspension concentrate
SCFAs	short-chain fatty acids
SCO	single cell oil
SD	sawdust
SmF	submerged fermentation
SNPs	single-nucleotide polymorphisms
SOD	superoxide dismutase
SPME	solid-phase microextraction
SPS	sanitary and phytosanitary standards
SSF	solid-state fermentation
T1D	type 1 diabetes
TAC	total antioxidant status
TBZ	thiabendazole
TC	total cholesterol
TCP	tricalcium phosphate
TGs	triglycerides
TLR	toll-like receptor
TNBS	trinitro-benzene-sulfonic acid
TNF	tumor necrosis factor
TNFR	tumor necrosis factor receptor
TOS	total oxidant status
TrMs	trichothecene mycotoxins
UC	ulcerative colitis
UL	ultra-low
USDA	United States Department of Agriculture
USEPA	United States Environmental Agency
UV	ultraviolet

VA	vascular-arbuscular
VA	veratryl alcohol
VCGs	vegetative compatibility groups
VOCs	volatile organic compounds
WB	wheat bran
WDG	water dispersible granules
WHO	World Health Organization
WP	wettable powder
WS	wheat straw
WTO	World Trade Organization
XD	xylose dehydrogenase
XOS	xylo-oligosacharides
XR	xylose reductase
ZO-1	zonula occludens-1

Preface

Mycology has been known for centuries, dealing with various groups of filamentous and lichen-forming fungi, molds, mushrooms, slime molds, yeasts, etc. Conventionally, these fungi were treated as pocket organisms under the Kingdom-Plantae since the era of the Linnaean classification system. Though the first proposal to separate fungi and treat them under an independent Kingdom 'Fungi' came way back in 1969, its global acceptance was not apparent for many decades which resulted in a lot of confusion among scholars as the conventional approach of classifying fungi continued. However, advances in molecular tools gradually changed the thought process and consensus on the status of fungi.

Several groups of fungi are now separated and classified under different kingdoms. As such, in the recent era, the field of fungal biodiversity, biology, and biotechnological applications has changed tremendously, and new avenues for the applications of fungi are being defined. Besides, their vital role in ecosystem functions, fungi touch, and influence every facet of human life. Therefore, mycology as a subject and its constituent organism 'fungi' have become essential and indispensable components of biodiversity on earth.

Interaction with the environment is important to many natural processes that occur in the biosphere. Being the most abundant eukaryotes on earth's biosphere, and the diverse nature of fungi play vital roles in various sectors. There is a great need to integrate conventional and biotechnological approaches in efficient applications of fungi in various sectors. Approaches of recovering fungi from various habitats/substrates in the new era needs to be redefined for the utilization of micro- and macrofungi in various sectors, like agriculture, food, healthcare, and industries. The purpose of bringing out this book is to bridge the knowledge gap between basic and applied aspects of mycology.

In the present book, we provide up-to-date glimpses on the strategic requirement of ex-situ conservation of native microorganisms in culture collections and their prospects. This fundamental aspect has drawn the attention of researchers/mycologists towards the flagship global program of the Convention on Biological Diversity (CBD). Information on various culture collections in India and abroad are also covered.

Since many microbes have neither been cultivated nor identified as the culture-based method has its limitations, an emerging area of metagenomic sequencing, amalgamated with environmental biology and genomics, facilitates a solution for overcoming these problems. In fact, this emerging approach has completely transformed the detection and characterization of complex microbial communities from the genetic material directly isolated from environmental samples. The historical perspectives of fungi and their role in eco-safety and warfare are another area of interest covered in this book.

Other chapters contain information valuable to various sectors. The developing interest and contribution to farming and expanded protection are topics of great interests. The use of beneficial microorganisms has turned out to be a boon for agriculture and forestry ecosystems. Plants hosted by AMF have shown various degrees of protective effect by mycorrhizal symbioses as a result of complex interactions between bacteria, AMF, and the plant. Myco-biocontrol offers an alluring option in contrast to the utilization of synthetic pesticides and are seen as less harmful to the climate.

Plant parasitic nematodes (PPNs) are the major threat to the agricultural and horticulture sectors worldwide causing huge economic losses. Plants infested with nematode often become targets for secondary infection by bacteria and fungi, making the condition even worse. Strategies are detailed out in this book to minimize these losses. In this regard, focus on the exploitation of macrofungi as a potential bio-control agent against PPNs and their diverse mechanism of nematophagy has been provided in the book.

An interesting overview of biocontrol technology is included in the book explaining how active ingredients such as atoxigenic isolates of *Aspergillus* species and other biological inoculants are selected, how compositions are designed and tested, and how biocontrol products are used effectively to combat the aflatoxin poisoning of food and feed crops.

The biodiversity of mushrooms is another area of interest included in this book. Mushrooms were said to be the "Food of the Gods," in ancient times by the Romans and Egyptians. They have been considered as versatile foods with accepted health benefits. Over the centuries, mushrooms have been used as medicine as well as food. Besides, mushrooms are emerging as novel resources of potential prebiotics, also considered as non-digestible functional ingredients facilitating the growth of helpful bacteria in the gut, which leads to several health benefits. This concept has drawn the attention of the nutraceutical industries and is scaling new heights as interest toward the identification of new molecules that can promote overall health gains momentum.

Currently, edible mushrooms have emerged as an excellent source of prebiotics as they contain numerous bioactive components, of which polysaccharides, β-(1→3)-D-glucans, and polysaccharide-peptide/protein complexes are of special significance. They contribute to improved gut barrier integrity, nutrient absorption, and gastrointestinal as well as systemic immunity. In this book, current knowledge on the effect of macrofungi on the improvement of gut dysbiosis and prevention of diabetes, osteoporosis, colitis, and metabolic disorders are very well-summarized, which would be helpful in translating the health benefits observed during an investigation into real-life outcomes.

Entomopathogenic fungi (EPF), especially *Cordyceps* and *Ophiocordyceps*, are the ecologically important indispensable constituents of coveted traditional Chinese medicines found in the Indian Himalayan region. The artificial cultivation of these fungi has always been a constraint. Therefore, various methods of artificial cultivation of *C. militaris* and *O. sinensis* are provided in this volume with a focus on cultural-environmental factors, biomass, and bioactive metabolite production, as well as details on other productivity constraints.

A review on the production of antibiotics and antimicrobial agents from filamentous fungi is also featured in this book. Globally, the healthcare system is facing challenges due to antimicrobial resistance exhibited by several classes of bacteria and fungi against existing antibiotics. It is thus the need of the hour to explore new compounds for new biomedical and industrial applications. In addition, recent developments in the fields of genomics and proteomics offer new opportunities to combine relevant information to extend the bio-manufacturing of recombinant proteins in filamentous fungi.

This book also features neem (*Azadirachta indica* A. Juss), a well-known plant containing about 400 chemicals that are utilized as a medicine in a variety of ways. However, search for key plant metabolites like azadirachtin, nimbin, nimbidin, etc., in endophytic mycoflora associated with neem plants is scaling new heights, with potential application in agriculture, pharmaceutical, and industrial sectors.

Prospects of fungi as producers of alternative fuels has been provided in detail as fungi possess excellent biosynthetic abilities due to their intricate metabolic pathways and their capability to utilize a range of substrates to derive nutrition for their survival and growth. In these processes, they produce a variety of primary and secondary metabolites which could be exploited in different industrial sectors. Besides being potent producers of hydrocarbons, fungi can replace the existing fuels for the transport industry as well as other applications for the betterment of mankind. Enhanced consumption of fossil fuels has led to the depletion of natural resources. Fungi, as well as their endophytic forms, have been implicated to be in great demand with the depletion of carboniferous reserves of fuels in nature. Similarly, the potential of fungi converting agro-residues into cellulase, bio-ethanol production, and other industrial applications are also well-illustrated in this book.

As such, this book contains comprehensive and up-to-date information on various fundamental and applied aspects of mycology. We are sure that researchers in various disciplines of science and those directly dealing with fungal biodiversity, biology, and bioprospecting would prefer to have this volume on their shelves and will find this volume highly beneficial. This volume will also be useful in the libraries of academic, research, and industrial institutions.

With great sincerity, the editors wish to acknowledge the authors of the chapters for their immense contribution to the completion of this volume. The editors also acknowledge the help and support taken from various books, resource persons, journals, and publishers whose publications have been used while preparing the manuscripts. Finally, we thank Apple Academic Press for accepting our proposal and bringing out this publication.

—Editors

CHAPTER 1

Indian Culture Collections and Fungal Biodiversity Conservation

RAHUL SHARMA[1] and ROHIT SHARMA[2]

[1]*Center for Biodiversity Exploration and Conservation (CBEC), Jabalpur, Madhya Pradesh, India*

[2]*National Center for Microbial Resources, National Center for Cell Science, Pashan, Pune, Maharashtra, India*

ABSTRACT

Fungal diversity has been a topic of great interest globally which has resulted in huge collections all over the world. Literature reveals that a large chunk of fungi known are not in culture either due to their non-cultivable nature or lack/loss of only available type-culture or the collection simply lack them. In spite of the fact that cultures collections have a major role in conserving the diversity of microbial strains as Germplasms. These collections provide a base for a sound pharmaceutical and biotechnological industry with a substantial share in the global economy. The authors believe that the lacuna has been on part of us. Microbiologist and mycologist fraternity in India were not able to communicate the importance of microbial culture collections (MCCs) to the policy makers. This further results into financial crunch to basic science, especially life sciences. A major bottleneck in India is the lack of awareness, expertise in conservation biology and funding. This chapter provides an overview and current status of MCCs in India and selected major collections in the world. Also, this chapter details out about the way forward in making concerted efforts in conserving the indigenous microbial treasure and their sustainable utilization by academia and industries.

Applied Mycology for Agriculture and Foods: Industrial Applications. Sanjay K. Singh, Deepak Kumar, Md. Shamim, & Rohit Sharma (Eds.)

1.1 INTRODUCTION

Fungi are versatile microbes that are so well adapted to various environment conditions that they comprise of forms that are part of 3 kingdoms of living organisms (Kirk et al., 2008), out of 7 recognized so far. Fungi lacking cell wall are placed in Protozoa, ones that have cellulosic cell wall are placed in Chromista and all those with chitinous cell wall come in Eumycota (True fungi) (Barr, 1992; Patterson & Sogin, 1992; Cavalier-Smith, 1993). These kingdoms have been established based on the type of mitochondria and sequence analysis of rDNA. These divisions or groupings of fungi reflect the evolutionary trend from simple aquatic lifestyle to their adaptation after they first colonized land several million years ago when life moved from water to land.

Fungal diversity has been a topic of interest to many (from plant pathologists to pharmaceutical industry) which resulted in huge collections all over the world. Around 0.12 million fungi have been described so far, although the total number estimated is about 1.5 million (Hawksworth, 1991, 2001; Kirk et al., 2001; Webster & Weber, 2007). The analysis of soil by direct sequencing of the ITS region and high throughput sequencing of the cloned fragments of two temperate sites in Duke Forest, North Carolina, USA suggests the 1.5 million figure is underestimated and close to around 3.5 to 5.1 million fungi are inhabiting the earth (O'Brien et al., 2005; Blackwell, 2011). However, fungal cultures in culture collections amount to one-third of all known fungal species on earth. A large portion of fungi known are not in culture either due to their non-cultivable nature or lack/loss of only available type-culture or the collection simply lacks them because the concept of culture collection first originated during 1900 in Prague since several fungi that had been described before 1900s are only known from herbarium specimen(s) or needs epi-or neo-typification with fresh collections due to loss of only available type (Hyde & Zhang, 2008). The world's biggest and best culture collections [viz. CBS (The Netherlands), ATCC (US), CABI (UK), UAMH (Canada), NBRC (Japan), etc.], have between 5,000 and 72,000 cultures.

Looking to the two lists of culture collections (Tables 1.1 and 1.2) the disparity is quite clear at the national and international stage, contrary to the fact that collections that are listed in Table 1.1 are from temperate countries and which have little indigenous biodiversity. The huge investments (continuous and long term) made by these countries on establishment and maintenance of culture collections give us an idea of the importance of microbes for man and its environment. In fact, these cultures collections have a major role in making these countries developed by way of establishing a base for a

TABLE 1.1 Status of Culture Collections for Fungi in the World* (Major 10 Alphabetically)

SL. No.	Acronym (WFCC No.)	Name	Country	Estd.	Holdings†	Staff#
1.	ATCC (1)	American Type Culture Collection	USA	1925	49,000	85
2.	BCCM/IHEM (642)	IHEM Biomedical Fungi and Yeasts Collection	Belgium	1982	15,000	11
3.	CABI (214)	CABI Genetic Resource Collection	UK	1910	28,000	23
4.	CBS (133)	Westerdijk Fungal Biodiversity Institute [Formerly Centraal Bureau voor Schimmelcultures (CBS-KNAW)]	The Netherlands	1904	72,000	29
5.	DSMZ (274)	German Collection of Microorganisms and Cell Cultures	Germany	1969	5,000	40
6.	FGSC (115)	Fungal Genetic Stock Center	USA	1961	24,000	1
7.	IBT (758)	IBT Culture Collection of Fungi	Denmark	–	35,500	26
8.	NBRC (825)	Biological Resource Center	Japan	–	15,145	41
9.	NRRL (97)	Agriculture Research Service Culture Collection	USA	1904	55,733	33
10.	UAMH (73)	University of Alberta Mold Herbarium	Canada	1933	11,750	3

*Data compiled from website of respective culture collection and WFCC.

#Staff includes scientific and technical.

†Filamentous fungi and yeasts.

TABLE 1.2 Status of Culture Collections of Fungi in India*

SL. No.	Acronym (WDCM No.)	Name	Located at	Estd.	Holding	Staff#	Funding
1.	ITCC (430)	Indian Type Culture Collection	New Delhi	1936	4,000	6	ICAR
2.	MCC (930)	Microbial Culture Collection	Pune	2008	117	35	DBT
3.	MTCC (773)	Microbial Type Culture Collection	Chandigarh	1986	3,020	12	CSIR
4.	NAIMCC (1060)	National Agriculturally Important Microbial Culture Collection	Kusmaur	2004	3,110	13	ICAR
5.	NCIM (3)	National Collection of Industrial Microorganisms	Pune	1951	3,700	4	CSIR
6.	NCCPF†	Nation Culture Collection of Pathogenic Fungi	Chandigarh	2010	2,500	4	ICMR
7.	NFCCI (932)	National Fungal Culture Collection of India	Pune	2008	5,000	6	DST
8.	NCDC (775)	National Collection of Dairy Cultures	Karnal	–	35**	6	ICAR

*Data compiled from website of respective culture collection and WFCC.

**This number is from WFCC, probably the number is slightly more since NCDC website now indicates 400 for bacteria, fungi, and yeasts.

#Scientific and technical.

†Not registered with WFCC.

sound pharmaceutical and biotechnological industry with a substantial share in the global economy. The authors feel that the lacuna was on part of us, microbiologist and mycologist in India who were not able to communicate and inform the importance of culture collections to the policy makers (who are mostly from administrative or political background) who allocate annual budgets to science and especially life science. Also, the growth of previously established culture collection to international standard did not occur, which also paved the way to budget cuts by funding agencies.

The microbial cultures have given us products with well recognized applications. One example can be had from the *Taq* polymerase (a DNA polymerase enzyme first extracted from the bacteria *Thermus aquaticus*), which is first applied for amplifying DNA sequence by Karry Mullis in 1986. He got a US patent on the enzymatic amplification of DNA in 1987, which was sold by Cetus (where K Mullis was employed) to Hoffman La-Roche for $330 million to the company (Mullis, 1994). Fungi, in specific, have also been of much importance to us. The pertinent fungal examples are Penicillin and Cyclosporine, the importance of which in saving human life (the figures in both cases are astonishing) surpasses their monetary or economic benefits. Penicillin, originally recovered from *Penicillium notatum* (current name *P. chrysogenum*) by Alexander Fleming is known to have saved millions of lives since its first discovery in 1929 (Fleming, 1929) and of Cyclosporine (recovered from *Tolypocladium inflatum*) in 1976 (Dreyfuss et al., 2003) which are now indispensable for surgeons the world over engaged in organ transplant. There are innumerable other examples of the role of microbes in changing the socio-economic status of a nation (in both directions-up and down) like the Ireland famine caused by the failure of potato crops due to fungus *Phytophthora infestans* resulting in 1.5 million deaths (Salaman, 1949). Also, three species of fungi *Penicillium chrysogenum* (=*P. notatum*) (discovery of Penicillin), *Neurospora crassa* (one gene, one enzyme theory) and *Saccharomyces cerevisiae* (discovery of key regulators of cell cycle in eukaryotes) are part of groundbreaking study that have won Nobel prizes in 1945, 1958, and 2001, respectively (Mathew et al., 2009). Therefore, the importance of culture collections can only be imagined in light of their beneficial and harmful effects of microbes in totality.

Cultivation, maintenance, and study of these groups of fungi require different expertise. Thus, a culture collection at least should have experts who could deal with minimum one group, i.e., one expert for each group, so that all are represented. In case of specialized groups, expert advice and assistance can be taken from outside the collection from India or abroad.

An essential component of culture collection is to provide fungal culture identification service and/or techniques associated with it viz., phenotypic characterization, sequencing, biochemical analysis. The culture collection can also generate some funds by providing these services on a charge basis, however for this, correct identification is a must. With the availability of DNA sequencing technology, authentic identification has become a norm rather an exception.

Relatively recently, there has been interest for searching the unknown fungal diversity of India, their conservation and also bioprospecting of industrially important ones. Here we discuss the status of culture collections of India, recent developments in tapping the fungal diversity of India, steps taken to achieve the same and also recently developed culture collections in India including implications or benefits to India due to such investments in culture collection.

1.2 NEED FOR CONSERVING FUNGAL DIVERSITY OF INDIA

India is a vast country with diverse habitats ideal for the survival of varied life forms including almost all groups of fungi. Some fungi are of direct use to mankind like mushrooms and yeasts (used in bread and wine making) while several others produce certain life-saving drugs that are extracted industrially viz. Penicillin (antibiotic), Cyclosporin (used in organ transplant).

While the biggest group of fungi economically relevant to man are the ones that cause severe diseases in crops plants rendering enormous losses to food production (Agrios, 2005). In fact, one-third of the crops are lost due to fungal and bacterial diseases worldwide and the situation in India is more severe since we have very miserable disease forecasting system since it requires prophylactic measures including pathogen identification during initial stages of disease incidence. All this requires expert identification services in the field supported with modern molecular techniques. A great number of fungal species are the cause of severe diseases in humans and animals, resulting in a great deal of agony to humans and animals (economic loss due to infection in life stock is one such example). A whole industry involving millions of euros of enterprise is engaged in controlling these infectious diseases caused by fungi. Several of these fungi are clinically very significant, for example, *Cladophialophora bantiana*, a fungus which has human and animal brain as its preferred niche and is mostly fatal (Gerrits & de Hoog, 1999; Horre & de Hoog, 1999; Ajantha & Kulkarni, 2011). Fungal cultures are essential for new drug to be tested against such pathogens, and also the identification of pathogens is usually done with the help of reference strains available in culture collections.

In India saving biodiversity has been more focused on plant and animal, until now. Microbes including fungi have been neglected for quite some time while assessing the impact on biodiversity and its conservation. Culture collections in India are the only alternative to conserving our microbial diversity, as the *in situ* conservation of all microbes is difficult except for those which have obligate associations with a particular plant (symbionts or biotrophs) or animal (Trichomycetes) or the ones that inhabit particular habitat like mushrooms (Bratton, 2003) There is a need to increase the capacity of present facilities generated by different funding agencies so as to preserve and conserve the fungal diversity of India. It will need continuous efforts (and funding) in a systematic manner for many years to achieve the same.

1.3 STATUS OF FUNGAL CULTURES IN INDIAN CULTURE COLLECTIONS

India is endowed with a large area with a variety of habitats. These are rich in microbial diversity. Nearly one-fourth (27,000 species) of the world's known fungal diversity have been reported from India (Manoharachary et al., 2005), however, the number of cultures deposited in our national culture collections does not represent even 10% of this diversity. The scenario has not changed from 2005 to 2014. Much of the greater works on fungal taxonomy and isolation of diverse fungal forms from different regions of India was done by earlier workers, which is evident from monographic accounts on fungi published by the Indian Agriculture Research Institute, New Delhi couple of decades ago [viz. *The Clavariaceae of India* by Thind (1961), *Indian Cercosporae* by Vasudeva (1963), *Mucorales of India* by Tandon (1968), *Hyphomycetes* by Subramanian (1971), and *The Myxomycetes of India* by Thind (1977)]. But somehow after 1990's the systematic exploration of fungi in India was gradually reduced, the reasons for this are varied including a pertinent one, that is the introduction of applied courses in undergraduate and postgraduate levels particularly the Biotechnology in colleges and universities which attracted most and best of the students, leaving few uninterested students to take up fungal systematics studies as chosen subject. However, after more than two decades, it is slowly becoming clear that India's biotechnology industry could not survive without germplasm of indigenous fungal and bacterial strains which are properly identified. This is because; indigenous strains are best suited for local applications including bioconversion, biocontrol, and even drug testing. Also, novel fungi and bacteria recently recovered from Indian soils (Table 1.3) will act as goldmine for functional genes, which will require only detection, annotation, and relevant application.

Until now, India has four prominent fungal culture collection facilities-one at IMTECH, Chandigarh, the MTCC (Microbial Type Culture Collection), second at IARI, New Delhi, the ITCC (Indian Type Culture Collection), third at NCL, Pune, NCIM (National Collection of Industrial Microorganisms), and the fourth at NBAIM, Kusmaur, the NAIMCC (National Agriculturally Important Microbial Culture Collection). The number of fungal species (and not strains) preserved in these four collections when added together does not represent fungal flora of a country with such enormous fungal diversity even if we include cultures from several other minor culture collections of India (15 are registered with WFCC, World Federation of Culture Collection, for fungi).

It was felt for long that there is an urgent need to map, catalog, and preserve the fungal wealth of our country through systematic surveys and taxonomic projects with the help of experts of various fields (Sharma, 2013). Recently, Vidhate et al. (2013) stressed the need for a focused approach to recover insect pathogenic fungi or the entomopathogens from soil where they mentioned about an improved technique (Maranaho & Santiago-Alvarez, 2003) for isolating wild types of this group. They also mentioned about Richard Humber's ARS collection, which is certainly, a good example and a roadmap before us to begin with and to lay a sound foundation for our country's biocontrol program for insect pests. However, whenever we start any program for testing of bio-control agents in the field, we must always keep in mind a bio-containment facility. It should always be included in the budget provisions to prevent anthropogenic inoculum built-up and spread in nature that might cause insect mortality of beneficial insects or unwarranted infections in humans and other animals. For example, the most commonly used insect bio-control fungus *Beauveria bassiana*, which is seemingly non-pathogenic to humans, may cause keratitis (Sachs et al., 1985; Ishibashi et al., 1987) or pulmonary infections in animals (Fromtling et al., 1979; Gonjalez et al., 1995). Also, another very commonly used biocontrol agent found in soil which is used indiscriminately in our country (without regulatory trials performed in containment facilities) belongs to genus *Trichoderma* whose species (prominent ones include-*T. harzianum, T. koningii, longibrachiatum, T. pseudokoningii*, and *T. viride*) are all known to cause various kind of infections in immune-compromised patients, often fatal (de Hoog et al., 2000). Since there is considerable rise in population of individual with reduced

TABLE 1.3 List of Participating Institutes (Sharma & Shouche, 2014) in DBT's Recently Concluded Mega-Project on Exploration of Bacterial Diversity in India Which Yielded 0.15 Million Bacterial Strains*

SL. No.	Institutes	Principal Investigators	Niches/Area Explored
1.	National Environment Engineering Research Institute, Nagpur	Dr. Hemant J. Purohit (Project Coordinator)	Effluent treatment plants
2.	MS Swaminathan Research Foundation, Chennai	Dr. Sudha Nair and Dr. V. K. Prabhavati	Mangroves and eastern ghats
3.	Institute of Genomics and Integrated Biology, Delhi	Dr. V. C. Kalia	River sediments
4.	Delhi University, Delhi	Prof. Rup Lal	Hot springs of Himachal Pradesh
5.	Guru Nanak Dev University, Amritsar	Prof. B. S. Chadha	Wetland ecosystems of Northwest India
6.	Institute of Life Science, Bhubaneswar	Dr. S. Das	Extremophiles from Orissa, Bihar, and Bengal
7.	National Institute of Oceanography, Goa	Dr. N. Ramaiah	Marine sediments
8.	Institute for Biodiversity and Sustainable Development, Imphal	Dr. O. N. Tiwari	Northeast
9.	National Center for Cell Science	Dr. Yogesh Shouche	Insect gut and soil from Western Ghats
		Industry Involved	
1.	Piramal Life Sciences Ltd.	Dr. Arun Balakrishnan and Dr. Saji George	Screened around 0.2 million bacterial isolates obtained during the project for anti-infective, anticancer, anti-inflammation, and anti-diabetic activities

*National Center for Cell Science was the central collection which preserved all the generated bacterial cultures (ca. 0.15 million) making one of the largest repositories of bacterial cultures in the world.

immunity in India (country stands third in the world with 2.1 million patients (UNAID, 2014), the uncontrolled use of these agents put life of several such individuals at risk. Since most of India's medical hospitals (except a very few) are poorly equipped to handle or even detect mycotic infections (Liu, 2011) other than the usual ones (dermatophytes) that cause visible symptoms on the skin. A sound infrastructure on testing the feasibility or safety of probable fungal biocontrol agents (BCA) is very much needed. Currently, India has only one such facility situated at Indian Agriculture Research Institute, New Delhi, the "Phytotron facility" but is very small (when you have a large number and variety of organism to investigate) as compared to the containment facilities in developed countries to handle such pathogens.

Certain funding agencies from India are now taking the lead in recovering the enormous fungal diversity from various Indian habitats (in a short time span) as has been envisaged for bacteria a few years ago resulting in recovery (of more than 0.2 million bacterial cultures) and final preservation of about 1,35,000 strains from 9 niches at MCC, Pune (Sharma & Shouche, 2014) (*see* Table 1.3).

1.4 THE THREE NEW CULTURE COLLECTIONS PRESERVING FUNGI!

The first one of the three new culture collections established (in 2008) is the National Fungal Culture Collection of India (NFCCI), Pune funded by Department of Science and Technology (DST) Government of India, of which the first author was part until 2013 as fungal molecular taxonomist. It is placed at Agharkar Research Institute (ARI), Pune. This culture collection preserves all groups of fungi except medically significant ones for which the second collection specializes. NFCCI currently holds >5,000 fungal strains and is equipped with staff to identify and characterize fungi by morphological as well as molecular means. There are 6 scientific staff with expertise on microfungi and molecular diagnostics. It provides services, like deposit of cultures under general deposit, supply of cultures, and identification of fungi. ARI also houses internationally recognized Ajrekar mycological herbarium (AMH) to preserve dried fungal material including lichen specimens. The herbarium has an impressive collection of 30,000 lichen specimens (Anonymous, 2011), several of which are type material. Their research area includes taxonomy, systematics, conservation and applications of fungi including endophytes. The details can be found at www.nfcci.aripune.org and one can contact the coordinator/curator of the national facility at nfcci.ari@gmail.com

The second collection established (in 2010) is the National Culture Collection of Pathogenic Fungi (NCCPF), Chandigarh, funded by Indian Council of Medical Research; Government of India. It is placed at Post Graduate Institute of Medical Education and Research (PGIMER), Chandigarh. NCCPF currently holds 2,500 fungal strains and is equipped with staff to identify and characterize fungi by morphological as well as molecular means. There are 4 scientific staff, with expertise on clinical zygomycetes, drug resistance in pathogenic fungi, and molecular diagnostics, typing, and epidemiology of pathogenic fungi. The services of the collection include deposit (with two categories: general and safe deposit), supply and identification of medically important fungal cultures. Their research area includes medical mycology. The details can be found at www.nccpf.com and can contact The Incharge at arunaloke@hotmail.com.

The third collection microbial culture collection (MCC) (Sharma & Shouche, 2014), Pune funded by Department of Biotechnology, Government of India is the last to be created (in 2009), has the smallest acronym but in fact, is the biggest in infrastructure and bacterial collection (1,35,000 bacterial strains), to support preservation of indigenous fungal wealth. It is placed at National Center for Cell Science (NCCS), Pune. MCC currently hold 117 fungal strains in their general deposit category, all of which have been identified by molecular sequencing and is equipped with staff to characterize fungi by morphological and molecular means. They have 16 scientific, 19 technical staff, and among one of the culture collections which providing the widest range of services. The services include general safe and IDA deposit, 16rRNA gene and ITS sequencing services, phylogenetic tree construction, MALDI-TOF, fatty acid methyl ester (FAME) analysis, DNA-DNA hybridization (DDH), etc. (Sharma & Shouche, 2014). The collection also pursues research primarily on molecular microbiology (gut, extremophiles viz., Lonar lake, Soldar lake, etc.), bacterial genome sequencing, barcoding of organisms (which includes plants, animals, and microbes) using DNA sequence data, fungal taxonomy, bacterial taxonomy, etc. MCC follows international standards and practices essential for a culture collection and is striding truly to be recognized globally for its services and collection. The details can be found at www.nccs.res.in/mcc and can contact The Service Coordinator at mcc@nccs.res.in.

An important feature of a good culture collection is the number of type (or ex-type) cultures it holds in its collection. Type cultures are references strains that are used in taxonomic studies and phylogenetic assignments of new forms. Any typical strain of a species can act as a reference strain (of that species) in taxonomic studies, but type-designated strains are greatly sort after. The value and service of a regional (national) culture collection mainly depends on the maintenance of regional fungal flora. However, there is always scope for betterment and enrichment of collection with strains from other collections (from overseas) in exchange. MCC has been working on increasing the number of type cultures in its collection. It has been done by procuring type cultures from taxonomists (directly) and/or from national and international collections viz., DSMZ, CBS, BCCM.

1.5 WHERE DO WE GO FROM HERE?

If we see some very well managed culture collection of US or Europe, one thing is clear, India has to has to go a long way and should be running and not walking, simply because all these collections have grown up to their present stature (in size and reputation) almost all taking a century (100 years' time). And above all, most of these collections have gained a free inflow of cultures from around the world either by their own staff working in remote locations (like Africa or Latin Americas) or by mycologist from countries like India or elsewhere interested in safe deposition of a novel find. All this coupled with some finest and most dedicated selfless mycologist were part of these collections until the turn of the millennia when Biodiversity convention altered the scenario altogether. The change predominantly happened because of the commercial interest of pharmaceutical and drug companies causing mycologist to turn entrepreneurs and run after patents (of immense value) and thereby losing their selfless character seen in naturalist. What does the history of three of world biggest and best culture collection like ATCC, CABI, and CBS tell us? They all have acquired this immense treasure after painstaking efforts for more than 100 years. India has both an advantage and disadvantage in starting seriously, so late. We mean "seriously" because ITCC was already established in 1936, but we did not take it anywhere further from there. No increase in staff, facilities, etc. Thus, we did not preserve our treasure the way it should have, and lost India a valuable >70 years. India has the advantage that it already has the guidelines from world known collections which are a result of their experiences in handling cultures for numerous decades (Smith & Onions, 1994; Crous et al., 2009), while a little disadvantage that we cannot collect microbial wealth from outside India's borders freely or send for taxonomic purposes unlike in earlier times when convention on Biological Diversity did not existed (Zedan, 2000). It is a matter of concern to many researchers working on taxonomy and systematics and has been raised on different platforms. Indian biodiversity scientists are currently at crossroads due to development agenda of our Government, which certainly is very essential for the common man for livelihood, calling for us to move very rapidly in scanning the entire geographic area, in limited time frame, otherwise undiscovered fungi will be permanently buried under concrete buildings to be recovered dead later by paleontologists, similar to ones recovered from a ninth century Longobard abbess exhumed from a monastery in Pavia, Italy (Caretta & Piontelli, 1998). The first author is also witness to one such event recently, when he tried to collect soil from Kalyan railway station, a second time, which previously yielded a new species of *Gymnoascus* (Sharma & Singh, 2013). The soil collection site, a railway platform in this case, was completely renovated (with concrete) with no sign of soil.

1.6 CONCLUSION

Looking to the global scenario of climate change and loss of habitats, not only by anthropogenic factors like pollution or development (expanding industrial or residential infrastructure) but also by natural factors like Tsunami, earthquakes, and floods causing irreparable damage to the life forms. By establishing these culture collections in a short time span, India has moved forward looking to the concerns that loss of microbial diversity is inevitable if timely actions are not taken (Sharma, 2003, 2013). What we can do now is to have a consortium of all the Indian culture collections (just in line with European Culture Collection Organization (ECCO)), so that they are interlinked for exchange of information and/or cultures so that redundancy does not take place at the cost of those fungi that are not represented even once in any of these collections. There can be an annual meeting of all culture collections of India to exchange ideas and information. It will provide a platform for future collaboration and developments in microbial taxonomy among members. It will also help to discuss problems faced by taxonomists related to convention of biological diversity (CBD). Projects such as barcode of all microbial species described from India or genome sequencing of important ones can be jointly undertaken. In this way Indian culture collections will also increase its capacity, even experts which are distributed among different centers, i.e., all collections do not need to have all group experts. Thus, together they can act as a source of authentic reference strains to industries and research organizations. Now the country has acquired sufficient infrastructure to preserve its fungal wealth, the only impetus is to recover this wealth (microbes present in Indian soils). Initiatives involving several institutions being in the pipeline since DBT recently invited proposals from mycologist in India for a coordinated megaproject similar to bacteria. Suryanarayanan & Gopalan (2014) recently proposed a roadmap for going about recovering fungal wealth which included exploring 8 habitats and utilization of active workforce of students. We partly agree with their suggestion and would add further that, of these 8 (now 9) microhabitats which were originally proposed by Hawksworth et al. (1997) we should explore all the 31 niches identified by them then, and updated later to 36 niches (Table 1.4) with the help of para-taxonomists (Hyde et al., 2000) using better techniques, for example particle filtration technique yielded 10 times more species than moist chamber technique in a study on leaf litter of two tree species in Puerto Rico (Polishook et al., 1996). Another approach will be the group-wise (taxonomic) exploration of fungi carried out by specialist mycologists having expertise on a particular group like Trichomycetes, Entomopthorales (*Conidiobolus* sp.), Mildews, Rusts, Smuts, Lichens, etc. We foresee a brighter future for systematic fungal diversity exploration and conservation in India and predict that the coming years will see a surge of novel fungal forms in all domains of organisms called "the fungi" from this country!

1.7 A WORD FOR THE FUTURE!

India does not need many collections; one is enough like CBS to excel and lead the world in fungal biodiversity conservation and most importantly to utilize the microbial wealth for the benefit of mankind and the planet. India has at least 15 culture collections registered with WFCC preserving fungi (with the sum total of type strains in few hundreds) while The Netherlands has only 1 culture collection registered with WFCC that preserve fungi, i.e., CBS having most of the type strains of cultured fungi on earth. Most of the genome sequencing projects completed or going on around the world (especially at Joint Genome Institute with funding from the Department of Energy, USA) are based on cultures with strains from CBS. However, recently initiated mega genome sequencing project funded by National Science Foundation, US which will sequence 1,000 fungal genomes that have representatives from all major fungal lineages (www.1000fungalgenomes.org) in a collaborative effort of 4 culture collections from US and CBS from The Netherlands. With the recent success of India's Mars mission costing a mere (as they say!) ₹400 crores (50 million euros), our country certainly needs an earth mission also, of equal proportion (at the least) to recover and characterize fungal genomes for immediate application and sustainable future of our country.

TABLE 1.4 Principle Niches and Microhabitats Identified by Hawksworth & Mueller (2005)

SL. No.	Principle Niches	Microhabitats
1.	Living vascular plants	• Biotrophs and necrotrophs of leaves, stems, flowers, fruits, seeds, roots, etc. • Commensals on bark and leaves (especially lichen forming fungi) • Endophytes of leaves, stems, bark, and roots • Secondary colonizers of dead attached tissues and leaf spots, etc. • Mycorrhizas (endo-, ecto-, eicoid, orchid, etc.) • Leaf surfaces • Nectar • Resin
2.	Dead vascular plants	• Saprobes on wood, bark, and litter • Burnt plant tissues • Saprobes on submerged and inundated plants • Pollen in water samples
3.	Non-vascular plants	• Algae (marine, terrestrial, and freshwater) • Bryophytes
4.	Fungi	• Biotrophs, necrotrophs, and saprobes of other fungi • Lichenicolous fungi • Myxomyceticolous fungi
5.	Vertebrates	• Skin, feathers, hair, bone, etc. • Dung • Nest, liars, etc. • Ruminant guts • Fish scales and guts
6.	Invertebrates	• Biotrophs and necrotrophs • Arthropods exoskeletons • Arthropods and annelid guts • Nematodes • Insect nests
7.	Rock	• Lichens • Epilithic fungi • Endolithic fungi
8.	Soil	• Surface • Soil cores
9.	Water	• Foam • Streams, permanent, and temporary ponds • Litter and wood immersed in sea- and freshwater • Plants (e.g., bromeliads)

ACKNOWLEDGMENTS

The authors thank Dr. D. L. Hawksworth, the untiring crusader of fungal biodiversity cause; Dr. K. D. Hyde for inspiration and Dr. R. C. Rajak for constant encouragement. We thank science funding agencies in India (DBT, DST, and ICMR) for their efforts in establishing the three new culture collections, enabling India to build sufficient infrastructure for conservation and utilization of its fungal wealth. We thank Dr. N. D. Sharma for his fruitful comments on our manuscript.

KEYWORDS

- **biodiversity**
- **conservation**
- **convention on biological diversity**
- **culture collection**
- **fungal diversity**
- **fungi**
- **microbes**
- **mycologist fraternity**
- **policymakers**

REFERENCES

Agrios, G., (2005). *Plant Pathology* (5th edn., p. 952). Academic Press.

Ajantha, G. S., & Kulkarni, R. D., (2011). *Cladophialophora bantiana*, the neotropic fungus, a mini review. *J. Clin. Diagn. Res., 5*(6), 1301–1306.

Anon., (2010–2011). *Annual Report* (p. 262). Department of Science & Technology, Government of India, New Delhi.

Barr, D. J. S., (1992). Evolution of kingdoms of organisms from the perspective of a mycologist. *Mycologia, 84*, 1–11.

Blackwell, M., (2011). The fungi: 1, 2, 3…5.1 million species? *Am. J. Bot., 98*(3), 426–438.

Bratton, J. H., (2003). *Habitat Management to Conserve Fungi – A Literature Review* (p. 20). CCW Natural Science report No. 03/10/1, Natural Science Group, Wales, UK.

Caretta, G., & Piontelli, G., (1998). Preserved ascomatal and other fungal structures on the remains of a ninth century Longboard abbess exhumed from a Monastery in Pavia, Italy, *Mycopathologia, 140*, 77–83.

Cavalier-Smith, T., (1993). Kingdom protozoa and its 18 phyla. *Microbiol. Rev., 57*, 953–994.

Crous, P. W., Verkley, G. J. M., Groenewald, J. Z., & Samson, R. A., (2009). *Fungal Biodiversity* (p. 270). CBS Laboratory Manual Series 1, Utrecht, Central Office for Fungal Cultures.

De Hoog, G. S., Guarro, J., Gene, J., & Figueras, M. J., (2000). *Atlas of Clinical Fungi* (p. 1126). Centraal Bureau voor Schimmel cultures, The Netherlands.

Dreyfuss, M., Haerri, E., Hoffman, H., Kobel, H., Pache, W., & Tscherter, H., (2003). Cyclosporin A and C–New metabolites from *Tolypocladium polysporum* (Link ex pers.) Rifai. *Europ. J. App. Microbiol., 3*, 125.

Fleming, A., (1929). On the antibacterial action of cultures of a *Penicillium*, with special reference to their use in the isolation of a *B. influenze*. *Brit. J. Exp. Path., 10*(3), 226–236.

Fromtling, R. A., Kosanke, S. D., Jensen, J. M., & Bulner, G. S., (1979). Fatal *Beauveria bassiana* infection in a captive American alligator. *J. Am. Vet. Med. Ass., 175*, 934–936.

Gerrits, V. D. E. A. H. G., & De Hoog, G. S., (1999). Variability and molecular diagnostics of the neurotropic species, *Cladophialophora bantiana*. *Stud. Mycol., 43*, 151–162.

Gonzalez, C. J. F., Espejo, S. J., & Barcena, A. M. C., (1995). Mycotic pulmonary disease by *Beauveria bassiana* in a captive tortoise. *Mycoses, 38*, 167–169.

Hawksworth, D. L., & Mueller, G., (2005). Fungal community: Their diversity and distribution. In: Dighton, J., White, J. F., & Oudemans, P., (eds.), *The Fungal Community: Its Organization and Role in the Ecosystem* (p. 30). CRC Press, Taylor & Francis Group, Boca Raton, USA.

Hawksworth, D. L., (1991). The fungal dimension of biodiversity: Magnitude, significance and conservation. *Mycol. Res., 95*, 641–655.

Hawksworth, D. L., (2001). The magnitude of fungal diversity: 1.5 million species estimate revisited. *Mycol. Res., 105*(12), 1422–1432.

Hawksworth, D. L., Minter, D. W., Kinsey, G. C., & Cannon, P. F., (1997). Inventorying a tropical fungal biota: Intensive and extensive approaches. In: Janardhan, K. K., Rajendran, C., Natrajan, K., & Hawksworth, D. L., (eds.), *Tropical Mycology* (pp. 29–50). Enfield, New Hampshire, Science Publishers.

Horré, R., & De Hoog, G. S., (1999). Primary cerebral infections which were caused by melanized fungi: A review. *Stud. Mycol., 43*, 176–193.

Hyde, K. D., & Zhang, Y., (2008). Epitypification: Should we epi-typify? *J. Zhejiang Univ. Science B, 9*(10), 842–846.

Hyde, K. D., Ho, W. H., Taylor, J. E., & Hawksworth, D. L., (2000). Estimating the extent of fungal diversity in the tropics. In: Raven, P. H., & Williams, T., (eds.), *Nature and Human Society-The Quest for a Sustainable World* (pp. 156–175). National Academic Press Washington DC.

Ishibashi, Y., Kaufman, Ichinoe, M., & Kagawa, S., (1987). The pathogenicity of *Beauveria bassiana* in the rabbit cornea. *Mykosen, 30*, 115–126.

Kirk, P., Cannon, P. F., David, J. C., & Stalpers, J. A., (2001). *Ainsworth & Bisby's Dictionary of Fungi* (9th edn., p. 655). CAB International, Surrey, UK.

Kirk, P., Cannon, P. F., Minter, D. W., & Stalpers, J. A., (2008). *Dictionary of Fungi* (10th edn, p. 771). CAB International, Wallingford, UK.

Liu, D., (2011). *Molecular Detection of Human Fungal Pathogens* (p. 932). CRC Press, Taylor & Francis Group, Boca Raton, USA.

Manoharachary, C., Sridhar, K., Singh, R., Adholeya, A., Suryanarayanan, T. S., Rawat, S., & Johri, B. M., (2005). Fungal biodiversity: Distribution, conservation and prospecting of fungi from India. *Curr. Sci., 89*(1), 58–71.

Maranhao, E. A. A., & Santiago-Alvarez, C., (2003). Occurrence of entomopathogenic fungi in soils from different parts of Spain. *IOBC WPRS Bulletin, 26,* 59–62.

Mathew, F., Jonathan, P., & Volk, T., (2009). Nobel prize winning fungi: An educational poster for teaching about mycology. *Inoculum, 60*(3), 16.

Mullis, K. B., (1994). PCR and scientific invention: The trial of DuPont vs Cetus. In: Mullis, K., et al., (eds.), *The Polymerase Chain Reaction* (pp. 427–441). Birkhauser, Berlin.

O'Brien, B. L., Parrent, J. L., Jackson, J. A., Moncalvo, J. M., & Vilgalys, R., (2005). Fungal community analysis by large-scale sequencing of environmental samples. *Appl. Environ. Microbiol., 71,* 5544–5550.

Patterson, D. J., & Sogin, M. L., (1992). Eukaryotic origins and protistan diversity. In: Hartman, H., & Matsuno, K., (eds.), *The Origin and Evolution of the Prokaryotic and Eukaryotic Cell* (pp. 13–46). World Scientific, Singapore.

Polishook, J. D., Bills, G. F., & Lodge, D. J., (1996). Microfungi from decaying leaves of two rain forest trees in Puerto Rico. *J. Ind. Microbiol., 17,* 284–294.

Sachs, S. W., Baum, J., & Mies, J., (1885). *Beauveria bassiana* keratitis. *Br. J. Opthalmol., 69,* 548–550.

Salaman, R. N., (1949). *The History and Social Influence of Potato.* Cambridge University Press, Cambridge.

Sharma, A., & Shouche, Y., (2014). Microbial culture collection (MCC) and international depository authority (IDA) at the national center for cell science, Pune. *Ind. J. Microbiol., 54*(2), 129–133.

Sharma, N. D., (2003). Inventorying biodiversity: Myopic planning. *Curr. Sci., 84*(5), 617, 618.

Sharma, N. D., (2013). Inventorying microbial diversity: New challenges for India. *Curr. Sci., 104*(10), 1280, 1281.

Sharma, R., & Singh, S. K., (2013). A new species of *Gymnoascus* with verruculose ascospores. *IMA Fungus, 4*(2), 177–186.

Smith, D., & Onions, A. H. S., (1994). *Preservation and Maintenance of Living Fungi* (p. 132). CAB International, UK.

Suryanarayanan, T. S., & Gopalan, V., (2014). Crowdsourcing to create national repositories of microbial resources: Fungi as a model. *Curr. Sci., 106*(9), 1196–1200.

UNAID, (2014). *The Gap Report* (p. 422). Geneva, Switzerland.

Vidhate, R., Ghormade, V., Kulkarni, S., Mane, S., Chavan, P., & Despande, M. V., (2013). Mission mode collections of fungi with special reference to entomopathogens and mycopathogens. *Kavaka, 41,* 33–42.

Webster, J., & Weber, R. W. S., (2007). *Introduction to Fungi* (p. 841). Cambridge University Press. Cambridge.

Zedan, H., (2000). *Sustaining Life on Earth-How the Convention on Biological Diversity Promotes Nature and Human Well-Being* (p. 20). Secretariat of the Convention on Biological Diversity, World Trade Centre, Montreal, Quebec, Canada.

CHAPTER 2

Fungal Metagenomics: An Emerging Approach to Determine the Imperceptible Facets of Fungal Diversity

SHIWALI RANA[1], MANISH KUMAR[2], DEEPAK K. MAURYA[1], and SANJAY K. SINGH[1]

[1]*National Fungal Culture Collection of India (NFCCI), Biodiversity and Paleobiology Group, MACS' Agharkar Research Institute, Pune, Maharashtra, India*

[2]*Amity University, Maharajpura, Dang, Gwalior, Madhya Pradesh, India*

ABSTRACT

Microbes are known to produce various metabolites of great industrial interest. Unfortunately, many microbes have neither been identified nor cultivated as the culture-based method has its limitations. An emerging area of metagenomic sequencing, amalgamated with environmental biology and genomics, facilitates a solution for overcoming these problems. It has completely transformed the detection and characterization of complex microbial communities from the genetic material isolated from environmental samples. Moreover, the characterization reveals functional diversity of microbial entities through culture independent assessment. Applications include studying environmental microbial community dynamics, discovering novel genes, enzymes, pathways, and bioactive molecules with entirely new or improved biochemical functions along with diagnosis and monitoring of pathogens. However, significant challenges persist in the functional annotation. In future, innovation and enhancement in technologies and methodologies, will lead to less sequencing cost. Furthermore, data integration using many technological platforms can

Applied Mycology for Agriculture and Foods: Industrial Applications. Sanjay K. Singh, Deepak Kumar, Md. Shamim, & Rohit Sharma (Eds.)

better understand metagenomes and subsequently characterize complex microbiomes and manipulate communities to accomplish better results for agriculture, environmental sustainability, and health. This chapter highlights the procedures for extracting, amplifying, sequencing, and storing fungal DNA from environmental samples, various applications of metagenomics approaches and the role of bioinformatician in the metagenomic study.

2.1 INTRODUCTION

Metagenomics, also known as ecogenomics, environmental genomics, or community genomics, refers to studying the genetic material which has been directly recovered from the environmental samples and provides an understanding of the functional potential of the population. Traditional single genomics relied upon the singularizing individual from complex microbial mixtures, cultivation of the clonal cultures, and sequencing specific cloned genes in order to know the diversity associated with a particular sample. Various types of selective media are used for the isolation of microbes. Careful selection of different culture conditions can help recover a tremendous range of microbes. However, microbiologists noticed a significant discrepancy in the actual number of organisms under the microscope and the factually recovered number in laboratory conditions (Mulcahy-O'Grady & Workentine, 2016). Furthermore, it was found that the immense majority of microbial biodiversity remained unexplored using such cultivation-based methods. Moreover, in the case of epidemiological studies, studying one isolate at a time approach is successful if a single genotype causes any disease. However, if we select and study single isolate, it may mask the possibility that in a highly similar population, yet different genotypes might be responsible for a particular disease (Fricke et al., 2011).

On the other hand, metagenomics offers a new vision for viewing the hidden diversity of the microbial world, which can potentially revolutionize our perception of the whole living world. It catalogs by sequencing all house-keeping, functional genes and whole genomes from a diverse community at once in one go. The cost of DNA sequencing continues to fall; metagenomics will allow environmental microbiology to be explored at a much larger scale and more details. These days shotgun sequencing or PCR-directed sequencing is largely used to achieve unbiased samples from all the members of any natural environmental community. Much interest arose in metagenomics when discoveries showed that the majority of microbes had earlier remained unnoticed with the culture-dependent approach. Metagenomic approaches are prevalent these days in large-scale studies for the genomics applications. They carve a way to study in-depth the taxonomy along with functional composition structure and associated processes of microbial communities from environmental, agricultural, and clinical settings.

In this process, whole genomic DNA is extracted from the environmental samples, irrespective of its composition, followed by their characterization by whole/segment sequencing using different platforms. The allotment of individual sequence reads or assembled sequence contigs to particular taxonomic groups is carried out by using sophisticated bioinformatics tools. Various tools exist that provide species identification in metagenomic samples. To understand the metagenomics data sets and their analysis, it requires a large statistical component. The sequence data needs to be evaluated on the basis of relative abundances rather than based on the absolute presence or absence of data (Fricke et al., 2011). The reduction of the cost and ever-increasing sequence read lengths obtained from high-throughput sequencing platforms has transformed it into a more accessible and significant powerful tool. rRNA gene amplicons sequencing from almost hundreds of samples can be finished parallelly. These finalized datasets can be used to understand the abundance as well as the taxonomic profiling of the microbial populations and species within a sample (Mulcahy-O'Grady & Workentine, 2016).

Omics technologies contributed to gain knowledge about fungal diversity, functionality, and community composition along with the discovery of novel fungal lineages to a greater extent. These technologies help to understand their ecological role, evolution, and biotechnological potential too and especially from poorly explored ecosystems (Vargas-Gastélum & Riquelme, 2020). This chapter is an effort to elaborate and explain the procedures, applications of metagenomics to help researchers working on fungal communities' diversity and dynamics.

2.2 DIVERSITY

The diversity of microorganisms on the planet is still poorly understood. It has been estimated that nearly 1.5 million fungal species belonging to a diverse group of organisms that functions as pathogens, symbionts, and decomposers in ecosystems are present; still, only 5% of the total estimated number has yet been described and discovered (Hawksworth, 2001). Furthermore, many unculturable fungal species inhabiting the phyllosphere, rhizosphere, and other niches are thought to represent a small fraction of the large unknown diversity. This calls for the need for culture-independent technologies commonly referred to as metagenomics in order to assess the true biodiversity of fungi in nature and for exploring microbial communities, especially "the uncultured majority" (O'Brien et al., 2005). The very first metagenomic study using NGS was done using pyrosequencing to study two different samples. They analyzed the data and observed significant differences between the metabolic potential of the two microbial populations/ communities (Edward et al., 2006).

2.2.1 FUNGI IN THE DEEP-SEA ECOSYSTEM

The deep-sea biosphere nearly covers almost 65% of the Earth's surface. The current knowledge about the fungi belonging to the marine environment, especially deep-sea, is scarce compared to the terrestrial environment. For many decades, most information on the fungal diversity (aerobic and anaerobic) of the deep sea was based on culture-dependent techniques. The majority of fungi in extreme marine environments have yet not been revealed due to unfavorable culture conditions in the laboratory. Recent advances in high-throughput next-generation sequencing platforms have revealed that the deep-sea ecosystem encompasses great fungal diversity despite exhibiting very harsh conditions such as high hydrostatic pressure, low temperatures, low organic matter, high salinity, and absence of light. Using high throughput platforms, various new groups of fungi have been identified, which are mostly referred to as "unidentified" or "unclassified." These unidentified groups often represent a very high proportion of the analyzed data, suggesting that the deep-sea harbors a largely unknown fungal community. Better characterization of the fungi belonging to deep-sea environment is essential in order to resolve their participation in various ecological processes and potential biotechnological importance. This has led to a rapid increment in the count of studies based on culture-independent approaches (Vargas-Gastélum & Riquelme, 2020). Microbial ecology studies in the past two decades have revealed significant prokaryotic diversity in hydrothermal vents. However, fungi were unknown from this extreme environment until a few years ago. Many discovered fungi from hydrothermal vents are good siderophore producers, indicating their role in biomineralization processes and seafloor alteration (Connell et al., 2009).

2.2.2 ENDOPHYTIC FUNGI

Endophytic fungi have come out as an essential source of many industrially important metabolites. Such ability of the endophytes results from their complex and continuous interactions with the host and its microbiome. Modern genomic studies involving metagenomics help us to understand the plant microbiomes and complex interactions. It can help to unravel the unidentified areas of endophytic study and improve agricultural management by plant growth promotion, biocontrol, and bioremediation (Kaul et al., 2016). The metagenomics approach helps to unveil uncultured endophytes' capability (Dinsdale et al., 2008); to reveal the information beyond individual taxa. Comparative metagenomics is an effective potential tool to study functional diversity between endophytes of the same host or else different host. Toju et al. (2013) described the fungal community composition of root-associated fungi in a temperate forest of Japan. They studied the coexistence of mycorrhizal fungi as well as endophytic fungi in the roots of different plant species. Coexistence indeed involves the complex interactions between different ecotypes, which can be studied using metaproteomics, metatranscriptomics, or metaproteogenomics approach. The high number of sequences with no homologs in the public databases is a major constrain in metagenomic studies.

2.2.3 SOIL

Almost a 1,000 million cells comprising different microbes inhabit a handful of soil. The fungal biomass accounts for 2–5 tha^{-1} in the soil of temperate grasslands. Soil microbial communities have been found to be of great interest to agriculture as they can degrade the soil's organic matter and make nutrients available to plants. These complex interactions are often mediated by molecular signals that are exchanged between the microbes, microbes and animals, and microbes and plants. Numerous microbial species have been studied from soil using molecular techniques, predominantly amplicon sequencing, and metagenomics. However, these in situ interactions are still poorly unknown because soil metatranscriptomics and soil proteomics has technical challenges (Nannipieri, 2020).

2.2.4 AIR

The 21st century has seen a tremendous increase in studies inspecting the effect of atmospheric particulate matter such as industrial pollutants, dust, and microorganisms, which can significantly affect human health, climate, and agriculture (Lacey & Crook, 1988). A lot of studies have been conducted in hospitals because immunocompromised patients are the ones who are most vulnerable to infection. Many undescribed pathogenic fungi have been identified in different studies where harmless fungal invasions could have resulted in a life-threatening situation (Iwen et al., 1994; Anderson et al., 1996). For the first time, Metagenomics came into play in aerosolic research in a study by Tringe et al. where researchers determined the airborne microorganism's composition in two indoor shopping centers in Singapore. The study found that the indoor populations were different from the outdoors; Most of the biogenic aerosols were prokaryotic, with the eukaryotic population being tiny but significant (between 0.26 and 2.0%) (Tringe et al., 2008). We are already aware of many human diseases, such as asthma and aspergillosis, severe life-threatening effects of black fungi in COVID patients, which are being caused or intensified by the presence of the aerosolic fungal spores (O'Gorman, 2011; Pringle, 2013); therefore, the need of the hour calls for metagenomics in order to know the taxonomic position of the causative microbes. Research in this particular area is developing very quickly, and it can be of great benefit to humankind from the newest available tools and well-suited metagenomic approaches.

2.3 MEDICINE

The microbes, which live in and on us have a great impact on our health and are linked to numerous diseases, which include infections and auto-immune disorders. In order to diagnose and treat such disorders, accurate identification, as well as characterization of the causative agents, is of utmost necessity. The metagenomic approach can provide sequencing of the entire nucleic acid component of the sample, thus giving an accurate snapshot of the microbial community as well as their potential (Maurya et al., 2021). There are many clinical applications that can benefit from such data, such as rapid identification of pathogens, diagnosis, and treatment of gastrointestinal disorders, microbial dysbiosis with intestinal disorders, problems by skin microbiome that may lead to skin disorders, and many other diseases which are associated with microbes, and the identification of antibiotic-resistance genes. Most disorders and diseases are not caused by a single pathogen but usually emerge as a result of many complex ecological interactions which occur within the microbiota.

2.4 UNDERSTANDING ANTIMICROBIAL RESISTANCE

The imbalance of the microflora in the gut is one of the major predisposing factors for antimicrobial resistant organisms such as *Klebsiella*, *Escherichia coli*, and vancomycin-resistant *Enterococci* (De La Cochetière et al., 2008; Bhalla et al., 2003; Ubeda et al., 2010). A healthy microbiome is one of the best defense mechanisms to fight against these organisms. Specific communities and functions which provide colonization resistance against antimicrobial resistance organisms can be identified. Traditional methods for the detection of antibiotic resistance were performed on isolated organisms. In context to it, the metagenome can provide a comprehensive picture of the whole community (Schmieder & Edwards, 2012). In the data set, genetic determinants of resistance are figured and used to predict the resistance patterns. In the metagenomics approach, all known determinants can be identified even if they are not present in the pathogen. However, they may be transferred because of the highly mobile nature of these genes. Mykrobe predictor (Bradley et al., 2015) is used to identify different allele types, such as indels and genes associated with antibiotic resistance and single-nucleotide polymorphisms (SNPs) using de Bruijn graphs. Databases such as the antibiotic resistance gene database (ARDB) (Liu & Pop, 2009) and comprehensive antibiotic-resistance database (CARD) (McArthur et al., 2013) are important for proper interpretation of resistance levels of a sample on the basis of gene content (Mulcahy-O'Grady & Workentine, 2016).

2.5 ROUTINE METHODS USED IN METAGENOMIC ANALYSIS

Recent advances in the sequencing of DNA directly from environmental samples such as soil, water, etc., offer enormous opportunities for biodiversity estimation and monitoring, especially where the collection and identification of whole organisms is not possible. However, there are many methods for extracting, amplifying, sequencing, and storing DNA from environmental samples which are discussed in subsections.

2.5.1 DNA EXTRACTION

The first step of a metagenomic experiment is similar to any other culture-independent method to extract the sample's DNA. The choice of extraction protocol to be followed is a crucial step and will finally influence downstream results. One can use any of the kits available by various manufacturers depending on the sample to be processed. There are specific kits available to extract DNA from sediment, plant tissue, soil, blood, tissue, feces, leaf, litter, water, and ice.

If DNA extraction of a large number of samples is not feasible, then samples can be pooled and mixed. However, wherever it is possible, pooling should be avoided; it can mask a significant proportion of microbial community which could be detectable, particularly organisms distributed with more significant spatial heterogeneity (Manter et al., 2010). Analysis of individual samples is recommended when taxon richness and diversity are to be studied.

2.5.2 AMPLIFICATION

Mass amplification of the genetic markers, also known as DNA barcodes using PCR, followed by high-throughput DNA sequencing in order to identify different organisms in a given sample is commonly known as 'metabarcoding.' The analysis of amplicons or metabarcoding is very distinct from the metagenomics approaches. The latter directly analyzes the genomes within an environmental sample and typically does not target individual genes for analysis.

Using ITS region as universal DNA barcode markers for fungi has been formalized after testing the potential of four markers (ITS, LSU, SSU, and *rpb1*). The internal transcribed spacer region (which includes ITS1, 5.8S, and ITS2) is the most accepted barcoding region for several fungi; however, other regions like 18S nuclear ribosomal small subunit rRNA gene have also been used but found to have fewer hypervariable regions thus not able to differentiate between closely related taxa (Chu et al., 2016; Schoch et al., 2012). ITS has superior species resolution for many taxonomic groups and also shows intra-specific differentiation.

The ITS copy number per fungal cell is roughly more than 250, making it an ideal target gene region for fungal identification, significantly where the concentration of environmental DNA (eDNA) recovered is low (Nilsson et al., 2009). SSU has been found to possess a low evolution rate compared to the ITS region, having lower variation in the identification of taxa, making it desirable for identifying fungi at species and strain level (Bruns et al., 1991). Although LSU performs similar to the ITS region for identifying fungi at the generic level, the ITS region allows the identification at the species level because of the sequence variability. Still, LSU has been found to perform better for fungal taxonomic identification as well as phylogeny (Porras-Alfaro et al., 2013).

To sequence the entire ITS1–5.8S–ITS2 region is not possible using Illumina; therefore, one needs to choose to amplify either ITS1 or ITS2. Both the regions have been widely used for metabarcoding. ITS2 has more data available in the GenBank sequence database than ITS1. The downstream gene region ITS2 (28S) is found to be much more variable compared to ITS1 (18S and 5.8S) (Nilsson et al., 2009; Schoch et al., 2012). The use of LSU barcoding is less because of the smaller database of reference sequences and many misidentifications.

The gITS7 and fITS7 forward primers in combination with the reverse primer ITS4 have been found to amplify the ITS2 region for Ascomycota and Basidiomycota (Ihrmark et al., 2012; White et al., 1990). ITS4 is highly recommended as the reverse primer. Based on the specificity and inclusivity of the primers, universal fungal specific primers ITS1F, ITS2, fITS7, gITS7, and ITS4 are recommended (Table 2.1).

TABLE 2.1 Details of Primers Used for PCR Amplification

Types of Markers	Primer name	Primer Sequences (5′-3′)	References
SSU	SSU817F	TTAGCATGGAATAATRRAATAGGA	Borneman & Hartin (2000)
	5.8S	CGCTGCGTTCTTCATCG	
	5.8SR	TCGATGAAGAACGCAGCG	
ITS	ITS1F	CTTGGTCATTTAGAGGAAGTAA	Gardes & Bruns (1993)
	ITS2	GCTGCGTTCTTCATCGATGC	White et al. (1990)
	ITS4	TCCTCCGCTTATTGATATGC	
	gITS7	GTGARTCATCGARTCTTTG	Ihrmark et al. (2012)
	fITS7	GTGARTCATCGAATCTTTG	
LSU	LR3 F	GTCTTGAAACACGGACC	Hopple & Vilgalys (1994)
	LR5 R	TCCTGAGGGAAACTTCG	
	LROR	ACCCGCTGAACTTAAGC	
rpb1	RPB1-Af	GARTGYCCDGGDCAYTTYGG	http://www.ddbj.nig.ac.jp/
	RPB1-Ac	CCNGCDATNTCRTTRTCCATRTA	
	RPB1-Cr		

Earth microbiome project (EMP) offers clear instructions in order to analyze microbial communities in different types of environmental samples.

Detailed protocols for the extraction of the genomic DNA and how to perform ITS1 amplicon sequencing libraries are available on their website (Gilbert et al., 2014). There arises a problem when one tries to follow standard protocols laid by EMP is that which molecular marker (i.e., whole ITS region, ITS1, or ITS2) should be used for identification purposes. The researcher has to follow Illumina ITS amplicon protocol even it is not the choice.

The other approach is to use primers such as ITS5 and ITS4 to amplify the entire ITS1 and ITS 2 region and sequence it is using Nanopore Platform.

The amplicon must be 'purified' in order to get rid of the remaining primers, PCR enzymes, and salts. A large variety of PCR purification approaches have been used, with AMPure XP being the most commonly adopted method for high-throughput sequencing studies.

2.5.3 SEQUENCING DEPTH AND INSTRUMENTATION

There are many platforms like 454 pyrosequencing, Illumina, Ion Torrent, PacBio, and Oxford Nanopore where the mixed amplicons are sequenced. PacBio and Oxford Nanopore companies are capable of providing long reads, which can likely improve metagenomic datasets, especially de novo assembly, but these are very expensive. Currently, Illumina is the most popular and favored choice for acquiring the read depth needed for better sample coverage. Presently, Illumina's sequencers provide good read quality and price-per-base. The choice of read depth highly depends on the experimental design and budget (Mulcahy-O'Grady & Workentine, 2016).

2.5.4 ANALYSIS

There is no standard protocol for the analysis of metagenomic data. Many new methods are published almost weekly. Good analysis requires a skilled analyst. There are quite a number of tools available, with Kraken (Wood et al., 2014) being the most popular, followed by MetaPhlAn (Segata et al., 2012) because of its speed and accuracy. The analysis of nucleotide sequence big data requires specific expertise with the application of high-performance computing system (HPCs) at different workbenches like CLC genomics and many others. In order to know the functional composition, the reads are usually searched against protein databases such as KEGG. A number of web-based graphical user interface tools, such as MG-RAST (Meyer et al., 2008), CAMERA (Sun et al., 2011) and MEGAN (Huson et al., 2011) are also available. MEGAN uses visualizing annotation results from BLAST searches for functional or taxonomic dendrogram, and it also makes the study of particular taxonomic or functional group visually very easy. Some pipelines have been designed in order to accept long reads derived from Sanger and 454/Roche sequencing while others like QIIME (Caporaso et al., 2010) and MG-RAST (Glass et al., 2010), which can directly accept short-read data generated from the SOLiD, Illumina/Solexa, and Ion Torrent platforms. MGRAST almost requires 75 bp or longer reads for similarity analysis or gene prediction which provides functional classification and taxonomic binning.

Metagenome assembly requires considerable work to be done using specialized assemblers such as Meta-IDBA (Peng et al., 2011). Further, the metagenomics contigs are binned into groups using software like CONCOCT (Alneberg et al., 2014). Manual analysis and binning correction can be done with a tool like Anvi'o (Eren et al., 2015). Contigs further need to be annotated and subsequently followed by functional analysis. Newer tools and algorithms allow the analysis of metagenomic samples till strain level (Luo et al., 2015; Nayfach & Pollard, 2015; Mulcahy-O'Grady & Workentine, 2016). The taxonomic classification depends mainly on the database which is used. During the quantification and identification of fungal communities, different technical aspects should be studied to obtain excellent and reliable data from which one can formulate conclusions. This is also required to carry out comparative analyzes among various data sets.

For the fungal taxonomic studies, a lot of databases are available such as FungiDB (Stajich et al., 2012), PHYMYCO-DB (Mahé et al., 2012), UNITE (Kõljalg et al., 2005), ITSone DB (Santamaria et al., 2012), and ITS2 Database (Schultz et al., 2006; Selig et al., 2008; Koetschan et al., 2010, 2012). FungiDB (Stajich et al., 2012) is a database for functional and genomic data. PHYMYCO-DB (Mahé et al., 2012) is a manually curated bank used for fungal phylogenetic analyzes from *TEF1-α* and SSU rRNA markers derived from GenBank. For Fungi, UNITE database (Nilsson et al., 2018) is the most widely used, as it contains 2,480,043 ITS sequences from the International Nucleotide Sequence Databases (NCBI, DDBJ, EMBL) and is a curated database, and lacks any cryptic sequences (Vargas-Gastélum & Riquelme, 2020) thus helps in improved identification from the environmental samples (Kõljalg et al., 2005; Abarenkov et al., 2010). It also has PlutoF, where users can store, manage, and conduct analysis. ITSone DB (Santamaria et al., 2012) has 4,05,433 ITS1 sequences specific to fungal taxonomy. ITS2 Database (Schultz et al., 2006; Selig et al., 2008; Koetschan et al., 2010,

2012) is a web interface that enables taxon sampling, secondary structure prediction, sequence-structure-based alignment, and tree reconstruction includes sequences from eukaryotic taxa. It has 2,88,044 ITS2 sequences. Noteworthy, is the fact that some databases, such as ITSone DB, UNITE, and ITS2, have a high percentage of sequences annotated as uncultured environmental samples. There is a need for further studies which focuses on fungal genomics as well as metagenomics to increase taxonomic knowledge and improve the classification.

2.6 ROLE OF BIOINFORMATICS IN METAGENOMICS

Advances in computational biology and sequencing have significantly increased the capability for the exploration of the functional and taxonomic compositions of different microbial communities (Van den Bogert et al., 2019). Decreasing sequencing costs, the large size of datasets in studies, and the high depth of sequencing have led to studies with more statistical power and subsequently led to change in OTU tables and functional profiles for machine learning (Pasolli et al., 2016). Many new metagenomic applications call for bioinformatics and data science challenges that vary from mere strain-level resolution in different community profiles to finally integrating these large datasets for machine learning purposes. Therefore, it requires efficient and novel computational tools and skilled bioinformatics analysts as most datasets require customized analysis, and there is no formal set analysis protocol available. However, many new tools these days are intended to be user-friendly in order to remove complex analysis. Moreover, the interpretation of the output should not be made blindly. One needs to have a good understanding of the tools so that one knows it's limitation to acquire precise results from the metagenomic data. New sequencing technologies capable of sequencing single-molecule and capable of producing very long reads are emerging. This calls for the need for a skilled bioinformatician as existing tools might not work on such new data (Figure 2.1).

2.7 CONCLUSION

The development of high throughput sequencing has helped us to obtain large datasets and represents a very cost-efficient alternative to existing conventional techniques. Metagenomics has made it possible to do a significant global analysis of the fungal communities associated with plant systems (pathogens, endophytes, and mycorrhizae), free-living saprobes, and fungi from extreme environments. It generates quantifiable and unbiased profiles of mixed samples needed for a biodiversity survey of any environmental sample. Metagenomics has enormous potential to help understand microbial biodiversity, their geographical distribution, and the ecological roles of fungi. It also holds a tremendous promising role in microbial diagnostics and research. The field of metagenomics is presently under active development. The changes in the metagenomic landscape are making it possible to increase the amount of available sequence data making it possible to fully characterize microbial communities of low complexity. All of these applications of metagenomics rely on the usage of sophisticated computational tools, and this calls for the need for skilled bioinformatics support in order to implement and use metagenomics in all dimensions.

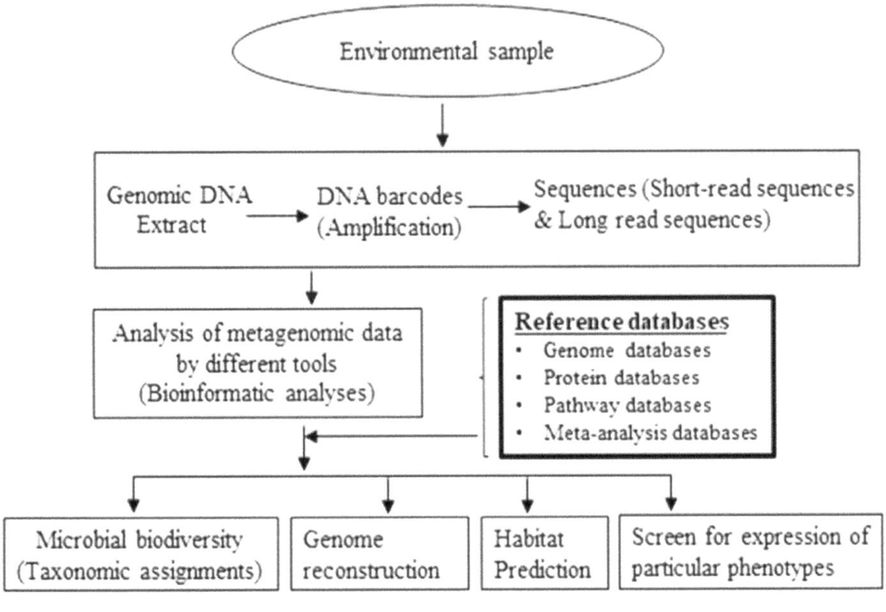

FIGURE 2.1 Flow diagram of metagenomic and bioinformatic analyzes in fungal microbial ecology, and reference databases lay foundations for various bioinformatic analyzes.

Source: Hiraoka et al. (2016).

DNA-based biodiversity assessment using metagenomics is not in widespread usage, but it has many potential benefits, as it does not require taxonomic expertise, which is scarce worldwide. The sampling can be done by non-specialists. The detection of microbes from eDNA is beneficial as it allows the detection of transient organisms, mostly missed by traditional observational sampling. The DNA sequence analysis of large sample numbers using metabarcoding approaches is becoming a more cost-competitive method for biodiversity and biosecurity monitoring as sequencing technologies advance. With reducing costs, long-read sequencing technologies will be commonly used for microbiome studies leading to enhanced taxonomic resolution with full-length marker gene sequencing and improved functional analyzes, which calls for more contiguous metagenome assemblies.

ACKNOWLEDGMENTS

Authors thank Director Agharkar Research Institute (ARI), Pune and Savitribai Phule Pune University (SPPU) for providing facilities. Shiwali Rana acknowledges University Grant Commission (UGC), New Delhi for granting Senior Research Fellowship (SRF).

KEYWORDS

- **antibiotic resistance gene database**
- **DNA analysis**
- **eco-genomics**
- **environmental sample**
- **metagenomics**
- **microbial community**

REFERENCES

Abarenkov, K., Nilsson, R. H., Larsson, K., Alexander, I. J., Eberhardt, U., et al., (2010). The UNITE database for molecular identification of fungi-recent updates and future perspectives. *New Phytologist, 186*, 281–285.

Alneberg, J., Bjarnason, B. S., De Bruijn, I., Schirmer, M., Quick, J., Ijaz, U. Z., Lahti, L., Loman, N.J., Andersson, A.F. & Quince, C. (2014). Binning metagenomic contigs by coverage and composition. *Nature Methods*, *11*(11), 1144–1146.

Anderson, K., Morris, G., Kennedy, H., Croall, J., Michie, J., et al., (1996). Aspergillosis in immunocompromised paediatric patients: Associations with building hygiene, design, and indoor air. *Thorax, 51*, 256–261.

Bhalla, A., Pultz, N. J., Ray, A. J., Hoyen, C. K., Eckstein, E. C., & Donskey, C. J., (2003). Anti-anaerobic antibiotic therapy promotes overgrowth of antibiotic-resistant, gram-negative *Bacilli* and vancomycin-resistant *Enterococci* in the stool of colonized patients. *Infection Control & Hospital Epidemiology, 9*, 644–649.

Borneman, J., & Hartin, R. J. (2000). PCR primers that amplify fungal rRNA genes from environmental samples. *Applied and Environmental Microbiology*, *66*(10), 4356–4360.

Bradley, P., Gordon, N. C., Walker, T. M., Dunn, L., Heys, S., Huang, B., et al., (2015). Rapid antibiotic-resistance predictions from genome sequence data for *Staphylococcus aureus* and *Mycobacterium tuberculosis*. *Nature Communications,* 1.

Bruns, T. D., White, T. J., & Taylor, J. W., (1991). Fungal molecular systematics. *Annual Review of Ecology, Evolution and Systematics, 22*, 525–564.

Caporaso, J. G., Kuczynski, J., Stombaugh, J., Bittinger, K., Bushman, F. D., et al., (2010). QIIME allows analysis of high-throughput community sequencing data. *Nature Methods, 7*, 335, 336.

Chu, H., Wang, C. Y., Wang, H. H., Chen, H., & Tang, M., (2016). Pine wilt disease alters soil properties and root associated communities in *Pinus tabulaeformis* forest. *Plant and Soil, 404*, 237–247.

Connell, L., Barrett, A., Templeton, A., & Staudigel, H., (2009). Fungal diversity associated with an active deep-sea volcano: Vailulu'u seamount, Samoa. *Geomicrobiology Journal, 8*, 597–605.

De La Cochetière, M. F., Durand, T., Lalande, V., Petit, J. C., Potel, G., & Beaugerie, L., (2008). Effect of antibiotic therapy on human fecal microbiota and the relation to the development of *Clostridium difficile*. *Microbial Ecology, 3*, 395–402.

Dinsdale, E. A., Edwards, R. A., Hall, D., Angly, F., Breitbart, M., Brulc, J. M., et al., (2008). Functional metagenomic profiling of nine biomes. *Nature, 452*, 629–632.

Edwards, R. A., Rodriguez-Brito, B., Wegley, L., Haynes, M., Breitbart, M., et al., (2006). Using pyrosequencing to shed light on deep mine microbial ecology. *BMC Genomics, 7*, 57.

Eren, A. M., Esen, Ö. C., Quince, C., Vineis, J. H., Morrison, H. G., Sogin, M. L., & Delmont, T. O., (2015). Anvi'o: An advanced analysis and visualization platform for 'omics data. *Peer J*, e1319.

Fricke, W., Cebula, T., & Ravel, J., (2011). Genomics. In: Budowle, B., Schutzer, S. E., Roger, G. B., Keim, P. S., & Morse, S. A., (eds.), *Microbial Forensics* (2nd edn., pp. 479–492). Academic Press.

Gardes, M., & Bruns, T. D., (1993). ITS primers with enhanced specificity for basidiomycetes-application to the identification of mycorrhizae and rusts. *Molecular Ecology, 2*(2), 11–118.

Gilbert, J. A., Jansson, J. K., & Knight, R., (2014). The earth microbiome project: Successes and aspirations. *BMC Biology*, (1).

Glass, E. M., Wilkening, J., Wilke, A., Antonopoulos, D., & Meyer, F., (2010). *Using the Metagenomics RAST Server (MG-RAST) for Analyzing Shotgun Metagenomes*. Cold Spring Harbor Protocols: PDB.

Hawksworth, D. L., (2001). The magnitude of fungal diversity: The 1.5 million species estimate revisited. *Mycological Research, 12*, 1422–1432.

Hiraoka, S., Yang, C. C., & Iwasaki, W., (2016). Metagenomics and bioinformatics in microbial ecology: Current status and beyond. *Microbes and Environments, 16024*.

Hopple, J. S., & Vilgalys, R., (1994). Phylogenetic relationships among coprinoid taxa and allies based on data from restriction site mapping of nuclear rDNA. *Mycologia, 86*, 96–107.

Huson, D. H., Mitra, S., Ruscheweyh, H. J., Weber, N., & Schuster, S. C., (2011). Integrative analysis of environmental sequences using MEGAN4. *Genome Research, 9*, 1552–1560.

Ihrmark, K., Bodeker, I. T. M., Cruz-Martinez, K., Friberg, H., Kubartova, A., Schenck, J., Strid, Y., et al., (2012). New primers to amplify the fungal ITS2 region – evaluation by 454-sequencing of artificial and natural communities. *FEMS Microbiology Ecology, 82*, 666–677.

Iwen, P. C., Davis, J. C., Reed, E. C., Winfield, B. A., & Heinrichs, S. H., (1994). Airborne fungal spore monitoring in a protective environment during hospital construction, and correlation with an outbreak of invasive aspergillosis. *Infection Control & Hospital Epidemiology, 15*, 303–306.

Kaul, S., Sharma, T., & Dhar, M. K., (2016). "Omics" tools for better understanding the plant–endophyte interactions. *Frontiers in Plant Science, 7*, 955.

Koetschan, C., Förster, F., Keller, A., Schleicher, T., Ruderisch, B., et al., (2010). The ITS2 database III-sequences and structures for phylogeny. *Nucleic Acids Research, 38*, 275–279.

Koetschan, C., Hackl, T., Müller, T., Wolf, M., Förster, F., et al., (2012). ITS2 database IV: Interactive taxon sampling for internal transcribed spacer 2 based phylogenies. *Molecular Phylogenetics and Evolution, 63*, 585–588.

Kõljalg, U., Larsson, K. H., Abarenkov, K., Nilsson, R. H., Alexander, I. J., et al., (2005). UNITE: A database providing web-based methods for the molecular identification of ectomycorrhizal fungi. *New Phytologist, 166*, 1063–1068.

Lacey, J., & Crook, B., (1988). Fungal and actinomycete spores as pollutants of the workplace and occupational allergens. *Annals of Occupational Hygiene*, 32, 515–533.

Liu, B., & Pop, M., (2009). ARDB-antibiotic resistance genes database. *Nucleic Acids Research, (Database)*, D443–D447.

Luo, C., Knight, R., Siljander, H., Knip, M., Xavier, R. J., & Gevers, D., (2015). ConStrains identifies microbial strains in metagenomic datasets. *Nature Biotechnology, 10*, 1045–1052.

Mahé, S., Duhamel, M., Le Calvez, T., Guillot, L., Sarbu, L., et al., (2012). PHYMYCODB: A curated database for analyses of fungal diversity and evolution. *PLoS One, 7*, e43117.

Manter, D. K., Weir, T. L., & Vivanco, J. M., (2010). Negative effects of sample pooling on PCR-based estimates of soil microbial richness and community structure. *Applied and Environmental Microbiology, 7*, 2086–2090.

Maurya, D. K., Kumar, A., Chaurasiya, U., Hussain, T., & Singh, S. K., (2021). Modern era of microbial biotechnology: Opportunities and future prospects. In: Solanki, M. K., Kashyap, P. L., Ansari, R. A., & Kumari, B., (eds.), *Microbiomes and Plant Health: Panoply and Their Applications* (pp. 317–343). Academic Press. ISBN 9780128197158.

McArthur, A. G., Waglechner, N., Nizam, F., Yan, A., Azad, M. A., Baylay, A. J., et al., (2013). The comprehensive antibiotic resistance database. *Antimicrobial Agents and Chemotherapy, 57*, 3348–3357.

Meyer, F., Paarmann, D., D'Souza, M., Olson, R., Glass, E., Kubal, M., et al., (2008). The metagenomics RAST server – a public resource for the automatic phylogenetic and functional analysis of metagenomes. *BMC Bioinformatics, 1*.

Mulcahy-O'Grady, H., & Workentine, M. L., (2016). The challenge and potential of metagenomics in the clinic. *Frontiers in Immunology, 7*, 29. doi: 10.3389/fimmu.2016.00029.

Nannipieri, P., (2020). Soil is still an unknown biological system. *Applied Sciences, 11*, 3717.

Nayfach, S., & Pollard, K. S., (2015). Population Genetic Analyses of Metagenomes Reveal Extensive Strain-Level Variation in Prevalent Human-Associated Bacteria. bioRxiv:031757. 10.1101/031757.

Nilsson, R. H., Larsson, K. H., Taylor, A. F. S., Bengtsson-Palme, J., Jeppesen, T. S., Schigel, D., et al., (2018). The UNITE database for molecular identification of fungi: Handling dark taxa and parallel taxonomic classifications. *Nucleic Acids Research, (D1)*, D259–D264.

Nilsson, R. H., Ryberg, M., Abarenkov, K., SjÃkvist, E., & Kristiansson, E., (2009). The ITS region as a target for characterization of fungal communities using emerging sequencing technologies. *FEMS Microbiology Letters, 1*, 97–101.

O'Brien, H. E., Parrent, J. L., Jackson, J. A., Moncalvo, J. M., & Vilgalys, R., (2005). Fungal community analysis by large-scale sequencing of environmental samples. *Applied and Environmental Microbiology, 9*, 5544–5550.

O'Gorman, C. M., (2011). Airborne *Aspergillus fumigatus* conidia: A risk factor for aspergillosis. *Fungal Biology Reviews, 25*, 151–157.

Pasolli, E., Truong, D., Malik, F., Waldron, L., & Segata, N., (2016). Machine learning meta-analysis of large metagenomic datasets: Tools and biological insights. *PLOS Computational Biology, 12*(7), e1004977.

Peng, Y., Leung, H. C. M., Yiu, S. M., & Chin, F. Y. L., (2011). Meta-IDBA: A de novo assembler for metagenomic data. *Bioinformatics, 13*, i94–i101.

Porras-Alfaro, A., Liu, K. L., Kuske, C. R., & Xie, G., (2013). From genus to phylum: Large-Subunit and internal transcribed spacer rRNA operon regions show similar classification accuracies influenced by database composition. *Applied and Environmental Microbiology, 3*, 829–840.

Pringle, A., (2013). Asthma and the diversity of fungal spores in air. *PLOS Pathogens, 9*, e1003371.

Santamaria, M., Fosso, B., Consiglio, A., De Caro, G., Grillo, G., et al., (2012). Reference databases for taxonomic assignment in metagenomics. *Briefings in Bioinformatics, 13*, 682–695.

Schmieder, R., & Edwards, R., (2012). Insights into antibiotic resistance through metagenomic approaches. *Future Microbiology, 1*, 73–89.

Schoch, C. L., Seifert, K. A., Huhndorf, S., Robert, V., Spouge, J. L., Levesque, C. A., et al., (2012). Fungal barcoding. Nuclear ribosomal internal transcribed spacer (ITS) region as a universal DNA barcode marker for fungi. *Proceedings of the National Academy of Sciences of the United States of America, 109*, 6241–6246.

Schultz, J., Müller, T., Achtziger, M., Seibel, P. N., Dandekar, T., & Wolf, M., (2006). The internal transcribed spacer 2 database--a web server for (not only) low level phylogenetic analyses. *Nucleic Acids Research, 34*, 704–707.

Segata, N., Waldron, L., Ballarini, A., Narasimhan, V., Jousson, O., & Huttenhower, C., (2012). Metagenomic microbial community profiling using unique clade-specific marker genes. *Nature Methods, 8*, 811–814.

Selig, C., Wolf, M., Müller, T., Dandekar, T., & Schultz, J., (2008). The ITS2 database II: Homology modelling RNA structure for molecular systematics. *Nucleic Acids Research, 36*, 377–380.

Stajich, J. E., Harris, T., Brunk, B. P., Brestelli, J., Fischer, S., et al., (2012). FungiDB: An integrated functional genomics database for fungi. *Nucleic Acids Research, 40*, 675–681.

Sun, S., Chen, J., Li, W., Altintas, I., Lin, A., et al., (2011). Community cyberinfrastructure for advanced microbial ecology research and analysis: The CAMERA resource. *Nucleic Acids Research, 39*, 546–551.

Toju, H., Yamamoto, S., Sato, H., Tanabe, A. S., Gilbert, G. S., & Kadowaki, K., (2013). Community composition of root-associated fungi in a Quercus-dominated temperate forest: "codominance" of mycorrhizal and root-endophytic fungi. *Ecology and Evolution, 3*, 1281–1293.

Tringe, S. G., Zhang, T., Liu, X., Yu, Y., Lee, W. H., et al., (2008). The airborne metagenome in an indoor urban environment. *PLoS One, 3*, e1862.

Ubeda, C., Taur, Y., Jenq, R. R., Equinda, M. J., Son, T., Samstein, M., et al., (2010). Vancomycin-resistant *Enterococcus* domination of intestinal microbiota is enabled by antibiotic treatment in mice and precedes bloodstream invasion in humans. *Journal of Clinical Investigation, 12*, 4332–4341.

Van, D. B. B., Boekhorst, J., Pirovano, W., & May, A., (2019). On the role of bioinformatics and data science in industrial microbiome applications. *Frontiers in Genetics, 10*, 721.

Vargas-Gastélum, L., & Riquelme, M., (2020). The mycobiota of the deep sea: What omics can offer. *Life, 11*, 292.

White, T. J., Bruns, T. D., Lee, S. G., & Taylor, J. W., (1990). Amplification and direct sequencing of fungal ribosomal RNA genes for phylogenetics. In: Innis, M. A., Gelfand, D. H., Sninsky, J. J., & White, T. J., (eds.), *PCR Protocols: A Guide to Methods and Applications* (pp. 315–322). New York, Academic Press.

Wood, D. E., & Salzberg, S. L., (2014). Kraken: Ultrafast metagenomic sequence classification using exact alignments. *Genome Biology, 3*, R46.

CHAPTER 3

Role of Fungi in Eco-Safety and Warfare

KHIROOD DOLEY[1], SANJAY K. SINGH[2], and MAHESH BORDE[1]

[1]*Department of Botany, Savitribai Phule Pune University, Pune, Maharashtra, India*

[2]*National Fungal Culture Collection of India (NFCCI), Agharkar Research Institute, Pune, Maharashtra, India*

ABSTRACT

Biological warfare involves employment of toxins of biological origin that are intentionally used for causing epidemic, disease, to eradicate or debilitate humans' lives and even animals and plants may undergo collateral damage. The approach for toxin inoculation may be via bacteria, virus, fungi, plants, etc. However, the desired effect similar to conventional warfare may not be obtained as biological agent does not show immediate effects because of their incubation period which leads to short or long-time delay for desired results. But they are advantageous in terms of acquirement, production, culture, economical, easy availability, etc. Therefore, it is an attractive option to terrorists or groups involved in war or proxies around the world. The biological agents possess potentials which are equivalent to conventional or chemical warfare agents. As these agents also produce obstruction of main cellular metabolism and fatalities that may ultimately causes death. In most cases, these toxins have acted very quickly when given in low but lethal doses. The fungus if developed as biological warfare agent seems to be more disastrous because of non-availability of any vaccines or cure to it which is evident in case of *Coccidioides* spp. Therefore, it has become an essential factor to anticipate biological warfare or bioterrorism an evident phenomenon for future vigilance and timely detection of precise agents will

Applied Mycology for Agriculture and Foods: Industrial Applications. Sanjay K. Singh, Deepak Kumar, Md. Shamim, & Rohit Sharma (Eds.)

be crucial for the masses. Also, the role of medical health authorities will have to participate significantly during biological warfare otherwise it can prove to be disastrous with mass casualties, disease, disabilities, death, chaos among civil population and disruption of economy of a state or world as well. It collectively brings about death, disease to large population along with social and political disorder. Hence, this chapter attempts to review various aspects of biological warfare and underline their successful evaluation in the process of bio-safety level.

3.1 INTRODUCTION

Biological warfare has been referred to deliberate use of biological weapons which may pathogenically strains of microbes or agents that causes death or disease in living organism or have potential to destroy environment or ecological balance (Tewari et al., 2013). Since ancient times humans have been utilizing several weapon technologies for caustic as well as for predation and survival. The ingredients of bioweapon may be biomolecules that possess toxic properties, which are products of organisms such as bacteria, viruses, fungi, plants, insects, and animals. Also, these molecules may bring about severe detrimental effects on the health of other organisms via inhalation, injection, ingestion or absorption (Dorner & Rummel, 2015). Among various biological agents that exists today, the relatively scarce mention of fungal species raise serious issue as to whether the human pathogenic fungi incur threat to standing agriculture only (Rogers et al., 1999). Despite mycotoxins which is a fungal product being reported to possess qualities similar to other potential biological warfare agents (Bennett & Klich, 2003). Therefore, it is now evident that pathogenic fungi have the potential to be used as biological warfare agents. Even the US biological warfare programs have demonstrated experiments that show unsuspecting potentials of *Aspergillus fumigates* spores (Riedel, 2004).

Nevertheless, at present, bioterrorism is of huge concern as it can prove to be catastrophic because of its potential to cause casualties to larger populations using very little technology (Zhang et al., 2014). The production, development, research, and management of biological agent's quality and delivery through proper means or instrument also plays vital role which was observed during decision of World Health Assembly (1996) and accident of Sverdlovsk (Fenner, 1996; Meselson et al., 1994). In addition, it might cause huge disruption in society as a whole, besides bringing death, disease or related disabilities (Hamburg, 1999). It can mark the rate of mortality and suffering together with a societal and political breakdown of any state. Therefore, at present it poses grave danger to 21st century and it becomes imperative to be conscious of the biological consequences in case of war or war-like situation (Kasdorf, 2011). Even at present COVID-19 is being projected by conspiracy theories as possible biological agent of warfare being used and the devastating potential of the COVID-19 pandemic in the lives of individuals, families, communities, countries, and the world is a true fact that microbes could be equally destructive and terrifying than their nuclear weapon counterparts. And it seems really difficult to manage than nuclear ones (Nie, 2020). Hence, it has become evident to take preventive measures beforehand and to establish effective neutralization by related experts and medical practitioners which can be accomplished via accurate, reliable detection by surveillance system, hospitals, laboratories, and other related public or private health departments. For this to achieve, a subservient staff, efficient data analysis systems, and appropriate laboratory infrastructure during an emergency are the essentials for a proper response which we witnessed during the COVID-19 situation in the year 2019.

3.2 HISTORY OF BIOLOGICAL WARFARE

Historically, the employment of biological agents for warfare is a very old concept which may date back to 600 BC and there exists many references to it which makes them difficult to trace (Eitzen & Takafuji, 1997; Stefan, 2004). Previously, scientific or microbial technology were not that advanced as of today which makes it difficult to ascertain exact origin of biological warfare or any information related to it. But biological warfare did exist militarily before the 20th century in the form of intentionally infecting very source of food and water, use of toxic micro-organisms or animals in weapons and use of biologically inoculated bodies of humans. The contemporary microbiology was noticeable at the beginning of the 19th century, marked by an initial understanding of biological warfare scientifically. Thus, it paved the way due to the work done by Louis Pasteur's and Robert Koch's theoretical and practical advancement in understanding of microbiology which ultimately began to offer researchers with the options of systematic isolating and mass multiplication of specific pathogens with their controlled dissemination (Barras & Greub, 2014). During World War I, Germany in particular is reported to be working largely on secret arsenal development of biological origin largely to infect animal fodder with *Bacillus anthracis* or *Burkholderia mallei* against enemy along with France who worked on much limited scale (Geissler & Moon, 2004). Therefore, in 1925 prohibition of biological and chemical warfare via the '*Geneva Protocol for the Prohibition of the Use in War of Asphyxiating, Poisonous or Other Gases, and of Bacteriological Methods of Warfare*' came into existence to cease major threat associated with it. However, it failed to provide any further prohibition in terms of its exploration and development of biological warfare agents which became evident when major developments were recognized in several European nations, USA and Russia until 1975 (Christopher et al., 1999). Noticeably, in 1932, the well-known Japanese laboratory set-up known as Unit 731 (the Army Epidemic Prevention Research Laboratory) where possible biological warfare agents that caused several diseases viz., anthrax, botulism, bubonic plague, cholera, tularemia, smallpox, and several venereal diseases were inoculated on prisoners and various effects was studied by giving no treatment at all. Even several human pathogenic agents such as *B. anthracis, Shigella* species, *Salmonella species*, *Vibrio cholera* and *Yersina pestis* were used against Chinese, Soviet Union and Mongolian military as well as civilian areas that were not under direct control by the Japanese government which cost huge number of human lives (Harris, 1994; Eitzen & Takafuji, 1997). However, evaluation of their effectiveness is questionable. The United States of America formed the US War Research Service in 1942 at Maryland and other parts of the country to counter Germans (Nazis) who were also experimenting with some biological agents. It is well known that when Second World War got over, US government completely reprieved prosecution of Japanese Unit 731 leaders for war crimes committed just to gain their experimental knowledge which can be co-related with USA's biological warfare program (Brody et al., 2014). During Cold-War era (1950–1980), several accusations of using biological warfare agents were reported (Carus, 1998). Therefore, there was growing apprehension amongst several nations concerning the epidemiological risks as well as the futility of the Geneva Protocol (1925). Thus, WHO demanded another one under the name of "Convention on the Prohibition of the Development, Production, and Stockpiling of Bacteriological (Biological) and Toxin Weapons and on their Destruction (BWC)," which came into existence in 1972, which included the US, the UK, and Soviet government along with several hundreds of nations all over the world and it came into force in 1975. Since then, it has been successively reviewed and largely it prohibits and mentions viz.: (i) the possession of biological agents except for 'prophylactic, protective, or other peaceful purposes; (ii) the development of technologies intended for the dispersal of biological agents for offensive military purposes; and (iii) the destruction of existing stocks (John & Courtland, 2006). Still, BWC convention was proved to be not enough for deterring countries to prohibit biological weapons development programs which became evident in famous Soviet Union's unintentional release of anthrax spores from a secret military research facility at Sverdlovsk that caused significant loss of human lives in hundreds in the year 1979. Even Iraqi government under the rule of Saddam Hussein is suspected to support the initiation of biological weapon development plan that which included possible candidature organisms such as *B. anthracis* and several viruses but none of them were put into service during the Gulf War (Noah et al., 2002). At present scenario, the threat pertaining to the use of biological agents as weapon still remains. For instance, in 1984, international contamination was observed in restaurant at Oregon by the supporters of pseudo-Buddhist Rajneeshee cult which infected 751 persons with *Salmonella typhimurium* in order to get control of local government by poisoning civic representatives (Carus, 2000). In 2001, in a span of just 3 months, 23 incidences related to bioterrorism occurred in USA in which letters contaminated with spores of anthrax were mailed through postal service to target television personalities, senator, and others which costed their lives and caused hospitalization (Stefan, 2004). Hence, the well preparedness in terms of clinical microbiologist for the identification of possible biological warfare agents still remains a challenge (Zhang et al., 2014). The early detection and therapy accordingly remain to be a principle to minimize the casualties. The modern development of techniques based on nucleic acid sensors may provide much more accurate data which might need complex sample preparations (Thavaselvam & Vijayaraghavan, 2010).

3.3 BIOLOGICAL WARFARE AGENTS

There are speculations of biological weapon's existence for mass destruction of life, and assumptions are that they are secretly being developed. Some of the biological as well as other weapons that came into notice are discussed as in subsections.

3.3.1 ETHNIC BOMBS

Advancement in genetic approaches for the development of biological weapons has led to target of certain or specific group of populations based on ethnicity, which has generated serious questions regarding its use. There are reports that the South African government were developing a genetically engineered biological weapon that would target the indigenous black population only (Purkitt & Burgess, 2002). In near future, there is possibility to envisage genetically engineered viruses or toxin-producing gene in bacterium that might activate or induce or regulate by gene product or signaling related to specific receptor binding that will determine particular or specific ethnic characteristic of certain population (Batra, 2000; Appel, 2009).

3.3.2 *FUNGI AS BIOWEAPON*

3.3.2.1 *MYCOTOXINS*

During fungal metabolism, there is the production of by-products known as mycotoxins which are considered to be the contributory agents for undesirable health effects in humans and animals (Ciegler & Bennett, 1980). As a result, a fungus that produces mycotoxins, as well as the mycotoxins themselves, is considered to cause potential problems to health and economy of the country. Several mycotoxins from the genera of fungi have been reported such as aflatoxins, fumonisins, satratoxins, and trichothecenes (Ciegler & Bennett, 1980). Justification of using fungi as an agent of biological origin in war is currently limited to *Coccidioides immitis* (Shannon, 2004). And several other fungi which are not pathogens producing toxins (Cole et al., 2003). And fungi that are human pathogenic, can cause disease but at a very slow rate which significantly reduces its ability to affect a larger population during war-like situation. Moreover, Paterson (2006) interestingly mentioned that an allergic response is not enough to be considered something as weapons. He also questioned the capability of *Aspergillus flavus* spraying onto crops to destroy for causing feminine or destructions similar to a war. On the other hand, the using *Fusarium oxysporum* for destruction of coca plants which result into reduction of cocaine production is noteworthy observation (de Vries, 2000).

3.3.2.2 *AFLATOXINS*

The death of thousands of turkeys due to turkey X disease was related to peanut consumption which was contaminated with molds (Blout, 1961). Finally, it resulted into isolation and characterization of aflatoxins. Generally, aflatoxins have been reported to be produced by *Aspergillus* species namely *A. flavus* and *A. parasiticus* even though other *Aspergillus* genera were also reported to be the producers of aflatoxins (Divakara et al., 2014). They are ubiquitous in its distribution among several crops that are cultivated for the consumption of humans and animals (Chandra Nayaka et al., 2013). The effect of aflatoxins exposure due to inhalation has been purified as AFB1 in animal experiments and formation of DNA adducts in the liver of rat (Zarba et al., 1992). Even, pulmonary and immune system suppression in rats and mice has been reported (Jakab et al., 1994). International Agency for Research on Cancer (1993) have described the AFB1 and AFG1 (naturally occurring aflatoxins) as Class I carcinogenic agent to human and linked it with in liver. The advantage of using aflatoxins is that it is relatively simple to produce and can lower combative effectiveness of enemy forces by compelling them into extra protective gears. Moreover, the disease known as aflatoxicosis caused by aflatoxins shows very unspecific clinical manifestations which makes it hard to diagnose.

3.3.2.3 CITRININ

Earlier, citrinin was first reported to be isolated from *Penicillium citrinum* (Hetherington & Raistrick, 1931), afterwards, it was identified in several species of *Penicillium* and *Aspergillus* (*Aspergillus terreus* and *Aspergillus niveus*) as well as in some strains of *Penicillium camemberti* and *Aspergillus oryzae* (Manabe, 2001). Currently, there are several other citrinin producers which include *A. awentil, A. ostianus, A. fumigatus, A. niveus, A. awamori* and *A. parasiticus* (Li et al., 2010). It also has been concerned to contribute to porcine nephropathy. Citrinin has been found to act as a nephrotoxin in all animal experiments which varied among several species. Even if citrinin is regularly related with human foods, the unidentified role it may play in human health is noteworthy.

3.3.2.4 RICIN

Ricin is a phytotoxin that used to be present in seeds of *Riccinus communis*. It was first reported by Hermann Still mark who did its extraction from castor seeds. In plants and bacteria, the ricin protein exhibited toxic properties that make use of conservative transport mechanism as a mode of entry into the host intracellular fluid (Yermakova et al., 2014). It was found that ricin was the main reason for erythrocytes agglutination and precipitation of serum proteins. The toxic property of ricin exhibited apoptosis pathways, alteration of membrane structure and function, direct cell membrane damage, inhibition of protein synthesis and release of cytokine inflammatory (Al-Tamimi & Hegazi, 2008). Clinically, the symptoms shown due to ricin toxin depends on its entry into host organism such as if it is ingested then gastrointestinal bleeding and renal necrosis occurs (Janik et al., 2019). If it enters through the respiratory tract, then fever accompanied by cough and progressive respiratory shortage and stiffness in the chest may be observed. In more severe cases, pulmonary edema, hypotension, and vascular collapse may follow (Aggarwal et al., 2017). The most important aspect of this toxication is that it hardly any treatment is available making it a suitable tool for bioterrorism by employing it in the particular mode of aerosol and by water and food poisoning (Audi et al., 2005). Therefore, it is featured in the list of US Centers for Disease Control (CDC) as a second highest main concerned bio-threatening agent (Shea & Gottron, 2013).

3.3.2.5 TRICHOTHECENE MYCOTOXINS (TRMS)

Trichothecene mycotoxins (TrMs) are a group of toxins known as T2 that several fungi are involved in its production viz., *Fusarium*, *Myrothecium*, *Stachybotrys*, *Cephalosporium*, *Verticimonosporium*, *Trichoderma*, etc. (Cole et al., 2003). It was reported that trichothecenes was used in the 1964 against Yemen by Egyptian (or Russian) (Mostrom & Raisbeck, 2012). Also, it was combined with mustards in attacks that are chemical in nature during the war between Iran & Iraq (1983–1984) (Ember et al., 1984). The Laos was showered with yellow liquid of gluey nature that mimicked rain when it was released from the sky during air-attack which famously became known as "yellow rain" (Venkataramana et al., 2015). The material involved in the attack consisted of yellow pigment that is said to cause symptoms which includes blistering, vomiting, bloody diarrhea (Haig, 1982) and even causes death within hours. Even when plant samples from affected areas exhibited unusual incidence of trichothecene mixtures. In addition, studies have shown that TrM which occurs naturally is found to induce diseases such as food associated diseases due molds, *Stachybotrys atra* (*S. alternans*) on animal feed, cotton lung disease which when cotton dust of *Dendrochilum toxicum* is inhaled. TrMs in another way is cytotoxic to eukaryotic cells due to its ability to inhibit of the process of protein and DNA synthesis. It has the ability to bind ribosomes or mitochondria which come about by the facility available in the form of bidirectional and free movement across the plasma membrane. In addition, TrMs inhibits succinic dehydrogenase activity which causes interference in the electron transport chain. TrM is readily absorbed from intestinal or pulmonary mucous membranes which show its lipophilic characteristics. When an individual gets exposed to TrMs, the severity of disease is affected by several factors such as nutritional status of the host, damage to liver, infections to intestine and route of toxin administration. The heightened exposure to TrMs can result in conditions such as immunosuppression and central nervous system toxicity. Manifestations such as irritation, local cutaneous necrosis and inflammation in combination with the formation of vesicles and bullae occur when dermal tissues are exposed to TrMs. Whereas eye exposure may cause symptoms such as tearing, burning sensation and blurred vision which could continue for weeks. And when there is chronical exposure, it causes alimentary toxic aleukia (ATA) which has four different stages viz., leukopenia, agranulocytosis, necrotic angina, hemorrhagic rash, sepsis, exhaustion of the bone marrow, bleeding from nose, throat, and gums and fever (Balaji-Mood et al., 2014).

In addition, the use of fungi such as *Fusarium oxysporum* has been perceived to cause significant loss of livestock and standing crops if used by terror organizations which will be a constant threat to several countries and has global implications (Arora et al., 2002; Puroshotum et al., 2009).

3.3.3 COCCIDIOIDES

Coccidioides immitis is reported to be the most virulent, endemic and systemic fungi which is considered as pathogen of humans as well as animals. *Coccidioides immitis* and *C. posadasii* are molds which are considered to be the pathogenic ones that pre-dominantly found as saprobes in regions of the south-western US, Mexico, and some parts of Central as well as South America having arid and alkaline soil. In artificial laboratory condition it can readily culturable and it exists as mold naturally that produces spores (arthroconidia) that separates in a particular fashion through dis-articulation from the parent mycelium (Dixon, 2001). These spores are small enough to infect a mammalian lung therefore inhalation of these infectious spores can result in respiratory disease in humans known as coccidioidomycosis or San Joaquin Valley fever. Human exposures are these asymptomatic infections accounts for about 60%, while 40% results in exhibition of manifestations that are similar to pulmonary, influenza to a quite unusual fatal ones (CDC, 2009). The earliest mention of cases among human during warfare was reported among Argentinian soldiers in the 1890s (Posada, 1892). Since then, the military personnel of the USA have been reported to be affected by this mycosis disease. During World War II, U.S. military as well as air force and allied service personnel suffered from *Coccidioides* infections when they were undergoing training in desert and it was firstly reported by Smith in the 1940s (Smith, 1958). Hence, during warfare Coccidioidomycosis might be used and it remains to be a significant menace to overall fitness and preparedness of any standing army personnel deployed at several military bases or exercise areas which are situated near to *Coccidioides* prevalent areas (Crum-Cianflone, 2007). Since, last decade several substantial preventive measures like the development of vaccines have been undertaken against San Joaquin Valley fever (Yoon & Clemons, 2013). However, so far there is currently no clinically available vaccine available against any fungal organism. On the whole, the fight against *Coccidioides* still remains to date (Van Dyke et al., 2019).

3.4 CHARACTERISTICS OF BIOLOGICAL WARFARE AGENTS

As far as biological weapons or biological warfare agents are concerned, it may include micro-organisms or toxins of biological origin that may cause disease or death in humans, animals, and plants (Suzanne & Etizen, 2000). The biological weapons mainly aim at destabilization and complete destruction of economic development and solidity of a country or state (DaSilva, 1999). They are highly capable of causing severe infectious disease and extensive illness that may cause death, chaos, and alarming situations which are economic in delivering to target areas. Thus, they are being called "Poor Man's Nuclear Arsenal" (Lawrence & Dennis, 2001).

3.5 CONSTRAINTS OF USING BIOLOGICAL WARFARE AGENTS

The BWA are difficult to maintain, and quality may vary in containment area or during the growth and harvesting of agents. In addition, if the pathogen is airborne then there is huge risk of escape into the nearby atmosphere. For instance, in 1979 the spores of anthrax got released accidentally from a Soviet military research facility in Sverdlovsk that caused significant loss of human lives of over 100. It also has been referred to as "biological Chernobyl." Moreover, the exact number of victims is still unknown (Meselson et al., 1994). Therefore, a biological warfare agent requires proper storage condition and is difficult to control if released. In addition, there are constraints in delivering it as a weapon in a conventional way, therefore other modes are preferred viz., food, water, aerosol sprays, etc. (Reshma et al., 2004).

3.6 GENETICALLY MODIFIED BIOLOGICAL AGENTS

As modern-day genetic engineering is advancing rapidly hence in the future, it could possibly conceal dangerous virus inside a seemingly harmless bacterium. In nature, we can observe this phenomenon where a bacteriophage inserts its genetic material into chromosomes or plasmids of bacteria and afterwards cause infection to other hosts by reappearance (Clark & Pazdernik, 2016). Also, there is speculation about the transformation of various strains which could carry more virulence or yields more toxins than previously existing ones. And this possibly can be generally developed through economically well-off countries. Also, due to the fact that modern advancement in the field of genetics would contribute to the development of such type of strains in the near future or may be developed already in disguise (Paterson, 2006).

3.7 BIOSAFETY CONCERNS

Nowadays, due to technological advancement, a great variety of biological agents could potentially be used for biological warfare. However, providentially only a few agents can be efficiently disseminated successfully. Given the fact that they cannot be prevented by vaccinations makes them more lethal. Also, the incidences of biological agent use not common which makes clinical laboratories unprepared once it becomes reality. Therefore, institutions dealing with such agents should be well prepared and be familiar with techniques while dealing in precautionary measures (Nulens & Voss, 2002). On the other hand, the requisite efforts necessary for fungal biological warfare agent's production requires much less complex technique and equipment when compared to other BW agent such as virological agents, still it has become imperative to develop effective preemptive measures in terms of epidemiology or clinical spectrum, laboratory research and analysis, disease surveillance, efficient medical response groups and their management systematically (Patwa, 2020). Further, prevention can be achieved by increasing awareness with regard to the clinical manifestations which can overwhelming during pandemic situation and actions would be necessary to reduce further spread of potentially dangerous toxins and it will definitely play its significant role in understanding the mechanisms associated with biotoxins. Therefore, sufficient knowledge of the fungal toxin and proper storage technique is essential to avoid possible contamination and further health and economic implication of fungal related poisoning is unavoidable. Currently, the data regarding potential fungi that can be used as a mode of weapon against human being is an important issue for which greater preparedness may be required against these threats. Earlier, the lack of threat perception posed by fungi in case of biowarfare and bioterrorism suggest itself because of previous non-communication, lack of historical use or developments in this area (Casadevall & Pirofski, 2006). At present, the emphasis on several other strategies that involves monitoring via intelligence inputs, checking on the capabilities of enemies and even terror organizations as to what kinds of resources available with them for likely use of biological weapon in future is also an indication which once becomes reality then it will prove to be very horrible bio-technological blow to any nation or state (DiSilva, 1999).

3.8 CONCLUSION

Historical data suggests successive use and development of biological weapons under disguise at large with the support of state sponsored agencies locally and globally despite serious implications associated with it. Moreover, nothing is deterring states from actually ceasing the research and development of biological weapons, notwithstanding its previous attempts of ceasing the programs via international conventions or protocols. Hence, there is presence of constant threats in the form of bioterrorism or biological warfare and biosafety, which can prove devastating for lives as a whole.

ACKNOWLEDGMENTS

The authors thank the authority of Savitribai Phule Pune University (SPPU) and Agharkar Research Institute (ARI), Pune and for providing the necessary facilities.

CONFLICTS OF INTEREST

The authors declare no conflict of interest.

KEYWORDS

- **alimentary toxic aleukia**
- **biological warfare agent**
- **biosafety**
- **bioterrorism**
- **Centers for Disease Control**
- **deoxyribonucleic acid**
- **ethnic bombs**
- **fungi**
- **trichothecene mycotoxins**

REFERENCES

Aggarwal, R., Aggarwal, H., & Chugh, P. K., (2017). Medical management of ricin poisoning. *Journal of Medical and Allied Science, 7*, 82–86.

Al-Tamimi, F. A., & Hegazi, A. E., (2008). A case of castor bean poisoning. *Sultan Qaboos University Medical Journal, 8*, 83–87.

Appel, M. J., (2009). Is all fair in biological warfare? The controversy over genetically engineered biological weapons. *Journal of Medical Ethics, 35*, 429–432.

Arora, D. R., & Gautum, V., & Arora, B., (2002). Biological warfare: Bioterrorism. *Indian Journal of Medical Microbiology, 20*, 6–11.

Audi, J., Belson, M., Patel, M., Schier, J., & Osterloh, J., (2005). Ricin poisoning: A comprehensive review. *JAMA, 294*, 2342–2351.

Balaji-Mood, M., Moshiri, M., & Etemad, L., (2014). Bio warfare and terrorism: Toxins and other mid-spectrum agents. *Encyclopedia of Toxicology, 1*, 503–508. http://dx.doi.org/10.1016/B978-0-12-386454-3.00589-3.

Barras, V., & Greub, G., (2014). History of biological warfare and bioterrorism. *Clinical Microbiology and Infection, 20*, 497–502. https://doi.org/10.1111/1469-0691.12706.

Batra, H. V., (2000). International response to BW threat and the global scenario. In: Patil, C. S., (ed.), *XXIV National Congress of Indian Association of Medical Microbiologists*. Department of Microbiology, JNMC, Belgaum.

Bennett, J. W., & Klich, M., (2003). Mycotoxins. *Clinical Microbiology Reviews, 16*, 497–516.

Blout, W. P., (1961). Turkey "X" disease. *Turkeys, 9*, 52.

Brody, H., Leonard, S. E., Nie, J. B., & Weindling, P., (2014). U.S. responses to Japanese wartime inhuman experimentation after World War II: National security and wartime exigency. *Cambridge Quarterly of Healthcare Ethics, 23*, 220–230.

Carus, W. S., (1998). *Bioterrorism and Biocrimes: The Illicit Use of Biological Agents in the 20th Century.* Washington, DC: National Defense University.

Carus, W. S., (2000). The Rajneeshees (1984), In: Tucker, J. B., (ed.), *Toxic Terror: Assessing Terrorist Use of Chemical and Biological Weapons* (pp. 115–137). MIT Press, Cambridge.

Casadevall, A., & Pirofski, L. A., (2006). The weapon potential of human pathogenic fungi. *Medical Mycology, 44*, 689–696.

Centers for Disease Control and Prevention (CDC), (2009). Increase in coccidioidomycosis –California, 2000–2007. *MMWR Morb. Mortal Wkly Rep., 58*, 105–109.

Chandranayaka, S., Venkata, R. M., Udayashankar, A. C., Niranjana, S. R., Mortensen, C. N., & Prakash, H. S., (2013). Chemical and molecular methods for detection of toxigenic fungi and their mycotoxins from major food crops. *Laboratory Protocols in Fungal Biology, 1*, 73–90.

Christopher, G. W., Cieslak, T. J., Pavlin, J. A., & Eitzen, E. M., (1999). Biological warfare: A historical perspective. In: Lederberg, J., (ed.), *Biological Weapons: Limiting the Threat* (pp. 17–35). Cambridge, MA: The MIT Press.

Ciegler, A., & Bennett, J. W., (1980). Mycotoxins and mycotoxicoses. *Bioscience, 30*, 512–515.

Clark, D. P., & Pazdernik, N. J., (2016). Biological warfare: Infectious disease and bioterrorism. In: Clark, D. P., & Pazdernik, N. J., (eds.), *Biotechnology* (2nd edn., pp. 687–719). Academic Cell. http://dx.doi.org/10.1016/B978-0-12-385015-7.00022-3.

Cole, R. A., Jarvis, B. B., & Schweikert, M. A., (2003). *Handbook of Secondary Metabolites* (pp. 199–560). Academic Press, New York, NY, USA.

Crum-Cianflone, N. F., (2007). Coccidioidomycosis in the U.S. military. *Ann. NY Acad. Sci., 1111*, 112–121. doi: 10.1196/annals.1406.001.

DaSilva, E. J., (1999). Biological warfare, bioterrorism, biodefence and the biological and toxin weapons convention. *Electronic Journal of Biotechnology, 2*, 109–139. http://www.ejb.org/content/vol2/issue3/full/2/ (accessed on 13 January 2023).

Divakara, S. T., Aiyaz, M., Hariprasad, P., Chandranayaka, S., & Niranjana, S. R., (2014). *Aspergillus flavus* infection and aflatoxin contamination in sorghum seeds and their biological management. *Archives of Phytopathology and Plant Protection, 1*, 1–16.

Dixon, D. M., (2001). *Coccidioides immitis* as a select agent of bioterrorism. *Journal of Applied Microbiology, 91*, 602–605.

Dorner, B., & Rummel, A., (2015). Preface biological toxins—Ancient molecules posing a current threat. *Toxins, 7,* 5320, 5321.

Eitzen, E. M., & Takafuji, E. T., (1997). Historical overview of biological warfare. In: Sidell, F. R., Takafuji, E. T., & Franz, D. R., (eds.), *Medical Aspects of Chemical and Biological Warfare* (pp. 415–423). Office of the Surgeon General, Department of the Army, Walter Reed Army Medical Center.

Ember, L. R., Sorenson, W. G., & Lewis, D. M., (1984). Charges of toxic arms use by Iraq escalate. *Chem. Eng. News, 62*, 16–18.

Fenner, F., (1996). Poxviruses. In: Knipe, D. M., Howley, P. M., Griffin, D. E., Lamb, R. A., Martin, M. A., Roizman, B., & Straus, S. E., (eds.), *Fields Virology* (3rd edn., pp. 2673–2695). Philadelphia: Lippincott-Raven Publishers.

Geissler, E., & van Courtland, J. E. (2004). In: Geissler, E., & Van Moon, J. E. V. C., (eds.), *Biological and Toxin Weapons Research, Development and Use from the Middle Ages to 1945*. Oxford University Press, Oxford.

Haig, A. M., (1982). *Chemical Warfare in Southeast Asia and Afghanistan: Report to the Congress from Secretary of State Haig AM, Special Report No. 98*. Washington, DC: US Department of State.

Hamburg, M. A., (1999). Addressing bioterrorist threats: Where do we go from here? *Emerging Infectious Diseases, 5*, 564, 565.
Harris, S. H., (1994). *Factories of Death*. Japanese biological warfare–1932–1945, and the American cover-up. New York: Routledge.
Hetherington, A. C., & Raistrick, H., (1931). Studies in the biochemistry of microorganisms. Part XIV. On the production and chemical constitution of a new yellow coloring matter, citrinin, produced from glucose by *Penicillium citrinum* Thom. *Philosophical Transactions of Royal Society of London, Ser. B, 220*, 269–295.
IARC, (1993). *Some Naturally Occurring Substances: Food Items and Constituents, Heterocyclic Aromatic Amines, and Mycotoxins, IARC Monographs on the Evaluation of Carcinogenic Risks to Humans* (Vol. 56). Lyon: International Agency for Research on Cancer.
Jakab, G. J., Hmieleski, R. R., & Zarba, A., (1994). Respiratory aflatoxicosis: Suppression of pulmonary and systemic host defenses in rats and mice. *Toxicology and Applied Pharmacology, 125*, 198–205.
Janik, E., Ceremuga, M., Saluk-Bijak, J., & Bijak, M., (2019). Biological toxins as the potential tools for bioterrorism. *International Journal of Molecular Sciences, 20,* 1181.
John, M., & Van, C. E., (2006). In: Wheelis, M., & Dando, M., (eds.), *Deadly Cultures: Biological Weapons Since 1945* (pp. 9–46). Cambridge, MA: Harvard University Press.
Kasdorf, B., (2011). EPUB 3: Not your father's EPUB. *Information Standards Quarterly, 23*, 4.
Lawrence, C. M., & Dennis, L. K., (2001). Basic considerations in infectious diseases. In: *Harrison's Principles of Internal Medicine* (15th edn., Vol. 1, pp. 763, 764).
Li, Y. N., Wang, Y. Y., Zheng, Y. Q., & Guo, Y. H., (2010). Preparation and characterization of the high specificity monoclonal antibodies against citrinin. *Progress in Biochemistry and Biophysics, 37,* 1248–1253.
Manabe, M., (2001). Fermented foods and mycotoxins. *Mycotoxins, 51*, 25–28.
Meselson, M., Guillemin, J., Hugh-Jones, M., Langmuir, A., Popova, I., Shelokov, A., & Yampolskaya, O., (1994). The Sverdlovsk anthrax outbreak of 1979. *Science, 266*, 1202–1208.
Mostrom, M. S., & Raisbeck, M. F., (2012). Trichothecenes. In: Gupta, R. C., (ed.), *Veterinary Toxicology* (2nd edn., pp. 1239–1265). Academic Press.
Nie, J. B., (2020). *In the Shadow of Biological Warfare: Conspiracy Theories on the Origins of COVID-19 and Enhancing Global Governance of Biosafety as a Matter of Urgency Bioethical Inquiry*. https://doi.org/10.1007/s11673-020-10025-8.
Noah, D. L., Huebner, K. D., Darling, R. G., & Waeckerle, J. F., (2002). The history and threat of biological warfare and terrorism. *Emergency Medicine Clinics of North America, 20*, 255–271.
Nulens, E., & Voss, A., (2002). Laboratory diagnosis and biosafety issues of biological warfare agents. *Clinical Microbiology and Infection, 8*, 455–466.
Paterson, R. R. M., (2006). Fungi and fungal toxins as weapons. *Mycol. Res., 110*, 1003–1010. doi: 10.1016/j.mycres.2006.04.004.
Patwa, J., & Flora, S. J. S., (2020). *Medical Management of Diseases Associated with Biological Warfare*, 151–172. https://doi.org/10.1016/B978-0-12-812026-2.00008-6.
Posada, A., (1892). Uno nuevo caso de micosisfungoidea con psorospermias. *Ann. Circ. Med. Argentino, 15,* 585–596.
Purkitt, H. E., & Burgess, S., (2002). South Africa's chemical and biological warfare program: A historical and international perspective. *Journal of Southern African Studies, 28*, 229–253.
Reshma, A., Shukla, S. K., Dharmani, S., & Gandhi, A., (2004). Biological warfare – an emerging threat. *Journal of the Physicians of India, 52*, 733–738.

Riedel, S., (2004). Biological warfare and bioterrorism: A historical review. *Proceedings Baylor University Medical Center, 17*, 400–406.

Rogers, P., Whitby, S., & Dando, M., (1999). Biological warfare against crops. *Scientific American, 280*, 70–75.

Shannon, M., (2004). Management of infectious agents of bioterrorism. *Clinical Pediatric Emergency Medicine, 5*, 63–71.

Shea, D., & Gottron, F., (2013). *Ricin: Technical Background and Potential Role in Terrorism*. CRS Report for Congress: Washington, DC, USA, RS2138.

Smith, C. E., (1958). Coccidioidomycosis. In: Colonel, J. B. J., (ed.), *Preventive Medicine in World War II: Communicable Diseases Transmitted Chiefly through Respiratory and Alimentary Tracts* (Vol. IV, pp. 285–316).

Stefan, R., (2004). Biological warfare and bioterrorism: A historical review. *Proc. Bayl. Univ. Med. Cent., 17*, 400–406.

Suzanne, R. W., & Col. Edward, M. E., (2000). Hazardous material exposure. In: *Emergency Medicine* (5th edn., pp. 1209–1214).

Tewari, A. K., Rashi, Wadhwa, G., Sharma, S. K., & Jain, C. K., (2013). BIRS: bioterrorism information retrieval system. *Bioinformation, 9*, 112–115.

Thavaselvam, D., & Vijayaraghavan, R., (2010). Biological warfare agents. *Journal of Pharmacy and Bioallied Sciences, 2*, 179.

Van, D., M. C. C., Thompson, G. R. 3rd., Galgiani, J. N., & Barker, B. M., (2019). The rise of Coccidioides: Forces against the dust devil unleashed. *Frontiers in Immunology, 10*, 2188.

Venkataramana, M., Chandranayaka, S., Prakash, H. S., & Niranjana, S. R., (2015). Mycotoxins relevant to biowarfare and their detection. In: Gopalakrishnakone, P., (ed.), *Biological Toxins and Bioterrorism, Toxinology* (pp. 1–22). Springer, Dordrecht.

Yermakova, A., Klokk, T. I., Cole, R., Sandvig, K., & Mantisa, N. J., (2014). *Antibody-Mediated Inhibition of Ricin Toxin Retrograde Transport.* mBIO5: E00995-13.

Yoon, H. J., & Clemons, K. V., (2013). Vaccines against *Coccidioides*. *Korean J. Intern. Med., 28*, 403–407. http://dx.doi.org/10.3904/kjim.2013.28.4.403.

Zarba, A., Hmieleski, R., & Hemenway, D. R., (1992). Aflatoxin B1-DNA adduct formation in rat liver following exposure by aerosol inhalation. *Carcinogenesis, 13*, 1031–1033.

Zhang, X., Kuca, K., Dohnal, V., Dohnalova, L., Wu, Q., & Wu, C., (2014). Military potential of biological toxins. *Journal of Applied Biomedicine, 12*, 63–77.

CHAPTER 4

Revisiting the Biodiversity and Ecosystem Functioning of Arbuscular Mycorrhizal Fungi Under Different Agricultural Management Practices and Environmental Stresses

DIPANTI CHOURASIYA[1], REENA BUADE[2], RAHUL GAJGHATE[3], ANIL PRAKASH[2], MANJU M. GUPTA[4], and MAHAVEER P. SHARMA[1]

[1]*Microbiology Section, ICAR–Indian Institute of Soybean Research, Indore, Madhya Pradesh, India*

[2]*Department of Microbiology, Barkatullah University, Bhopal, Madhya Pradesh, India*

[3]*Crop Improvement Division, ICAR–Indian Grassland and Fodder Research Institute, Jhansi, Uttar Pradesh, India*

[4]*Sri Aurobindo College, Delhi University, Malviya Nagar, New Delhi, India*

ABSTRACT

Arbuscular mycorrhizal fungi (AMF) form a mutualistic symbiosis with the roots of more than 80% of plants. These fungi provide numerous benefits to the plants, *viz.*, improved uptake of mineral nutrients, plant growth, soil carbon sequestration, and confers resistance to biotic and abiotic stresses in plants. AMF functioning and diversity is immensely altered by chemical fertilizers, soil disruption, and cropping pattern. Admittance of appropriate crop and soil management practices for maintaining a high population

Applied Mycology for Agriculture and Foods: Industrial Applications. Sanjay K. Singh, Deepak Kumar, Md. Shamim, & Rohit Sharma (Eds.)

and functioning of indigenous resident AMF would help in sustaining plant productivity under different ecosystems. Studies have indicated that the application of organic fertilizers with reduced tillage practices under particular crop sequences and selecting the efficient AMF strain are acting as potential drivers for AMF functioning. This chapter focuses on: (i) the status of distribution of functional diversity of AMF in agroecosystems; (ii) crop and soil management practices as key drivers affect the AMF diversity; and (iii) cautions for commercial exploitation of AMF biodiversity for application in sustaining the plant productivity.

4.1 INTRODUCTION

Arbuscular mycorrhizal fungi (AMF), members of subphylum *Glomeromycotina*, are the most common association among crop plants (Spatafora et al., 2016). These fungi support plant growth by providing various kind of benefits to plant, such as increased nutrient uptake (Bagyaraj et al., 2015), heavy metal tolerance (Hildebrandt et al., 2007: Abdelhameed et al., 2019), soil carbon sequestration (Rillig, 2004; Parihar et al., 2020; Agnihotri et al., 2021), altered rate of water movement (Auge, 2001) alleviate moisture stress (Bharti et al., 2017; Mathimaran et al., 2017) and also influence soil microbes (Drigo et al., 2010; Gupta et al., 2018). In addition, AMF secretes a sticky substance called glomalin (an insoluble glue-like substance protein) (Rillig & Mummey, 2006), which improves soil structure, water retention and promotes soil aggregation resulting in improved soil health (Willis et al., 2013; Gupta, 2020; Agnihotri et al., 2021).

Numerous studies and critical reviews have suggested the beneficial effect of AMF in increasing plant growth and enhancing the productivity of many crop plants (Hijri, 2016; Ryan & Graham, 2018; Rillig et al., 2019; Gupta & Abbott, 2020), including production of wheat (Ryan & Graham, 2018), maize (Chen et al., 2014), yam (Lu et al., 2015), tomato (Latef & Chaoxing, 2011; Bona et al., 2017), potato (Hijri, 2016), cotton (Gao et al., 2020), cape gooseberry (Ramírez-Gómez et al., 2019) and soybean (Sharma et al., 2012, 2016). Besides improving the plant growth parameters, AMF can also enhance crop quality parameters by stimulating the production of health promoting phytochemicals in edible plants (Sbrana et al., 2014; Rouphael et al., 2015; Gupta & Abbott, 2020). To cite an example, Zeng et al. (2014) reported increased in contents of sugar, soluble solids, vitamin C, total phenolics, flavonoids, and minerals elements due to *Glomus versiforme*, which improves the quality of citrus fruit. Further, the AMF response in crop plants is reported to vary with AMF strain therefore a well-adapted and competitive AMF strains have to be selected and exploited for the growth promotion and biocontrol abilities (Hart et al., 2017).

The AMF fungal diversity, functioning, and efficiency in agroecosystems have been reported to be influenced due to crop and soil management practices (Agnihotri et al., 2017, 2021; Lombardo et al., 2019; Wang et al., 2015) and environmental stresses like rainfall patterns (Hazard et al., 2013) and temperature (Dumbrell et al., 2011). The intensive uses of agronomic practices such as monoculture cropping, tillage practices, chemical fertilizers, biocides although has sustained the crop productivity but poses negative impact on the diversity of native AMF inhabiting in these ecosystems (Hazard et al., 2013). The use of conservation tillage practices (zero and minimum tillage), crop sequences involving AM-host crops promotes population of native AMF (Sharma et al., 2012) and organic farming promotes to greater AMF diversity (Oehl et al., 2004). Whereas arable field showed low taxonomic diversity of AMF (Helgason et al., 1998) but negatively correlated with the richness of AMF species (Oehl et al., 2003; Hijri et al., 2006). It has also demonstrated that a plant growth stage affects the community composition and diversity of AMF (Jansa et al., 2002; Buade et al., 2020). The AMF community and diversity has been also influenced due to land use types. Grasslands showed higher AMF diversity and abundance in AMF taxa than forest and arable lands (Xu et al., 2017).

In the present chapter, we aim to revisit and reassess the influence of different agricultural management practices and environmental stresses on the biodiversity and ecosystem functioning of AMF in an ecosystem. The status of AM fungal diversity (both taxonomic and functional) in agroecosystems is discussed with reference to several soil and crop management practices as main drivers and the risks and cautions before commercial exploitation of AMF biodiversity for application in sustaining the plant productivity.

4.2 STATUS OF AMF DIVERSITY AND DISTRIBUTION UNDER AGROECOSYSTEMS

AMFs are one of the most widespread important components of the soil biota influencing the soil nutrient cycling and functioning of agricultural systems. The biodiversity of AMF includes the species variability (taxonomic), genes (genetic) and the ecosystems they involve (functional) (Gupta et al., 2019).

Generally, the AMF biodiversity is being determined based on morphometric features of spores. Therefore, as a polyphasic approach, molecular studies employing ITS/18S rRNA sequences supplemented with spore morphology are being advocated (Redecker et al., 2013). Based on comparison drawn from the classifications given by Goto et al. (2012); and Redecker et al. (2013), a current status of taxonomic classification of AMF is depicted in Figure 4.1. As per the current status, about 300 AMF species were reported from all over the world. During the last 10 years about 50 AMF species were determined based on metagenomic methods and expected may be more (Öpik et al., 2013). AMF belonging to 10 out of 11 families such as *Acaulosporaceae, Ambisporaceae, Archaesporacea, Claroidoglomeraceae, Diversisporaceae, Gigasporaceae, Glomeraceae, Pacisporaceae, Paraglomeraceae,* and *Sacculosporaceae* and from which 323 AMF species have been defined (Opik et al., 2013). However, studies pertaining to AM fungal communities associated with different agroecosystems are limited.

The diversity of AM fungal communities has been studied across different agroecosystems (Table 4.1), which range from conventional soybean-based cropping systems (Buade et al., 2020) to tropical cowpea agro-ecosystems of African region (Johnson et al., 2013). Under temperate conditions, the diversity of AMF in calcareous grassland field contaminated with phosphate was detected to be a total of 6 species belonging to the genus *Glomus* (Renker et al., 2005). AMF genus *Glomus* was found to be most widely distributed and predominant in many ecosystems (Sýkorová et al., 2007) and 45% species described, the *Rhizophagus irregularis* was described as a common strain and found mostly in the agroecosystems. On the other hand, *Gigaspora* and *Sclerocystis* spp. have been reported to be more common in tropical soils but less in disturbed agroecosystems (Siqueira et al., 1989). AMF genera *Scutellospora,* the members of the *Gigasporaceae* family were observed to be more predominant in sandy soils (Cuenca & Lovera, 2010; Chaudhary et al., 2016). Currently, many reports on the AMF community have been published under different environmental conditions (Öpik et al., 2006; Krüger et al., 2009; Lin et al., 2019) where some of AMF species are rare occurrence and are unique. For example, in forest, *Sacculospora baltica;* in grassland, *Acaulospora nivalis, A. scrobiculata, Gigaspora brohultii, Paraglomus laccatum* and *Racocetra* spp.; in horticultural agroecosystems, *Funneliformis monosporum, Glomus clavisporum, G. vesiculiferum*, and *Rhizoglomus aggregatum*, and in crop agroecosystems, *Ambispora gerdemannii, Dominikia aurea, G. ambisporum, Rhizoglomus microaggregatum* and *Sclerocystis* spp. were observed which may or may not be beneficial to the plants (Fyson & Oaks, 1991; Johnson et al., 1992).

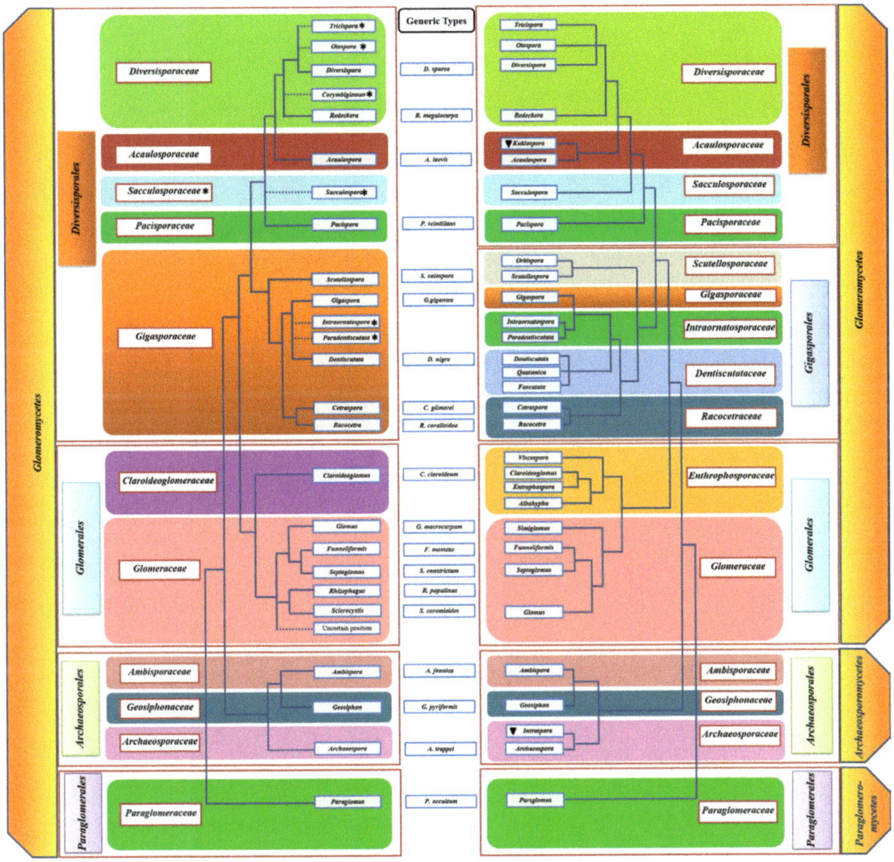

FIGURE 4.1 Current status of AMF classification. The genera assigned to uncertain position are indicated by dotted lines; asterisks indicate insufficient evidence and validated awaited; inverted triangles indicate taxa already rejected in previous publication.
Source: Adapted from: Goto et al. (2012); Redeker et al. (2013); Oehl et al. (2011).

The dynamics of AMF communities assessed across the soybean growth stages showed changes in the AMF colonization pattern and diversity throughout its phenological growth stages (Buade et al., 2020). They reported a high dominance of AMF spp. mainly *Glomus* morphotypes as compared to *Gigaspora* and *Acaulospora.* They also reported a total of 40 AMF morphotypes such as *Glomus aggregatum*, *Rhizophagus fasciculatum*, *Glomus coronatum*, *Glomus etunicatum*, *Glomus* spp. as most frequent and abundant irrespective of growth stages (Buade et al., 2020).

It is evident that the majority of modern intensive agricultural practices, such as frequent tillage practices, monocultures, luxuriant uses of fertilizer and

TABLE 4.1 A Brief Account of AMF Species Reported Under Different Agroecosystems

Agroecosystems	AMF Dominant Species Description	References
Continuous cropping soybean system	*Funneliformis mossae* and *Glomus*.	Cui et al. (2018)
Wheat agro-ecosystem in Southern Chile	26 AMF species *Acaulospora*.	Castillo et al. (2016)
Cropland, grassland, and forest Southern-Central zone of Chile	66 species of *Glomeromycota*.	Castillo et al. (2016)
Araucaria nursery seedling	*Glomus* was dominated and *Acaulospora scrobiculata, Denticulata heterogama, Glomus spinoliferum* and unidentified species of genera *Gigaspora* and *Glomus*.	Vilcatoma-Medina et al. (2018)
Mouth of lake Budi, Chile	Low AMF species richness (only 5)	Medina et al. (2015)
Organic tomato in a Mediterranean site	58 AMF species genera were *Glomus* and *Funneliformis*, which accounted for 27.8 and 25% of total species. Five species *Funneliformis geosporum, F. mossae, G. badium, Septoglomus costrictum*, and *Claroideoglomus lutem* were evenly distributed.	Njeru et al. (2015)
No-till and in temperate grasslands	38 AMF species	Maurer et al. (2014)
Tropical Atlantic Forests	50 AMF species belonging to 15 genera were recorded. *Acaulospora* spp. and *Glomus* spp. predominated, accounting for 52% of total species.	Pereira et al. (2014)
Conventional agri-practices (soybean-based cropping system)	40 AMF morphotypes were observed from which species, *viz. Glomus aggregatum, Rhizophagus fasciculatum, Glomus coronatum, Glomus etunicatum, Glomus* sp. (76.92%).	Buade et al. (2020)
Tropical agroecosystems of African region (cowpea)	15 AMF species belonging to 8 genera (*Gigaspora, Sculellospora, Racocetra, Acaulospora, Funneliformis, Rhizophagus, Glomus*, and *Claroideoglomus*) and 4 families.	Johnson et al. (2013)
Cape gooseberry (*Physalis perviana* L.) crops grown in rainy and dry season	46 species, grouped in 16 genera, *Glomus* species were dominated only four species (*G. macrocarpum, G sp12, Diversipora celata*, and *Kuklospora colombiana*).	Ramírez-Gómez et al. (2019)
Conventional agri-practices (maize-soybean-based cropping system)	*Glomus-f-Glomeraceae, Paraglomus*, and *Gigaspora*.	Zhang et al. (2020)

chemical pesticides although increases the plant yield in the short period but decreases the biodiversity involved in the ecosystem services (Sendek et al., 2019). The community of native AMF spp. in African tropical agroecosystems associated with the rhizosphere of cowpea was found to be highly influenced due to soil edaphic-climatic and agricultural practices (Johnson et al., 2013). It was observed that across all the zones, the soils were marginal and deficient in available P and K, the population of indigenous AMF spore density was not altered significantly. From all the locations, a total of 15 AMF morphospecies covering eight genera *viz., Gigaspora, Scutellospora, Racocetra, Acaulospora, Funneliformis, Rhizophagus, Glomus*, and *Claroideoglomus* were detected out of which *Glomeraceae* was the most prevalent (92%). Moreover, AMF diversity indices such as evenness and diversity indices were adversely affected with rainfall, available phosphorus, and other soil management practices such as tillage, and soil disturbance (Johnson et al., 2013).

4.3 DRIVERS OF AMF BIODIVERSITY AND ECOLOGICAL FUNCTIONS IN AGROECOSYSTEMS

4.3.1 CROPPING SYSTEMS AND TILLAGE PRACTICES

Crop rotation modifies the population of soil microflora, including AMF by changing the availability of soil nutrients and interactions among them. Several studies have demonstrated the positive influence of crop rotation on mycorrhizal fungal communities for example sunflower-sugarcane (Oehl et al., 2004; Hijri et al., 2006; Ambrosano et al., 2010). Bakhshandeh et al. (2017) showed that AMF plays a key role in crop rotation systems by decreasing the dependency on fertilizers and it can reduce the use of chemical fertilizers by 50% without compromising the agriculture production (Srivastava et al., 2017). Most agricultural crops can perform better when colonized by compatible/specific AMF species (Mathimaran et al., 2007), however a critical perspective to this has been put forward recently by Ryan & Graham (2018); Rillig et al. (2019). Different crops influence the AMF diversity (Johnson et al., 2003) for example crop rotation favors the species diversity of AMF spore communities in the rhizosphere of soybean (Mathimaran et al., 2007). The stage of soybean also influences the AMF community and population with a higher population at the maturation stage (Buade et al., 2020). The AM colonization increased in soybean was planted before the maize, whereas the colonization increases in maize when soybean was sown with rhizobial inoculations (Sanginga et al., 1999). Similarly, irrespective of tillage systems, the inclusion of maize in the soybean rotation showed comparatively more infectivity potential of local AM fungi, which has supported higher soybean yield (Sharma et al., 2012).

It is evident from a number of studies that agricultural soils have low diversity mostly inhabited by Glomus species than the natural ecosystems (Menéndez et al., 2001; Chen et al., 2014). This could be due to adoption of monocropping practices that have low diversity of hosts (Oehl et al., 2003). On the other hand, the continuous cropping systems showed better results than the monocropping. In the earlier studies, it was found that forage rotation, conserve AMF diversity where alfalfa-winter cereal found abundance of *Scutellospora* and *Funneliformis* was dominated in frequent tillage practices (Pellegrino et al., 2020). Turrini et al. (2016) showed that AMF are more abundant in cover crops and were represented by *Acaulospora avernata*, while maize roots colonize by *Funneliformis mosseae*, *Rhizophagus intraradices*, and *Glomus* spp. Hence, host represents a strong driver in agroecosystems influencing the AMF community dynamics and functioning.

Every type of tillage practices, e.g., digging, stirring or overturning disrupt the hyphal network and negatively affect the mycorrhizal symbiosis (Agnihotri et al., 2017). AMF spore density, species richness and diversity were increased in reduced till and affected by land use type (Säle et al., 2015). Under no-tillage and reduced till found to alter the AMF community, increased AMF spore density, species richness and diversity in maize and bean crop as compared to tilled soils (Higo et al., 2020) whereas in tilled soil, the members of family *Glomeraceae* species were found to have higher abundance (Rosendahl et al., 2009) because it maintains random hyphal connections after soil disruption whereas *Gigasporaceae* family does not rejuvenate from hyphal fragments. Soil tillage alters the significant reduction of colonized roots because of breaking up their hyphal network (McGonigle & Miller, 1996). In maize, higher mycorrhizal root colonization was observed in no-tilled and ridge-tilled plots as compared to the conventional-tilled plots (McGonigle & Miller, 1996). AMF community structure significantly differed between conventional and no-till treatments, *Acaulospora* spp. showed higher affinity towards the conventional management whereas *Glomus* and *Gigaspora* spp. were correlated with no-till management (Jansa et al., 2003; Castillo et al., 2006; Gottshall et al., 2017). Since under no-till system mycorrhizal hyphal network and other fungal structures are not disrupted and hence accelerates mycorrhizal functioning and nutrient uptake in soil (Rillig et al., 2018; Ryan & Graham, 2018). Tillage disrupts the AMF extraradical hyphal network and spore abundance results into decline inoculation potential of the soil (Galvez et al., 2001; Sharma et al., 2012; Agnihotri et al., 2021).

Douds et al. (1995) found variation in *Glomus* species where *Glomus etunicatum* occurred with higher frequency in tilled soil and *Glomus occultum* spores were higher in no tillage soils. Hence, it becomes logical to identify the dominant and stabilized AMF community composition under reduced tillage practices for promoting the AMF biomass and overall crop productivity or performance.

4.3.2 DROUGHT, SALINITY, AND HEAVY METALS

AMF play a crucial role in drought mitigation (Mathimaran et al., 2017; Sharma et al., 2020) because of extended hypha and aquaporin protein (Bárzana et al., 2014; Zhang et al., 2018). The impacts of drought on AMF diversity and abundance have been reported and showed enhanced inoculation responses to soil ecosystems strong to cope up with drought (Sallah et al., 2002; Oyewole et al., 2017; Sendek et al., 2019). The AMF inoculation (*Funneliformis mosseae*) under drought modulates the fungal communities in the citrus plantations where more abundance of specific fungal communities (*Sordariomycetes*) in roots was observed and that may be playing role in enhancing tolerance to drought (He et al., 2019). Whereas another study suggested richness of barley plant genotypes and AMF on the alleviation of drought did not mitigate detrimental drought effects on the plant and AMF (Sendek et al., 2019).

Besides, drought tolerance, several studies have suggested that AM symbiosis play an important role in overcoming the salinity stress and yield enhancement in plants (Evelin et al., 2019; Li et al., 2020). However, studies on the diversity of AMF under saline conditions are limited. The occurrence of AMF assessed in two salt marshes (sodium salt in Island-Atlantic Coast, the Netherlands and potassium carbonate marshy site in Schreyahn, Northern Germany) reported a low diversity in roots at both sites (Wilde et al., 2009). However, in the rhizosphere soil both the sites exhibited a higher AMF biodiversity and across both the sites *Glomus geosporum* was found to be the predominant species in root and soil samples of the *Aster tripolium* and *Puccinellia* spp. at both saline sites (Wilde et al., 2009). A high AMF diversity in saline area was observed comprising 33 AMF species from 11 genera from the rhizosphere of two food legumes (Liu et al., 2017). They observed *Acaulospora* and *Glomus*, as dominant species where *Septoglomus constrictum* was the most abundant species inhabiting in the saline soils. Very recently, Samba-Mbaye et al. (2020) studied the AMF communities associated with the rhizosphere of *Vachellia nilotica* growing in salt-affected soils of six different sites of the central region of Senegal. It was observed that salinity was the main factor which impacted negatively on AMF development particularly in dry season than in wet season. AM morphotypes identified from salinity area belonged to the genus *Glomus, Gigaspora*, and *Acaulospora*. On the other hand, the desert soils of two halophytes showed the typical AMF structures inside roots and 26 molecularly distinct AMF taxa were recovered from soil and root DNA (Lumini et al., 2020). Numerous research reported inoculation of AMF in host plants confers salinity tolerance by increasing AMF colonization, and thus increases biomass in *Trigonella foenum-graecum* (Evelin et al., 2012), *Oryza sativa* (Porcel et al., 2015) and *Chrysanthemum morifolium* (Wang et al., 2018).

Now a days AMF inoculation has become a promising technology to enhance plant tolerance against the different environmental stress, e.g., metal-contaminated soils (Gong & Tian, 2019; Dhalaria et al., 2020). There are many reports signifying the role of the AM symbiosis in the remediation of stressed ecosystems, although the underlying mechanisms are not yet fully understood (Gaur & Adholeya, 2004). The long-term use of sewage sludge in soil contributed to heavy metal levels, which influences the AMF. In contaminated soil, six different AMF ecotypes were observed showing consistent differences in tolerance levels to heavy metals. AMF ecotypes showed a wide range of tolerance index from very sensitive to relative high tolerance to heavy metals in relation to the presence of heavy metals in soil (Del Val et al., 1999). The existence of AMF in nickel hyper accumulating plant species was found naturally on metal rich soils. It offers the possibilities of using heavy metal hyper accumulating plants along with AMF for phytoremediation strategies (Turnau & Mesjasz-Przybylowicz, 2003; Gamalero et al., 2009). Moreover, many phosphate-based fertilizers contaminated agricultural lands with high deposition of cadmium in soil where use of AMF, through their extensive network of hyphae improve the P uptake efficiently as well as have buffering effect on the cadmium uptake, thus reduces the toxic effect of cadmium in plant growth and development (Rivera-Becerril et al., 2002; Nziguheba & Smolders, 2008; López-Millán et al., 2009). AMF *viz., Glomus mosseae* and *Rhizophagus irregularis* facilitates to translocate heavy metals in shoot (Zaefarian et al., 2013; Ali et al., 2015), and even some AMF spp. effectively restrict the high accumulation of heavy metals (Na, Mn, Mg, and Fe) in roots (Bati et al., 2015).

4.3.3 CHEMICAL PESTICIDES, FERTILIZERS, AND ORGANIC MANURES

The benefits of AMF are higher in systems where the inputs are very low (Martinez & Johnson, 2010). Long-term heavy usages of phosphatic and nitrogenous fertilizers alter the diversity of AMF and can inhibit mycorrhizal colonization and growth (Santos et al., 2006; Wilson et al., 2009; Chen et al., 2014). There is a strong impact of nitrogen fertilizers on AMF assemblage as compared to tillage where members of *Diversisporaceae* and *Entrophospo-raceae* were found to be more prevalent (Borriello et al., 2012). The spore population of *Glomus* spp. was not affected significantly due to mineral (N and P) and organic fertilizers. However, the population of *Acaulospora* and *Scutellospora* species was enhanced in soils that received organic fertilizers (Oehl et al., 2004; Harinikumar & Bagyaraj, 1989). In general, organic manures promoted AMF population and communities. The comparative analysis of organic farming versus conventional farming systems revealed that organic systems inhabit high AMF spores and its species diversity than the conventional system (Oehl et al., 2004). The study further reported that long-term conventional farming systems showed a reduced number of AMF spores mainly belonging to species of *Acaulospora* and *Scutellospora*. On contrary, AMF sporulation and diversity enhanced in conventional farming than the organic farming (Purin et al., 2006). The organic practice with no-tilled plots showed increased AMF-glomalin production, diversity, and community stability over the conventional practice. Conventional (chemical inputs) showed the prevalence of *Acaulospora* spp., while *Glomus* spp. *and Gigaspora* spp. were associated with no-till management (Gottshall et al., 2017). Further, they suggested that integrated agricultural systems consisting of organic management with conservation tillage improve the AM fungal benefits to crops and soils. The choice of chemical pesticides needs to be cautiously selected as some fungicides (Calonne et al., 2011), herbicides (Zaller et al., 2014), nematicides (Bakhtiar et al., 2001) applied at field application rate has been shown to reduce the functioning and population of the AMF. Although the reported effects of pesticides on AMF symbiosis are varied, and these effects depend on many components, such as active substance ingredients, mode of action, application, and doses (Hage-Ahmed et al., 2019). However, information is limited, particularly on the impact of fungicides on AMF biomass when applied at threshold level. The impact of biological pesticides *viz.*, azadirachtin, terpenes pyrethrum and spinosad, on AMF was assessed where pesticide affects the structure of the intraradical AM fungal community. Among the biological pesticides spinosad, pyrethrum, and terpenes did not affect the structure and colonization ability of the AM fungal community where as azadirachtin resulted in a selective inhibition of the AMF species and shifts in the AM fungal community. Carbendazim not promoted AM colonization rather decreased colonization and the community structure of indigenous AM fungi (Ipsilantis et al., 2011). Fungicides *viz.,* azoxystrobin, pencycuron, and flutolanil were not detrimental to *Rhizophagus irregularis* (Buysens et al., 2015). Jin et al. (2013) studied the consequences of systemic and contact fungicides on AMF where AM colonization and host plant growth were inhibited by former and only slightly affected by the latter. A brief account of reports on factors influencing AMF has been provided in Table 4.2.

4.4 SCOPE OF COMMERCIAL EXPLOITATION OF AMF BIODIVERSITY IN SUSTAINABLE AGRICULTURE

All over the world, the commercial exploitation of AMF biodiversity is being practiced through their conservation in different gene banks (Agnihotri et al., 2018). For application, the commercial exploitation of AMF in sustainable agriculture is being undertaken either through management of ex situ production of stabilized AMF or through management of native AMF through long-term adoption of crop management practices (Gupta & Abbott, 2020). For example, the AMFs of organic systems can be mass-produced and inoculated back to the systems. AMF infectious propagules include spores, arbuscules, vesicles, hyphae, and infected root pieces, and each of these can serve as inoculum components and can be used for their mass multiplication (Biermann & Linderman, 1983; Klironomos & Hart, 2002). Although the bulk production of desired AMF inoculum is a critical issue therefore, mass production of quality production of AMF inocula at affordable price is need of an hour. Constraints in their application in fields along with several risk minimization strategies have been summarized recently by Gupta & Abbott (2020). Various techniques have been used for mass production of AMF inocula such as hydroponics, aeroponic, substrate-based pot cultures (Agnihotri et al., 2021), nutrient film technique, *in-vitro* cultivation, and on farm production (Douds et al., 2006; Ijdo et al., 2011; Sharma et al., 2011; Schlemper & Stürmer, 2014). AMF producing and selling industries after their inception in 1970 have reached to establishment phase in 2021 (Figure 4.2). However, their application by farmers for securing harvest benefit is still not achieved. For commercial AM fungi production, the focal issue remains quality assurance. To achieve this, the only way is to yield pure, non-contaminated AM inoculum on transformed roots in axenic

TABLE 4.2 Factors Affecting the Performance of AMF Species on Various Crop Plants

Factor	AMF Species	Crop	Inference	References
Phosphorus level	*Rhizophagus irregularis*	Cotton	Enhanced photosynthesis rate, P concentration in the cotton biomass, plant growth, boll number per plant, maturity of fiber, increased yield.	Gao et al. (2020)
Nitrogen level	*Glomus intraradices*	*Plantago lanceolata, Trifolium repens L.*	Increased shoot dry weight, root dry weight, shoot, and root N content.	Thirkell et al. (2016)
Salt tolerance	*Rhizophagus intraradices*	*Euonymus maackii*	Increased photosynthesis, nutrient uptake, activating the antioxidant enzyme system.	Li et al. (2020)
Salt tolerance	*Glomus etunicatum*, *Glomus intraradices*, *Glomus mosseae*	*Cucumis sativus* L.	Increased biomass, photosynthetic pigment synthesis, and enhanced antioxidant enzymes.	Hashem et al. (2018)
Heat	*Rhizophagus irregularis*, *Funneliformis mosseae*, *Funneliformis geosporum*	*Triticum aestivum* L.	Increased grain number, nutrient allocation, and nutrient composition in root.	Cabral et al. (2016)
High temperature	*Rhizophagus intraradices*, *Funneliformis mosseae*, *F. geosporum*	*Zea mays*	Increased leaf length, plant height, leaf number, chlorophyll a, photosynthetic rate, stomatal conductance, and transpiration rate.	Mathur & Jajoo (2014)
Heavy metal stress (cadmium)	*Glomus monosporum*, *G. clarum*, *Gigaspora nigra*, and *Acaulospora laevis*	*Trigonella foenum-graecum* L.	Increased antioxidant enzymes activities and malondialdehyde content.	Abdel-Hameed & Rabab (2019)
Heavy metal stress (cadmium and zinc)	*Rhizophagus irregularis*	*Cajanus cajan* L.	Improved root biomass, nutrient status (P, N, Mg, Fe.), and proline biosynthesis.	Garg & Singh (2018)
Salinity	*Glomus mosseae*	*Leymus chinensis*	Increased colonization rate, seedling weight, water contents, and both P and N.	Lin et al. (2017)
Bioprotection	*Glomus deserticola*; *Gigaspora gigantean*, *Funneliformis mosseae* or *Rhizophagus irregularis*	Cowpea, rice	Protection against charcoal rot, rice blast fungus.	Oyewole et al. (2017); Campo et al. (2020)

conditions. But not all fungal species can be cultivated successfully in axenic conditions due to various issues. Conventionally, AMF are being multiplied in "substrate-based pot cultures," in association with plant roots (Coelho et al., 2014) and can be produced on-farm by farmers using indigenous or introduced AMF isolates (Douds et al., 2008, 2010). A native AMF inoculum that is locally suited to soil conditions performs better in providing the plants and soil with numerous benefits (Douds et al., 2010; Hart et al., 2017).

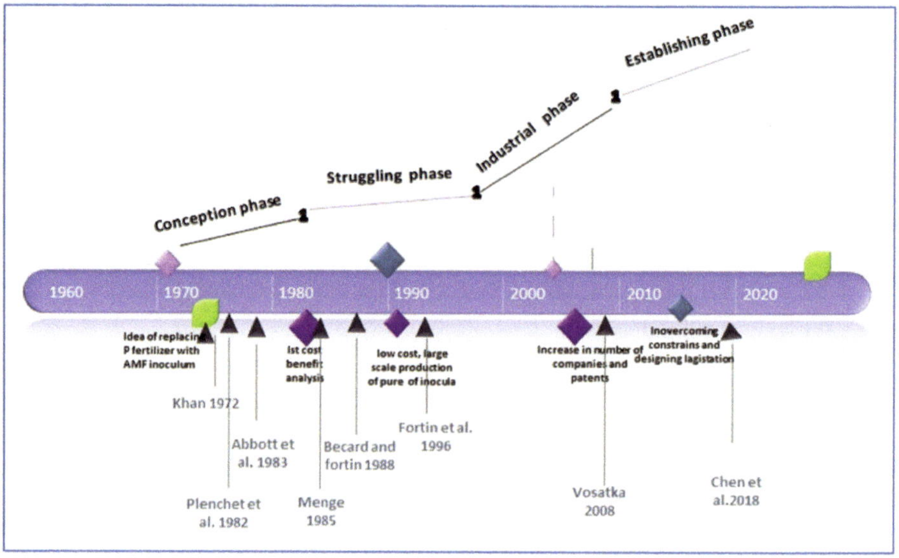

FIGURE 4.2 Diagrammatic representation of AMF establishment stages.

Although the success of AM production in any system may vary with the type of AMF and host species used (McGonigle & Fitter, 1990). Hetrick & Bloom (1986) observed that *Glomus fasciculatum* spore development was influenced by the host plant, they investigated the influence of five host plants, including tomato, Sudan grass and red clover, on spore production and colonization development of AMF. As a result of identification of stabilized AMF community from the agroecosystems needs a gestation period where such AMF communities can be exploited through farming and commercialization interventions. Hence, the appropriate native AMF and their production in sufficient quantity can be achieved through application in agriculture systems, because native adapted AMF isolates appears to be physiologically, ecologically, and genetically stronger against the different environmental stresses along with hosts (de Oliveira et al., 2017).

4.5 CONCLUSION AND FUTURE STRATEGIES

The main aim of the sustainable plant production system is to maintain soil biodiversity equilibrium including AMF abundance for performing various soil and ecological functions to maintain long-term plant and soil productivity of agroecosystems. AMF functioning and population is greatly affected by the use of chemical fertilizers, mono cropping and soil disturbance. Adoption of potential hosts along with stabilized and well acclimatized AMF to support high populations of indigenous AMF would help in identifying functionally superior AMF as biofertilizers (Sharma et al., 2012). AMF "friendly" agricultural methods such as minimum tillage, crop rotation, and inter and mixed cropping practices all contribute to long-term plant production sustainability. However, the plant families *Brassicaceae* and *Chehnopodiaceae* species should be discouraged as they reduced and do not usually form mycorrhizal symbiosis. Therefore, the selection of an appropriate host plants will always stay an important aspect in not only supporting mycorrhizal population inside soil but also directs the AMF composition. The assessment of abundance and diversity of AMF in the agricultural field is an important component of mycorrhizal technology for effective utilization of mycorrhizal fungal inoculum for increasing the crop productivity (Solaiman et al., 2014). Each AMF-host plant partnership is typical not only in morphology but also in the extraradical mycelium, nutrient uptake and transfer capacity (Lee et al., 2013). Thus, more detailed and extensive studies are needed for predicting the functional performance of species or genera (Cavagnaro et al., 2015). For ecological perspective, a correct understanding of the local AMF biodiversity and their beneficial impacts should be unraveled to sustain the agricultural production systems (Hart et al., 2017). As a future strategy, some of the areas, which are, still need to be relooked and targeted to achieve the millennium development goals:

- More detailed studies on deciphering the AMF community and diversity are needed in stressed environment of different agroecosystems.
- Region specific studies should be targeted on identifying the most prevalent species to undertake their mass production.
- Irrespective of ecological regions, identify the potential driving factors governing the AMF diversity.
- Capacity building in studying the AMF diversity using correct identification modules.
- Mass production and its quality assessment of AMF always remains a thrust area of AM technology.

ACKNOWLEDGMENTS

This work is part of in-house AMF mass production approved project under the Institute's research advisory committee and ICAR-AMAAS network subproject. The authors are grateful to the Director, ICAR-Indian Institute of Soybean Research, Indore, India, for providing necessary infrastructure facilities is gratefully acknowledged.

KEYWORDS

- **agroecosystem**
- **arbuscular mycorrhizal fungi**
- **biodiversity**
- **biofertilizer**
- **cropping systems**
- **environmental stress**
- **fungi**

REFERENCES

Abdelhameed, R. E., & Metwally, R. A., (2019). Alleviation of cadmium stress by arbuscular mycorrhizal symbiosis. *International Journal of Phytoremediation, 21*(7), 663–671.

Agnihotri, R., Bharti, A., Ramesh, A., Prakash, A., & Sharma, M. P., (2021). Glomalin related protein and C16:1ω5 PLFA associated with AM fungi as potential signatures for assessing the soil C sequestration under contrasting soil management practices. *European Journal of Soil Biology, 103*, 103286. https://doi.org/10.1016/j.ejsobi.2021.10328.

Agnihotri, R., Maheshwari, H. S., Sharma, S. K., & Sharma, M. P., (2018). Biobank for conservation of arbuscular mycorrhizal (AM) fungi. In: Sharma, S. K., & Varma, A., (eds.), *Microbial Resource Conservation: Soil Biology* (Vol. 54, pp. 199–221). Springer, Cham.

Agnihotri, R., Ramesh, A., Singh, S., & Sharma, M. P., (2017). Impact of agricultural management practices on mycorrhizal functioning and soil microbiological parameters under soybean-based cropping systems. In: Rakshit, A., Chirakuzhyil, P., Harikesh, A., Singh, B., & Ghosh, S., (eds.), *Adaptive Soil Management: From Theory to Practices* (pp. 301–322). Springer Nature Singapore Pte Ltd., Gateway East, Singapore 189721.

Ali, N., Masood, S., Mukhtar, T., Kamran, M. A., Rafique, M., Munis, M. F. H., & Chaudhary, H. J., (2015). Differential effects of cadmium and chromium on growth, photosynthetic

activity, and metal uptake of *Linum usitatissimum* in association with *Glomus intraradices*. *Environmental Monitoring and Assessment, 187*(6), 311.
Ambrosano, E. J., Azcón, R., Cantarella, H., Ambrosano, G. M. B., Schammass, E. A., Muraoka, T., Trivelin, P. C. O., et al., (2010). Crop rotation biomass and arbuscular mycorrhizal fungi effects on sugarcane yield. *Scientia Agricola, 67*(6), 692–701.
Auge, R. M., (2001). Water relations, drought and vesicular-arbuscular mycorrhizal symbiosis. *Mycorrhiza, 11*, 3–42.
Bagyaraj, D. J., Sharma, M. P., & Maiti, D., (2015). Phosphorus nutrition of crops through arbuscular mycorrhizal fungi. *Current Science, 108*, 1288–1293.
Bakhshandeh, S., Corneo, P. E., Mariotte, P., Kertesz, M. A., & Dijkstra, F. A., (2017). Effect of crop rotation on mycorrhizal colonization and wheat yield under different fertilizer treatments. *Agriculture, Ecosystems and Environment, 247*, 130–136.
Bakhtiar, Y., Miller, D., Cavagnaro, T., & Smith, S., (2001). Interactions between two arbuscular mycorrhizal fungi and fungivorous nematodes and control of the nematode with *fenamifos*. *Applied Soil Ecology, 17*(2), 107–117.
Bárzana, G., Aroca, R., Bienert, G. P., Chaumont, F., & Ruiz-Lozano, J. M., (2014). New insights into the regulation of aquaporins by the arbuscular mycorrhizal symbiosis in maize plants under drought stress and possible implications for plant performance. *Molecular Plant-Microbe Interactions, 27*(4), 349–363.
Bati, C. B., Santilli, E., & Lombardo, L., (2015). Effect of arbuscular mycorrhizal fungi on growth and on micronutrient and macronutrient uptake and allocation in olive plantlets growing under high total Mn levels. *Mycorrhiza, 25*(2), 97–108.
Bharti, A., Garg, S., Prakash, A., & Sharma, M. P., (2017). *Contribution of AMF in the Remediation of Drought Stress in Soybean Plants.* New India Publishing Agency, New Delhi, India.
Biermann, B., & Linderman, R. G., (1983). Use of vesicular-arbuscular mycorrhizal roots, intraradical vesicles and extraradical vesicles as inoculum. *New Phytologist, 95*, 97–105.
Bona, E., Cantamessa, S., Massa, N., Manassero, P., Marsano, F., Copetta, A., Lingua, G., et al., (2017). Arbuscular mycorrhizal fungi and plant growth-promoting *pseudomonads* improve yield, quality and nutritional value of tomato: A field study. *Mycorrhiza, 27*(1), 1–11.
Borriello, R., Lumini, E., Girlanda, M., Bonfante, P., & Bianciotto, V., (2012). Effects of different management practices on arbuscular mycorrhizal fungal diversity in maize fields by a molecular approach. *Biology and Fertility of Soils, 48*(8), 911–922.
Buade, R., Chourasiya, D., Prakash, A., & Sharma, M. P., (2020). *Changes in Arbuscular Mycorrhizal Fungal Community Structure in Soybean Rhizosphere Soil Assessed at Different Growth Stages of Soybean.* Agriculture Research. https://doi.org/10.1007/s40003-020-00481-4.
Buysens, C., De Boulois, H. D., & Declerck, S., (2015). Do fungicides used to control *Rhizoctonia solani* impact the non-target arbuscular mycorrhizal fungus *Rhizophagus irregularis*? *Mycorrhiza, 25*(4), 277–288.
Calonne, M., Fontaine, J., Debiane, D., Laruelle, F., Grandmougin, A., & Lounes-Hadj, A. S., (2011). Side effects of the sterol biosynthesis inhibitor fungicide, propiconazole, on a beneficial arbuscular mycorrhizal fungus. *Communications in Agricultural and Applied Biological Sciences, 76*(4), 891–902.
Castillo, C. G., Rubio, R., Rouanet, J. L., & Borie, F., (2006). Early effects of tillage and crop rotation on arbuscular mycorrhizal fungal propagules in an ultisol. *Biology and Fertility of Soils, 43*(1), 83–92.

Cavagnaro, T. R., Bender, S. F., Asghari, H. R., & Van, D. H. M. G., (2015). The role of arbuscular mycorrhizas in reducing soil nutrient loss. *Trends in Plant Science, 20*(5), 283–290.

Chaudhary, V. B., Rúa, M. A., Antoninka, A., Bever, J. D., Cannon, J., Craig, A., Duchicela, J., et al., (2016). MycoDB, a global database of plant response to mycorrhizal fungi. *Scientific Data, 3*, http://doi.org/10.1038/sdata.2016.28.

Chen, X., Song, F., Liu, F., Tian, C., Liu, S., Xu, H., & Zhu, X., (2014). Effect of different arbuscular mycorrhizal fungi on growth and physiology of maize at ambient and low temperature regimes. *The Scientific World Journal*, 1–7. doi: 10.1155/2014/956141.

Chen, Y. L., Zhang, X., Ye, J. S., Han, H. Y., Wan, S. Q., & Chen, B. D., (2014). Six-year fertilization modifies the biodiversity of arbuscular mycorrhizal fungi in a temperate steppe in inner Mongolia. *Soil Biology and Biochemistry, 69*, 371–381.

Coelho, I. R., Pedone-Bonfim, M. V. L., Silva, F. S. B., & Maia, L. C., (2014). Optimization of the production of mycorrhizal inoculum on substrate with organic fertilizer. *Brazilian Journal of Microbiology, 45*, 1173–1178.

Cuenca, G., & Lovera, M., (2010). Seasonal variation and distribution at different soil depths of arbuscular mycorrhizal fungi spores in a tropical *Sclerophyllous* shrub land. *Botany, 88*, 54–64.

De Oliveira, J. R. G., De Resende, G. M., De Melo, N. F., & Yano-Melo, A. M., (2017). Symbiotic compatibility between arbuscular mycorrhizal fungi (autoctone or exotic) and three native species of the *Caatinga* in different phosphorus levels. *Acta Scientiarum. Biological Sciences, 39*(1), 59–69.

Del Val, C., Barea, J. M., & Azcon-Aguilar, C., (1999). Diversity of arbuscular mycorrhizal fungus populations in heavy-metal-contaminated soils. *Applied and Environmental Microbiology, 65*(2), 718–723.

Dhalaria, R., Kumar, D., Kumar, H., Nepovimova, E., Kuča, K., Torequl, I. M., & Verma, R., (2020). Arbuscular mycorrhizal fungi as potential agents in ameliorating heavy metal stress in plants. *Agronomy, 10*(6), 815. doi: 10.3390/agronomy10060815.

Douds, D. D., Nagahashi, G., & Hepperly, P. R., (2010). On-farm production of inoculum of indigenous arbuscular mycorrhizal fungi and assessment of diluents of compost for inoculum production. *Bioresource Technology, 101*, 2326–2330. doi: 10.1016/j.biortech.2009.11.071.

Douds, Jr. D. D., Galvez, L., Janke, R. R., & Wagoner, P., (1995). Effect of tillage and farming system upon populations and distribution of vesicular-arbuscular mycorrhizal fungi. *Agriculture, Ecosystems and Environment, 52*(2, 3), 111–118.

Douds, Jr. D. D., Nagahashi, G., Pfeffer, P. E., Reider, C., & Kayser, W. M., (2006). On-farm production of AM fungus inoculum in mixtures of compost and vermiculite. *Bioresource Technology, 97*(6), 809–818.

Douds, Jr. D. D., Nagahashi, G., Reider, C., & Hepperly, P. R., (2008). Choosing a mixture ratio for the on-farm production of AM fungus inoculum in mixtures of compost and vermiculite. *Compost Science & Utilization, 16*(1), 52–60.

Drigo, B., Pijl, A. S., Duyts, H., Kielak, A. M., Gamper, H. A., Houtekamer, M. J., Boschker, H. T., et al., (2010). Shifting carbon flow from roots into associated microbial communities in response to elevated atmospheric CO_2. *Proceedings of the National Academy of Sciences, 107*(24), 10938–10942.

Dumbrell, A. J., Ashton, P. D., Aziz, N., Feng, G., Nelson, M., Dytham, C., Fitter, A. H., & Helgason, T., (2011). Distinct seasonal assemblages of arbuscular mycorrhizal fungi revealed by massively parallel pyrosequencing. *New Phytologist, 190*, 794–804. https://doi.org/10.1111/j.1469-8137.2010.03636.x.

Evelin, H., Devi, T. S., Gupta, S., & Kapoor, R., (2019). Mitigation of salinity stress in plants by arbuscular mycorrhizal symbiosis: Current understanding and new challenges. *Frontiers in Plant Science, 10*, 470.
Evelin, H., Giri, B., & Kapoor, R., (2012). Contribution of *Glomus intraradices* inoculation to nutrient acquisition and mitigation of ionic imbalance in NaCl-stressed *Trigonella foenum-graecum*. *Mycorrhiza, 22*(3), 203–217. doi: 10.1007/s00572-011-0392-0.
Fyson, A., & Oaks, A., (1991). Promotion of maize growth by legume soil factors. In: *The Rhizosphere and Plant Growth* (p. 370). Springer, Dordrecht.
Galvez, L., Douds, D. D., Drinkwater, L. E., & Wagoner, P., (2001). Effect of tillage and farming system upon VAM fungus populations and mycorrhizas and nutrient uptake of maize. *Plant and Soil, 228*(2), 299–308.
Gamalero, E., Lingua, G., Berta, G., & Glick, B. R., (2009). Beneficial role of plant growth promoting bacteria and arbuscular mycorrhizal fungi on plant responses to heavy metal stress. *Canadian Journal of Microbiology, 55*(5), 501–514.
Gao, X., Guo, H., Zhang, Q., Guo, H., Zhang, L., Zhang, C., Gou, Z., et al., (2020). Arbuscular mycorrhizal fungi (AMF) enhanced the growth, yield, fiber quality and phosphorus regulation in upland cotton (*Gossypium hirsutum* L.). *Scientific Reports, 10*(1), 1–12.
Gaur, A., & Adholeya, A., (2004). Prospects of arbuscular mycorrhizal fungi in phytoremediation of heavy metal contaminated soils. *Current Science, 86*, 528–534.
Gong, X., & Tian, D. Q., (2019). Study on the effect mechanism of arbuscular mycorrhiza on the absorption of heavy metal elements in soil by plants. *IOP Conference Series: Earth and Environmental Science, 267*(5), 052064. doi: 10.1088/1755-1315/267/5/052064.
Goto, B. T., Da Silva, G. A., De Assis, D. M., Silva, D. K. A., Souza, R. G., Ferreira, A. C. A., Jobim, K., et al., (2012). *Intraornatospora* (*Gigasporales*), a new family with two new genera and two new species. *Mycotaxon, 119*, 117–132.
Gottshall, B., & Cooper, M., & Emery, S., (2017). Activity, diversity and function of arbuscular mycorrhizae vary with changes in agricultural management intensity. *Agriculture, Ecosystems and Environment, 241*, 142–149. 10.1016/j.agee.2017.03.011.
Gupta, M. M., & Abbott, L. K., (2020). Exploring economic assessment of the arbuscular mycorrhizal symbiosis. *Symbiosis*. doi: 10.1007/s13199-020-00738-0.
Gupta, M. M., (2020). Arbuscular mycorrhizal fungi – the potential soil health indicators. In: Giri, B., & Verma, A., (eds.), *Soil Health* (pp. 183–195). Springer, Cham.
Gupta, M. M., Aggarwal, A., & Asha, (2018). From mycorrhizosphere to rhizosphere microbiome: The paradigm shift. In: Giri, B., Prasad, R., & Varma, A., (eds.), *Root Biology* (pp. 487–500). Springer, Cham.
Gupta, M. M., Chourasiya, D., & Sharma, M. P., (2019). Diversity of arbuscular mycorrhizal fungi in relation to sustainable plant production systems. In: Satyanarayana, T., Johri, B. N., & Das, S. K., (eds.), *Microbial Diversity in Ecosystem Sustainability and Biotechnological Applications: Microbial Diversity in Normal and Extreme Environments* (Vol. 1, pp. 167–186). Springer Singapore.
Hage-Ahmed, K., Rosner, K., & Steinkellner, S., (2019). Arbuscular mycorrhizal fungi and their response to pesticides. *Pest Management Science, 75*(3), 583–590. doi: 10.1002/ps.5220.
Harinikumar, K. M., & Bagyaraj, D. J., (1989). Effect of cropping sequence, fertilizers and farmyard manure on vesicular-arbuscular mycorrhizal fungi in different crops over three consecutive seasons. *Biology and Fertility of Soils, 7*, 173–175.
Hart, M. M., Antunes, P. M., & Abbott, L. K., (2017). Unknown risks to soil biodiversity from commercial fungal inoculants. *Nature Ecology and Evolution, 1*(4), 1.

Hazard, C., Gosling, P., Van, D. G. C. J., Mitchell, D. T., Doohan, F. M., & Bending, G. D., (2013). The role of local environment and geographical distance in determining community composition of arbuscular mycorrhizal fungi at the landscape scale. *The ISME Journal, 7*(3), 498–508.

He, J. D., Wu, Q. S., & Zou, Y. N., (2019). Effects of mycorrhiza and drought stress on the diversity of fungal community in soils and roots of trifoliate orange. *Biotechnology, 18*, 32–41.

Helgason, T., Daniell, T. J., Husband, R., Fitter, A. H., & Young, J. P. W., (1998). Ploughing up the wood-wide web? *Nature, 394*(6692), 431.

Hetrick, B. A. D., & Bloom, J., (1986). The influence of host plant on production and colonization ability of vesicular-arbuscular mycorrhizal spores. *Mycologia, 78*(1), 32–36.

Higo, M., Tatewaki, Y., Iida, K., et al., (2020). Amplicon sequencing analysis of arbuscular mycorrhizal fungal communities colonizing maize roots in different cover cropping and tillage systems. *Scientific Reports, 10*, 6039. https://doi.org/10.1038/s41598-020-58942-3.

Hijri, I., Sýkorová, Z., Oehl, F., Ineichen, K., Mäder, P., Wiemken, A., & Redecker, D., (2006). Communities of arbuscular mycorrhizal fungi in arable soils are not necessarily low in diversity. *Molecular Ecology, 15*(8), 2277–2289.

Hijri, M., (2016). Analysis of a large dataset of mycorrhiza inoculation field trials on potato shows highly significant increases in yield. *Mycorrhiza, 26*(3), 209–214.

Hildebrandt, U., Regvar, M., & Bothe, H., (2007). Arbuscular mycorrhiza and heavy metal tolerance. *Phytochemistry, 68*, 139–146. https://doi.org/10.1016/j.phytochem.2006.09.023.

Ijdo, M., Cranenbrouck, S., &Declerck, S., (2011). Methods for large-scale production of AM fungi: Past, present, and future. *Mycorrhiza, 21*, 1–16. doi: http://doi.org/10.1007/s00572-010-0337-z.

Ipsilantis, I., & Samourelis, C., & Karpouzas, D., (2011). The impact of biological pesticides on arbuscular mycorrhizal fungi. *Soil Biology and Biochemistry, 45*. 10.1016/j.soilbio.2011.08.007.

Jansa, J., Mozafar, A., Anken, T., Ruh, R., Sanders, I. R., & Frossard, E., (2002). Diversity and structure of AMF communities as affected by tillage in a temperate soil. *Mycorrhiza, 12*, 225–234. doi: 10.1007/s00572-002-0163-z.

Jansa, J., Mozafar, A., Kuhn, G., Anken, T., Ruh, R., Sanders, I. R., & Frossard, E. J. E. A., (2003). Soil tillage affects the community structure of mycorrhizal fungi in maize roots. *Ecological Applications, 13*(4), 1164–1176.

Jin, H., Germida, J. J., & Walley, F. L., (2013). Suppressive effects of seed-applied fungicides on arbuscular mycorrhizal fungi (AMF) differ with fungicide mode of action and AMF species. *Applied Soil Ecology, 72*, 22–30.

Johnson, D., Vandenkoornhuyse, P., Leake, R. J., Gilbert, L., Booth, R. E., Grime, J. P., Young, J. P. W., & Read, D., (2003). Plant communities affect arbuscular mycorrhizal fungal diversity and community composition in grassland microcosms. *New Phytologist, 161*, 503–515. 10.1046/j.1469-8137.2003.00938.x.

Johnson, J. M., Houngnandan, P., Kane, A., Sanon, K. B., & Neyra, M., (2013). Diversity patterns of indigenous arbuscular mycorrhizal fungi associated with rhizosphere of cowpea (*Vigna unguiculata* (L.) Walp.) in Benin, West Africa. *Pedobiologia, 56*(3), 121–128. doi: 10.1016/j.pedobi.2013.03.003.

Johnson, N. C., Tilman, D., & Wedin, D., (1992). Plant and soil controls on mycorrhizal fungal communities. *Ecology, 73*(6), 2034–2042.

Klironomos, J. N., & Hart, M. M., (2002). Colonization of roots by arbuscular mycorrhizal fungi using different sources of inoculum. *Mycorrhiza, 12*, 181–184. doi: 10.1007/s00572-002-0169-6.

Krüger, M., Stockinger, H., Krüger, C., & Schüßler, A., (2009). DNA-based species level detection of *Glomeromycota*: One PCR primer set for all arbuscular mycorrhizal fungi. *New Phytologist, 183*(1), 212–223.

Latef, A. A. H. A., & Chaoxing, H., (2011). Arbuscular mycorrhizal influence on growth, photosynthetic pigments, osmotic adjustment and oxidative stress in tomato plants subjected to low temperature stress. *Acta Physiologiae Plantarum, 33*(4), 1217–1225.

Lee, E. H., Eo, J. K., Ka, K. H., & Eom, A. H., (2013). Diversity of arbuscular mycorrhizal fungi and their roles in ecosystems. *Mycobiology, 41*(3), 121–125.

Li, Z., Wu, N., Meng, S., Wu, F., & Liu, T., (2020). Arbuscular mycorrhizal fungi (AMF) enhance the tolerance of *Euonymus maackii* Rupr. at a moderate level of salinity. *PLoS One, 15*(4), e0231497. doi: 10.1371/journal.pone.0231497.

Lin, T. C. J., Wang, P. H., & Lin, W. R., (2019). Changes of mycorrhizal fungal community occurring during the natural restoration after the Chi-Chi earthquake in Taiwan. *Symbiosis, 77*, 177–184. https://doi.org/10.1007/s13199-018-0582-z.

Liu, H., Wang, Y., & Tang, M., (2017). Arbuscular mycorrhizal fungi diversity associated with two halophytes *Lycium barbarum* L. and *Elaeagnus angustifolia* L. in Ningxia, China, *Archives of Agronomy and Soil Science, 63*(6), 796–806. 10.1080/03650340.2016.1235783.

Lombardo, L., Palese, A. M., Grasso, F., Duffy, D. H., Briccoli, B. C., & Xiloyannis, C., (2019). Mechanical tillage diversely affects glomalin content, water stable aggregates and AM fungal community in the soil profiles of two differently managed olive orchards. *Biomolecules, 9*(10), 639.

López-Millán, A. F., Sagardoy, R., Solanas, M., Abadía, A., & Abadía, J., (2009). Cadmium toxicity in tomato (*Lycopersicon esculentum*) plants grown in hydroponics. *Environmental and Experimental Botany, 65*(2, 3), 376–385.

Lu, F. C., Lee, C. Y., & Wang, C. L., (2015). The influence of arbuscular mycorrhizal fungi inoculation on yam (*Dioscorea* spp.) tuber weights and secondary metabolite content. *PeerJ, 3*, e1266.

Lumini, E., Pan, J., Magurno, F., Huang, C., Bianciotto, V., Xue, X., Balestrini, R., & Tedeschi, A., (2020). Native arbuscular mycorrhizal fungi characterization from saline lands in arid oases, Northwest China. *Journal of Fungi, 6*, 80.

Martinez, T. N., & Johnson, N. C., (2010). Agricultural management influences propagule densities and functioning of arbuscular mycorrhizas in low-and high-input agroecosystems in arid environments. *Applied Soil Ecology, 46*(2), 300–306.

Mathimaran, N., Ruh, R., Jama, B., Verchot, L., Frossard, E., & Jansa, J., (2007). Impact of agricultural management on arbuscular mycorrhizal fungal communities in Kenyan ferralsol. *Agriculture, Ecosystems and Environment, 119*(1, 2), 22–32.

Mathimaran, N., Sharma, M. P., Raju, M. B., & Bagyaraj, D. J., (2017). Arbuscular mycorrhizal symbiosis and drought tolerance in crop plants. *Mycosphere, 8*, 361–376. doi: 10.5943/mycosphere/8/3/2.

McGonigle, T. P., & Fitter, A. H., (1990). Ecological specificity of vesicular–arbuscular mycorrhizal associations. *Mycological Research, 94*, 120–122.

McGonigle, T. P., & Miller, M. H., (1996). Mycorrhizae, phosphorus absorption, and yield of maize in response to tillage. *Soil Science Society of America Journal, 60*(6), 1856–1861.

Menéndez, A. B., Scervino, J. M., & Godeas, A. M., (2001). Arbuscular mycorrhizal populations associated with natural and cultivated vegetation on a site of buenos Aires province, Argentina. *Biology and Fertility of Soils, 33*(5), 373–381.

Nziguheba, G., & Smolders, E., (2008). Inputs of trace elements in agricultural soils via phosphate fertilizers in European countries. *Science of the Total Environment, 390*(1), 53–57.

Oehl, F., Sieverding, E., Ineichen, K., Mäder, P., Boller, T., & Wiemken, A., (2003). Impact of land use intensity on the species diversity of arbuscular mycorrhizal fungi in agroecosystems of central Europe. *Applied and Environmental Microbiology, 69*(5), 2816–2824.

Oehl, F., Sieverding, E., Mäder, P., Dubois, D., Ineichen, K., Boller, T., & Wiemken, A., (2004). Impact of long-term conventional and organic farming on the diversity of arbuscular mycorrhizal fungi. *Oecologia, 138*(4), 574–583.

Oehl, F., Sieverding, E., Palenzuela, J., Ineichen, K., & Da Silva, G. A., (2011). Advances in *Glomeromycota* taxonomy and classification. *IMA Fungus, 2*, 191–199.

Öpik, M., Davison, J., Moora, M., & Zobel, M., (2013). DNA-based detection and identification of *Glomeromycota*: The virtual taxonomy of environmental sequences. *Botany, 92*(2), 135–147.

Öpik, M., Moora, M., Liira, J., & Zobel, M., (2006). Composition of root-colonizing arbuscular mycorrhizal fungal communities in different ecosystems around the globe. *Journal of Ecology, 94*(4), 778–790.

Oyewole, O. B., Olawuyi, J. O., Odebode, C. A., & Abiala, A. M., (2017). Influence of arbuscular mycorrhiza fungi (AMF) on drought tolerance and charcoal rot disease of cowpea. *Biotechnol. Reports, 14*, 8–15.

Parihar, M., Rakshit, A., Meena, V. S., Gupta, V. K., Rana, K., Choudhary, M., Tiwari, G., et al., (2020). The potential of arbuscular mycorrhizal fungi in C cycling: A review. *Archives of Microbiology, 202*, 1581–1596. https://doi.org/10.1007/s00203-020-01915-x.

Pellegrino, E., Gamper, H. A., Ciccolini, V., & Ercoli, L., (2020). Forage rotations conserve the diversity of arbuscular mycorrhizal fungi and soil fertility. *Frontiers in Microbiology, 10*, 2969.

Porcel, R., Redondo-Gómez, S., Mateos-Naranjo, E., Aroca, R., Garcia, R., & Ruiz-Lozano, J. M., (2015). Arbuscular mycorrhizal symbiosis ameliorates the optimum quantum yield of photosystem II and reduces non-photochemical quenching in rice plants subjected to salt stress. *Journal of Plant Physiology, 185*, 75–83. doi: 10.1016/j.jplph.2015.07.006.

Purin, S., Filho, O. K., & Sturmer, S. L., (2006). Mycorrhizae activity and diversity in conventional and organic apple orchards from Brazil. *Soil Biology Biochemistry, 38*, 1831–1839.

Ramírez-Gómez, M., Pérez-Moncada, U., Serralde-Ordoñez, D., Peñaranda-Rolón, A., Roveda-Hoyos, G., & Rodriguez, A., (2019). Diversity of arbuscular mycorrhizal fungi communities associated with cape gooseberry (*Physalis peruviana L.*) crops. *Agronomía Colombiana, 37*(3), 239–254. doi: 10.15446/agron.colomb.v37n3.74008.

Redecker, D., Schüßler, A., Stockinger, H., Stürmer, S. L., Morton, J. B., & Walker, C., (2013). An evidence-based consensus for the classification of arbuscular mycorrhizal fungi (*Glomeromycota*). *Mycorrhiza, 23*, 515–531. 10.1007/s00572-013-0486-y.

Renker, C., Blanke, V., & Buscot, F., (2005). Diversity of arbuscular mycorrhizal fungi in grassland spontaneously developed on area polluted by a fertilizer plant. *Environmental Pollution, 135*(2), 255–266.

Rillig, M. C., (2004). Arbuscular mycorrhizae, glomalin, and soil aggregation. *Canadian Journal of Soil Science, 84*(4), 355–363.

Rillig, M. C., Aguilar-Trigueros, C. A., Camenzind, T., Cavagnaro, T. R., Degrune, F., Hohmann, P., & Yang, G., (2019). Why farmers should manage the arbuscular mycorrhizal symbiosis. *New Phytologist.* doi: 10.1111/nph.15602.

Rillig, M. C., Lehmann, A., Lehmann, J., Camenzind, T., & Rauh, C., (2018). Soil biodiversity effects from field to fork. *Trends in Plant Science, 23*(1), 17–24.

Rivera-Becerril, F., Calantzis, C., Turnau, K., Caussanel, J. P., Belimov, A. A., Gianinazzi, S., et al., (2002). Cadmium accumulation and buffering of cadmium induced stress by arbuscular mycorrhiza in three *Pisum sativum* L. genotypes. *Journal of Experimental Botany, 53*, 1177–1185.

Rosendahl, S., McGee, P., & Morton, J. B., (2009). Lack of global population genetic differentiation in the arbuscular mycorrhizal fungus *Glomus mosseae* suggests a recent range expansion, which may have coincided with the spread of agriculture. *Molecular Ecology, 18*, 4316–4329.

Rouphael, Y., Franken, P., Schneider, C., Schwarz, D., Giovannetti, M., Agnolucci, M., De Pascale, S., et al., (2015). Arbuscular mycorrhizal fungi act as biostimulants in horticultural crops. *Scientia Horticulturae, 196*, 91–108.

Ryan, M. H., & Graham, J. H., (2018). Little evidence that farmers should consider the abundance or diversity of arbuscular mycorrhizal fungi when managing crops. *New Phytologist, 220*(4), 1092–1107.

Säle, V., Aguilera, P., Laczko, E., Mäder, P., Berner, A., Zihlmann, U., Van, D. H. M. G. A., & Oehl, F., (2015). Impact of conservation tillage and organic farming on the diversity of arbuscular mycorrhizal fungi. *Soil Biology and Biochemistry, 84*, 38–52.

Sallah, P. Y. K., Obeng-Antwi, K., & Ewool, M. B., (2002). Potential of elite maize composites for drought tolerance in stress and non-drought environments. *African Crop Science Journal, 10*(1), 1–9.

Samba-Mbaye, R. T., Anoir, C. M., Diouf, D., Kane, A., Diop, I., Assigbete, K., Tendeng, P., & Sylla, S. N., (2020). Diversity of arbuscular mycorrhizal fungi (AMF) and soils potential infectivity of *Vachellia nilotica* (L.) PJH hurter Mabb. rhizosphere in Senegalese salt-affected soils. *African Journal of Biotechnology, 19*(7), 487–499.

Sanginga, N., Thottappilly, G., & Dashiell, K., (1999). Effectiveness of rhizobia nodulating recent promiscuous soybean selections in the moist savanna of Nigeria. *Soil Biology Biochemistry, 32*, 127–133.

Santos, J. C., Finlay, R. D., & Tehler, A., (2006). Molecular analysis of arbuscular mycorrhizal fungi colonizing a semi-natural grassland along a fertilization gradient. *New Phytologist, 172*(1), 159–168.

Sbrana, C., Avio, L., & Giovannetti, M., (2014). Beneficial mycorrhizal symbionts affecting the production of health-promoting phytochemicals. *Electrophoresis, 35*(11), 1535–1546.

Schlemper, T. R., & Stürmer, S. L., (2014). On farm production of arbuscular mycorrhizal fungi inoculum using lignocellulosic agrowastes. *Mycorrhiza, 24*, 571–580. doi: 10.1007/s00572-014-0576-5.

Sendek, A., Karakoç, C., Wagg, C., Domínguez-Begine, J., Martucci Do, C. G., Van, D. H. M. G. A., Naz, A. A., Locher, A., et al., (2019). Drought modulates interactions between arbuscular mycorrhizal fungal diversity and barley genotype diversity. *Science Reports, 9*, 9650. https://doi.org/10.1038/s41598-019-45702-1.

Sharma, M. P., Grover, M., Chourasiya, D., Bharti, A., Agnihotri, R., Maheshwari, H. S., Pareek, A., et al., (2020). Deciphering the role of trehalose in tripartite symbiosis among rhizobia, arbuscular mycorrhizal fungi, and legumes for enhancing abiotic stress tolerance in crop plants. *Frontiers in Microbiology, 11*, 509919.

Sharma, M. P., Jaisinghani, K., Sharma, S. K., & Bhatia, V. S., (2012). Effect of native soybean rhizobia and am fungi in the improvement of nodulation, growth, soil enzymes

and physiological status of soybean under microcosm conditions. *Agricultural Research, 1*(4), 346–351. https://doi.org/10.1007/s40003-012-0038-2.
Sharma, M. P., Reddy, U. G., & Adholeya, A., (2011). Response of arbuscular mycorrhizal fungi on wheat (*Triticum aestivum* L.) grown conventionally and on beds in a sandy loam soil. *Indian Journal of Microbiology, 51,* 384–389. doi: 10.1007/s12088-011-0134-1.
Sharma, M. P., Sandeep, S., Sharma, S., Aketi, R., & Bhatia, V. S., (2016). Co-inoculation of resident AM fungi and soybean rhizobia enhanced nodulation, yield, soil biological parameters and saved fertilizer inputs in vertisols under microcosm and field conditions. *Soybean Research. 14,* 39–53.
Siqueira, J. O., Colozzi-Filho, A., & Oliveira, E., (1989). Occurrence of vesicular-arbuscular mycorrhizae in agro- and natural ecosystems in the state of minas Gerais. *Brazilian Agricultural Research, 24,* 1499–1506.
Solaiman, Z. M., Abbott, L. K., & Varma, A., (2014). *Mycorrhizal Fungi: Use in Sustainable Agriculture and Land Restoration* (Vol. 41). Heidelberg: Springer.
Spatafora, J. W., Chang, Y., Benny, G. L., Lazarus, K., Smith, M. E., Berbee, M. L., Bonito, G., et al., (2016). A phylum-level phylogenetic classification of zygomycete fungi based on genome-scale data. *Mycologia, 108*(5), 1028–1046.
Srivastava, P., Saxena, B., & Giri, B., (2017). Arbuscular mycorrhizal fungi: Green approach/technology for sustainable agriculture and environment. In: *Mycorrhiza-Nutrient Uptake, Biocontrol, Ecorestoration* (pp. 355–386). Springer, Cham.
Sýkorová, Z., Ineichen, K., Wiemken, A., & Redecker, D., (2007). The cultivation bias: Different communities of arbuscular mycorrhizal fungi detected in root from the field, from bait plants transplanted to the field, and from a greenhouse trap experiment. *Mycorrhiza, 18,* 1–14.
Turnau, K., & Mesjasz-Przybylowicz, J., (2003). Arbuscular mycorrhiza of *Berkheya coddii* and other Ni-hyperaccumulating members of *Asteraceae* from ultramafic soils in South Africa. *Mycorrhiza, 13*(4), 185–190.
Turrini, A., Sbrana, C., Avio, L., Njeru, E. M., Bocci, G., Bàrberi, P., & Giovannetti, M., (2016). Changes in the composition of native root arbuscular mycorrhizal fungal communities during a short-term cover crop-maize succession. *Biology Fertility Soils, 52*(5), 643–653.
Wang, H., Boutton, T., Xu, W., Hu, G., Jiang, P., & Bai, E., (2015). Quality of fresh organic matter affects priming of soil organic matter and substrate utilization patterns of microbes. *Scientific Reports, 5*(1), 1–13.
Wang, Y., Wang, M., Li, Y., Wu, A., & Huang, J., (2018). Effects of arbuscular mycorrhizal fungi on growth and nitrogen uptake of *Chrysanthemum morifolium* under salt stress. *PLoS One, 13,* e0196408. doi: 10.1371/journal.pone.0196408.
Wilde, P., Manal, A., Stodden, M., Sieverding, E., Hilderbrandt, U., & Bothe, H., (2009). Biodiversity of arbuscular mycorrhizal fungi in roots and soils of two salt marshes. *Environmental Microbiology, 11,* 1548–1561.
Willis, A., Rodrigues, B. F., & Harris, P. J., (2013). The ecology of arbuscular mycorrhizal fungi. *Critical Reviews in Plant Sciences, 32*(1), 1–20.
Wilson, G. W., Rice, C. W., Rillig, M. C., Springer, A., & Hartnett, D. C., (2009). Soil aggregation and carbon sequestration are tightly correlated with the abundance of arbuscular mycorrhizal fungi: Results from long-term field experiments. *Ecology Letters, 12*(5), 452–461.
Xu, M., Li, X., Cai, X., Li, X., Christie, P., & Zhang, J., (2017). Land use alters arbuscular mycorrhizal fungal communities and their potential role in carbon sequestration on the Tibetan plateau. *Scientific Reports, 7*(1), 1–11.

Zaefarian, F., Rezvani, M., Ardakani, M. R., Rejali, F., & Miransari, M., (2013). Impact of mycorrhizae formation on the phosphorus and heavy-metal uptake of Alfalfa. *Communications in Soil Science and Plant Analysis, 44*(8), 1340–1352.

Zaller, J. G., Heigl, F., Ruess, L., & Grabmaier, A., (2014). Glyphosate herbicide affects belowground interactions between earthworms and symbiotic mycorrhizal fungi in a model ecosystem. *Scientific Reports, 4*, 5634.

Zeng, L., Li, J., Liu, J., & Wang, M., (2014). Effects of arbuscular mycorrhizal (AM) fungi on citrus fruit quality under nature conditions. *Southwest China Journal of Agricultural Sciences, 27*(5), 2101–2105.

Zhang, T., Hu, Y., Zhang, K., Tian, C., & Guo, J., (2018). Arbuscular mycorrhizal fungi improve plant growth of *Ricinus communis* by altering photosynthetic properties and increasing pigments under drought and salt stress. *Industrial Crops and Products, 117*, 13–19.

CHAPTER 5

Role of Fungi in Biocontrol of Diseases in Cereal Crops

MD. SHAMIM[1], DEEPAK KUMAR[2], MAHESH KUMAR[1], DEEPTI SRIVASTAVA[3], SANTOSH KUMAR[4], TUSHAR RANJAN[5], and V. B. JHA[6]

[1]*Department of Molecular Biology and Genetic Engineering, Dr. Kalam Agricultural College, Kishanganj (Bihar Agricultural University), Sabour, Bhagalpur, Bihar, India*

[2]*R&D Division, Nextnode Bioscience Pvt. Ltd., Kadi, Gujarat, India*

[3]*Integral Institute of Agricultural Science and Technology, Integral University, Dasauli, Lucknow, Uttar Pradesh, India*

[4]*Department of Plant Pathology, Mandan Bharti Agricultural College (Bihar Agricultural University), Sabour, Bhagalpur, Agwanpur Saharsa, Bihar, India*

[5]*Department of Molecular Biology and Genetic Engineering, Bihar Agricultural University, Sabour, Bhagalpur, Bihar, India*

[6]*Department of Plant Breeding and Genetics, Dr. Kalam Agricultural College, Kishanganj, (Bihar Agricultural University), Sabour, Bhagalpur, Bihar, India*

ABSTRACT

Fungal antagonists participate as an important key role in managing plant diseases caused by pathogens in plants. These fungi are employed as biocontrol agents (BCAs) for important diseases all over the world.

Applied Mycology for Agriculture and Foods: Industrial Applications. Sanjay K. Singh, Deepak Kumar, Md. Shamim, & Rohit Sharma (Eds.)

There are about 300 fungal antibodies in 13 classes, and 113 generations have been documented along with targeted pathogens and associated plant diseases. As only organisms can provide commercially adequate levels of disease management, their assimilation with additional control systems is expected to provide better strength and efficiency. It is also an attractive approach than the chemical fungicide that should be in line with the present executive approach that may limit the effectiveness of enemy types. The use of antagonists must be accompanied by other control strategies for biological control to be effective. An effective biocontrol based on a combination of several competing and non-competitive competitors has several advantages: in addition to the wide range of functions, they increase efficiency, greater reliability and allow for reduced application times and treatment costs. As indicated by the expanding number of commercial treatments available or manufactured, fungal exploitation to control pests and invertebrates is on the rise.

5.1 INTRODUCTION

The possible use of fungi as biological control agents for plant pathogens has enlarged significantly because the fungus has a higher reproductive rate (sexually and asexually), shorter production time and more direct. Furthermore, in the absence of a host, they can live in the environment and switch their pathogenesis to propagation, ensuring stability. Many fungal species have mechanisms that allow them to effectively defend plants from different diseases originated by pathogenic fungi (Thambugala et al., 2020). The most commonly used method of combating plant-induced losses is the common use of pesticides on plants, with the aim of eliminating or reducing the severity of disease phenotypes. However, it is becoming increasingly clear that long-term use of chemical pesticides can have many harmful effects. For example, many pesticides can lead to serious and permanent harm to humans and have been shown to cause serious damage to the general ecosystem by affecting unintended factors, such as pollen, and soil pollution and water systems (Law et al., 2017; Pimentel et al., 1993). These non-target effects can also extend to reduce the diversity of microorganisms in the soil, which can also disperse viruses from the competition and increase the chances of infection (Jacobsen & Hjelmso, 2014). The use of chemical pesticides is also hampered by the emergence of antimicrobial resistance. In the same way that we deal with the problem of modern medicine due to antimicrobial resistance, so do we also experience a decline in the effectiveness of pesticide antimicrobials due to phytopathogen resistance (Lucas et al., 2015). As a result of the side effects and side effects of using chemical pesticides to control plant diseases, research is beginning to focus on finding alternative solutions to pathogenic infections. It has long been known that soil-related microorganisms can contribute to plant as well as soil health, as exposure to certain strains of bacteria can lead to a reduction in the incidence and severity of plant diseases (Weller et al., 2002; Schlatter et al., 2017). In addition, certain solvents from the roots of the plant microbiome create a variety of secondary metabolites that can slow down pathogen in both *in vitro* and *in vivo* condition (Coombs et al., 2004; Mendes et al., 2013).

5.2 HISTORY OF FUNGI AS BIOCONTROL AGENT

During the study of fluid culture between *Penicillium glaucum* and bacteria, Roberts (1874) proved the hostile behavior of germs and coined the term antagonism in microbiology. By mixing soil with microbes that were assumed to have antagonistic qualities, Hartley (1921) conducted the first attempt at direct biological control of plant bacteria. He incorporated forest nursery soil with 13 opposing fungi to control fishing by *Pythium debaryanum* (Baker, 1987; Gupta & Sharma, 2014). Weindling (1932, 1934) described the ability of *Trichoderma lignorum* (*T. viride*) to control plant-pathogenic fungi by mycoparasitism and reported the first use of a known antimycotic antagonist in the control of plant diseases (Baker, 1987). Weindling (1941) discovered that Trichoderma species generated gliotoxin, a toxic antimycotic that killed plant bacteria including *Rhizoctonia solani* and *Sclerotinia americana*. This was the first time a known antimycotic antagonist was used to treat plant diseases (Baker, 1987; Howell, 2003). A. Fleming's discovery of penicillin in 1928, as well as its refinement and usage in pharmaceutical manufacture, rekindled interest in antimicrobial agents (Baker, 1987).

The Arbuscular mycorrhizae (AMF) fungus is a symbiotic fungus prominent in the roots and soil of agricultural plants (Tahat et al., 2010). Mycorrhizal fungi association varies greatly in structure and function, but Arbuscular Mycorrhizae (AM) is the most common contact (Harrier, 2001). Two major types of mycorrhizae fungi most commonly in the cereal plant play a role in controlling disease in grain plants. Endo-mycorrhizae is a fungus group that is found in many agricultural plants and protects the environment from soil-borne diseases (Smith & Read, 2008). These fungi were reported across the several species of living things on earth and are found in many kinds of plants (wheat, corn, rice, grapes, beans, and cotton) and flowering species, petunias, and lilies) (Peterson et al., 2004). Arbuscular Mycorrhizae fungi is a binding of biotrophs that feed the products of the owner of a living plant and those fungi are not selected by those who can handle them. The host plant obtains mineral nutrients and new fungus mycelium from outside the root zone, while AMF receives a carbon-derived picture from the host (Smith & Read, 1997). In the other ways, ecto-mycorrhizal fungus (ECM) forms a thick layer of membranes within the spaces between the root cortex cells and the canal around the root of the feeder that acts as a connector for transporting nutrients from the plant to the fungus and vice versa (Kumar & Satyanarayana, 2002). Ectomycorrhizal fungi can only surround the living cells in the catch roots, not enter in the plant cells. The ectomycorrhizal fungus produces a wide mycelium that can be used to transmit nutrients straight from rotting leaves (Suverch et al., 1991).

5.3 ARBUSCULAR MYCORRHIZAE FUNGI (AMF) AS PROTECTION AGAINST PATHOGENS (BIO-CONTROL AGENTS)

Microbes (e.g., *Pseudomonas fluorescens*, *Bacillus subtilis*, etc.) and fungi (e.g., AMF, *Trichoderma*, etc.) compete with plant germs for nutrients and the environment by creating antibiotics, destroying bacteria, or diminishing invasive plant resistance. These microbes have been used in biocontrol for pathogens (Berg et al., 2007). Excessive use of chemicals to manage diseases endangers modern crop production systems (Dehne, 1982). Beneficial microorganisms are currently being studied as one of the various techniques of management that has the ability to safeguard the bacteria that carry plant soil (Brimmer & Boland, 2003; Mukerji et al., 2002). Several studies have looked into the protective impact of mycorrhizal symbioses against root pathogenic fungi (Caron, 1989; Dehne, 1982). Reduction of disease in plants hosted by AMF is a result of the release of complex interactions between bacteria, AMF, and the plant (Harrier & Watson, 2004). The AMF symbiosis has been demonstrated to minimize infections in the soil (Azcon-Aguilar et al., 2002). The increase in *Phytophthora parasitica* decreased significantly when the tomato root was colonized by *Glomus mosseae* and *P. parasitica* compared to non-mine tomato roots (Cordier et al., 1996). Trotta et al. (1996) discovered that phosphate produced by AMF may help to reduce *P. parasitica* damage in tomatoes. The presence of AMF successfully slows Ganoderma Boninense's ability to infect and destroy the palm oil plant, and the seedlings were *G. boninense* resistant (Rini, 2001). AMF did not show indirect contact with soil pathogen through antagonism, mycoparasitism, and or antibiotic (Harrier & Watson, 2004). Various methods have been reported to define bio control by AMF including chemical changes in plant tissues, microbial changes in the rhizosphere, genetic status, anatomical changes in cells, changes in root system morphology and stress reduction (Hooker et al., 1994).

5.4 *GLOMUS* SPECIES

Many species of *Glomus* have been found to control soil-borne diseases (Olowe et al., 2018) such as *Aphanomyces, Cylindrocladium, Macrophomina, Phytophtora, Rhizoctonia, Sclerotinium,* and *Verticillium* (Harrier & Watson, 2004; Oyewole et al., 2017). *Glomus clarum* and *G. deserticola* has been reported to reduce the risk of necrosis caused by *Rhizoctonia solani* in cowpea, and *Pythium aphanidermatum* in pepper and pawpaw (Odebode et al., 1997; Abdel-Fattah & Shabana, 2002; Olawuyi et al., 2013).

Control of fungal biology is an exciting and rapidly evolving field of research that contributes to plant production, animal and human health and food production. The attraction of fungi as biocontrol agents (BCAs), in particular, is due to their common availability, high level of host clarity, destruction of the keeper, persistence, good dispersion efficiency, cultural freedom and laboratory retention. The success story of fungi, especially rusts, in biocontrol systems is well documented, including the rapid weed control of *Chondrilla juncea* by *Puccini achondrillina* (Hasan, 1972; Cullen et al., 1973; Hasan & Wapshere, 1973; Emge et al., 1981). There are several challenges remain reported in the study, development, and marketing of final fungal control agents, ranging from deciphering critical biological data to social and economic concerns. There has been significant improvement in a number of areas, but it is critical to integrate and convey these new results (Button et al., 2000). Fungal control agents can be used when chemical pesticides are banned (e.g., organochlorines) or extracted (e.g., methyl bromide) or when pesticides have become resistant to common pesticides.

5.5 MECHANISMS OF FUNGI MEDIATED BIO-CONTROLS

Biological control is the result of many forms of communication between living organisms. Researchers have recently concentrated on demonstrating approaches that operate in a range of experimental settings. In all cases, the bacteria are resistant to the presence and activity of other substances. The various biocontrol modalities of contraindication occur in a wide range that is directly related to the interaction of interspecies interactions and interaction specifications. Hyper-parasitism by binding to parasites of a plant pathogen can be considered as the most direct form of resistance due to the activities of non-other organisms.

5.5.1 DIRECT ANTAGONISM

The direct opposite or lysis of the death of a pathogen by other microorganisms is called hyper-parasitism and fungal fungus in other fungi known as mycoparasites (Baker & Cook, 1974): (a) concomitant contraindications; (b) a mixed argument; and (c) no contradiction, with no sense at all. However, the understanding of my coparasitism has greatly improved these days. The fungus *Ampelomyces quisqualis* occurs in Erysiphales (powdery mildews) which is a natural deuteromycete hyper-parasite contributes to the formation of pycnidia (fruit-bearing bodies) within powdery mildew hyphae, conidiophores, and cleistothecia (closed bodies of powdery mildew). For more than 50 years, *Ampelomyces quisqualis* has been the focus of various studies into powdery mildew treatment. *Trichoderma lignorum (T. viride)* suppresses *Rhizoctonia solani* hyphae and also suggests that the presence of *Trichoderma* spores may assist regulate citrus seedling decline (Lo, 1997). Trichoderma species' mycoparasitic power against the most important plant pathogens enables the development of biocontrol techniques (Harman et al., 2004). There are several insect-borne plant diseases, some of which attack sclerotia (e.g., *Coniothyrium minitans*) and others which attack living hyphae (e.g., *Coniothyrium minitans*) (e.g., *Pythium oligandrum*) and, one fungal virus can be attacked by many examples ofhyperparasites, *Acremonium alternatum, Acrodontium crateriforme, Ampelomyce squisqualis, Cladosporium oxysporum*, and *Gliocladium virens* are just a few of them that has the potential to reduce the incidence of powdery mildew (Kiss, 2003). Microbial predation is highly widespread, in contrast to hyperparasitism and the pathogen is unclear and often provides uncontrolled levels of disease control.

5.5.2 ANTIBIOSIS

Antibiosis is the process of disinfection of antibacterial agents by antagonist fungi suppresses or kills pathogenic fungi in the newly developed environment. Most fungi are able to hide one or more computers and secondary metabolites through the action of antibiotics, which are often associated with certain stages of behavioral differentiation and are associated with an active growth phase. Interestingly, some secondary fungal metabolites can alter plant growth, sporulation, and hyphal extension, are exploited in agricultural applications (Keller, 2005). Menendez & Godas (1998) reported on the bio-control study of *Trichoderma harzianum* directed at *Sclerotinia sclerotiorum* – a soil-carrying plant that attacks many economically important plants, such as soybean and studied the antibiotic *T. harzianum* against plant pathogen, it is thought that beneficial effect was due to mycoparasitism and similar competition (Inbar et al., 1996; Ghisalberti, 2002). In another example, without the close connection between the hyphae of *Trichoderma* spp. and with *Fusarium moniliforme*/ *Aspergillus flavus* on coculturing, hyphal infiltration was absent, suggesting that mycoparasitism was not the only cause of the observed inhibitory effects (Calistru, 1997).

Therefore, metabolites are produced by *Trichoderma* sp. (e.g., mutations, extracellular enzymes, and/or antibiotics) were considered potential risk factors for antibiotic resistance. It has proven to be effective in a variety of situations and in the long-term prevention of pathogenic fungal sclerotia. Szekeres et al. (2005) reviewed the antagonistic metabolites produced by *Trichoderma* sp. Specific metabolites, amphipathic polypeptides, i.e., peptaibols and peptaibiotics and their physicochemical and biological properties of these antibiotics include lipid membrane disruption, antibacterial activity, and implantation plant resistance is also reported. Some biocontrol manufacturers exhibit edible behavior under limited nutrient conditions. However, such activity is usually not demonstrated under normal growth conditions. For example, some species of *Trichoderma* produce large amounts of enzymes directed to the fungal wall. Peptaibols and peptaibiotics are amphipathic linear polypeptides. These substances are composed of 5–20 amino acids and are usually produced in micro-organic compounds that contain gram-positive antibodies and fungi – acting in harmony with cell-wall-degrading enzymes (CWDEs) to slow down bacterial and fungal growth and stimulates plant resistance to bacteria (Wiest et al., 2002; Szekeres et al., 2005). Peptaibols and peptaibiotics with abnormal amino acid content are the result of non-ribosomal biosynthesis. The major active enzymes are known called peptide synthetases that bind these molecules is a thio-template carrier machine that carries many of the primary precursors, which can be N-methylated, acylated or reduced (Szekeres et al., 2005).

Peptaibols and peptaibiotic show interesting physicochemical and biological properties, including the formation of pores in the lipid membrane. Excessive use of peptaibols creates an immune response and reduces susceptibility to mosaic virus (Wiest et al., 2002). Peptaibol synthetase from *T. virens* in has recently been purified, and the same type, made, will facilitate studies of this gene and its contribution to biocontrol. An in-depth review of antibiotics and the production of secondary *Trichoderma* metabolites provided by Hutchinson (1999); and Hanson & Howell (2002) explained the importance of secondary metabolites (antibiotic activity) in the antagonistic resistance of *Trichoderma* spp. However, there seems to be a general consensus on the combined effect of the combination of two substances (enzymes and antibiotic compounds) (Schirmbock, 1994).

5.5.3 COMPETITION

Hunger is the most common cause of death of microorganisms, so that competition for nutrient depletion leads to the control of biological fungal phytopathogens (Chet et al., 1997). Celar (2003) directed research on genetically modified genes found in phytopathogenic and antagonistic fungi. Previous research has reaffirmed Blakeman's (1978) finding those nutrient deficiencies in microorganisms, especially in those living in soils and vegetation, lead to open competition for nutrients between microorganisms (Sivan & Chet, 1986; Lewis, 1991). The type of competition is fungistatic (inhibitory) and there will be a slight change in the inoculum strength of the rhizosphere near the foci of infection before and after the competitive biological control strategy is implemented. The potential for effective fungal competition in biological control is therefore slim.

It has been discovered that biocontrol based on competition for uncommon but critical micronutrients, such as iron, which is severely limited to the rhizosphere, is influenced by soil pH. Iron is present in highly oxidized soils in the ferric state (Lindsay, 1979), which is insoluble in water (pH – 7.4) and has a concentration as low as 10–18 M. This concentration is insufficient to support microbial development, usually require concentrations close to 10–6 M. In order to make it in such a hostile environment, living organisms have been found to produce fibers comprising iron-containing siderophores that have a high affinity for iron ore from microbial origin. The most filamentous fungi, in turn, take iron as an important mineral for their function and are less susceptible to iron deficiency, and especially fungi produce low-density irons-iron specific chelators, called siderophores, which include natural iron (Eisendle et al., 2004). After that, metal is obtained from ferri-siderophore companies through certain extraction processes. In *Aspergillus fumigatus* and *Aspergillus nidulans*, siderophore biosynthesis is poorly regulated by a carbon source (Eisendle et al., 2004). In *Ustilago maydis*, genetic products related to iron deficiency affect plant growth (McIntire et al., 2004). Some *Trichoderma* BCAs produce highly effective siderophores that test iron and stop the growth of other fungi (Chet & Inbar, 1994). As a result, depending on the availability of iron, soil composition determines Pythium biocontrol performance by *Trichoderma*. In addition, *T. harzianum* T35 regulates *Fusarium oxysporum* by competing with the colonization of the rhizosphere and nutrients, and biocontrol is effective as gene expression decreases (Tjamos et al., 1992). Biocontrol of phytopathogens such as Botrytis cinerea, a significant pathogenic compound, has been demonstrated to be dependent on competition. Increased efficiency of iron detection of microbes is thought to be the cause of their strong grip binding to plant roots and aids in the transport of predatory organisms to potential areas of infection. Therefore, the cited examples have confirmed the competitive advantage of fungal growth nutrients as bio-control agents.

5.5.4 MYCOPARASITISM

Mycoparasitism occurs when one fungus is present in close contact with another where it receives a portion of or all of its nutrients without reciprocating benefits (Lewis et al., 1989). Depending on their mode of parasitism, mycoparasites are divided into two major groups, biotrophs and necrotrophs (Barnett & Binder, 1973). Biotrophs are considered to be those organisms that are able to absorb nutrients from living cells and are identified by the fact that the infected tissue does not die. This type of insect relationship is physically limited, and parasites seem to be very compatible with this lifestyle. Biotrophic mycoparasites tend to have a very limited processing range and produce specialized nutrients. Necrotrophic or destructive mycoparasites kill a trapped cell shortly before or after the attack and absorb nutrients from dying or dead cells. Necrotrophic mycoparasites, unlike biotrophic parasites, tend to be more aggressive, have a wide range of handling and are less well known for their parasitism. The antagonistic activity of necrotrophic mycoparasites is caused by the production of antibiotics, toxins or hydrolytic enzymes (Chet et al., 1997). Mycoparasitism is a common phenomenon *in vitro* and *in vivo* and its involvement in the control of biological diseases has been extensively reviewed (Ayers & Adams, 1981; Lumsden, 1981; Manocha, 1991; Nigam & Mukerji, 1992; Sharma et al., 1998; Wells, 1988).

Trichoderma hyphae has been shown to disable pathogenic fungi, viz. *Rhizoctonia, Sclerotium, Sclerotinia, Helminthosporium, Fusarium, Verticillium, Venturia, Pythium, Phytophthora, Rhizopus, Botrytis*, etc. (Beagle-Restaino & Papavizas, 1985; Chet et al., 1981; Kumar, 1993; Wolfehechel & Jensen, 1992). When mycoparasite has grown with its host carrying a dual culture, it grows towards the host and a typical branch pattern occurs. The state of recognition was made by binding the host agglutinin (lectin) to the carbohydrate residues in the cell walls of the Trichoderma spp. Hydrolytic enzymes such as β-l, 3-glucanase, cellulase, chitinase, and various other compounds of enzymes that assist BCAs in the entry of hypha cell center (Baker & Dickman, 1992). To achieve the desired effect from mycoparasitic biocontrol, physical, and nutritional environment should be good for agents to work. Many researchers have reported effective biocontrol in critical diseases using mycoparasitic agents in glasshouse experiments (Ayers & Adams, 1981; Sharma et al., 1998). Whipps et al. (1989) reported care by *Coniothyrium minitans* against *Sclerotinia sclerotiorum* disease in lettuce and edible plants with watery branches in glasshouse experiments. Although success in open space is rare, mycoparasitism has the potential to eradicate pathogens and remains an attractive strategy for controlling biodiversity.

5.6 FUNGAL BIOLOGICAL CONTROL AGENTS' COMMERCIALIZATION

Doing business with biological control agents is expensive because it entails a number of measures, including isolation from pure culture or enrichment of microorganism, identification, and performance, development of appropriate structures, mass production, successful product testing, storage stability, industrial co-operation, concern for human safety environment, registration, and marketing (Punja, 1997; Stirling & Stirling, 1997; Janisiewicz & Korsten, 2002; Montesinos, 2003). Many environmentally friendly products are marketed around the world to control fungal plant fungi and are often produced as granules (GR), liquid powders, fungi, and products containing strong liquids or oil using various mineral and organic carriers (Ardakani et al., 2009; Nega, 2014). Many microbial antagonists are patented and tested for commercial use (Schena et al., 2004; Nabi et al., 2017) and these substances are frequently recommended for plants (Albajes et al., 2000; Fravel, 2005; O'Brien, 2017).

5.7 APPLICATIONS OF BCAS IN COMBINATION WITH SYNTHETIC FUNGICIDES FOR PLANT FUNGAL PATHOGENS

Plant fungal infections are commonly treated with chemical agents such as Captan, dithiocarbamates, thiabendazole (TBZ), and imazalil (IMZ) (Lucas et al., 2015; Perez et al., 2016; Gupta, 2018). However, the extensive and indiscriminate use of fungicides in plant protection and post-harvest food preservation has resulted in the development of resistance to other fungicides, as well as negative consequences for humans, animals, and wildlife, resulting in negative environmental consequences (Gupta, 2018, 2019).

5.8 PLANT FUNGAL PATHOGEN BIOCONTROL: BIOTECHNOLOGICAL APPROACHES

The use of genomics, genetic engineering, and repeated DNA technologies have altered the important genetic makeup of fungi and their products for better use in disease management. Several investigations have been carried out to evaluate the genetic features of antifungal antibodies and their ability to improve biocontrol action (Janisiewicz & Korsten, 2002; Droby, 2006; O'Brien, 2017). Here are a couple such examples: (a) the addition of several enzyme genes – encoding into the *Trichoderma* virens genome has led to stress that has produced a combination of glucanases; and shows the highly advanced inhibition of *Pythium ultimum* (Oomycota, Chromista), *Rhizoctonia solani*, and *Rhizopus oryzae* (b). McDougal et al. (2012) introduced the genetic modification of *Cyclaneusma minus*, utilizing protoplasts produced by treatment with the enzyme Glucanex ™ as the cause of Cyclaneusma needle-cast. The *Cyclaneusma minus* was modified with a gene encoding green fluorescent protein (GFP), which was approved to identify several *Trichoderma*-capable biocontrol species of the disease. Interaction between *C. minus* and *Trichoderma* problems, at the point of contact where GFP is lost, determined the two-cultural process to be antifungal agents (c). Yakoby et al. (2001) produced modified pathogenicity mutants of the avocado fruit pathogen *Colletotrichum gloeosporioides* using insertion mutagenesis by inhibiting mediated integrated enzyme mutations and these mutants further can be used for C-anthracnose biological control.

5.9 CONCLUSIONS AND FUTURE PROSPECTS

If scientists are successful in developing relaxing grains and compatible mycelia, the usage of bio-control fungi will rise in the future. Studies related to the use of genes to increase fungal vitality are growing but are not as widely studied as germs and viruses. A better understanding of abiotic and biotic and fungal interactions is required to determine the appropriate dose and time of application. The application period must take into account the participants' categorization, as well as environmental and agricultural activities (e.g., avoid mold). By reducing the negative impact of pests and weeds and hence enhancing crop quality, fungal biological management can play an essential role in sustainable agriculture.

KEYWORDS

- **antimycotic antagonist**
- **arbuscular mycorrhizae**
- **biocontrol agent**
- **biocontrol agents**
- **cell-wall-degrading enzymes**
- **disease control**
- **ectomycorrhizal fungus**
- **green fluorescent protein**
- **mycoparasitism**

REFERENCES

Abdel-Fattah, G. M., & Shabana, Y. M., (2002). Efficacy of arbuscular mycorrhizal fungus (*Glomus clarum*) in protection of cowpea plants from root rot pathogen *Rhizoctonia solani*. *Zeitschrift fur Pflanzenkrankheiten und Pflanzenschutz. Journal of Diseases and Protection, 109*(2), 207–215.

Albajes, R., Gullino, M. L., Van, L. J. C., & Elad, Y., (2000). *Integrated Pest and Disease Management in Greenhouse Crops* (Vol. 14). Dordrecht, The Netherlands: Springer Science & Business Media.

Ardakani, S., Heydari, A., Khorasani, N., Arjmandi, R., & Ehteshami, M., (2009). Preparation of new biofungicides using antagonistic bacteria and mineral compounds for controlling cotton seedling damping-off disease. *J. Plant Prot. Res., 49*(1), 49–56.

Ayers, W. A., & Adams, P. B., (1981). Mycoparasitism and its application of biological control of plant diseases. In: Papavizas, G. C., (ed.), *Biological Control in Crop Production* (pp. 91–103). Allanueld, Granada.

Azcon-Aguilar, C., Jaizme-Vega, M. C., & Calvet, C., (2002). The contribution of arbuscular mycorrhizal fungi for bioremediation. In: Gianinazzi, S., Schuepp, H., Barea, J. M., & Haselwandter, K., (eds.), *Mycorrhizal Technology in Agriculture: From Genes to Bioproducts* (pp. 187–197). Birkhauser Verlag, Berlin, ISBN-10: 0-89054-245-71.

Baker, K. F., (1987). Evolving concepts of biological control of plant pathogens. *Annu. Rev. Phytopathol., 25*(1), 67–85.

Baker, R., & Dickman, M. B., (1992). Biological control with fungi. In: Blain, F., (ed.), *Soil Microbial Ecology: Application in Agricultural and Environmental Management* (pp. 275–305). Metting, Jr.

Bakers, K. F., & Cook, R. J., (1974). *Biological Control of Plant Pathogens* (p. 433). Am. Phytopathol. Soc. St. Paul MN.

Barnett, H. L., & Binder, F. L., (1973). The fungal host-parasite relationship, *Ann. Rev. Biochem., 50*, 207–292.

Beagle-Restaino, I. E., & Papavizas, G. C., (1985). Biological control of *Rhizoctonia* stem canker and black scurf of potato. *Phytopathol., 75*, 560–564.

Berg, G., Grosch, R., & Scherwinski, K., (2007). Risk assessment for microbial antagonists: Are there effects on non-target organisms? *Gesunde Pflanzen, 59,* 107–117.

Blakeman, J. P., (1978). Microbial competition for nutrients and germination of fungal spores. *Ann. Appl. Biol., 89*, 151–155.

Brimmer, T., & Boland, G. J., (2003). A review of the non-target effects of fungi used to biologically control plant diseases. *Agric. Ecosyst. Environ., 100*, 3–16.

Butt, T. M., & Copping, L., (2000). Fungal biological control agents. *Pesticide Outlook, 11,* 186–191.

Calistru, C., McLean, M., & Berjak, P., (1997). *In vitro* studies on the potential for biological control of *Aspergillus flavus* and *Fusarium moniliforme* by *Trichoderma* species 1. Macroscopical and microscopical observations of fungal interactions. *Mycopathologia, 139,* 115–121.

Caron, M., (1989). Potential use of mycorrhizae in control of soil-borne diseases. *Can. J. Plant Pathol., 11,* 177–179.

Celar, F., (2003). Competition for ammonium and nitrate forms of nitrogen between some phytopathogenic and antagonistic soil fungi. *Biol. Control, 28,* 19–24.

Chet, I., & Inbar, J., (1994). Biological control of fungal pathogens. *Appl. Biochem. Biotechnol., 48,* 37–43.

Chet, I., Hadar, Y., Elad, Y., Kalan, J., & Henis, Y., (1997). Biological control of soil borne plant pathogens by *Trichoderma harzianum.* In: Schippers, B., & Grams, W., (eds.), *Soil-borne Plant Pathogens* (pp. 585–590). Academic Press, London.

Chet, I., Harman, G. E., & Baker, R. I., (1981). *Trichoderma hamatum:* Its hyphal interaction with *Rhizoctonia solani* and *Pythium* spp. *Microb. Ecol. 7,* 29–38.

Chet, I., Inbar, J., & Hadar, I., (1997). Fungal antagonists and mycoparasites. In: Wicklow, D. T., & Söderström, B., (eds.), *The Mycota IV: Environmental and Microbial Relationships* (pp. 165–184). Springer-Verlag, Berlin.

Cook, R. J., (1993). Making greater use of introduced microorganisms for biological control of plant pathogens. *Annu. Rev. Phytopathol., 31,* 53–80.

Coombs, J. T., Michelsen, P. P., & Franco, C. M. M., (2004). Evaluation of endophytic actinobacteria as antagonists of *Gaeumannomyces graminis* var. *tritici* in wheat. *Biol. Control., 29*, 359–366.

Cordier, C., Gianinazzi, S., & Gianinazzi-Pearson, V., (1996). Colonization patterns of root tissues by *Phytophthora nicotianae* var. *parasitica* related to reduced disease in mycorrhizal tomato. *Plant Soil, 185*, 223–232.

Cullen, J. M., Kable, P. F., & Catt, M., (1973). Epidemic spread of a rust imported for biological control. *Nature, 244*, 462–464.

Dehne, H. W., (1982). Interaction between vesicular-arbuscular mycorrhizal fungi and plant pathogens. *Phytopathology, 72*, 1115–1119.

Djonovic, S., Vittone, G., Mendoza-Herrera, A., & Kenerley, C. M. (2007). Enhanced biocontrol activity of *Trichoderma virens* transformants constitutively co-expressing β-1,3- and β-1,6-glucanase genes. *Molecular Plant Pathology, 8*(4), 469–480. https://doi.org/10.1111/j.1364-3703.2007.00407.x.

Droby, S., (2006). Biological control of postharvest diseases of fruits and vegetables: Difficulties and challenges. *Phytopathol. Pol., 39*, 105–117.

Eisendle, M., Oberegger, H., Buttinger, R., Illmer, P., & Haas, H., (2004). Biosynthesis and uptake of siderophores are controlled by the PacC-mediated ambient-pH regulatory system in *Aspergillus nidulans*. *Euk. Cell., 3*, 561–565.

Emge, R. G., Melching, J. S., & Kingsolver, C. H., (1981). Epidemiology of *Puccinia chondnllina,* a rust pathogen for the biological control of rush skeleton weed in the United States. *Phytopathol., 7*, 839–843.

Fravel, D. R., (2005). Commercialization and implementation of biocontrol. *Annu. Rev. Phytopathol. 43*, 337–359.

Ghisalberti, E. L., (2002). Anti-infective agents produced by the *Hyphomycetes* genera *Trichoderma* and *Gliocladium*. *Curr. Med. Chem. Anti-Infect. Agents, 1*, 343–374.

Gupta, P. K., (2018). "*Toxicity of Fungicides,*" *in Veterinary Toxicology* (pp. 569–580). Cambridge, USA: Academic Press.

Gupta, P. K., (2019). Toxic effects of pesticides and agrochemicals. In: *Concepts and Applications in Veterinary Toxicology* (pp. 59–82). Cham: Springer.

Gupta, S. K., & Sharma, M., (2014). *Approaches and Trends in Plant Disease Management* (p. 429). India: Scientific Publishers.

Hanson, L. E., & Howell, C. R., (2002). Bio-control efficacy and other characteristics of protoplast fusants between *Trichoderma koningii* and *T. virens*. *Mycol. Res., 106*, 321–328.

Harman, G. E., Howell, C. R., Viterbo, A., Chet, I., & Lorito, M., (2004). *Trichoderma* species opportunistic, avirulent plant symbionts. *Nat. Rev. Microbiol., 2*, 43–56.

Harrier, L. A., & Watson, C. A., (2004). The potential role of arbuscular mycorrhizal (AM) fungi in the bioprotection of plants against soil-borne pathogens in organic and/or other sustainable farming systems. *Pest Manage. Sci., 60*, 149–157.

Harrier, L. A., (2001). The arbuscular mycorrhizal symbiosis: A molecular review of the fungal dimension. *J. Exp. Bot., 52*, 469–478.

Hartley, C., (1921). *Damping-Off in Forest Nurseries*. U.S. Dept. Agr. Bul., 934, 99.

Hasan, S., & Wapshere, A. J., (1973). The biology of *Puccinia chondrillina,* a potential biological control agent of skeleton weed. *Ann. Appl. Biol., 74*, 325–332.

Hasan, S., (1972). Specificity and host specialization of *Puccinia chondrillina. Ann. Appl. Biol., 72*, 257–252.

Hooker, J. E., Jaizme-Vega, M., & Alkinson, D., (1994). Biocontrol of plant pathogen using arbuscular mycorrhizal fungi. In: Gianinazzi, S., & Schhepp, H., (eds.), *Impact of Arbuscular Mycorrhizas on Sustainable Agriculture and Natural Ecosystems* (pp. 191–209). Birkhauser Verlag, Basle, Switzerland.

Howell, C. R., (2003). Mechanisms employed by *Trichoderma* species in the biological control of plant diseases: The history and evolution. *Current Horticultural Science*, 2115–2120.

Hutchinson, C. M., (1999). *Trichoderma virens*-inoculated composted chicken manure for biological weed control. *Biol. Control., 16*, 217–222.

Inbar, J., Menendez, A., & Chet, I., (1996). Hyphal interaction between *Trichoderma harzianum* and *Sclerotinia sclerotiorum* and its role in biological control. *Soil Biol. Biochem., 28,* 757–763.

Jacobsen, C. S., & Hjelmso, M. H., (2014). Agricultural soils, pesticides and microbial diversity. *Curr. Opin. Biotechnol., 27*, 15–20.

Janisiewicz, W. J., & Korsten, L., (2002). Biological control of postharvest diseases of fruits. *Ann. Rev. Phytopathol., 40*(1), 411–441.

Kasun, M., Thambugala, K. M., Daranagama, D. A., Phillips, A. J. L., Kannangara, S. D., & Promputtha, I., (2020). Fungi vs. fungi in biocontrol: An overview of fungal antagonists applied against fungal plant pathogens. *Front. Cell. Infect. Microbiol., 10,* 604923.

Keller, N. P., Turner, G., & Bennett, J. W., (2005). Fungal secondary metabolism—From biochemistry to genomics. *Nat. Rev. Microbiol., 3,* 937–947.

Kiss, L., (2003). A review of fungal antagonists of powdery mildews and their potential as biocontrol agents. *Pest Manag., Sci., 59,* 475–483.

Kumar, R. N., (1993). Biological control of damping off disease of chir pine. *Ind J. Mycol. Plant Pathol. 23*(3), 35–37.

Kumar, S., & Satyanarayana, T., (2002). Isolation of ectomycorrhizal fungi: Methods and techniques. In: Mukerji, K. G., Manoharachary, C., & Chamola, B. P., (eds.), *Techniques in Mycorrhizal Studies* (pp. 143–166). Kluwer Academic Publishers, London.

Law, J. W., Ser, H. L., Khan, T. M., Chuah, L. H., Pusparajah, P., Chan, K. G., Goh, B. H., & Lee, L. H., (2017). The potential of *Streptomyces* as biocontrol agents against the rice blast fungus, *Magnaporthe oryzae* (*Pyricularia oryzae*). *Front. Microbiol., 8*, 3.

Lewis, J. A., & Papavizas, G. C., (1991). Biocontrol of plant diseases: The approach for tomorrow. *Crop Prot., 10*, 95–105.

Lindsay, W. L., (1979). *Chemical Equilibria in Soils*. John Wiley and Sons, Inc., New York.

Lo, C. T., (1997). Biological control of turfgrass diseases using *Trichoderma harzanium. Plant Pro. Bull., 39,* 207–225.

Lucas, J. A., Hawkins, N. J., & Fraaije, B. A., (2015). The evolution of fungicide resistance. *Adv. Appl. Microbiol., 90*, 29–92.

Lumsden, R. D., (1981). Ecology of my co-parasitism. In: Wicklow, D. T., & Carrol, G. C., (eds.), *The Fungal Community: Its Organization and Role in the Ecosystem* (pp. 295–318). Marcel Dekker, New York.

Manocha, M. S., (1991). Physiology and biochemistry of biotrophic mycoparasitism, In: Arora, D. K., Rai, B., Mukerji, K. G., & Knudsen, G. R., (eds.), *Handbook of Applied Mycology* (Vol. I, pp. 273–300). Marcel Dekker, New York.

McDougal, R., Stewart, A., & Bradshaw, R., (2012). Transformation of *Cyclaneusma minus* with green fluorescent protein (GFP) to enable screening of fungi for biocontrol activity. *Forests, 3*(1), 83–94.

McIntyre, M., Nielsen, J., Arnau, J., Van, D. B. H., Hansen, K., & Madrid, S., (2004). *Proceedings of the 7th European Conference on Fungal Genetics*. Copenhagen, Denmark.

Mendes, R., Garbeva, P., & Raaijmakers, J. M., (2013). The rhizosphere microbiome: Significance of plant beneficial, plant pathogenic, and human pathogenic microorganisms. *FEMS Microbiol. Rev., 37*, 634–663.

Menendez, A. B., & Godeas, A., (1998). Biological control of *Sclerotinia sclerotiorum* attacking soybean plants. Degradation of the cell walls of this pathogen by *Trichoderma harzianum* (BAFC 742). *Mycopathologia, 142*, 153–160.

Montesinos, E., (2003). Development, registration and commercialization of microbial pesticides for plant protection. *Int. Microbiol., 6*(4), 245–252.

Mukerji, K. G., Manoharachary, C., & Chamola, B. P., (2002). *Techniques in Mycorrhizal Studies* (1st edn., pp. 285–296). Kluwer Academic Publishers., London-Netherlands, ISBN-10: 1402005326.

Nabi, S. U., Raja, W. H., Kumawat, K. L., Mir, J. I., Sharma, O. C., & Singh, D. B., (2017). Post-harvest diseases of temperate fruits and their management strategies-a review. *Int. J. Pure App. Biosci., 5*(3), 885–898.

Nega, A., (2014). Review on concepts in biological control of plant pathogens. *J. Biol. Agric. Healthcare, 4*(27), 33–54.

Nigam, N., & Mukerji, K. G., (1992). Biocontrol of powdery mildew. In: Mukerji, K. G., Tiwari, J. P., Arora, D. K., & Saxena, G., (eds.), *Recent Developments in Biocontrol of Plant Diseases* (pp. 17–22). Aditya Books Pvt. Ltd., New Delhi.

O'Brien, P. A., (2017). Biological control of plant diseases. *Australas. Plant Pathol., 46*(4), 293–304.

Odebode, A. C., Oladoye, O. A., & Osonubi, O., (1997). Effect of *Pythium aphanidermatum* and the arbuscular mycorrhizal fungus *Glomus deserticola* on disease severity and growth of pepper. *Int. J. Trop. Plant Dis., 15*, 85–92.

Olawuyi, O. J., Odebode, A. C., Oyewole, I. O., Akanmu, A. O., & Afolabi, O., (2013). Effect of arbuscular mycorrhizal fungi on *Pythium aphanidermatum* causing foot rot disease on pawpaw (*Carica papaya* L.) seedlings. *Arch. Phytopathol. Plant Prot., 40*, 185–193.

Oyewole, B. O., Olawuyi, O. J., Odebode, A. C., & Abiala, M. A., (2017). Influence of arbuscular mycorrhizal fungi (AMF) on drought tolerance and charcoal rot disease of cowpea. *Biotechnol. Rep., 14*, 8–15.

Perez, M. F., Contreras, L., Garnica, N. M., Fernández-Zenoff, M. V., Farías, M. E., Sepulveda, M., et al., (2016). Native killer yeasts as biocontrol agents of postharvest fungal diseases in lemons. *PLoS One, 11*(10), e0165590.

Peterson, R. L., Massicotte, H. B., & Melville, L. H., (2004). *Mycorrhizas: Anatomy and Cell Biology* (p. 173). NCR Research Press, Ottawa, Canada.

Pimentel, D., McLaughlin, L., Zepp, A., Lakitan, B., Kraus, T., Kleinman, P., Vancini, F., Roach, W. J., Graap, E., Keeton, W. S., et al., (1993). Environmental and economic effects of reducing pesticide use in agriculture. *Agric. Ecosyst. Environ., 46*, 273–288.

Punja, Z. K., (1997). Comparative efficacy of bacteria, fungi, and yeasts as biological control agents for diseases of vegetable crops. *Can. J. Plant Pathol., 19*(3), 315–323.

Rini, V. M., (2001). *Effect of Arbuscular Mycorrhiza on Oil Palm Seedling Growth and Development of Basal Stem Rot Disease Caused by Ganoderma Boninense*. Master Thesis, Universiti Putra Malaysia.

Roberts, W., (1874). Studies on biogenesis. *Philos. Trans. R. Soc Lond., 164*, 466.

Schena, L., Nigro, F., SoletiLigorio, V., Yaseen, T., Ippolito, A., & El Ghaouth, A., (2004). Biocontrol activity of bio-coat and biocure against postharvest rots of table grapes and sweet cherries. In: *V International Postharvest Symposium* (Vol. 682). Leuven, Belgium.

Schirmbock, M., Lorito, M., Wang, Y. L., Hayes, C. K., Arisan, A. I., Scala, F., Harman, G. E., & Kubicek, C. P., (1994). *Parallel Formation and Synergism of Hydrolytic Enzymes and Peptaibol Antibiotics, Molecular Mechanisms Involved in the Antagonistic Action of Trichoderma Harzianum Against Phytopathogenic Fungi*, 4364–4370.

Schlatter, D., Kinkel, L., Thomashow, L., Weller, D., & Paulitz, T., (2017). Disease suppressive soils: New insights from the soil microbiome. *Phytopathology, 107*, 1284–1297.

Sharma, M., Mittal, N., Kumar, R. N., & Mukerji, K. G., (1998). Fungi: Tool for plant disease management. In: Varma, A., (ed.), *Microbes: For Health, Wealth and Sustainable Environment*. Malhotra Publishing House, New Delhi.

Sivan, A., & Chet, I., (1986). Biological control of *Fusarium* spp. in cotton, wheat and muskmelon by *Trichoderma harzianum*. *J. Phytopathol., 116*, 39–47.

Smith, S. E., & Read, D. J., (1997). *Mycorrhizal Symbiosis* (2nd edn., p. 605). Academic Press, London, UK., ISBN-13: 978-0-12-652840-4.

Smith, S. E., & Read, D. J., (2008). Mineral nutrition, toxic element accumulation and water relations of arbuscular mycorrhizal plants. In: *Mycorrhizal Symbiosis* (3rd edn., pp. 145–148). Academic Press, London, ISBN-10: 0123705266.

Stirling, M., & Stirling, G., (1997). Disease management: Biological control. In: Brown, J., & Ogle, H., (eds.), *Plant Pathogens and Plant Diseases* (pp. 427–439).

Suverch, K., Mukerji, K. G., & Arora, D. K., (1991). Ectomycorrhizal: Handbook of applied mycology. In: Arora, D. K., Rai, B., Mukerji, K. G., & Kndson, G. R., (eds.), *Soil and Plants* (Vol. 1, pp. 187–215). Marcel Dekkler Inc., New York.

Szekeres, A., Leitgeb, B., Kredics, L., Zsuzsanna, A., Hatvani, L., Manczinger, L., & Vagvolgyi, C., (2005). Peptaibols and related peptaibiotics of *Trichoderma*. *Acta Microbiol. Immunol. Hung, 52*, 137–168.

Tahat, M. M., Kamaruzaman, S., & Othman, R., (2010). Mycorrhizal fungi as a biocontrol agent. *Plant Pathology Journal, 9, 198–207.*

Tjamos, E. C., Papavizas, G. C., & Cook, R. J., (1992). *Biological Control of Plant Diseases: Progress and Challenges for the Future.* Plenum Press, New York.

Trotta, A., Vanese, G. C., Gnavi, E., Fascon, A., Sampo, S., & Berta, G., (1996). Interaction between the soilborne root pathogen *Phytophthora nicotianae* var. *parasitica* and the arbuscular mycorrhizal fungus *Glomus mosseae* in tomato plant. *Plant & Soil, 185*, 199–209.

Weindling, R., (1934). Studies on lethal principle effective in the parasitic action of *Trichoderma lingorum* on *Rhizoctonia solani* and other soil fungi. *Phytopathology, 24*, 1153–1179.

Weindling, R., (1941). Experimental consideration of the mold toxins of *Gliocladium* and *Trichoderma*. *Phytopathology, 31*(11), 991.

Weller, D. M., Raaijmakers, J. M., Gardener, B. B., & Thomashow, L. S., (2002). Microbial populations responsible for specific soil suppressiveness to plant pathogens. *Ann. Rev. Phytopathol., 40*, 309–348.

Wells, H. D., (1988). *Trichoderma* as biocontrol agent. In: Mukerji, K. G., & Garg, K. L., (eds.), *Biocontrol of Plant Disease* (Vol. I, pp. 71–82). CRC Press, Florida.

Whipps, J. M., Budge, S. P., & Ebben, M. H., (1989). Effect of *Coniothyrium minitans* and *Trichoderma harzianum* on *Sclerotinia* disease of celery and lettuce in the glass house at a range of humidities. In: *Proc. CEC Joint Experts.* Meeting Cabrils, Spain.

Wiest, A., Grzegorski, D., Xu, B., Goulard, C., Rebuffat, S., Ebbole, D. J., Bodo, B., & Kenerley, C., (2002). Identification of peptaibols from *Trichoderma virens* and cloning of a peptaibol synthetase. *J. Biol. Chem., 277,* 20862–20868.

Wolfehechel, H., & Jensen, D. F., (1992). Use of *Trichoderma harzianum* and *Gliocladium virens* for biological control of post emergence damping off and root rot of cucumbers caused by *Pythium ultimum. J. Phytopathol. 136*(3), 221–230.

Yakoby, N., Zhou, R., Kobiler, I., Dinoor, A., & Prusky, D., (2001). Development of *Colletotrichum gloeosporioides* restriction enzyme-mediated integration mutants as biocontrol agents against anthracnose disease in avocado fruits. *Phytopathology, 91*(2), 143–148.

CHAPTER 6

Entomopathogenic Fungi: A Boon towards Organic Life Support Management

LOKNATH DESHMUKH[1] and SARDUL SINGH SANDHU[2]

[1]Fungal Biotechnology and Invertebrate Pathology Laboratory,
[2]Department of Biological Science, R.D. University, Jabalpur, Madhya Pradesh, India

ABSTRACT

The developing commercialization in the global world has prompted a lift in the far-reaching utilization of substance pesticides for crop security in agricultural green fields. It has not just added to an expansion in food creation, yet its poisonous and non-biodegradable character has additionally brought about unfriendly impacts on climate and non-targeted creatures. Entomopathogenic fungi (EPF) assume a vital part in the directive of pest populaces in nature, and agent species have been created as promising ecologically well-disposed mycoinsecticides. Taken along with endeavors toward the hereditary improvement of contagious destructiveness and stress obstruction, information on entomopathogenic growths and parasitic connection among host and insects additionally talked about with their behavioral changes and defensive response. Entomo-pathogenicity spectacles of model fungal agents, molecular advancement, and leading challenges are explored. The present review focuses on the need and use of EPF as potent biocontrol agents (BCAs) for organic lifestyle. Designing of effective and combinational formulations also mentioned. EPF and their secretion, as pest immune suppressor and the impact of biotechnology in the field of

Applied Mycology for Agriculture and Foods: Industrial Applications. Sanjay K. Singh, Deepak Kumar, Md. Shamim, & Rohit Sharma (Eds.)

biological insect pest management suitably reviewed. Different fungal spore modifications and metabolic formulations and its successful implementation for pest control and the strategies for the selection of suitable fungal agents are covered and diagrammatically represented. We have also discussed the most recent prospects of additional and offbeat utilization of EPF.

6.1 INTRODUCTION

Over the most recent couple of years, the continually expanding ecological contamination is a genuine concern. Consistently substances are being presented in the climate that adversely affects the equalization of the biological system and living creatures (Subbanna et al., 2019). Fungal parasites that possess the dirt are additionally touchy to toxins (e.g., herbicides, insect sprays, fungicides, and hefty metals) that cause diminished or postponed development, irregularities of cell metabolic pathways, and harm of cell structures (Stolarek et al., 2019). Filamentous growths speak to a significant transformative part of the eukaryotes. A few basal contagious ancestries are parasites of spineless creatures, suggesting that contagious arthropod associations establish antiquated ideal models for looking at pathogenesis. What's more, phylogenetic dispersion information unequivocally underpins the possibility that parasitic destructiveness towards insects autonomously emerged in a few contagious heredities, speaking to instances of joined development (Zheng et al., 2013). Such contagious creepy crawly communications have significant ramifications in ecological microbiology, environment balance, biodiversity, eukaryotic, and microorganism advancement, and insect and irritation control issues (Kirkland et al., 2005). Entomopathogenic growths are envisioning options in contrast to compound insect sprays. Undoubtedly, parasitic entomopathogens have been generally investigated as organic control agents of insect's vermin in endeavors to improve the manageability of harvest security (Watanabe et al., 2010). Parasitic entomopathogens have advanced some detailed relations with arthropods, plants, and different microorganisms. Basic boundaries for picking a contagious microorganism for its function in biocontrol incorporate the financially savvy manufacture of a steady, infective propagule that is fitting for use in the climate where the creepy crawly should be controlled. Productivity and cost are the two important boundaries that should be taken a gander at while contrasting the entomopathogens (biopesticides) with the traditional compound pesticides. Of the assessed 1.5–5.1 million types of parasites on the planet, around 1,00,000 have been portrayed (Blackwell et al., 2011). Of these, around 750–1,000 are parasitic entomopathogens put in more than 100 genera. However, in light of the quantity of obscure species uncovered by ongoing atomic phylogeny contemplates (Rehner et al., 2009), it is apparent that these assessments are low. Parasitic entomopathogens subsequently establish the biggest number of taxa that are insect microorganisms. Recognized 171 parasitic-based items utilized as biocontrol operators since the 1960s, the greater part of them dependent on entomopathogenic model organisms (de Faria & Wraight, 2007).

Notwithstanding viability, there are favorable circumstances in utilizing microbial control operators, for example, human security and other non-target living beings; pesticide build-ups are minimized in food and biodiversity would be expanded in oversaw biological systems (Hibbett et al., 2011). Creation of catalysts everywhere scale and some different metabolites in evaluation of expanding the entomopathogenic organisms' destructiveness, in the control of creepy crawlies and conceivably in certain sicknesses influencing plants, opens new possibilities so as to improve the entomopathogenic growths being used. This methodology of utilizing biocontrol agents (BCAs) rather than synthetic pesticides is by all accounts promising in the coming a very long time as it heads towards feasible rural practices and safeguarding environmental factors, which is the need of the current situation (O'Brien et al., 2005). Organic control agents have indicated a ton of guarantee as far as movement, however its viability is influenced by numerous variables, for example, biotic and nonbiotic factors, have plant, and at the degree of nematode pervasion. There is a compelling impulse to explain the substance of these elements to improve the general adequacy of these control agents alongside creating novel techniques to convey adequate inoculums at the objective destinations. Current strategies in biotechnology can possibly control attractive qualities of these entomopathogenic organisms to improve the general field movement (Sandhu et al., 2012).

6.2 DIVERSIFIED LIFE FORM OF MYCO-BIOCONTROL AGENTS (BCAS)

The Kingdom Fungi is one of the first gatherings of eukaryotic microorganisms in earthly environments (Mueller et al., 2007). There are just around 1,00,000 portrayed types of Fungi, which just comprises a small amount of its assortment, assessed to be somewhere in the range of 1.5 and 5 million species (Hawksworth et al., 1997). In light of sub-atomic clock dating techniques used to evaluate contagious advancement, parasitic presence on Earth fluctuates generally relying upon the fossils utilized for adjustment of the clock, and current assessments suggest that organisms developed around 0.5–1.5 billion years back (Berbee & Taylor, 2010). Fundamentally, one of the brand names of growths is their propensity to shape private collaborations and relationships with different gatherings of life on Earth. As per Hawksworth (Vega et al., 2005), 21% of all types of detailed growths are connected with green growth as lichens and 8% structure private alliance with plants as mycorrhiza, being this an exceptional case of such cozy relationship, which happen in rhizosphere (around roots), where arbuscular mycorrhizal (AM) and ectomycorrhizal parasite's structure relations with plants (Hawksworth et al., 1998). Hardly any living beings in earthly environments stay alive in nature in the total nonexistence of growths and consequently they are crucial parts in the upkeep of biological system wellbeing. Another gathering of extent is the Oomycetes. These are supposed water forms and fit into a far-off Kingdom (Stramenopila), all the more firmly related to brown, green growth (Samson et al., 1988). Nonetheless, it is reasonable to consult them with parasites as they were for some time respected to be growths and are environmentally fundamentally the same. Entomopathogenic parasites are discovered in the divisions of Zygomycota and Ascomycota, just as the Chytridiomycota (fossil growths) and Oomycota, which were prior ordered inside the organisms. A significant number of the taxa of entomopathogenic growths as of now under exploration either fit into the class Entomophthorales in the Zygomycota or the class Hyphomycetes in the Deuteromycota (Kamoun et al., 2003). Entomopathogenic fungi (EPF) being a potential natural control specialist essentially because of their high conceptive capacities, target explicit action, short age time, and resting stage or saprobic stage creating abilities that can guarantee their endurance for a more drawn-out time when no host available. An essential prerequisite for the utilization of an entomogenous parasite as a myco-biocontrol specialist is the vulnerability of the insect pests on one hand and the destructiveness of the growth then again. The last relies upon the choice of a strain with steady, explicit adequacy for an objective host. Henceforth there is a tremendous potential for hereditary improvement of organisms for myco-biocontrol (Sandhu et al., 2012).

6.3 ENTOMOPATHOGENIC FUNGI (EPF): AS PERSUASIVE BIOCONTROL AGENT

The entomogenous word has been originated from two Greek words, "*entomon*" which means creepy crawlies and "*genic*" which means emerging in. Hence, the etymological significance of entomogenous microorganism is "microorganisms which emerge in insects." The intensity of these entomogenous fungi in achieving a specific level of characteristic or microbial control of insect nuisances is legitimately identified with human government assistance which has pulled in the consideration of microbiologist, atomic scientists, and entomologists in the ongoing years (Sandhu et al., 2012). Entomopathogenic organisms perceive and taint insects through the spore grip and arrangement of appressoria that infiltrate the cuticle. In the wake of coming to the hemocoel (body cavity) of a creepy crawly, contagious fibers will switch into yeast-like cells that go through sprouting for quick proliferation and balance the insusceptible reaction of the hosts (Shang et al., 2015). For the contamination cycle to finish, dead insects must be either mycosed to deliver abiogenetic conidial spores or colonized to frame a fruiting body to yield sexual spores for the following disease. Then again, insect microbes, for example, *M. robertsii* with an expansive host reach can shape a root-rhizosphere relationship to move nitrogen from dead insects to a plant and secure carbon consequently and this is a methodology for long haul diligence in the dirt when potential hosts are missing (Behie et al., 2012). Traditional natural control has been characterized as 'the deliberate presentation of an outlandish organic control specialist for lasting foundation and long-haul pest control. This strategy has frequently been utilized to target arthropod insects and weeds that have been acquainted with new territories. It has prevalently focused on utilization of herbivores or arthropod parasitoids and hunters while it has been utilized moderately infrequently for entomopathogens (Eilenberg et al., 2001). Entomopathogenic contagious strains, including *Verticillium lecanii*, *Beauveria bassiana*, *Isaria fumosorosea*, and *Metarhizium anisopliae*, were utilized as the particular natural pesticides, which are ecologically cordial and can be utilized against many sucking pest bothers (Majeed et al., 2017). Spores sprouted subsequent to appending to the epidermis of the host insects, and the hyphae enter the body of the insects, which causes the passing of the host inside a couple of days (Mus et al., 2015). In expansion, these entomopathogenic growths have no or little destructiveness on warm blooded animals. Their residuals are target explicit and less helpless against obstruction advancement (Nasir et al., 2011). For the choice of appropriate and compelling entomopathogenic parasites the harmfulness factors assume significant job. Destructiveness relies upon a progression of complex sub factors including contagious spore hydrophobicity, spore germination extremity, adequacy of hydrolytic compounds and affectability to abiotic factors, which are at last fundamental for the advancement of productive mycoinsecticides or biocontrol operators. Figure 6.1 shows strategies of virulence factor for the selection of effective EPF (Hussain et al., 2014). For effectual strains selection in the agricultural field is a first step to develop strong biocontrol formulations. The keen information regarding genetic variations along with isolated EPF can be successively obtained by the use of ISSR markers (Vianna et al., 2020). Myco-biocontrol is an ecologically stable and powerful method for diminishing or moderating insect vermin and its belongings using common foes. Vermin-related harms bring about a substantial misfortune, roughly assessed to be US $10,000 million yearly in rural creation in the field and capacity in India. Myco-biocontrol is the utilization of organisms in natural cycles to bring down the insect thickness with the point of decreasing illness delivering action and subsequently crop harm (Chet et al., 1993). All gatherings of creepy crawlies might be influenced, and more than 700 types of growths have been recorded as microorganisms. A portion of these organisms have confined host ranges, for instance, *Aschersonia aleyrodes* taints just scale insects and whiteflies, while other contagious species have a wide host range, with individual disengages being more explicit to target insects. A few animal categories are facultative generalist microorganisms, for example, *Aspergillus* and *Fusarium*. In any case, most species are committed microorganisms, frequently very explicit and infrequently discovered (Sandhu et al., 2012). Chemical pest sprays have been normally used to control rural vermin, termites, and organic vectors, for example, mosquitoes and ticks. Notwithstanding, the unsafe effects of harmful synthetic insect sprays on the climate, the improvement of opposition in insects and vectors towards compound insect sprays, and public concern have driven broad exploration for options, particularly organic control operators, for example, growth and microorganisms. Entomopathogenic fungal strains, including *Beauveria*, *Metarhizium*, *Verticillium*, *Nomuraea*, *Trichoderma*, and *Cordyceps* were majorly utilized as the particular organic pesticides, which are naturally benevolent and can be utilized against many sucking creepy crawly insects. Figure 6.2 depicts the identifying spore modifications and life cycle of insect pest after fungal infection (Majeed et al., 2017).

6.3.1 *BEAUVERIA* SPP.

Beauveria bassiana is one of the most well-informed destructive entomopathogenic organisms having a place with the request of Hypocreales. They have a wide scope of insect colonization (Dogan et al., 2017). This parasitic strain is effortlessly gathered from the phylloplane of vegetation, just as from tainted insects and soil (Freed et al., 2011). As a bio-insect spray, the EPF *B. bassiana* has been created as one of the major new bioactive compounds for plant microorganism and creepy crawly insect control (Shin et al., 2017). A wide scope of restoratively or horticulturally critical strains of *Beauveria bassiana* has been obtained from different insects around the world. *B. bassiana* has no notable sexual cycle. Insect pests are contaminated by conidia (abiogenetic propagules) which hold fast to the host cuticle. Conidia fill in a climate with high dampness.

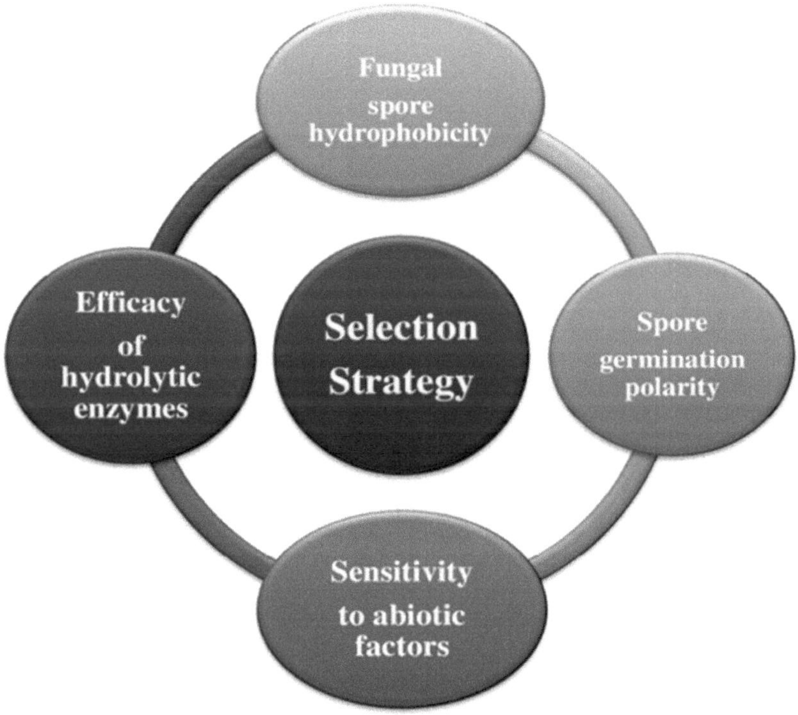

FIGURE 6.1 Strategies of virulence factor for the selection of potent entomopathogenic fungi.

The germ tubes emerging from the conidia penetrate the host cuticle and attack the hemocoel. A fruitful disease by *B. bassiana* is reliant fundamentally on different enzymatic exercises for corruption of proteins, chitin, and lipids in the creepy crawly integument (Koiri et al., 2017). Contagious conidia are delivered abiogenetically and turn into the premise of contamination in creepy crawly irritations of yields. Disease through conidia begins when they are appended to the host cuticle, at that point grow following the initiation of the enzymatic response and attacked the body of the creepy crawlies by germ tube, appressoria, and entrance stakes (Khan et al., 2012). Refined culture

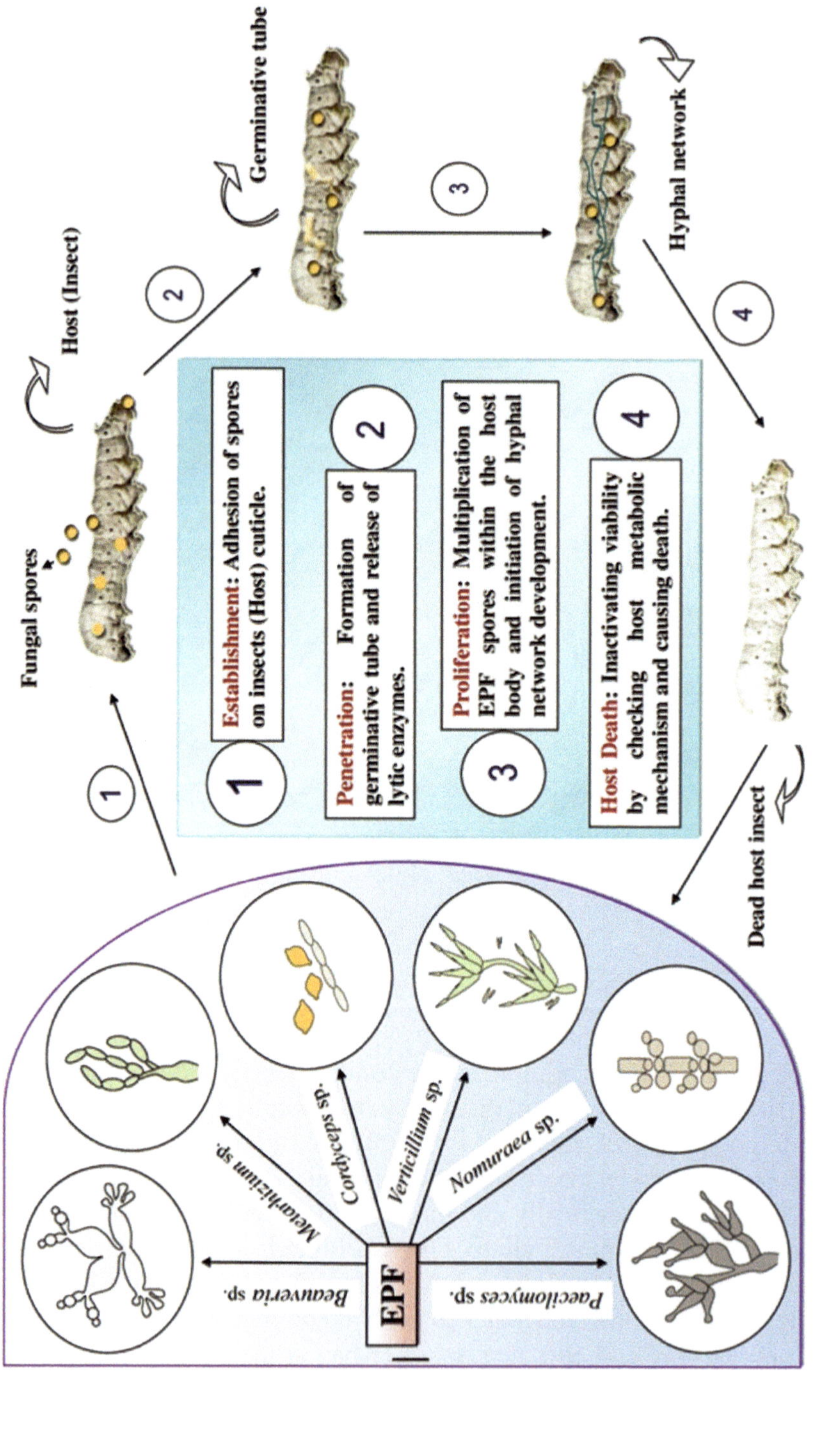

FIGURE 6.2 Identifying spore modifications and life cycle of insect pest after fungal infection.

filtrates of EPF, *B. bassiana*, diminished the conceptive pace of aphids (Kim et al., 2010) and forestalled taking care of the hatchling of *Spodoptera littoralis* and *Bemisia tabaci* (Wang et al., 2007). Expanded organism fixation diminished the quantity of grown-up parasitoids and furthermore adversely influenced its formative stages (Fazeli-Dinan et al., 2016). Filtrate culture contains numerous proteins like chitinase, lipases, and protease, and these catalysts help in the contamination cycle by corrupting the cuticle of creepy crawlies. The grouping of a chemical can be upgraded by the utilization of various added substances in the way of life media, as colloidal chitin (Kim et al., 2010).

6.3.2 METARHIZIUM SPP.

The overall mode of infection of *Metarhizium* can be classified into attachment, germination, appressorium development, entrance, expulsion, and sporulation. *M. anisopliae*, which is a generalist, is a potential biocontrol specialist of termites, mosquitoes, and cows ticks, while *M. acridum* is a potential biocontrol specialist of beetles and grasshoppers. Nonetheless, numerous difficulties, for example, the moderate method of activity and erratic consequences of the use of these growths, must be defeated to augment the adequacy of both *M. anisopliae* and *M. Acridum* (Yang et al., 2011). Studies have been directed to accelerate the method of activity by hereditarily designing both *M. anisopliae* and *M. acridum*. In any case, the viability of these hereditarily designed parasites in the field has not been distributed. The impacts of these growths on the impacts of non-target creatures and the climate ought to likewise be considered prior to utilizing them in the field. Beetles and grasshoppers are major rural irritations which have caused colossal rural and financial misfortunes. Broad examination on the utilization of *M. acridum* on insects and grasshoppers has been directed by the LUBILOSA program (Lutte Biologiquw contre les Locustes et les Sauteriaux) in Africa and CSIRO (Commonwealth Scientific and Industrial Research Organization) in Australia in the course of the most recent 10 years (Schneider et al., 2013). *M. acridum* is a pro, having a limited host range where it contaminates just beetles and grasshoppers. The possible utilization of *M. anisopliae* as a powerful natural control of termites has been accounted for. The viability of *M. anisopliae* is influenced by three factors: the strain of *M. anisopliae* utilized, the dose of *M. anisopliae* conidia, and the genera and types of termites which have an alternate vulnerability to a similar strain of *M. anisopliae*. The mortality of termites is portion subordinate, whereby expanding the measurement of conidia of *M. anisopliae* expands the mortality of termites. In particular, the most elevated insecticidal measurement of conidia is around 1×10^6 conidia/mL to 1×10^{10} conidia/mL. The most un-number of days taken to arrive at 100% was between three days and 21 days post-vaccination (Niassy et al., 2011).

6.3.3 VERTICILLIUM SPP.

The organism *Verticillium lecanii* is one of the individuals from Deuteromycetes and it very well may be utilized for crop assurance *V. lecanii* can develop on both living and dead materials. It is non-particular and can develop on all ordinary mycological media so far tried, for example Czapeck Dox, Malt concentrate, Sabouraud, and potato dextrose agars (PDAs), including a media containing chitin as the sole wellspring of carbon and nitrogen. It has the ability to deliver conidia on strong media; conversely, *V. lecanii* expects a semi-yeast morphology in fluid media (Schuler et al., 1991). *V. lecanii* is one of the most well-known and significant entomophagous Hyphomycetes parasites happened on coccids, aphids, thrips, Diptera, Homoptera, Hymenoptera, Lepidoptera, and insects and in all the climatic areas. Other significant substrates for *V. lecanii* are rusts and other parasites. It is an outcome of this propensity that the species is as often as possible disconnected from soil; it has likewise been separated from leaf litter of oak, debris, and birch, tea leaves, grain seed, pastry specialist's yeast, beet seed and blasting corn pieces (Alavo et al., 2002a). Strategies for presenting these growths incorporate splashing the spores and parts of the mycelium of the organisms onto trees which it is proposed to contaminate, integrating tainted material with trees which it is wanted to taint lastly, planting among the trees to be contaminated, little trees whose foliage is all around contaminated with different parasitic scale parasites, so the leaves of the little trees come into contact with those of the bigger ones (Koppert et al., 2015). The bio-examine method comprises in treating falsely raised aphids with the contagious spore on channel paper in Buchner pipe, by pouring delicately a known measure of the fitting spore suspension on them. After treatment, aphids were independently positioned on leaf circles in high stickiness test cells which were kept in bewilder confines at high moistness for the term of the trial. This test method gives a decent estimation of the pathogenicity of clusters of conidiospores of *V. lecanii*. The fluctuation of the bioassay results and the inconceivability of characterizing plainly qualities related to the harmfulness of a growth strain toward a particular insect animal group have been talked about and spores' improvement utilizing adjuvants was recommended. These days, a plan of *V. lecanii* is popularized under the name of Mycotal® just for use against Whitefly hatchlings. This item viability is supposed to be improved whenever applied along with adjuvant dependent on emulsifiable vegetable oil (Alavo et al., 2015).

6.3.4 NOMURAEA SPP.

Nomuraea sp. cause parasitic disease to edit insects in nature in a few nations around the world; over the most recent 20 years, this species has been seriously researched as an organic control operator of arthropods insects. The practicality of *Nomuraea rileyi* conidia stayed higher than 75% on the outside of *Anticarsia gemmatalis* corpses for 10 days after treatment, showing that *N. rileyi* conidia may persevere in the climate as an inoculum source (Sujii et al., 2002a). At any rate 32 creepy crawly species (for the most part lepidopteran) are powerless to *Nomuraea* sp. This species announced as causing epizootic flare-ups in caterpillars that harm numerous yields and fields. *N. rileyi* is extremely pathogenic to a few rural nuisances and is professed to be a proficient natural control operator. Cows ticks cause huge monetary misfortune because of microbe transmission, which can bring about harmed covers up and low weight gain. The financial misfortunes are assessed to arrive at 2 billion dollars every year in Brazil, making this species the main steers ectoparasite (Grisi et al., 2002). *Nomuraea rileyi* is a cosmopolitan animal groups tainting numerous noctuids, for example, *Helicoverpa armigera*, *Spodoptera litura*, *Tricoplusiani*, *Anticarsia gammatalis*, *Pseudoplusia* incorporates and has a potential for improvement into mycoinsecticide and happens in soils of different agro biological system (Namasivayam et al., 2015). *Nomuraea* shows significant mortality factor for some lepidopteran insects all through the world. Being created as a microbial creepy crawly control operator, the parasite is fit for causing awesome epizotic in caterpillar nuisances of cabbage, clover, soybean, potato, cotton, and so on (Sahayaraj et al., 2011).

6.3.5 TRICHODERMA SPP.

Trichoderma is a variety of soil-borne organisms with notable enemy of phytopathogen exercises. Instruments of activity incorporate rivalry, mycoparasitism, antibiosis, and host prompted foundational opposition. *Trichoderma* species contend with microbes mostly for supplements and natural specialties. Other than fast development and bountiful creation of spores, a few strains can orchestrate siderophores that repress the development of other parasites during rivalry (Ghorbanpour et al., 2018). Mycoparasitism characteristics of *Trichoderma* depend on the movement of cell divider corrupting catalysts emitted by the parasite after its hyphae curl around and additionally enter the microbe's hyphae cells. The utilization of *Trichoderma* as a natural control specialist of phytopathogens is accounted for against tobacco mosaic infection. Further examination assessed its latent capacity controlling plant pathogenic parasites and Oomycetes that prompted the improvement of *Trichoderma*-based business items (Harman et al., 2004). A trademark white cotton-like mycelium on contaminated tissues is symptomatically noticeable. At later phase of sickness advancement, endurance structures of the organism, named sclerotia, are shaped on has. Seed parts defiled with organism mycelia or sclerotia comprise the most widely recognized wellspring of spread of the microbe. A few strains of the family *Trichoderma* are being tried as choices in contrast to synthetic fungicides (Tuao et al., 2016). *Trichoderma viride* was demonstrated as a powerful biocontrol specialist against two parasitic microorganisms, *Fusarium oxysporum* sp. and *Pythium arrhenomanes*, contaminating soybean. *Trichoderma viride* is a biocontrol agent against soil-borne plant microorganisms and it can without much of a stretch colonize in plant rhizosphere and help to advance the plant development (Verma et al., 2007). The productivity of *Trichoderma* as biocontrol operators against contagious soil microorganisms and shows the need of creation and improvement of *Trichoderma*-based BCAs to fill in as a model for climate inviting biocontrol specialist (John et al., 2010).

6.3.6 CORDYCEPS SPP.

The entomopathogenic therapeutic mushroom *Cordyceps* species has been presenting wellbeing organically and pharmacologically for quite a long time. It harbors an assortment of bio-metabolites having far-going exercises (Deshmukh et al., 2020). The *Cordyceps* family incorporates numerous types of organisms, the vast majority of which are endoparasitoids on arthropods. The conveyance of these parasites is cosmopolitan, however many happen in areas, for example, Asia with a hot, damp atmosphere. These microorganisms of creepy crawly irritations are promising contender for use as organic control factors. A few mitotic or anamorphic species have gotten consideration as organic control operators of creepy crawly bothers. Up-and-comer species have generally originated from the Cordycipitaceae or Clavicipitaceae (Evans et al., 2011). Entomopathogenic growths *Cordyceps* sps. are all around described in regard to pathogenicity to a few insects and have been utilized as myco-biocontrol operators for organic control of biosphere horticulture bothers, particularly at high height. The insecticidal movement of methanolic concentrates of *C. militaris* and their principle part cordycepin on the precious stone back moth *Plutella xylostella* hatchlings was set up. Various agents have portrayed the improvement of the fruiting assemblages of the growth on insects after infusions or skin treatment with various sorts of spores. *Cordyceps militaris* (explicit to caterpillars), and *Ophiocordyceps unilateralis* sensu lato (explicit to formicine ants), just taint a restricted scope of creepy crawlies. Specifically, *O. unilateralis* is an animal categories complex that incorporates various species being profoundly have explicit, which is near a degree of one organism versus one insect animal types (de Bekker et al., 2014; Deshmukh et al., 2019a). *Cordyceps militaris* and most species assault the hatchlings and pupae of arthropods. *Ophiocordyceps* is the biggest variety of arthropod pathogenic growths (Dworecka-Kaszak, 2014). Entomopathogenic parasites, *Cordyceps javanica* (Hypocreales: Cordycipitaceae), in the past *Isaria javanica*, is a broadly utilized natural control operator of arthropod insects on vegetables, organic products, and decorative plants (Mascarin et al., 2018). Their obligately-executing life cycle implies that there is probably going to be solid choice weight for qualities that permit them to sidestep the impacts of the host invulnerable framework. *Cordyceps* conceivably go about as a silencer of the safe reaction during parasitic contamination of insects (Woolley et al., 2020).

6.4 BIO-PATHOGENICITY OF INSECTS BY ENTOMOPATHOGENIC FUNGI

EPF are microorganisms that cause lethal illnesses of arthropods. The disease cycle includes a few phases that comprise of direct contact of the parasite with the outside of the cuticle of the assaulted creepy crawly. The variables that decide the adequacy of the disease cycle incorporate lytic proteins, optional metabolites, and adhesins created by EPF. Due to their high insecticidal viability, these parasites are regularly utilized as biopesticides in natural cultivating. As the climate and farmlands are sullied with numerous mixes of anthropogenic roots (e.g., pesticides), the impacts of these harmful mixes on EPF and the components that influence their endurance in such a poisonous climate have been concentrated as of late (Anna et al., 2020).

Entomopathogenic growths contaminate insects by the direct entrance of the cuticle. In contrast to microbes or infections, they don't need to be ingested by an insect (Bilgo et al., 2018). Entomopathogenic parasites show a remarkable method of contamination; they come to the hemocoel through the cuticle or conceivably through the mouthparts. Ingested contagious spores don't sprout in the gut and are voided in the dung. The passing of the insect results from a mix of elements: mechanical harm coming about because of tissue intrusion, consumption of supplement assets and toxicosis, and creation of poison in the group of insects (Bhattacharyya et al., 2004). The greater part of entomopathogenic contagious species proceeds with normal advances incorporating Conidial Attachment with the Cuticle and producing two significant kinds of proteins, in particular, hydrophobins (whose layers deteriorate during the sporulation of spores) and adhesins (MAD1 and MAD2), which empower both close grips to the insect cuticle and response of the host by the parasitic microbe (Greenfield et al., 2014), Formation of an Infection Structure and the creation of cuticle debasing lytic chemicals including lipases, phospholipase C, proteolytic catalysts, proteases (subtilisin and trypsin), chitinases (Zhao et al., 2016) at that point at last entrance of the cuticle. At that point, after they start their infective cycle when spores are held on the integument surface, where the arrangement of the germinative cylinder starts, the growths begin discharging catalysts, for example, proteases, chitinases, quitobiases, lipases, and lipoxygenases (Amer et al., 2008). The pathogenesis brought about by EPF requires the inclusion of a few irresistible operators, the most significant of which are adhesins, lytic chemicals, and optional metabolites. The association of the parasitic poisons assume significant part in have passing. The activities of cytotoxins are recommended by cell disturbance preceding hyphae infiltration. Conduct indications, for example, halfway or general loss of motion, drowsiness, and diminished peevishness in mycosed creepy crawlies are predictable with the activity of neuromuscular poisons. The poisons have appeared to effect devious different insect tissues (Sandhu et al., 2000). The whole contamination measure is generally long and takes roughly 14 days after disease, yet the first side effects of disease as a rule happen around seven days post contamination (or much prior, contingent upon parasitic species). Subsequent to slaughtering the creepy crawly and utilizing all nourishment, the hyphae of the organism rise up out of the corpse of the host through openings in its body (mouth opening, rear-end) and through intersegmental territories. At that point, resting or infective spores are delivered, which permits the organism to spread and contaminate others (Skinner et al., 2014). Entomopathogenic growths emit bountiful low-sub-atomic weight natural mixes called auxiliary metabolites, particularly in light of ecological conditions. The quantity of created mixes proposes that they are essential both for keeping up the indispensable elements of the hosts and for viably tainting microbes by harming the sensory system or diminishing creepy crawly obstruction (Donzelli & Krasnoff, 2016).

6.5 HOST INSECT BEHAVIORAL CHANGES: A CAUSE OF FUNGAL ASSAIL

The cooperation between the microbe and attacking entomopathogenic contagious spores triggers a progression of safeguard components in the arthropod. These components are generally set off following the respectability of the cuticle is undermined by the contagious spores and includes complex biochemical collaborations. Different changes happen inside the host, while it sends the protection instruments related with cell and humoral safe reactions (Hussain et al., 2013). Contaminations intervened by wide host range entomopathogenic growths speak to original perceptions that prompted one of the principal germ speculations of infection and are an exemplary case of a co-developmental weapons contest between a microbe and target has. These organisms can parasitize defenseless has by means of direct entrance of the cuticle with the underlying and possibly deciding association happening between the contagious spore and the insect pest epicuticle (Fan et al., 2007). Entomogenous organisms have advanced instruments for attachment and acknowledgment of host surface signals that help direct a versatile reaction that incorporates the creation of hydrolytic, assimilatory, and additionally detoxifying chemicals including lipase/esterases, catalases, cytochrome P450s, proteases, and chitinase, particular irresistible structures, e.g., appressoria or penetrant cylinders; and optional and different metabolites that encourage disease (Sandhu et al., 2017). Beside resistant reactions, insects have advanced various components to keep microorganisms under control that incorporate the creation of (epi) cuticular antimicrobial lipids, proteins, and metabolites, shedding of the cuticle skin during improvement; and social natural variations, for example, prompted fever, tunneling, and prepping, just as conceivably enrolling the assistance of different organisms, all proposed to stop the microbe before it can penetrate the cuticle. Virulence and host-guard can be recognized as under steady equal particular weight, and the activity on a superficial level probably adds to marvels, for example, strain variety, have range, and the expanded destructiveness frequently noted once a destructive strain passage through a creepy crawly have (Zhang et al., 2011). Since the cuticle speaks to the principal purpose of contact and boundary between the parasite and the insect, the activity on the cuticle surface may speaks to the characterizing collaboration that at last can lead either to fruitful mycosis by the microbe or effective protection by the host (Pedrini et al., 2013). Information concerning the atomic components basic this cooperation can reveal insight into the environment and advancement of harmfulness and can be utilized for balanced plan methodologies at expanding the viability of entomopathogenic organisms for pest control in field applications (Ortiz-Urquiza & Keyhani, 2013).

An exact signaling mechanism encouraged by the existence of effectors and receptors at the host-microbe edge causes the spores to follow onto the larval cuticular layer. The contagious discharged effector molecules have a fluctuating capacity to adjust the physiology of their hosts. Cost like receptors (TLRs) assume a focal function in perceiving host cells and the resulting reactions to the microbial microorganisms (Kawai et al., 2011). Also, late ID of other non-TLR design acknowledgment receptors (PRRs, for example, C-type lectin receptors, RIG-I-like receptors, and NOD-like receptors assists with revealing insight into the intricacy of natural insusceptibility (Baral et al., 2017). The function of effectors is to confine the organism onto the insects' bodies. Effector triggered immunity may help shield the creepy crawly from have adjusted biotrophic contagious microorganisms. Not at all like in different creatures, *Cordyceps* is seldom followed/related with a hypersensitivity response with programmed cell death at the disease site. Besides, two unique sorts of PCDs (apoptosis and autophagy) needed for the finish of transformation have been recorded in various tissues of the insect pests (Fahrbach et al., 2012). A few insects (e.g., Drosophila melanogaster) have reaper, hid, and grim (RHG) protein apoptosis activators and IAP family proteins (inhibitors of apoptosis) with functions in ubiquitination in the muscle programmed cell death of insect. Inferable from the fruitful arriving of spores on to the cuticular surfaces, alteration of creepy crawlies' integuments happens, trailed by the germination of spores (Charnley et al., 2003), and finally germlings adhesion. Operating system have extraordinary mixes with insecticidal properties, (for example, short-chain unsaturated fats) which fill in as armoires against the hindrance managed by insect integuments and help break up the cuticular layer (Boucias et al., 1988). Furthermore, spore germination is quickened by the ownership of a few chemicals that help in penetrating the host boundary, all the while supporting growing spores. The larval integument involves chitin, proteins, lipids, different chemicals and phenolic mixes. Chemicals at larval integuments incorporate subtilisin, chymotrypsin, trypsin, metalloprotease, aminopeptidase, post proline dipeptidyl peptidase, post alanine peptidase, serine carboxypeptidase and zinc carboxypeptidase (Leger et al., 1991). During parasitic foundation inside hatchlings, compounds, for example, esterase and proteolytic catalysts (aminopeptidase, endoprotease, and carboxypeptidase) are discharged at first, trailed by the creation of chitinase and lipase at later periods (following hardly any long stretches) of contamination. Consequently, cuticle layer debasing compounds, for example, serine-protease (with fibrinolytic action) encoded by csp1 and csp2 qualities help to crumble the larval cuticular layer and encourage parasitic foundation onto the larval surface (Baral et al., 2017). Figure 6.3 indicates EPF and host insect pest interaction: Detailed portrayal of contagious fungal infection and host defense behavioral changes (Almudena et al., 2013: Sandhu et al., 2012).

6.6 PROGRESSION IN ENTOMOINFECTION

The developing commercialization everywhere on the world has prompted a lift in the far-reaching utilization of synthetic pesticides for crop assurance in agrarian fields. It has not just added to an expansion in food creation, yet its poisonous and non-biodegradable character has additionally brought about unfriendly consequences for climate and non-target life forms. In addition, the vast majority of the vermin have created obstruction against them. These disadvantages of regular pesticides have prompted an expansion in the requirement for the hunt of some novel, non-hurtful, eco-accommodating pesticides. Regular irritation control materials usually known as biocontrol operators are the most encouraging of them. BCAs incorporate microorganisms just as microorganisms. The microorganisms utilized are microscopic organisms, growths, infections, nematodes, and protozoan. The misuse of these characteristic and sustainable assets is fundamental for an effective biocontrol methodology (Sandhu et al., 2017). Molecular science and techniques have clarified pathogenic cycles in a few contagious BCAs including two of the most regularly applied entomopathogenic growths, *Metarhizium anisopliae* and Beauveria bassiana. In this survey, we portray how a mix of sub-atomic procedures including; distinguished and described qualities engaged with contamination, controlled the qualities of the microbe to improve biocontrol execution and permitted articulation of a neurotoxin from the scorpion *Androctonus australis* (Federici et al., 2008). The total sequencing of four model types of entomopathogenic parasites including *B. bassiana* and *M. anisopliae* have been achieved. Inclusion of these genomes

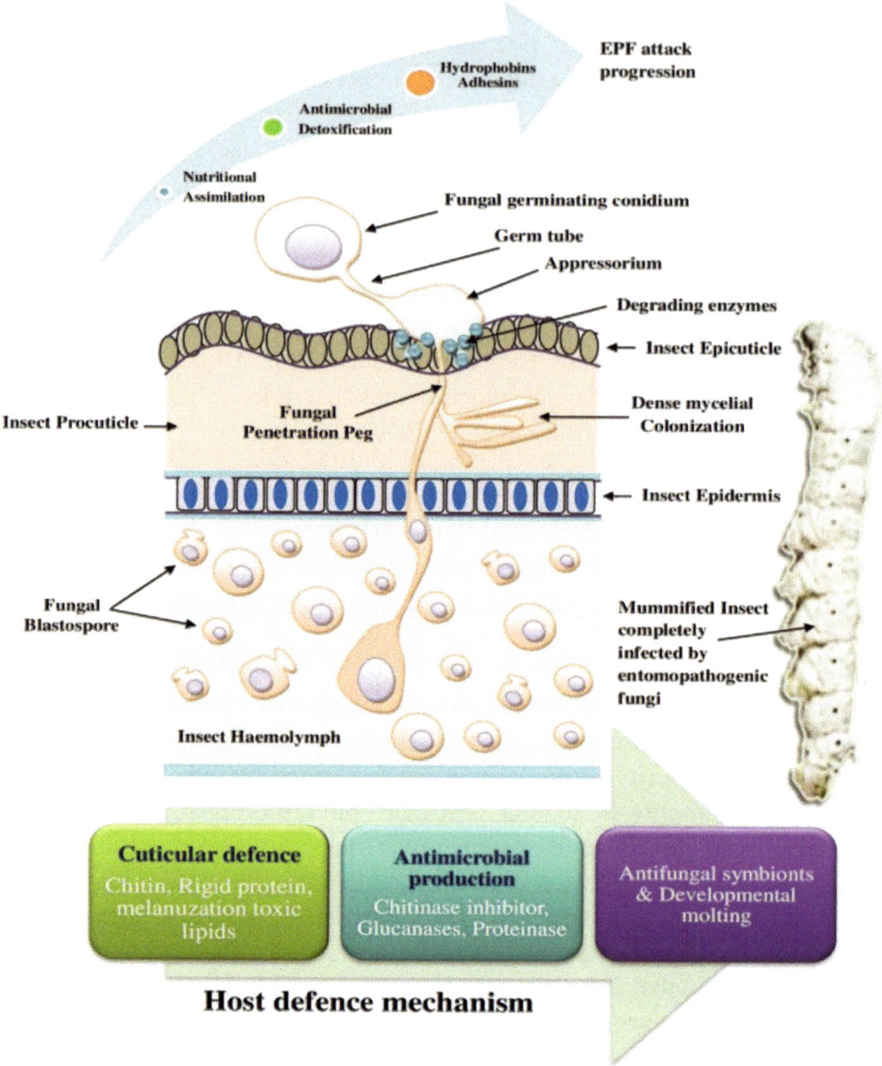

FIGURE 6.3 Entomopathogenic fungi and host insect interaction: Detailed depiction of fungal infection and host defensive behavioral changes.

will help decide the character, cause, and advancement of characteristics required for assorted ways of life and host exchanging. Such information joined with the exactness and pliability of atomic methods will permit plan of various microorganisms with various techniques to be utilized for various biological systems and maintain a strategic distance from the chance of the host creating opposition (Raymond et al., 2017). Exhausting examines detailed most developed strategy to segregate entomopathogenic organisms. The nucleotide grouping of the different area (d1/d2) at the distal finish of the 26S ribosomal RNA (rRNA) quality and the inner deciphered spacer 1 (ITS1) district was then gotten for each refined disconnect after DNA extraction. Polymerase chain response (PCR) intensifications with 0.5 unit of Amplitaq DNA Polymerase (Invitrogen Life Technologies, Carlsbad, CA), 2.5 mM $MgCl_2$, 200 μM dNTPs, 0.5 μM each of forward and switch oligonucleotide groundworks, 1X response cradle, and 50 to 100 ng format DNA. All PCR responses were performed for 35 cycles, each comprising of a 30-sec denaturation venture at 94°C, a 30-sec strengthening venture at 50°C, and a 1-min expansion venture at 72°C. Intensification items were cleaned and outwardly altered at that point adjusted to the realized nucleotide arrangement for growth applied. Characters were affirmed when the nucleotide arrangement was 100% homologous with the known parasite (Ayala-Zermeno et al., 2011). On account of insects that didn't give outer indications of disease, quantitative PCR (qPCR) strategy used to affirm the parasitic contamination with positive controls. Conidial fixation was estimated minutely utilizing a hemocytometer, and afterward sequentially weakened 1:10 with sterile water to shape required arrangements. Decontaminated DNA was eluted and was performed utilizing a QIAgility PCR robot. Each example was run in 3 specialized reproduces. Particularity of every enhancement was affirmed with a softening bend of each amplicon from 60–95°C at 12°C/min. Both positive and negative controls were remembered for all runs (Bell et al., 2009). On the off chance that the creation and plan methods produced for *M. anisopliae* var. acridum are pertinent to different strains of *M. anisopliae* with various host ranges, at that point this leaves the need to improve adequacy as the important boundary staying to financially savvy biocontrol of numerous insects (Bagga et al., 2004). Besides, a large number of the new application strategies being created for entomopathogenic parasites include some sort of lure station where harmfulness and powerful determination are especially significant. Absence of viability is presumably inbuilt in microbes in light of the fact that a developmental offset will have created with their hosts so speedy murder, even at high portions, isn't versatile for the microorganism. In which case, cost-effective biocontrol will require moving qualities to the parasites (Gressel et al., 2001).

Researchers today are not restricted to the tremendous variety of peptide and protein successions that as of now happen in nature as they can infer engineered multifunctional qualities that are half and halves of various exercises. Intriguingly, the chitinases disconnected from *B. bassiana*, up to this point, seem to need chitin binding spaces. Fan et al. (2007a) developed a few *B. bassiana* cross breed chitinases where the chitinase was combined to chitin-restricting areas got from plant, bacterial, or creepy crawly sources. A half breed chitinase containing the chitin binding space from the silkworm *Bombyx mori* chitinase combined to the *B. bassiana* chitinase indicated the best capacity to tie to chitin contrasted with other half breed chitinases. Constitutive articulation of this half and half chitinase quality by *B. bassiana* diminished opportunity to death of insect has by 23% contrasted with the wild-type organism demonstrating that hereditary segments of the host insect fused into the parasitic microbe can expand destructiveness. A similar gathering has created manufactured chitinase with improved compound exercises through DNA rearranging (Fan et al., 2007b). The developing interest for lessening substance contributions to farming and expanded protection from insect sprays has given incredible force to the improvement of elective types of insect control. Myco-biocontrol offers an alluring option in contrast to the utilization of synthetic pesticides. Myco-biocontrol agents are normally happening life forms which are seen as less harmful to the climate. Their method of activity shows up minimal complex which makes it exceptionally impossible that obstruction could be created to a biopesticide. Past exploration has given some guarantee of the utilization of parasites as a particular pesticide. The current chapter refreshes us about the ongoing advancement in the field of myco-biocontrol of insect bothers and their conceivable instrument of activity to additional improve our comprehension about the natural control of insect pests (Sandhu et al., 2012).

6.7 STEP FORWARDING TREND (IMPROVEMENT)

Biocontrol operators for the most part don't perform all around ok under field conditions to contend with compound fungicides. Subsequently, at an atomic hereditary qualities level, endeavors to expand the biocontrol capacity of *Trichoderma* have been centered on expanding chitinase or proteinase action either by expanding the quantity of duplicates of the suitable qualities or by intertwining them with solid advertisers. Cell divider corrupting chemical movement, Determination of glucose oxidase action, unexpected change incited obstruction action can assume significant function to improve biocontrol possibility of entomopathogenic growths (Brunner et al., 2015). For distinguishing the improved strains of entomopathogenic organisms and their upgraded profitability, entrenched markers can be utilized (Deshmukh et al., 2019b). It has been discovered that the insecticidal cyclopeptide destruxins delivered by *Metarhizium* could be utilized by the growth to sidestep insect insusceptibility, and its capacity to create a poison is associated with have particularity. The foundation of an ideal model framework (e.g., the zombie insect), the procurement of genomic data, and the sending of the information and strategies of the studies of parasitic hereditary qualities, optional digestion, science, and insect physiology and nervous system science would help reveal the organic insider facts behind changes in insect larvae conduct (Wang et al., 2012). Biotechnology with regards to insect the board can be characterized as the controlled and intentional control of natural frameworks to accomplish effective insect pest control. Living creatures have advanced a tremendous range of natural abilities and by picking proper living beings with explicit capacity, it is conceivable to get important control of such insect species. Biotechnology can possibly add to economical organic components of coordinated vermin the executives (IPM), however biotechnology improvement to date has been aimed at more traditional models for Pest control advances (Dilawari et al., 2002). Up to this point, most obstruction reproducing has zeroed in on strategies wherein opposition depended on a solitary quality as these gave elevated levels of obstruction and furthermore the techniques were viable with rearing plans. Nonetheless, its quality for quality nature gives prospects to opposition breakdown through the development of obstruction breaking irritation genotypes. Biotechnology for insect pest the executives has somewhat been an ahead of schedule result of the securing of biotechnological expertise, which will have more generous ramifications for horticulture than just improved insect administrations (Gatehouse et al., 2008). Atomic markers have just been created and misused for quick and exact distinguishing proof of both old and recently developed strains of different creepy crawly pest species. They have additionally demonstrated valuable in understanding different other significant parts of the insect, for example, nature, hereditary variety and host explicitness in insects. The effective and wide-scale appropriation of hereditarily adjusted biotech crops worldwide has set up the capability of biotechnology in improved harvest creation. Be that as it may, the rise of creepy crawly protection from Bt-cotton once in a while raised worries on its restricted flexibility comparable to conceivable insect opposition advancement (Franz et al., 2011). The eventual fate of GM crops, in any case, depends upon the quest for new qualities, which, by acting in an unexpected way, could manage the cost of comparable or enhanced opposition in transgenic plants and has brought about the recognizable proof of various qualities from different sources. A large number of these when assessed have demonstrated critical potential for misuse in crop insurance. In this way, future patterns and prospects for biotechnological applications to intercede crop insurance against insects incorporate systems utilizing stacked qualities changed Bt-poisons, insect/scorpion toxin peptides, vegetative insecticidal proteins, lectins, endogenous opposition instruments just as novel methodologies (Singh et al., 2011). Then again, hereditary control to upgrade the viability of mycoinsecticides stays as a moderately ineffectively investigated region. Set against this foundation, future examination should zero in on the consolidation or over expression of destructiveness related qualities in entomopathogenic growths. These freak strains will encourage the improvement of another age of mycoinsecticides for arthropod insect control. In any case, while misusing such techniques, the advantages and dangers related to the appropriation of GM creepy crawly safe yields, particularly for non-industrial nations and asset helpless little holder ranchers should be remembered. As of now, most consideration is being centered on crucial qualities and metabolic pathways that are innate to the creepy crawly vermin and host-plant science. The host-insect relationship being a perplexing wonder, the ID of such qualities and the particular part of their capacity in this unpredictable digestion has consistently stayed a bulky assignment (Behura et al., 2006).

6.8 ENTOMOPATHOGENIC FUNGAL-BASED MICROBIAL CONSORTIUM: AN ECO-FRIENDLY APPROACH TO CONTROL INSECTS AND PESTS

The improvement of insect pest control estimates utilizing microorganisms, particularly entomopathogens, has achieved expanding consideration lately. Detailing of natural control specialist is a significant measure for practical farming. Different details of *N. rileyi* with biogel, oil, and hydrogel were arranged, and assessed for post-treatment ingenuity under various temperatures and the biocontrol potential against groundnut defoliator *Spodoptera litura* (Fab.) (*Lepidoptera Noctuidae*). Among the plans, the greatest pace of tirelessness was recorded in biogel. Contagious spores could hold the practicality in all the tried temperatures in biogel plan. Upgraded pesticidal movement was additionally recorded in a similar detailing (Namasivayam et al., 2014). Plan of natural control specialist is a significant model for practical farming. Definition can improve the item solidness and practicality may bring about consistency of field execution of numerous potential organic control operators. Definition of biocontrol items has been utilized against sicknesses (biofungicides), weeds (bioherbicides) and insect pests' control (bioinsecticides) (Sharma et al., 2004). A considerable lot of the BCAs have been defined with dried milk, powdered casein, gelatin, saponins, oils, cleansers, and so forth to the extent that microbial insect sprays are concerned, it is basic that the compound utilized should untimely the development or germination and that it ought not repress the fruitful foundation of the microorganisms (Tincilley et al., 2000). The advancement of stable definitions of opposing microorganisms and other biocontrol operators is vital to numerous nations, particularly those where means agribusiness is conspicuous, soil-borne illnesses are the fundamental issue, and fungicides are exorbitant. Plan and foundation of biocontrol operators are significant for their adequacy (Ardakani et al., 2010). The oil-based fluid definitions have no ecological results or hurtful poisonous consequences for treated plants or plant items since they are primarily made out of regular substances that are utilized as food-added substances or in assembling of beauty care products. All around the world, oil-based fluid definitions, for example, rearrange emulsions (water-in-oil type) are the most encouraging and generally fitting for plan of viable strains of entomopathogenic growths. Dry details, for example, mix of entomopathogenic parasitic conidia with Diatomaceous earth characteristic residue ought to be additionally engaged and created during the future examination because of the synergistic impact delivered between the segments of these mixes (Batta et al., 2016).

6.9 FOREMOST OBSTACLE TO PREVAIL OVER

Under *in vitro* conditions, most entomopathogenic organisms perform outstandingly. Significant levels of mortality can be watched for most insect-pest focuses under such controlled conditions. A deplorable outcome of such reports is that it sets up ridiculous assumptions about the presentation of these operators in the field. Varieties in temperature, moistness, and introduction to UV light are among a portion of the significant abiotic factors that can fundamentally affect the viability of EPF in field applications (Fernandes et al., 2007; Le Grand & Cliquet, 2013; Rangel et al., 2005; Thompson et al., 2006). Spore practicality and ingenuity are additionally significant contemplations with some ongoing advances in our comprehension of, at any rate, one atomic determinant included, i.e., a putative parasitic methyltransferase (Qin et al., 2014). In spite of the fact that not tended to in this audit, while plan, e.g., consideration of UV-protectants (safeguards), heat/moistness stabilizers, and so forth, has a solid potential for tending to a portion of these issues, costs, and unforeseen impacts related with detailing mixes can exceed benefits (Behle et al., 2009). Development substrate for the creation of the irresistible conidia has appeared to impact pathogenesis and stress obstruction, with millet grain or potentially consideration of salicylic corrosive promising contender for the creation of thermotolerant conidia (Rangel et al., 2012) and incorporation of hydrocarbons bringing about more destructive spores. In spite of the fact that requiring further affirmation, strains with expanded destructiveness have been created by means of protoplast combinations of *B. bassiana* with a toxinogenic strain of *B. sulfurescens* that created substantial half and halves that were steady in any event for one entry through a creepy crawly have (Crespo et al., 2002), and hyphal combinations happening during pairings of *B. bassiana* confines have been utilized to produce thermotolerant. Coordinated advancement by constant culture of parasitic cells under particular conditions, i.e., high temperature, has likewise come about in thermotolerant strains, despite the fact that the maximum capacity of this innovation still can't seem to be acknowledged (de-Crecy et al., 2009). The utilization of the nonmetabolizable simple, 2-deoxy-glucose, to create freaks hindered in glucose usage has likewise yielded strains with more prominent harmfulness, in spite of the fact that the sub-atomic nature of these freaks has not yet been described (Robledo-Monterrubio et al., 2009). A significant issue during utilization of entomopathogenic parasites as biopesticides is that on the off chance that the creation of the entomopathogenic growths requires an expansion of anti-microbials, at that point hints of that would end up in with the spores. This is on the grounds that the creation of entomopathogenic organisms without tainting by different microorganisms, particularly in enormous fermenters, can be tested. Such toxins can both meddle with the entomopathogenic properties of the parasites in question, just as become a wellbeing danger in their own right (Meikle et al., 2012). Detailing progresses regardless, consequences for stress opposition and additionally harmfulness have been moderately little, and these endeavors have not yet prompted any forward leaps in the utilization of entomopathogenic parasites as well as commercialization, and in a few cases, the created strains showed diminished wellness or different aggregates, e.g., decreased conidiation, that would be survived.

6.10 CONCLUSION

The developing interest for diminishing substance contributions to farming and expanded protection from insect sprays have given astonishing thrust to the improvement of elective types of insect pest control. Myco-biocontrol offers an alluring option in contrast to the utilization of synthetic pesticides. Myco-biocontrol operators are normally happening creatures which are seen as less harmful to the climate. Their method of activity shows up minimal complex which makes it profoundly impossible that obstruction could be created to a biopesticide. The opposition with microorganisms, parasitism and the creation of antifungal mixes are the main systems in biocontrol movement. Biopesticides dependent on microscopic organisms, infections, entomopathogenic parasites, and nematodes are frequently significant extension as plant assurance operators against a few creepy crawlies. Be that as it may, the utilization of entomopathogenic organisms as natural control operators for creepy crawly species has expanded worldwide consideration during the most recent couple of many years. Substance insect sprays assume a significant function in the control of plant harm and plant infections. In any case, broad utilization of these items has prompted the disturbance of biological systems in view of a few reasons, for example, passing of non-target species, collection of pesticide build-ups in the climate and food, and develop of pesticide obstruction in the objective species. Natural control is one of the options in contrast to substance pesticides, and it very well may be depicted as the restriction of the bounty of living life forms and their items by other living beings. Hunters, parasitoids, growths, and other useful life forms can be utilized for the biocontrol of creepy crawly bothers. Natural control speaks to a monetary and suitable option in contrast to the usage of substance insect sprays. Use of entomopathogenic parasites is prescribed to control plant plagues in agrarian yields as an environmentally and monetarily acceptable technique. Subsequently research enthusiasm for growth and use of biopesticides has additionally been developing with a definitive target of improving business creation and manageable usage of the biopesticides. In the most recent decade, broad and efficient examination has upgraded the adequacy of biopesticides while the procedures for their large scale manufacturing, stockpiling, transport, and application have immensely improved. In excess of 750 types of growths generally deuteromycetes and entomophthorales are pathogenic to insects. They are quite certain to pests and do not taint the host plants. It is shown that the transgenic utilization of biocontrol-related advertisers related with a suitably chosen heterologous gene is an incredible practice to perk up both biocontrol and the capacity of soilborne fungal biocontrol cause to foundationally instigate sickness obstruction against foliar pathogenic microbes.

ACKNOWLEDGMENT

Authors are grateful to the Bio-Design Innovation Center, Grant No.-F. No. 17-14/2014P.N.I, M.H.R.D., New Delhi, India, for availing basic study platform.

CONFLICT OF INTEREST

No conflicts declared from the authors concerning the publication of this chapter.

KEYWORDS

- **biocontrol formulation**
- **biocontrol improvement strategies**
- **entomopathogenic fungal agent**
- **entomopathogenic fungi**
- **insect-host interaction**
- **pest management**
- **polymerase chain response**
- **quantitative PCR**

REFERENCES

Alavo, T. B. C., (2015). The insect pathogenic fungus *Verticilium lecanii* (Zimm.) viegas and its use for pest control: A review. *J. Exp. Biol. Agricultl. Sci., 3*(4), 337–345.

Alavo, T. B. C., Sermann, H., & Bochow, H. (2002). Virulence of strains of the entomopathogenic fungus *Verticillium lecanii* to aphids: Strain improvement. *Arch. Phytopathol. Plant Protec. 34*(6), 379–398.

Amer, M. M., El-Sayed, T. I., Bakheit, H. K., Moustafa, S. A., et al., (2008). Pathogenicity and genetic variability of five entomopathogenic fungi against *Spodoptera littoralis. Res. J. Agricul. Biol. Sci., 4*(5), 354–367.

Ardakani, S. S., Heydari, A., Khorasani, N., & Arjmandi, R., (2010). Development of new bioformulations of *Pseudomonas fluorescens* and evaluation of these products against damping-off cotton seedlings. *J. Plant Pathol., 92*(1), 83–88.

Ayala-Zermeno, M. A., Reyes-Montes, M. R., Arroyo-Vazquez, E., et al., (2011). An *Isaria fumosorosea* SCAR marker for evaluation of soil, insect, and airborne samples. *Biocontrol. Sci. Technol., 21,* 1091–1102.

Bagga, S., Hu, G., Screen, S. E., & St. Leger, R. J., (2004). Reconstructing the diversification of subtilisins in the pathogenic fungus *Metarhizium anisopliae. Gene, 324*, 159–169.

Baral, B., (2017). Entomopathogenicity and biological attributes of Himalayan treasured fungus *Ophiocordyceps sinensis* (yarsagumba). *J. Fungi, 3*(4), 1–24.

Batta, Y. A., (2012). Invert emulsion: Method of preparation and application as proper formulation of entomopathogenic fungi. *Methods, 3*, 119–127.

Behie, S. W., Zelisko, P. M., & Bidochka, M. J., (2012). Endophytic insect-parasitic fungi translocate nitrogen directly from insects to plants. *Science, 336,* 1576, 1577.

Behle, R. W., Compton, D. L., Laszlo, J. A., & Shapiro-Ilan, D. I., (2009). Evaluation of soyscreen in an oil-based formulation for UV protection of *Beauveria bassiana* conidia. *J. Econ. Entomol., 102*, 1759–1766.

Behura, S. K., (2006). Molecular marker systems in insects: Current trends and future avenues. *Mol. Ecol., 15,* 3087–3113.

Bell, A. S., Blanford, S., Jenkins, N., Thomas, M. B., et al., (2009). Real-time quantitative PCR for analysis of candidate fungal biopesticides against malaria: Technique validation and first applications. *J. Invertebr. Pathol., 100*, 160–168.

Berbee, M. L., & Taylor, J. W., (2010). Dating the molecular clock in fungi, how close are we? *Fungal Biol. Rev., 24*, 1–16.

Bhattacharyya, A., Samal, A. C., & Kar, S., (2004). Entomophagous fungus in pest management. *Newsletter, 5*, 12.

Bilgo, E., Lovett, B., Leger, R. J. S., et al., (2018). Native entomopathogenic *Metarhizium* spp. from Burkina Faso and their virulence against the malaria vector *Anopheles coluzzii* and non-target insects. *Parasites Vectors, 11,* 11–16.

Blackwell, M., (2011). The fungi: 1,2,3. 5.1 million species? *American J. Botany, 98,* 426–438.

Boucias, D. G., Pendland, J. C., & Latge, J. P., (1988). Nonspecific factors involved in attachment of entomopathogenic deuteromycetes to host insect cuticle. *Appl. Environ. Microbiol., 54*, 1795–1805.

Brunner, K., Zeilinger, S., Ciliento, R., Sheridian, L. W., et al., (2005). Improvement of the fungal biocontrol agent *Trichoderma atroviride* to enhance both antagonism and induction of plant systemic disease resistance. *Appl. Environ. Microbiol.*, 3959–3965.

Charnley, A. K., (2003). Fungal pathogens of insects: Cuticle degrading enzymes and toxins. *Adv. Bot. Res., 40*, 241–321.

Chet, T., Schichler, H., Haran, S., & Appenheim, A. B., (1993). Cloned chitinase and their role in biological control of plant pathogenic fungi. In: *Proceedings of the International Symposium on Chitin Enzymology* (pp. 47, 48). Senigalia, Italy.

Crespo, R., Juarez, M. P., Dal Bello, G. M., Padin, S., Fernandez, G. C., & Pedrini, N., (2002). Increased mortality of *Acanthoscelides obtectus* by alkane grown *Beauveria bassiana. BioControl., 47*(6), 685–696.

de-Bekker, C., Quevillon, L. E., Smith, P. B., Fleming, K. R., Ghosh, D., et al., (2014). Species-specific ant brain manipulation by a specialized fungal parasite. *BMC Evol. Biol., 14*(166), 2–12.

de-Crecy, E., Jaronski, S., Lyons, B., Lyons, T. J., & Keyhani, N. O., (2009). Directed evolution of a filamentous fungus for thermotolerance. *BMC Biotechnol., 9*, 74.

de-Faria, M. R., & Wraight, S. P. (2007). Mycoinsecticides and mycoacaricides: A comprehensive list with worldwide coverage and international classification of formulation types. *Biol. Control. 43*, 237–256. http://dx.doi.org/10.1016/j.biocontrol.2007.08.001.

Deshmukh, L., Agrawal, D., & Sandhu, S. S., (2019b). Development of marker in the soft gold mushroom *Cordyceps* spp. for strain improvement. In: Kundu, R., & Narila, R., (eds.), *Advance in Plant & Microbial Biotechnology* (pp. 33–39). Singapore: Springer Nature.

Deshmukh, L., Sharma, A. K., & Sandhu, S. S., (2020). Contrive Himalayan soft gold *Cordyceps* species: A lineage of eumycota bestowing tremendous pharmacological and therapeutic potential. *Curr. Pharmacol. Rep., 6,* 155–166.

Deshmukh, L., Singh, R., & Sandhu, S. S., (2019a). Far ranging antimicrobial and free radical scavenging activity of Himalayan soft gold mushroom; *Cordyceps* sp. In: Sen, R., Mukherjee, S., Paul, R., & Narula, R., (eds.), *Biotechnology and Biological Sciences* (pp. 297–302). London, UK: Taylor & Francis Group.

Dilawari, V. K., Bons, M. S., Sethi, A., Kumar, P., & Gupta, V. K., (2002). RAPD-PCR for studying host specificity of *Trichogramma. Egg Parasitoid New, 14*, 9.

Dogan, Y. O., Hazir, S., Yildiz, A., et al., (2017). Evaluation of entomopathogenic fungi for the control of *Tetranychus urticae* (*Acari*: *Tetranychidae*) and the effect of *Metarhizium brunneum* on the predatory mites (*Acari*: *Phytoseiidae*). *Biol. Control, 111,* 6–12.

Donzelli, B. G. G., & Krasnoff, S. B., (2016). Molecular genetics of secondary chemistry in *Metarhizium* fungi. In: Lovett, B., & Leger, R. J. S., (eds.), *Advances in Genetics* (Vol. 94, pp. 365–436). Elsevier, Amsterdam.

Dworecka-Kaszak, B., (2014). *Cordyceps* fungi as natural killers, new hopes for medicine and biological control factors. *Annals Parasitol., 60*(3), 151–158.

Eilenberg, J., Hajek, A. E., & Lomer, C., (2001). Suggestions for unifying the terminology in biological control. *BioControl., 46,* 387–400.

Evans, H. C., Elliot, S. L., & Hughes, D. P., (2011). Hidden diversity behind the zombie-ant fungus *Ophiocordyceps unilateralis*: Four new species described from carpenter ants in Minas Gerais, Brazil. *PLoS One, 6*, e17024. doi: 10.1371/journal.pone.0017024 PMID: 21399679.

Fahrbach, S. E., Nambu, J. R., & Schwartz, L. M., (2012). Programmed cell death in insects. In: Gilbert, L. I., (ed.), *Insect Molecular Biology and Biochemistry*. Academic: Waltham, MA, USA.

Fan, Y. H., Fang, W. G., Guo, S. J., Pei, X. Q., et al., (2007a). Increased insect virulence in *Beauveria bassiana* strains over expressing an engineered chitinase. *Appl. Environ. Microbiol., 73*, 295–302.

Fan, Y., Fang, W., Xiao, Y., Yang, X., Zhang, Y., Bidochka, M. J., & Pei, Y., (2007b). Directed evolution for increased chitinase activity. *Appl. Microbiol. Biotechnol., 76*, 135–139.

Fazeli-Dinan, M., Talaei-Hassanloui, R., & Goettel, M., (2016). Virulence of the entomopathogenic fungus *Lecanicillium longisporum* against the greenhouse whitefly *Trialeurodes vaporariorum* and its parasitoid *Encarsia formosa. Int. J. Pest Manag., 62*, 251–260.

Federici, B. A., Bonning, B. C., & St. Leger, R. J., (2008). Improvement of insect pathogens as insecticides through genetic engineering. In: Hill, C., & Sleator, R., (eds.), *Patho Biotechnology* (pp. 15–40). Landes Bioscience, Austin.

Fernandes, E. K. K., Rangel, D. E. N., Moraes, A. M. L., Bittencourt, V. R. E. P., & Roberts, D. W., (2007). Variability in tolerance to UV-B radiation among *Beauveria* spp. isolates. *J. Invertebr. Pathol., 96*, 237–243.

Franz, G., & Robinson, A. S., (2011). Molecular technologies to improve the effectiveness of the sterile insect technique. *Genetica, 139*, 1–5.

Freed, S., Jin, F. L., & Ren, S. X., (2011). Phylogenetics of entomopathogenic fungi isolated from the soils of different ecosystems. *Pak. J. Zool., 43*, 417–425.

Gatehouse, J., (2008). Biotechnological prospects for engineering insect resistant plants. *Plant Physiol., 146,* 881–887.

Ghorbanpour, M., Omidvari, M., Abbaszadeh-Dahaji, P., et al., (2018). Mechanisms underlying the protective effects of beneficial fungi against plant diseases. *Biological Control., 117*, 147–157.

Greenfield, B. P. J., Lord, A. M., Dudley, E., & Butt, T. M., (2014). Conidia of the insect pathogenic fungus, *Metarhizium anisopliae*, fail to adhere to mosquito larval cuticle. *R. Soc. Open. Sci., 2*, 140–193.

Gressel, J., (2004). Potential failsafe mechanisms against the spread and introgression of transgenic hypervirulent biocontrol fungi. *Trends Biotechnol., 19,* 149–154.

Grisi, L., Massard, C. L., Borja, G. E. M., & Pereira, J. B., (2002). Impacto econômico das principais ectoparasitoses em bovinos no Brazil. *Hora Vet., 21*, 23–28.

Harman, G. E., Howell, C. R., Viterbo, A., Chet, I., & Lorito, M., (2004). *Trichoderma* species opportunistic, avirulent plant symbionts. *Nature Reviews Microbiol., 2,* 43–56.

Hawksworth, D. L., & Rossman, A. Y., (1997). Where are all the undescribed fungi? *Phytopathol., 87*, 888–891.

Hawksworth D. L. (1998). The consequences of plant extinctions for their dependent biotas: an overlooked aspect of conservation science. In *Rare, threatened and endangered floras of Asia and the Pacific Rim* (eds. Peng, C.-I. and Lowry II, P. P.), pp. 1–16. (*Academica sinica*: Taipei).

Hibbett, D. S., Ohman, A., Glotzer, D., et al., (2011). Progress in molecular and morphological taxon discovery in fungi and options for formal classification of environmental sequences. *Fungal Biol. Rev., 25*, 38–47.

Hussain, A., Li, Y. F., Cheng, Y., Liu, Y., Chen, C. C., & Wen, S. Y., (2013). Immune related transcriptome of *Coptotermes formosanus* shiraki workers: The defense mechanism. *PLoS One, 8*, e69543.

Hussain, A., Muhammad, R., Hassan, A., et al., (2014). Mycoinsecticides: Potential and future perspective. *Recent Pat. Food Nutr. Agric., 6*, 45–53.

John, R. P., Tyagi, R. D., Prevost, D., et al., (2010). Mycoparasitic *Trichoderma viride* as a biocontrol agent against *Fusarium oxysporum* f. sp. *adzuki* and *Pythium arrhenomanes* and as a growth promoter of soybean. *Crop Protection, 29*, 1452–1459.

Kamoun, S., (2003). Molecular genetics of pathogenic oomycetes. *Eukaryotic Cell, 2*, 191–199.

Kawai, T., & Akira, S., (2011). Toll-like receptors and their crosstalk with other innate receptors in infection and immunity. *Immunity, 34,* 637–650.

Khan, S., Guo, L., Shi, H., et al., (2012). Bioassay and enzymatic comparison of six entomopathogenic fungal isolates for virulence or toxicity against green peach aphids *Myzus persicae*. *Afr. J. Biotechnol., 11*, 14193–14203.

Kim, J. S., Roh, J. Y., Choi, J. Y., et al., (2010). Correlation of the aphicidal activity of *Beauveria bassiana* SFB-205 supernatant with enzymes. *Fungal Biol., 114,* 120–128.

Kirkland, B. H., Eisa, A., & Keyhani, N. O., (2005). Oxalic acid as a fungal acaricidal virulence factor. *J. Med. Entomol., 42,* 346–351.

Koiri, R. K., Naik, R. A., Rawat, D., Chhonker, S. K., & Ahi, J. D., (2017). Bioecological perspective of entomopathogenic fungi with respect to biological control. *J. Appl. Microb. Res., 1*(1), 7–14.

Koppert, B. V., (2015). *Mycotal*: *Verticillium Lecanii*-m. http://www.koppert.com (accessed on 13 January 2023).

Leger, R. J. S., Goettel, M., Roberts, D. W., & Staples, R. C., (1991). Prepenetration events during infection of host cuticle by *Metarhizium anisopliae*. *J. Invertebr. Pathol., 58*, 168–179.

Le-Grand, M., & Cliquet, S., (2013). Impact of culture age on conidial germination, desiccation and UV tolerance of entomopathogenic fungi. *Biocontrol Sci. Tech., 23*, 847–859.

Litwin, A., Nowak, M., & Rozalsk, S., (2020). Entomopathogenic fungi: Unconventional applications. *Rev. Environ. Sci. Biotechnol., 19*, 23–42.

Majeed, M. Z., Fiaz, M., Ma, C. S., et al., (2017). Entomopathogenicity of three muscardine fungi, *Beauveria bassiana, Isaria fumosorosea* and *Metarhizium anisopliae*, against the Asian citrus psyllid, *Diaphorina citri* Kuwayama (*Hemiptera*: *Psyllidae*). *Egypt. J. Biol. Pest Control., 27*, 211–215.

Mascarin, G. M., Alves, P. J. R., Fernandes, E. K. K., et al., (2018). Phenotype responses to abiotic stresses, asexual reproduction and virulence among isolates of the entomopathogenic fungus *Cordyceps javanica* (*Hypocreales*: *Cordycipitaceae*). *Microbiol. Res., 216*, 12–22.

Meikle, W. G., Mercadier, G., Guermache, F., & Bon, M. C., (2012). *Pseudomonas* contamination of a fungus-based biopesticide: Implications for honey bee (*Hymenoptera*: *Apidae*) health and varroa mite (*Acari*: *Varroidae*) control. *Biol. Control., 60*, 312–320.

Mueller, G. M., & Schmit, J. P., (2007). Fungal biodiversity: What do we know? What can we predict? *Biodivers. Conserv., 16*, 1–5.

Mustu, M., Demirci, F., Kaydan, M. B., & Ülgentürk, S., (2015). Laboratory assay of the effectiveness of the entomopathogenic fungus *Isaria farinosa* (Holmsk.) Fries (*Sordariomycetes*: *Hypocreales*) against the vine mealybug *Planococcus ficus* (Signoret) (*Hemiptera*: *Pseudococcidae*), even under the use of fungicides. *Int. J. Pest Manag., 61*, 264–271.

Namasivayam, S. K. R., & Arvind, B. R. S., (2015). Biocontrol potential of entomopathogenic fungi *Nomuraea rileyi* (f.) samson against major groundnut defoliator *Spodoptera litura* (fab.) lepidoptera; noctuidae. *Adv. Plants Agric. Res., 2*(5), 221–225.

Namasivayam, S. K. R., & Vidyasankar, A., (2014). Biocompatible formulation of potential fungal biopesticide *Nomuraea rileyi* (f.) samson for the improved post treatment persistence and biocontrol potential. *Nature Environment and Pollution Technology, 13*(4), 835–838.

Nasir, M., Hussain, M., Iqbal, B., et al., (2011). A review on whitefly control by the use of entomopathogenic fungi. In: *Proceedings of the 8th National Conference of Plant Pathology, University of Agriculture* (pp. 28–40). Faisalabad, Pakistan.

Niassy, S., Diarra, K., Ndiaye, S., & Niassy, A., (2011). Pathogenicity of local *Metarhizium anisopliae* var. *acridum* strains on *Locusta migratoria migratorioides* Reiche and Farmaire and *Zonocerus variegatus* Linnaeus in Senegal. *Afr. J. Biotechnol., 10*(1), 28–33.

O'Brien, H. E., Parrent, J. L., Jackson, J. A., et al., (2005). Fungal community analysis by large-scale sequencing of environmental samples. *Appl. Environ. Microbiol., 71*, 5544–5550.

Ortiz-Urquiza, A., & Keyhani, N. O., (2013). Action on the surface: Entomopathogenic fungi versus the insect cuticle. *Insects, 4*, 357–374.

Pedrini, N., Ortiz-Urquiza, A., Huarte-Bonnet, C., Zhang, S., et al., (2013). Targeting of insect epicuticular lipids by the entomopathogenic fungus *Beauveria bassiana*: Hydrocarbon oxidation within the context of a host-pathogen interaction. *Front. Microbiol., 4*, 24.

Qin, Y., Ortiz-Urquiza, A., & Keyhani, N. O., (2014). A putative methyltransferase, mtrA, contributes to development, spore viability, protein secretion, and virulence in the entomopathogenic fungus *Beauveria bassiana*. *Microbiol., 160*, 2526–2537.

Rangel, D. E. N., Braga, G. U. L., Anderson, A. J., & Roberts, D. W., (2005). Influence of growth environment on tolerance to UV-B radiation, germination speed, and morphology of *Metarhizium anisopliae* var. *acridum* conidia. *J. Invertebr. Pathol., 90*, 55–58.

Rangel, D. E. N., Fernandes, E. K. K., Anderson, A. J., & Roberts, D. W., (2012). Culture of *Metarhizium robertsii* on salicylic-acid supplemented medium induces increased conidial thermotolerance. *Fungal Biol., 16,* 438–442.

Raymond, J. St. L., & Wang, C., (2010). Genetic engineering of fungal biocontrol agents to achieve greater efficacy against insect pests. *Appl. Microbiol. Biotechnol., 85*, 901–907.

Rehner, S. A., (2009). Molecular systematics of entomopathogenic fungi. In: Stock, S. P., Vandenberg, J., Glazer, I., & Boemare, N., (eds.), *Insect Pathogens: Molecular Approaches and Techniques* (pp. 145–165). Wallingford: CABI.

Robledo-Monterrubio, M., Alatorre-Rosas, R., Viniegra-Gonzalez, G., & Loera, O., (2009). Selection of improved *Beauveria bassiana* (Bals.) Vuill. strains based on 2-deoxy-D-glucose resistance and physiological analysis. *J. Invertebr. Pathol., 101*, 222–227.

Sahayaraj, K., & Namasivayam, S. K. R., (2011). Field evaluation of three entomopathogenic fungi on groundnut pests. *Tropicultura, 29*, 143–147.

Samson, R. A., Evans, H. C., & Latge, J. P., (1988). *Atlas of Entomopathogenic Fungi*. Springer.

Sandhu, S. S., Rajak, R. C., & Hasija, S. K., (2000). Potential of entomopathogens for the biological management of medically important pest: Progress and prospect. In: *Glimpses in Plant Sci.* (pp. 110–117).

Sandhu, S. S., Sharma, A. K., Beniwal, V., et al., (2012). Myco-biocontrol of insect pests: Factors involved, mechanism, and regulation. *J. Pathog., 126819*, 01–10.

Sandhu, S. S., Shulka, H., Aharwal, R. P., Kumar, S., & Shukla, S. E., (2017). Efficacy of entomopathogenic fungi as green pesticides: Current and future prospects. In: Panpatte, D., Jhala, Y., Vyas, R., & Shetal, H., (eds.), *Microorganisms for Green Revolution. Microorganisms for Sustainability* (p. 6). Springer, Singapore.

Schneider, L., Silva, C., Pamphile, J., & Conte, H., (2013). Infection, colonization and extrusion of *Metarhizium anisopliae* (Metsch) Sorokin (*Deuteromycotina*: *Hyphomycetes*) in pupae of *Diatraea saccharalis* F. (*Lepidoptera*: *Crambidae*). *J. Entomol. Nematol., 5*, 1–9.

Schuler, T., (1991). *Verticillium lecanii (Zimmermann) Viegas (Hyphomycetales: Moniliaceae): History, Systematics, Distribution, Biology, and Application in Plant Protection* (Vol. 269, p. 154). Communications from the Federal Biological Institute for Agriculture and Forestry, Berlin-Dahlem Notebook.

Shang, Y., Feng, P., & Wang, C., (2015). Fungi that infect insects: Altering host behavior and beyond. *PLoS Pathog., 11*(8), 1005037.

Sharma, K., (2004). Bionatural management of pests in organic farming. *Agrobios. Newsl., 2*, 296–325.

Shin, T. Y., Bae, S. M., Kim, D. J., et al., (2017). Evaluation of virulence, tolerance to environmental factors and antimicrobial activities of entomopathogenic fungi against two-spotted spider mite, *Tetranychus urticae*. *Mycosci., 58*, 204–212.

Singh, A., Kumar, B., & Srivastava, A. K., (2011). Metabolic engineering of crops using RNA interference. *Pac. J. Mol. Biol. Biotechnol. 19*, 137–148.

Skinner, M., Parker, B. L., & Kim, J. S., (2014). Role of entomopathogenic fungi. In: Abrol, D. P., (ed.), *Integrated Pest Management* (pp. 169–191). Academic Press, Cambridge.

Stolarek, P., Rozalska, S., & Bernat, P., (2019). Lipidomic adaptations of the *Metarhizium robertsii* strain in response to the presence of butyltin compounds. *Biochim Biophys Acta-Biomembr., 1861*, 316–326.

Subbanna, A., Stanley, J., Venkateswarlu, V., et al., (2019). Toxicological prospects on joint action of microbial insecticides and chemical pesticides. In: Khan, M. A., & Ahmad, W., (eds.), *Microbes for Sustainable Insect Pest Management* (pp. 317–340). Springer, Cham.

Sujii, E. R., Carvalho, V. A., & Tigano, M. S., (2002a). Sporulation kinetics and viability of conidia of *Nomuraea rileyi* (Farlow) samson on corpses of the fall armyworm, *Anticarsia gemmatalis* Hübner (Lepidoptera: Noctuidae), under field conditions. *Neotrop. Entomol., 31,* 85–90.

Thompson, S. R., Brandenburg, R. L., & Arends, J. J., (2006). Impact of moisture and UV degradation on *Beauveria bassiana* (Balsamo) Vuillemin conidial viability in turfgrass. *Biol. Control., 39*, 401–407.

Tincilley, A., Easwaramoorthy, G., & Santhanalakshmi, G., (2000). Attempts on mass production of *Nomuraea rileyi* on various agricultural products and byproducts. *J. Biol. Contr., 18*(1), 33–40.

Tuao, G. C. A., & Pinto, J. M., (2016). Biocontrol of melon wilt caused by *fusarium oxysporum* Schlect f. sp. *melonis* using seed treatment with *Trichoderma* spp. and liquid compost. *Biol. Control., 97*, 13–20.

Vega, F. E., & Blackwell, M., (2005). *Insect-Fungal Associations: Ecology and Evolution.* Oxford University Press, Oxford.

Vianna, M. F., Pelizza, S., Russo, M. L., et al., (2020). ISSR markers to explore entomopathogenic fungi genetic diversity: Implications for biological control of tobacco pests. *J. Biosci., 45*, 136.

Wang, B., Kang, Q., Lu, Y., Bai, L., & Wang, C., (2012). Unveiling the biosynthetic puzzle of destruxins in *Metarhizium* species. *Proc. Natl. Acad. Sci. USA., 109*, 1287–1292.

Wang, C., & St Leger, R. J., (2007). The MAD1 adhesin of *Metarhizium anisopliae* links adhesion with blastopore production and virulence to insects, and the MAD2 adhesin enables attachment to plants. *Eukaryotic Cells, 6*, 808–816.

Watanabe, T., (2010). Pictorial atlas of soil and seed fungi. *Morphologies of Cultured Fungi and Key to Species* (3rd edn.). Boca Raton: CRC Press.

Woolley, V. C., Graham, R. T., Gillian, P., et al., (2020). Cordycepin, a metabolite of *Cordyceps militaris*, reduces immune-related gene expression in insects. *J. Invertebr. Pathol.*

Yang, M., Jin, K., & Xia, Y., (2011). MaFKS, a β-1,3-glucan synthase, is involved in cell wall integrity, hyperosmotic pressure tolerance and conidiation in *Metarhizium acridum. Curr. Genet., 57*, 253–260.

Zhang, S. Z., Xia, Y. X., Kim, B., & Keyhani, N. O., (2011). Two hydrophobins are involved in fungal spore coat rodlet layer assembly and each play distinct roles in surface interactions, development and pathogenesis in the entomopathogenic fungus, *Beauveria bassiana. Mol. Microbiol., 80*, 811–826.

Zhao, H., Lovett, B., & Fang, W., (2016). Genetically engineering entomopathogenic fungi. In: Lovett, B., & Leger, R. J. T., (eds.), *Advances in Genetics* (pp. 137–163). Elsevier, Amsterdam.

Zheng, P., Xia, Y. L., Zhang, S. W., & Wang, C. S., (2013). Genetics of *Cordyceps* and related fungi. *Appl. Microbiol. Biotechnol., 97*, 2797–2804.

CHAPTER 7

An Outlook of Nematophagous Fungi and the Underlying Mechanism of Nematophagy

ARGHYA NASKAR, KISHOR ROY, BAISHAKHI SANTRA, ANIK SARKAR, and KRISHNENDU ACHARYA

Molecular and Applied Mycology and Plant Pathology Laboratory, Department of Botany, University of Calcutta, Kolkata, West Bengal, India

ABSTRACT

Plant parasitic nematodes (PPNs) are a major threat to the agricultural sector worldwide. These tiny, parasitic nematodes can be considered a major hazard for crop plants. Statistical reports show huge economic losses faced in the agriculture and horticulture field. Plants infested with nematodes often become targets for secondary infection by bacteria and fungi, making the condition even worse. Increasing concerns of health over the usage of chemical methods have shifted our focus to finding alternative means of controlling nematodes. Many organisms act antagonistically against nematodes, among which the usage of fungi has been reviewed in this chapter. Due to its wide variety and distribution all over the world, fungi provide a good scope for research in this field. Different mechanisms of nematophagy are mentioned here. With the advancement of technology over the years, particularly in the field of molecular biology and omics studies, our understanding of nematophagous fungi has increased, making them potent bio-control agents. Further studies are required for identifying different fungi with nematophagous ability.

Applied Mycology for Agriculture and Foods: Industrial Applications. Sanjay K. Singh, Deepak Kumar, Md. Shamim, & Rohit Sharma (Eds.)

7.1 INTRODUCTION

Yield loss due to pathogenic organisms is of major concern in agriculture. Nowadays besides insect pests, nematodes are emerging as a serious threat towards the agricultural sectors throughout the world, especially in developing countries. Phytoparasitic nematodes cause huge worldwide economic loss, estimated to be around 125 million dollars annually. According to a recent survey done by AICRP (All India Coordinated Research Project) 2014–2015 on 19 field crops and 11 horticultural crops, economic loss due to plant parasitic nematode infestation is around Rs. 10,203,979 million per annum (Kumar et al., 2020). On an average 21% yield loss is caused by plant parasitic nematode infection in crops (Figure 7.1).

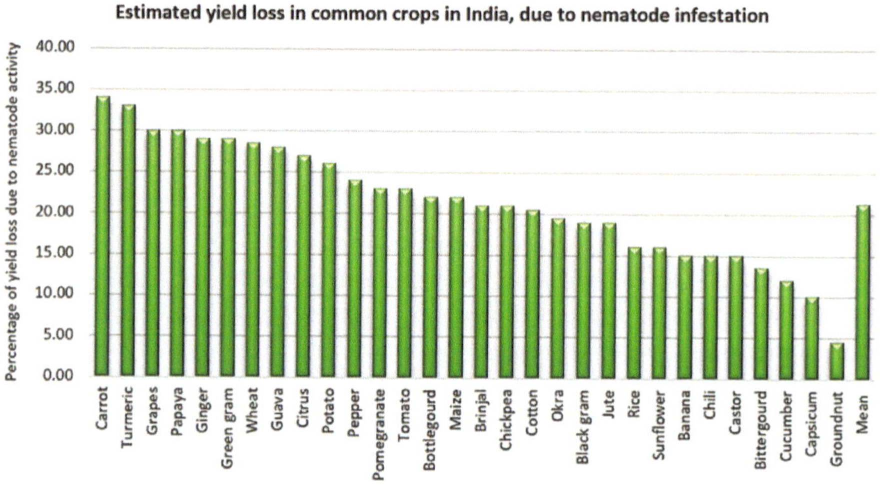

FIGURE 7.1 Estimated yield loss in common crops in India, due to nematode infestation.

More than 4,100 species are encountered to be having plant parasitic activity, of which *Meloidogyne* spp. are most dangerous. Root-knot disease in different crops caused by *Meloidogyne* spp. accounts for 75% of the total estimated loss amounting to Rs. 77,373.87 million. This estimation makes it clear that *Meloidogene* is the most important among all plant parasitic nematodes (PPNs). *Meloidogene graminicola* is the most common one, causing root-knot disease in rice and this particular species alone is responsible for yield loss of Rs. 23,272.32 million yearly (Mesa-Valle et al., 2020). Beside causing huge yield loss, nematodes are also responsible for the loss of vigor (Trudgill et al., 1992).

In a developing country like India, where more than 260.2 million people (26.1%) are still lying below the poverty line (Mehta & Shah, 2003), even a small fraction of yield loss in crop production can be a major factor of distress.

Nematodes are microscopic roundworms belonging to the distinct phylum Nematoda or Nemata under the Ecdysozoa. Traces regarding its origin dates back to the Cambrian period (600–500 million years ago). Fossil record of nematode *Palaeonema phyticus* in association with *Aglaophyton major* traces back to early Devonian period (416–396 million years ago). Indirect evidence suggests a marine origin of this phylum during the Cambrian period (Khan, 2015). Mostly, PPNs belong to one of the four orders viz., Tylenchida, Aphelenchida, Dorylaimida, and Triplonchida, constituting a very small fraction of the Phylum Nematoda.

Among the PPNs, the root-knot nematodes (*Meloidogyne* spp.) and the cyst nematodes (*Heterodera* spp. and *Globodera* spp.) are considered as primary models for understanding the host-parasitic relationship because of their prevalence and economic impact (Khan, 2015). As the name suggests, root-knot disease is characterized by the formation of knots or small gall like bodies on the roots, which reduce the absorbing capability of the plant and thus affecting the transportation of water and nutrients to the rest of the plant parts. Wilting especially in the afternoon is one of the prominent markers of the root-knot disease caused by nematodes (Ralmi et al., 2016). Cyst nematodes are associated with the development of cysts which are smaller in size than galls. Above-ground symptoms include stunting, discoloration, and wilting of the plant in strong sunlight (Davis & Tylka, 2000). Although the symptoms vary depending on the particular nematode species, the development of disease depends on the complex interaction among the host, pathogen, and the environment. Some of the common PPNs and their host plants are mentioned in Table 7.1.

There are several nematode-controlling strategies which are being practically applied for decades, of which the use of chemical nematicides is the most accepted one (Hildalgo-Diaz & Kerry, 2008). To reduce the nematode infestation cultural nematode management methods including cultivation of antagonistic crops, trap crops and resistant cultivars, crop rotation in seasons, soil solarization, bio-fumigation, destruction of residual crop roots and flooding before plantation are used in combinations or in addition to chemical nematicides (Ntalli & Caboni, 2017; Ouiza & Samira, 2017; Trivedi & Barker, 1986). Although these cultural practices play a major role in integrated pest management (IPM), the majority of these methods come with limitations and serious drawbacks (Hildalgo-Diaz & Kerry, 2008). In terms of efficacy against nematodes, the place of chemical nematicides comes first, but most of these are extremely toxic. The extreme toxicity of these chemical nematicides can be considered as major threat towards animals and human beings. Chemical nematicides such as methyl bromide, dibromo chloropropane, fosthiazate are being frequently used but most of them are getting banned in several countries (Chen et al., 2020). These broad spectrum nematicides affects our environment by contaminating the ground water and help in depletion of ozone layer (Hussain et al., 2017).

TABLE 7.1 Some Common Plant Parasitic Nematode and Their Host Plants

Nematode Genus	Host Plant	References
Meloigogyne	Almost all fruit crops and cereals	Anwar & Mckenry (2010)
Ditylenchus	Potato, garlic, onion	Poirier et al. (2019)
Rotylenchus	Boxwood, cotton, castor, papaya	Davis & Webster (2005); Khan & Khan (1973)
Globodera	Potato	Ingham & Phillips (2015)
Heterodera	Maize, wheat, carrot, beans, potatoes, tobacco	Dong & Opperman (1997); Escobar-Avila et al. (2018); Ringer et al. (1987)
Aphelenchoides	Rice, strawberry	Franklin (1950); Khan et al. (2012)
Trichodorus	Corn, but beets, celery, cabbage, and tomatoes	Allen (1957)
Anguina	Wheat, rye	Koshy & Swarup (1971); Stynes & Bird (1980)
Pratylenchus	Walnut, cherry, strawberry, apple, potato	Koen (1967); Pinochet et al. (1992)
Radopholus	Citrus, rice, potato	Feldman & Hanks (1968); Wuyts et al. (2007)

Keeping the negative impacts and drawbacks of conventional control strategies in mind, the developing world is looking forward to environment-friendly and hazard-less sustainable control strategies having high efficacy against pathogens.

Biological control strategies are emerging as an eco-friendly tool in countering plant pathogens. Eilenberg et al. (2001) defined biological control as 'the use of living organisms to suppress the population density or impact of a specific pest organism, making it less abundant or less damaging than it would otherwise be.' Decades of research established the use of antagonistic microorganisms or their metabolic products as a potential weapon against pathogens including PPNs. Many bacteria including human pathogenic bacteria such as *Staphylococcus aureus, Pseudomonas aeruginosa* and many strains of *Bacillus* have the ability to infect nematodes and are commonly utilized as biocontrol agents (Liang et al., 2019). Several varieties of cry proteins obtained from *Bacillus thuringiensis* show the significant reduction of *Meloidogyne incognita* population in tomato roots (Li et al., 2007). Apart from bacteria, microfungi hold an important position as biocontrol agents (BCAs). Because of their constant association in the rhizosphere, antagonistic fungal strains can act on nematodes by several means (Siddiqui & Mahmood, 1996). In the recent decade, advancement in molecular biology and omics science helped us to decipher the molecular mechanisms behind the host parasitic interactions among the nematodes and associated nematophagous fungi (Soares et al., 2018) In this study, we will be focusing on the exploitation of macrofungi as a potential bio-control agent against PPNs and their diverse mechanism of nematophagy.

7.2 NEMATOPHAGOUS FUNGI

Nematophagous fungi are often called 'carnivorous fungi' or 'nematode hunting fungi' or 'nematode destroying fungi' which have antagonistic relationships with nematodes. They are cosmopolitan in distribution and mostly facultative parasites and can survive saprophytically on decaying organic matter present in soil. However, the nematophagous fungi are less competitive in soil thus making their predatory nature even more important for their survival. The parasitic behavior of fungi on nematodes may have evolved over time as a source for nutrients in deficient soil or as a source for spore dispersal to ensure continuation of generation. Whatever the driving cause may have been, the parasitic behavior of fungi on nematodes also made them a potential biocontrol agent for nematodes. The nematode trapping behavior was first identified in *Arthrobotrys oligospora* (Fresenius, 1863). Around 700 fungi have been demonstrated to date with nematophagous ability under Chytridiomycota, Zygomycota, and Ascomycota, Basidiomycota (Li et al., 2015). According to their mode of action, these fungi are classified into four groups: Nematode trapping fungi, endoparasitic fungi, egg, and female parasitic fungi, and toxin producing fungi (Jansson & Lopez-Llorca, 2001). Taxonomy of some nematophagous fungi and their mode of action are summarized in subsections (Table 7.2). The infection of nematodes by the nematophagous fungi involves a series of steps including recognition, adhesion, penetration, and digestion (Dong & Zhang, 2006).

TABLE 7.2 Taxonomy of Some Nematophagous Fungi and Their Mode of Action

Phylum	Anamorph	Teliomorph	Mode of Action
Chitridiomycota	–	*Catenaria*	Endoparasitic
		Olpidium	Endo parasitic and egg/female parasitic
Zygomycota	–	*Cystopage*	Nematode-trapping
		Stylopage	Nematode-trapping
		Helicocephalum	Nematode trapping, egg/female parasitic
		Rhopalomyces	Egg and female parasitic
Ascomycota	*Arthrobotrys*	*Orbilia*	Nematode-trapping
	Dactylellina		
	Drechslerella		
	Duddingtonia		
	Gamsylella		
	Monacrosporium		
	Harposporium	*Podocrella*	Endoparasitic
	Drechmeria		
	Haptocillium		
	Meria		
	Paecilomyces	–	Egg/female parasitic
	Pochonia	*Cordyceps*	
	Lecanicillium		
	Trichoderma	*Hypocreae*	
	Dactylella	*Orbilia*	
Basidiomycota	*Nematoctonus*	*Hohenbuehelia*	Nematode trapping and endoparasitic
	–	*Stropharia*	Nematode-trapping
	–	*Peniophorella*	
	–	*Pleurotus*	Toxin-Producing
	–	*Coprinus*	

7.2.1 NEMATODE TRAPPING FUNGI

As the name suggests, the fungi belonging to this group develop special hyphal traps to entrap nematodes. These fungi have very low host specificity and can entrap all moving stages of nematodes (Pramer, 1964). More than 200 species of fungi belonging to Ascomycota, Basidiomycota, and Zygomycota forms adhesive traps to capture free living nematodes in the soil. Different groups of fungi can produce different types of nematode trapping devices like adhesive nets, adhesive hyphae, adhesive branches, adhesive knobs, non-constricted rings and constricted rings (Table 7.3).

Antimicrobial and nematicidal compounds like linoleic acid (*Arthrobotrys oligospora, A. conoides*) or pleurotin (*Nematoctonus robustus, N. concurrens*) were also reported to be secreted by nematode trapping fungi (Anke et al., 1995). The majority of trapping fungi colonize bulk soil and wait for the nematode to make contact. Production of secondary attractive compounds for nematodes is reported in *A. superba* which attract second juvenile stages of *Meloidogyne incognita* nematode (Hallmann et al., 2009). Some fungi prefer colonizing the rhizosphere of plants to increase their chances of trapping PPNs on their way towards the root. For example, *A. oligospora* populations were seen growing in abundance in the rhizosphere of tomato and barley plants (Bordallo et al., 2002).

Ulzurrun & Hsueh (2018) described different stages involved in the trapping process utilized by *A. oligospora* to trap *Caenorhabditis elegans* adult nematodes. The interaction involves several sequential events including attraction of nematodes, prey recognition, triggering of trap development, trap production, adhesion of nematodes to the traps, penetration of nematodes and finally digestion and uptake of nutrients.

The formation of traps in trapping fungi is induced chemically by small peptides secreted by the nematodes (Nordbring-Hertz, 1973, 1977). The infection starts with the recognition of the host (nematode) including attraction and host chemotaxis towards fungal hyphae or traps. However, the identity of these compounds is still not clear. The underlying mechanism of trap formation during nematophagy was extensively studied in *Arthrobotrys oligospora* (Niu & Zhang, 2011). The nematode *Neoaplectana glaseri* releases a small low molecular weight peptide 'nemin' which triggers nematode trapping fungi to differentiate into three dimensional networks to trap the nematodes (Pramer & Stoll, 1954). Recognition of the host, i.e., nematode by the nematophagous fungi *A. oligospora* mediated by the interaction between fungal surface lectin with sugar molecules viz., galNAc (Nordbring-Hertz & Mattiasson, 1979), AoL (Rosen et al., 1992), Aofle A (Liu et al., 2020), etc., present on the cuticular surface of the nematodes. This cell-to-cell interaction and a group of volatile compounds (Wang et al., 2018), nitrate (Liang et al., 2016) and autophagy (Chen et al., 2013) make the saprophytic form of fungi to pathogenic stage and helps in adhesion. After successful adhesion with exerting mechanical force and by excreting different extracellular enzymes like serine proteases (Yang et al., 2007; Wang et al., 2006), collagenase and chitinases (Jansson et al., 1985) the fungal penetration hyphae enter into the interior of the nematodes, obtaining nutrients for their luxurious growth and reproduction.

TABLE 7.3 Some Important Nematode Trapping Fungi and Their Associated Trapping Structures

Trapping Apparatus	Species
Adhesive nets	*Arthrobotry oligospora,*
	Arthrobotrys irregularis,
	Arthrobotrys robustus,
	Arthrobotrys superba,
	Dactylaria zhongdianensis,
	Duddingtonia flagrans,
	Geniculifera perpasta,
	Monacrosporium cinophagum
	Monacrosporium elegans
	Monacrosporium psychrophilum,
	Monacrosporium eudermatum
Adhesive hyphae	*Cystopage cladospora,*
	Dactylaria lobata,
	Helicocephalum oligosporum,
	Peniophorella praetermissia,
	Stropharia rugosoannulata,
	Stylopage hadra,
	Stylopage leihypha
Adhesive branches	*Monacrosporium gephyropagum*
Adhesive knobs/non-constricting rings	*Arthrobotrys haptotyla,*
	Dactylaria candida,
	Dactylaria ellipsospora,
	Dactylellina haptophyla,
	Dactylellina sichuanensis,
	Dactylellina varietas, Monacro,
	Monacrosporium ellipsosporum,
	Monacrosporium haptotylum
Constricting rings	*Arthrobotrys brochopage,*
	Dactylella bembicodes,
	Drechslerella anchoria,
	Drechslerella brochopage
	Monacrosporium bembicodesr

When a nematode touches the inner ring wall, inflation of the ring cells results in enclosure and trapping of nematode. The mechanism is still not known. It has been observed that the application of mild heat, pressure, and calcium ions stimulates the swelling of ring cells. Upon entrance of the nematode, the pressure activates the G-proteins of the cells and a consequent rise of calcium ion concentration in the cytoplasm. Calcium ions activate the calmodulin and the water channels open. The rapid entrance of water leads to cell inflation and enclosure of the traps. Calmodulin was seen to be the key factor in regulating ring closure (Chen et al., 2001; Wyss et al., 1992). In the case of *Drechslera stenobrocha* the trap formation and nematode capturing capacity was induced by abscisic acid and reverse in the case of compromised nitric oxide production (Xu et al., 2011). It has been demonstrated that *Arthrobotrys oligospora* also produces a Serine protease PII belonging to the subtilisin family (Åhman et al., 1996). Serine proteases produced by other trapping fungi such as Mlx produced by *Arthrobotrys microscaphoides*, DsI produced by *A. shizishanna* has also been characterized and sequenced (Wang et al., 2006).

7.2.2 ENDOPARASITIC FUNGI

Fungi belonging to this group are endoparasites of host nematodes, usually having a broad nematode host range (Viene et al., 2006). These fungi are mostly obligate parasites, spending the entire vegetative life cycle within their infected host and rarely saprophytic in nature (Lopez-Llorca et al., 2008). Reports of distribution of this group indicate a fairly cosmopolitan nature but few species are confined to either tropical or temperate areas (Gams & Zare, 2003). Endoparasitic fungus infects nematodes using spores which are either motile (zoospores) or non-motile (conidia) (Table 7.4). The spores get attached to the cuticular skin of nematodes or sometimes the spores are ingested by the nematodes. Some of the extensively studied endoparasitic fungi include *Drechmeria balanoides, D. coniospora, Nematoctonus sp., Hirstella rohssiliensis*, and *Harposporium anguillulae*.

Nematodes having wide mouth openings, which facilitate conidial ingestion are much more prone to endoparasitism (Gams & Zare, 2003). *Drechmeria* produces teardrop shaped conidia coated with sticky layer and radiating fibrils (Jansson, 1993; Saikawa, 1982). Sticky conidia get attached to the chemosensory organs near the mouth region of the nematodes, disabling their ability to respond to all chemotactic attraction sources including the host plants (Jansson & Nordbring-Hertz, 1983). The spores germinate and insert its content inside the nematode by means of penetration tube which moves forward through an infection vesicle formed within the cuticle layers (Dijksterhuis et al., 1994; Sjollema et al., 1993). In the case of *D. conidiospora*, direct penetration of conidia through the cuticle is the only way of infection and the conidia do not germinate in the intestine if ingested by the nematode (Jansson, 1994). After attachment of the adhesive conidia to the cuticle of the nematode, it develops an appressoria and presses against the cuticle. With strong mechanical force and extracellular enzyme action, the fungal penetration hyphae invade the cuticle of the nematode and enter the interior part followed by vigorous growth. It disturbs the metabolism of nematode. These fungi uptake nutrients from the nematode which leads to the death of the organism (Jansson et al., 1985; Zhang et al., 2016; Zuckerman et al., 1988).

TABLE 7.4 Infective Propagule of Some Endoparasitic Nematophagous Fungi

Infective Propagules	Species
Zoospores	*Catenaria anguillulae,*
	Catenaria vermiformis,
	Meristacrum asterospermum,
	Olpidium vermicola
Adhesive spores	*Hirsutella rhossiliensis,*
	Nematoctonus concurrens,
	Nematoctonus leiosporus,
	Verticillum balanoides
Adhesive conidia	*Drechmeria conidiospore,*
	Haptocillium bacterosporum,
	Haptocillium balanoides,
	Meristacrum asterospermum,
	Maria conidiospora
Integrated spores	*Harposporium anguillulae,*
	Harposporium bysmatosporum,
	Harposporium leptospira,
	Spirogyromyces vermicola

Drechmeria conidiospore produces up to 10,000 conidia, whereas *H. rhossoliensis* forms 100–1,000 conidia per infected nematode (Nordbring-Hertz et al., 2006). Investigations revealed that application of *H. rhossoliensis*

decreases cyst nematode (*Heterodera glycines*) infection in soyabean *and D. conidiospore* significantly reduces root-knot nematode (*Meloidogyne* spp.) in many crops including tomato and other plants under the family, Solanaceae (Jansson et al., 1985).

In a lab experiment, *Hirsutella rhossiliensis* killed the nematode *Ditylenchusdipsaci* in four days and the *Meloidogyne incognitia* juveniles in two days (Cayrol et al., 1986). In the absence of nematode host as a food source, the population of *H. rhossiliensis* drastically decreased. The conidia produced by this group are infectious only when attached to a conidiophore (Timper et al., 1991). A new extracellular alkaline protease (Hasp) has been identified from *H. rhossiliensis*. Isolated and purified Hasp was found effective in killing soybean cyst nematode *Heterodera glycines* (Wang et al., 2009).

A different style of infection has been observed in *Catenaria anguillulae.* Motile uniflagellate zoospores of *C. anguillulae* swim towards the natural openings (mouth, anus, excretory pores) of nematodes and after making contact with the cuticle, shows amoeboid movement (Jansson & Thiman, 1992). A sticky material covering the zoospores helps in adherence. Infection is achieved with the help of a penetration peg which breaks and digests the nematode cuticle within 24 hours. At later stages, the hyphae develop sporangium containing zoospores that can infect new nematodes after their liberation (Wyss et al., 1992).

7.2.3 EGG PARASITIC/CYST PARASITIC FUNGI

Egg parasitic or, cyst parasitic fungi are those which infect the non-motile stages of nematodes, eggs or, cysts by forming hyphal structure and appressoria at the tip, which invades the eggshell and digests the whole content of egg of both mature and immature eggs or, the fungi secrete some extracellular enzymes which completely or, partially degrade the outer shell layer of egg or cyst which further ease up the penetration of hyphae and destroys nematodes in the egg or juvenile stage (Nordbring-Hertz et al., 2006; Kerry, 1988). Among different fungi showing nematode egg/cyst parasitic activity, mostly studied species includes *Pochonia chlamydosporia, Paecilomyces lilacinum, Metapochonia suchlasporia, Lecanicillium lecanii, Nematophthora gynophila* and *Catenaria auxilaris* (Jatala, 1986; Moosavi et al., 2010).

The eggshell or cysts are the rigid and organized structure composed of mainly chitin, protein, and lipid (Wharton, 1980). The eggshell is composed of three different layers: the outermost vitelline layer, the middle chitinous layer, and the innermost lipid layer. The thickness of each layer differs from species to species. For example, the vitelline layer is comparatively thinner than the chitinous layer in *Meloidogyne javanica.* A vitelline layer of eggshell derives from the fertilized oocyte remains a unit membrane like structure and became thicker when the egg is fully formed. The middle chitinous layer is the thickest layer which provides structural strenght to the eggshell, mostly composed of non-helicoidal chitin and protein elements. The innermost lipoprotein layer forms a barrier to permeability (Bird & Mcclure, 1976; Mcclure & Bird, 1976; Wharton & Jenkins, 1978).

The interaction between egg/cyst parasitic fungi and the nematode egg/cyst takes place in 4 stages: attraction, recognition, penetration, and digestion. Fungi like *Pochonia chlamydosporia, Paecilomyces lilacinus and Colletotricum gloeosporioides* first form an appressorium to infect nematode eggs. Appressoria secrete certain enzymes which facilitate the penetration of hyphae into the eggs (Lopez-Llorca et al., 2008). Apart from this, *P. clamydosporia* also produces many secondary metabolites such as radiciol, pseurotin A and pochonin which facilitate the recognition of nematode host. Another group of secondary metabolites, aurovertin (A-S) is found to be produced in several egg parasitic fungi, which helps in recognition of host in soil. Aurovertin B binds to the beta subunit of F_1-ATPase in the inner mitochondrial membrane, ultimately blocking the function of ATPase (Wang et al., 2015). Aurovertin D shows the most toxicity at the larval stage of *C. elegans.* The adhesion on nematode egg takes place mainly by a certain glycoprotein, lectin-protein or, GLEYA-protein which remain present on the appressoria (Monteiro et al., 2017). Certain extracellular enzymes secreted by fungi are involved in the penetration process. The eggshell layer or the cyst outer layer remains the major barrier between the host and the fungus. Fungus overcome these barriers by secreting VCP1 (subtilisin-like serine protease of 33 kDa) which degrades the vitelline layer and exposes the chitinous layer of egg. VCP1 is functionally related to pr1 protein (Morton et al., 2003).

7.2.4 TOXIN PRODUCING FUNGI

Fungi having antagonistic relationships with nematodes often produce some toxic compounds which can paralyze and eventually kill nematodes (Table 7.5). These toxic compounds can either be nematicidal or nematostatic in nature. The toxin producing fungi usually belong to Ascomycota and

Basidiomycota (Degenkolb & Vilcinskas, 2016). Most wood-rot fungi, which have the ability to feed on wood by decaying it with certain enzyme, often produce toxic substances which shows toxicity to nematodes (Soares et al., 2018).

TABLE 7.5 Important Nematocidal Toxin-Producing Fungi

Toxin-Producing Fungi	Toxin	References
Nematocnus robustus	Dihydropleurotinic acid	Stadler et al. (1994)
	Pleurotin	
	Leucopleurotin	
Pleurotus pulmonarious	S-coriolic acid	
	Linoleic acid	
	p-Anisaldehyde	
	1-(4-methoxyphenyl)-1,2-propanediol	
	2-Hydroxy-(4′-methoxy)-propiophenone	
Pleurotus ostreatus	Trans-2-decenedioic acid	Kwok et al. (1992)

One of the popular edible mushrooms, *Pleurotus* has been reported to release nematotoxic compounds from their hyphae. *Pleurotus ostreatus* produces a toxic substance having structural similarity with trans-2-decenedioic acid, that quickly immobilizes nematodes and finally leads to shrinkage of the nematode (Kwok et al., 1992). Several investigations have revealed the molecular mechanism of nematotoxin produced by *P. ostreatus* which paralyze *C. elegans* adult nematodes within a minute, when it comes in contact. When the sensory cilia of *C. elegans* come in contact with the fungus, it releases the toxin which paralyzes the nematodes and gradually causes a massive influx of calcium ions in head cell and shows hyper concentration of calcium ions in pharyngeal and other body cells leading to the death of the nematodes by damaging the neuronal cells. Forward genetic approaches showed that, mutant species of *C. elegans* which are unable to respond in ciliogenesis can escape themselves from *Pleurotus* nematotoxin mediated paralysis. It may be due to the fact that cilia of nematodes induce the release of nematotoxin at the initial stage of interaction (Lee et al., 2020).

Another species of *Pleurotus*, i.e., *P. eryngii* produces a toxin which is effective against *Panagrellus* sp. larvae (Marlin et al., 2019). Nematicidal toxins like S-coriolic acid, linoleic acid, panisaldehyde, and p-anisyl alcohol secreted from *P. pulmonarius* are also found effective in *in vitro* studies (Stadler et al., 1994). Culture filtrate of *P. flabellatus, P. florida and P. sajor-caju* showed nematotoxic effect (Palizi et al., 2009; Shariat et al., 1994).

Crude extract of some species of *Pleurotus*, e.g., *P. citrinopileatus* and *P. florida* showed broad spectrum nematocidal activity (Khan et al., 2014).

Further molecular study is required to discover new nematocidal compounds and the pathway behind the synthesis of these compounds.

7.2.5 SPECIAL STRUCTURES

In addition to the above-mentioned mechanisms, some fungi produce special attacking hyphal devices such as spiny balls in *Coprinus comatus* (Luo et al., 2007) and acanthocytes in *Stropharia* sp. (Luo et al., 2006) which help in adhesion and penetration of the thick cuticle of nematodes. The genus *Hyphoderma* from the order Polyporales produces a special type of attacking device having terminal globular cells surrounded by bands of spines, which resembles a medieval weapon named 'morning star' (Liou & Tzean, 1992; Tzean & Liou, 1993). These structures damage the cuticle of the nematodes by applying mechanical force using the sharp projections allowing extravasation of the inner contents of the nematode body (Soares et al., 2018).

7.3 CONCLUSION

The present study has described the various aspects of nematophagous fungi, including their effects on nematodes, mechanism of nematicidal and nematostatic activity. By looking at the global scenario of nematode induced yield loss, it is required to protect economically sound food crops from the PPNs for securing food security. The fungal diversity of the world is still to be discovered. According to Hawksworth & Lücking (2017), only about 5% of the fungi have been described among 3.8 million species. First of all, proper identification of these fungi is required by different microscopic and molecular approaches. So, future research must be directed in the field of fungal taxonomy. To check the negative impacts of different nematicidal or nematostatic chemicals on the environment and organisms, the alternative way is nowadays emerging. Nematophagous fungi can potentially be considered as suitable alternative to protect numerous economically important crops from the devastating effects of nematodes. Understanding the different biochemical and molecular events during this host-parasite interaction will be helpful for future workers in various ways to design the protection tools. In recent years nematophagous fungi have been extensively studied to exploit them as BCAs against PPNs, but due to lack of knowledge on the ecological perspective of these organisms, we met with very little success. Proper identification and transparent understanding of this host parasite interaction will be helpful in commercializing nematophagous fungi as an alternative of chemical nematicides. The study of signal transduction pathways during nematode and nematophagous fungi interactions through different sophisticated approaches like transcriptomics and proteomics must be undertaken to dissect the underlying mechanisms. The organic amendment and biological control using nematophagous fungi can be a great tool in suppression of PPNs, replacing the pre-existing control strategies like using resistant crop cultivars, crop rotation, using chemical nematicides. So, taking nematophagous fungi as an alternative tool can be portrayed as a sustainable approach in countering nematode infestation for a better future.

KEYWORDS

- **hyphal devices**
- **integrated pest management**
- **molecular approaches**
- **nematophagous fungi**
- **pathogenic organisms**
- **plant parasitic nematodes**
- **polyporales**

REFERENCES

Åhman, J., Ek, B., Rask, L., & Tunlid, A., (1996). Sequence analysis and regulation of a gene encoding a cuticle-degrading serine protease from the nematophagous fungus *Arthrobotrys oligospora*. *Microbiol., 142*(7), 1605–1616.

Allen, M. W., (1957). A review of the nematode genus *Trichodorus* with descriptions of ten new species 1). *Nematol., 2*(1), 32–36.

Anke, H., Stadler, M., Mayer, A., & Sterner, O., (1995). Secondary metabolites with nematicidal and antimicrobial activity from nematophagous fungi and ascomycetes. *Can. J. Bot., 73*(S1), S932-S939.

Anwar, S. A., & Mckenry, M. V., (2010). Incidence and reproduction of *Meloidogyne incognita* on vegetable crop genotypes. *Pak. J. Zool., 42*(2), 135–141.

Barron, G. L., (1992). Lignolytic and cellulolytic fungi as predators and parasites. In: Carroll, G. C., & Wicklow, D. T., (eds.), *The Fungal Community: Its Organization and Role in the Ecosystem*, (2nd edn., Vol. 9, p. 311). Marcel Dekker.

Bird, A. F., & Mcclure, M. A., (1976). The tylenchid (*Nematoda*) eggshell: Structure, composition and permeability. *Parasitol., 72*(1), 19–28.

Bordallo, J. J., Lopez-Llorca, L. V., Jansson, H. B., Salinas, J., Persmark, L., & Asensio, L., (2002). Colonization of plant roots by egg-parasitic and nematode-trapping fungi. *New Phyt., 154*(2), 491–499.

Cayrol, J. C., Castet, R., & Samson, R. A., (1986). Comparative activity of different *Hirsutella* species towards three plant parasitic nematodes. *Rev. Nématol., 9*(4), 412–414.

Chen, J., Li, Q. X., & Song, B., (2020). Chemical nematicides: Recent research progress and outlook. *J. Agric. Food Chem., 68*, 12175–12188.

Chen, T. H., Hsu, C. S., Tsai, P. J., Ho, Y. F., & Lin, N. S., (2001). Heterotrimeric G-protein and signal transduction in the nematode-trapping fungus *Arthrobotrys dactyloides*. *Plant., 212*, 858–863.

Chen, Y. L., Gao, Y., Zhang, K. Q., & Zou, C. G., (2013). Autophagy is required for trap formation in the nematode-trapping fungus *Arthrobotrys oligospora*. *Environ. Microbiol. Rep., 5*(4), 511–517.

Davis, E. L., & Tylka, G. L. (2021). Soybean cyst nematode disease. *The Plant Health Instructor*. DOI: 10.1094/PHI-I-2000-0725-02. *Updated 2021*.

Davis, R. F., & Webster, T. M., (2005). Plant pathology and nematology: Relative host status of selected weeds and crops for *Meloidogyne incognita* and *Rotylenchulus reniformis*. *J. Cotton Sci., 9*(1), 41–46.

Degenkolb, T., & Vilcinskas, A., (2016). Metabolites from nematophagous fungi and nematicidal natural products from fungi as an alternative for biological control. Part I. metabolites from nematophagous ascomycetes. *Appl. Microbiol. Biotechnol., 100*(9), 3799–3812.

Dijksterhuis, J., Veenhuis, M., Harder, W., & Nordbring-Hertz, B., (1994). Nematophagous fungi: Physiological aspects and structure-function relationships. *Adv. Microb. Physiol., 36*, 111–143.

Dong, K., & Opperman, C. H., (1997). Genetic analysis of parasitism in the soybean cyst nematode *Heterodera glycines*. *Genetics, 146*(4), 1311–1318.

Dong, L. Q., & Zhang, K. Q., (2006). Microbial control of plant-parasitic nematodes: A five-party interaction. *Plant Soil., 288*, 31–45.

Eilenberg, J., Hajek, A., & Lomer, C., (2001). Suggestions for unifying the terminology in biological control. *BioCont., 46*, 387–400.

Escobar-Avila, I. M., López-Villegas, E. Ó., Subbotin, S. A., & Tovar-Soto, A., (2018). First report of carrot cyst nematode *Heterodera carotae* in Mexico: Morphological, molecular characterization and host range study. *J. Nematol., 50*(2), 229–242.

Feldman, A. W., & Hanks, R. W., (1968). Phenolic content in the roots and leaves of tolerant and susceptible citrus cultivars attacked by *Radopholus similis*. *Phytochem., 7*, 5–12.

Franklin, M. T., (1950). Two species of *Aphelenchoides* associated with strawberry bud disease in Britain. *Appl. Biol., 37*(1), 1–10.

Fresenius, G., (1863). *Beitrage zur Mykologie (Contributions to Mycology)*. The British Library.

Gams, W., & Zare, R., (2003). *A taxonomic review of the Clavicipitaceous Anamorphs Parasitizing Nematodes and Other Microinvertibrates*. In: James, F. W., Charles, W. B., Nigel, L. H., & Joseph, W. S., (eds.), *Clavicipitalean Fungi: Evolutionary Biology, Chemistry, Biocontrol and Cultural Impacts* (p. 17). Marcel-Dekker.

Hallmann, J., Keith, G. D., & Sikora, R., (2009). Biological control using microbial pathogens, endophytes and antagonists. In: Perry, R. N., Moens, M., & Starr, J. L., (eds.), *Root-Knot Nematodes* (p. 380). CABI.

Hawksworth, D. L., & Lücking, R., (2017). Fungal diversity revisited: 2.2 to 3.8 million species. In: Heitman, J., Howlett, B. J., Crous, P. W., Stukenbrock, E. H., James, T. Y., & Gow, N. A. R., (eds.), *The Fungal Kingdom* (p. 79). American Society for Microbiology, Washington, DC.

Hildalgo-Diaz, L., & Kerry, B. R., (2008). Integration of biological control with other methods of nematode management. In: Ciancio, A., & Mukerji, K. G., (eds.), *Integrated Management and Biocontrol of Vegetable and Grain Crops Nematodes* (Vol. 4, No. 1, p. 29). Springer.

Hussain, M., Zouhar, M., & Ryšánek, P., (2017). Comparison between biological and chemical management of root knot nematode, *Meloidogyne hapla*. *Pakistan J. of Zool., 49*(1), 205–210.

Jansson, H. B., & Lopez-Llorca, L. V., (2001). *Biology of nematophagous fungi*. In: Misra, J. K., & Horn, B. W., (eds.), *Trichomycetes and Other Fungal Groups* (p. 145). Science publishers Inc.

Jansson, H. B., & Nordbring, H. B., (1983). The endoparasitic nematophagous fungus *Meria coniospora* infects nematodes specifically at the chemosensory organs. *J. Gen. Microbiol., 129*(4), 1121–1126.

Jansson, H. B., & Nordbring-Hertz, B., (1988). Infection events in the fungus-nematode system. In: Poinar, G. O., & Jansson, H. B., (eds.), *Diseases of Nematodes* (Vol. 2, p. 59). CABI.

Jansson, H. B., & Thiman, L., (1992). A preliminary study of chemotaxis of zoospores of the nematode-parasitic fungus *Catenaria anguillulae*. *Mycologia, 84*(1), 109–112.

Jansson, H. B., (1993). Adhesion to nematodes of conidia from the nematophagous fungus *Drechmeria coniospora*. *J. Gen. Microbiol., 139*(8), 1899–1906.

Jansson, H. B., (1994). Adhesion of conidia of *Drechmeria coniospora* to *Caenorhabditis elegans* wild type and mutants. *J. Nematol., 26*(4), 430–435.

Jansson, H. B., Jeyaprakash, A., & Zuckerman, B. M., (1985). Control of root-knot nematodes on tomato by the endoparasitic fungus *Meria coniospora*. *J. Nematol., 17*(3), 327–329.

Jansson, H. B., Jeyaprakash, A., & Zuckerman, B. M., (1985). Differential adhesion and infection of nematodes by the endoparasitic fungus *Meria coniospora (Deuteromycetes)*. *Appl. Environ. Microbiol., 49*(3), 552–555.

Jatala, P., (1986). Biological control of plant-parasitic nematodes. *Ann. Rev. Phytopathol., 24*(17), 453–489.

Kerry, B., (1988). Fungal parasites of cyst nematodes. *Agric. Ecos. Environ., 24*(1–3), 293–305.

Khan, A., Iqbal, M., Hussain, S., Pakhtunkhwa, K., & Pakhtunkhwa, K., (2014). Organic control of phytonematodes with *Pleurotus* species. *Pakistan J. Nematol., 32*(2), 155–161.

Khan, F. A., & Khan, A. M., (1973). Studies on the reniform nematode, *Rotylenchulus reniformis*. I. Host range and population changes. *Indian J. Nematol., 3*, 24–30.

Khan, M. R., (2015). Nematode diseases of crops in India. In: Awasthi, L. P., (ed.), *Recent Advances in the Diagnosis and Management of Plant Diseases* (p. 183). Springer India.

Khan, M. R., Handoo, Z. A., Rao, U., Rao, S. B., & Prasad, J. S., (2012). Observations on the foliar nematode, *Aphelenchoides besseyi*, infecting tuberose and rice in India. *J. Nematol., 44*(4), 391–398.

Koen, H., (1967). Notes on the host range, ecology and population dynamics of *Pratylenchus brachyurus*. *Nematologica, 13*(1), 118–124.

Koshy, P. K., & Swarup, G., (1971). Distribution of *Heterodera avenae, H. zeae, H. cajani, and Anguina tritici* in India. *Indian J. Nematol., 1*, 106–111.

Kumar, V., Khan, M. R., & Walia, R. K., (2020). Crop loss estimations due to plant-parasitic nematodes in major crops in India. *Natl. Acad. Sci. Lett.* [Online], https://link.springer.com/article/10.1007/s40009-020-00895-2 (accessed on 13 January 2023).

Kwok, O. C. H., Plattner, R., Weisleder, D., & Wicklow, D. T., (1992). A nematicidal toxin from *Pleurotus ostreatus* NRRL 3526. *J. Chem. Ecol., 18*(2), 127–136.

Lee, C. H., Chang, H. W., Yang, C. T., Wali, N., Shie, J. J., & Hsueh, Y. P., (2020). Sensory cilia as the Achilles heel of nematodes when attacked by carnivorous mushrooms. *PNAS, 117*(11), 6014–6022.

Li, J., Zou, C., Xu, J., Ji, X., Niu, X., Yang, J., Huang, X., & Zhang, K. Q., (2015). Molecular mechanisms of nematode-nematophagous microbe interactions: Basis for biological control of plant-parasitic nematodes. *Annu. Rev. Phytopathol., 53*, 67–95.

Li, X. Q., Wei, J. Z., Tan, A., & Aroian, R. V., (2007). Resistance to root-knot nematode in tomato roots expressing a nematicidal *Bacillus thuringiensis* crystal protein. *Plant Biotech. J., 5*, 455–464.

Liang, L. M., Zou, C. G., Xu, J., & Zhang, K. Q., (2019). Signal pathways involved in microbe-nematode interactions provide new insights into the biocontrol of plant-parasitic nematodes. *Philos. Trans. R. Soc. B., 374*, 1–8.

Liang, L., Liu, Z., Liu, L., Li, J., Gao, H., Yang, J., & Zhang, K. Q., (2016). The nitrate assimilation pathway is involved in the trap formation of *Arthrobotrys oligospora*, a nematode-trapping fungus. *Fungal Genetics and Biol., 92*, 33–39.

Liou, J. Y., & Tzean, S. S., (1992). Stephanocysts as nematode-trapping and infecting propagules. *Mycologia, 84*(5), 786–790.

Liu, M., Cheng, X., Wang, J., Tian, D., Tang, K., Xu, T., Zhang, M., et al., (2020). Structural insights into the fungi-nematodes interaction mediated by fucose-specific lectin AofleA from *Arthrobotrys oligospora. Int. J. Biol. Macromol., 164*, 783–793.

Lopez-Llorca, L. V., Maciá-Vicente, J. G., & Jansson H. B., (2008). Mode of action and interactions of nematophagous fungi. In: Ciancio, A., & Mukerji, K. G., (eds.), *Integrated Management and Biocontrol of Vegetable and Grain Crops Nematodes* (Vol. 4, No. 1, p. 51–71). Springer.

Luo, H., Li, X., Li, G., Pan, Y., & Zhang, K., (2006). Acanthocytes of *Stropharia rugosoannulata* function as a nematode-attacking device. *Appl. Environ. Microbiol., 72*(4), 2982–2987.

Luo, H., Liu, Y., Fang, L., Li, X., Tang, N., & Zhang, K., (2007). *Coprinus comatus* damages nematode cuticles mechanically with spiny balls and produces potent toxins to immobilize nematodes. *Appl. Environ. Microbiol., 73*(12), 3916–3923.

Marlin, M., Wolf, A., Alomran, M., Carta, L., & Newcombe, G., (2019). Nematophagous *Pleurotus* species consume some nematode species but are themselves consumed by others. *Forests, 10*(5), 1–11.

Mcclure, M. A., & Bird, A. F., (1976). The tylenchid (*Nematoda*) eggshell: Formation of the eggshell in *Meloidogyne javanica. Parasitol., 72*, 29–39.

Mehta, A. K., & Shah, A., (2003). Chronic poverty in India: Incidence, causes and policies. *World Dev., 31*(3), 491–511.

Mesa-Valle, C. M., Garrido-Cardenas, J. A., Cebrian-Carmona, J., Talavera, M., & Manzano-Agugliaro, F., (2020). Global research on plant nematodes. *Agronomy, 10*(1148), 1–11.

Monteiro, T. S. A., Lopes, E. A., & Evans, H. C., (2017). Interactions between *Pochonia* chlamydosporia and nematodes. In: Manzanilla-López, R. H., & Lopez-Llorca, L. V., (eds.), *Perspectives in Sustainable Nematode Management Through Pochonia chlamydosporia Applications for Root and Rhizosphere Health* (p. 77). Springer.

Moosavi, M. R., Zare, R., Zamanizadeh, H. R., & Fatemy, S., (2010). Pathogenicity of *Pochonia* species on eggs of *Meloidogyne javanica. J. Invert. Pathol., 104*, 125–133.

Morton, C. O., Mauchline, T. H., Kerry, B. R., & Hirsch, P. R., (2003). PCR-based DNA fingerprinting indicates host-related genetic variation in the nematophagous fungus *Pochonia chlamydosporia*. *Mycol. Res., 107*(2), 198–205.
Niu, X. M., & Zhang, K. Q., (2011). *Arthrobotrys oligospora*: A model organism for understanding the interaction between fungi and nematodes. *Mycology, 2*(2), 59–78.
Nordbring-Hertz, B., & Mattiasson, B., (1979). Action of a nematode-trapping fungus shows lectin-mediated host-microorganism interaction. *Nature, 281*, 477–479.
Nordbring-Hertz, B., (1973). Peptide-induced morphogenesis in the nematode-trapping fungus *Arthrobotrys oligospora*. *Physiol. Plant., 29*(2), 223–233.
Nordbring-Hertz, B., (1977). Nematode-Induced morphogenesis in the predacious fungus *Arthrobotrys oligospora*. *Nematologica, 23*(4), 443–451.
Nordbring-Hertz, B., Jansson, H. B., & Tunlid, A., (2006). Nematophagous fungi. *eLS.*, 1–11.
Ntalli, N., & Caboni, P., (2017). A review of isothiocyanates biofumigation activity on plant parasitic nematodes. *Phytochem. Rev., 16*(5), 827–834.
Ouiza, D. Z., & Samira, S., (2017). Effects of soil solarization and sincocin treatments for controlling *Meloidogyne Javanica* (*Nematoda: Meloidogynidae*) in eggplant (*Solanum melongena* L.). *PONTE Int. Sci. Res. J., 73*(12/1), 271–279.
Palizi, P., Goltapeh, E. M., Pourjam, E., & Safaie, N., (2009). Potential of oyster mushrooms for the biocontrol of sugar beet nematode (*Heterodera schachtii*). *J. Plant Protect. Res., 49*(1), 27–33.
Pinochet, J., Verdejo, S., Soler, A., & Canals, J., (1992). Host range of a population of *Pratylenchus vulnus* in commercial fruit, nut, citrus, and grape rootstocks in Spain. *J. Nematol., 24*(4S), 693–698.
Poirier, S., Dauphinais, N., Van, D. H. H., Véronneau, P. Y., Bélair, G., Gravel, V., & Mimee, B., (2019). Host range and genetic characterization of *Ditylenchus dipsaci* populations from Eastern Canada. *Plant Dis., 103*(3), 456–460.
Pramer, D., & Stoll, N., (1959). Nemin: A morphogenic substance causing trap formation by predaceous fungi role of myocardial catecholamines in cardiac contractility. *Sci., 129*, 966, 967.
Pramer, D., (1964). Nematode-trapping fungi. *Sci., 144*, 382–388.
Ralmi, N. H. A. A., Khandaker, M. M., & Mat, N., (2016). Occurrence and control of root knot nematode in crops: A review. *AJCS, 10*(12), 1649–1654.
Ringer, C. E., Sardanelli, S., & Krusberg, L. R., (1987). Investigations of the host range of the corn cyst nematode, *Heterodera zeae*, from Maryland. *Ann. Appl. Nematol., 1*, 97–106.
Rosen, S., Ek, B., Rask, L., & Tunlid, A., (1992). Purification and characterization of a surface lectin from the nematode-trapping fungus *Arthrobotrys oligospora*. *J. Gen. Microbiol., 138*, 2663–2672.
Saikawa, M., (1982). An electron microscope study of *Meria coniospora*, an endozoic nematophagous hyphomycete. *Can. J. Bot., 60*, 2019–2023.
Shariat, H. S., Farid, H., & Kavianpour, M., (1994). A study of anthelmintic activity of aqueous extract of *Pleurotus eryngii* on *Syphacia obvelata* and *Hymenolepis nana*. *J. Sci. I. R. Iran, 5*(1, 2), 19–22.
Siddiqui, Z. A., & Mahmood, I., (1996). Biological control of plant parasitic nematodes by fungi: A review. *Biores. Tech., 58*, 229–239.
Sjollema, K. A., Dijksterhuis, J., Veenhuis, M., & Harder, W., (1993). An electron microscopical study of the infection of the nematode *Panagrellus redivivus* by the endoparasitic fungus *Verticillium balanoides*. *Mycol. Res., 97*(4), 479–484.

Soares, F. E. F., Sufiate, B. L., & Queiroz, J. H., (2018). Nematophagous fungi: Far beyond the endoparasite, predator and ovicidal groups. *Agric. Nat. Res., 52*, 1–8.

Speijer, P. R., & Sikora, R. A., (1991). *Influence of a Complex Disease Involving Pratylenchus Goodeyi and a Non-Pathogenic Strain of Fusarium oxysporum on Banana Root Health* (p. 231). CABI.

Stadler, M., Mayer, A., Anke, H., & Sterner, O., (1994). Fatty acids and other compounds with nematicidal activity from cultures of basidiomycetes. *Planta Med., 60*(2), 128–132.

Stynes, B. A., & Bird, A. F., (1980). *Anguina Agrostis*, the vector of annual rye grass toxicity in Australia. *Nematol., 26*(4), 475–490.

Timper, P., Kaya, H. K., & Jaffee, B. A., (1991). Survival of entomogenous nematodes in soil infested with the nematode-parasitic fungus *Hirsutella rhossiliensis* (*Deuteromycotina*: *Hyphomycetes*). *Biol. Cont., 1*, 42–50.

Trivedi, P., & Barker, K., (1986). Management of nematodes by cultural practices. *Nematropica, 16*(2), 213–236.

Trudgill, D. L., Kerry, B. R., & Phillips, M. S., (1992). Integrated control of nematodes (With particular reference to cyst and root knot nematodes). *Nematologica, 38*, 482–487.

Tzean, S. S., & Liou, J. Y., (1993). Nematophagous resupinate basidiomycetous fungi. *Phytopathol., 83*(10), 1015–1020.

Ulzurrun, V. G., & Hsueh, Y. P., (2018). Predator-prey interactions of nematode-trapping fungi and nematodes: Both sides of the coin. *Appl. Microbiol. Biotech., 102*, 3939–3949.

Viaene, N., Coyne, D. L., & Kerry, B. R., (2006). Biological and cultural management. In: Perry, R. M., & Moens, M., (eds.), *Plant Nematology*. CABI.

Wang, B. L., Chen, Y. H., He, J. N., Xue, H. X., Yan, N., Zeng, Z. J., Bennett, J. W., et al., (2018). Integrated metabolomics and morphogenesis reveal volatile signaling of the nematode-trapping fungus *Arthrobotrys oligospora*. *Appl. Environ. Microbiol., 84*(9), 1–19.

Wang, B., Liu, X., Wu, W., Liu, X., & Li, S., (2009). Purification, characterization, and gene cloning of an alkaline serine protease from a highly virulent strain of the nematode-endoparasitic fungus *Hirsutella rhossiliensis*. *Microbiol. Res., 164*, 665–673.

Wang, R. B., Yang, J. K., Lin, C., Zhang, Y., & Zhang, K. Q., (2006). Purification and characterization of an extracellular serine protease from the nematode-trapping fungus *Dactylella shizishanna*. *Let. Appl. Microbiol., 42*, 589–594.

Wang, Y. L., Li, L. F., Li, D. X., Wang, B., Zhang, K., & Niu, X., (2015). Yellow pigment aurovertins mediate interactions between the pathogenic fungus *Pochonia chlamydosporia* and its nematode host. *J. Agric. Food Chem., 63*, 6577–6587.

Wharton, D. A., & Jenkins, T., (1978). Structure and chemistry of the eggshell of a nematode (*Trichuris suis*). *Tissue and Cell, 10*(3), 427–440.

Wharton, D., (1980). Nematode eggshells. *Parasitol., 81*, 447–463.

Wuyts, N., Lognay, G., Verscheure, M., Marlier, M., Waele, D. D., & Swennen, R., (2007). Potential physical and chemical barriers to infection by the burrowing nematode *Radopholus similis* in roots of susceptible and resistant banana (*Musa* spp.). *Plant Pathol., 56*, 878–890.

Wyss, U., Voss, B., & Jansson, H. B., (1992). *In vitro* observations on the infection of *Meloidogyne incognita* eggs by the zoosporic fungus *Catenaria anguillulae*. *Fundam. Appl. Nematol., 15*(2), 133–139.

Xu, L. L., Lai, Y. L., Wang, L., & Liu, X. Z., (2011). Effects of abscisic acid and nitric oxide on trap formation and trapping of nematodes by the fungus *Drechslerella stenobrocha* AS6.1. *Fungal Biol., 115*, 97–101.

Yang, J., Li, J., Liang, L., Tian, B., Zhang, Y., Cheng, C., & Zhang, K. Q., (2007). Cloning and characterization of an extracellular serine protease from the nematode-trapping fungus *Arthrobotrys conoides*. *Arch. Microbiol., 188*, 167–174.

Yang, J., Liang, L., Zhang, Y., Li, J., Zhang, L., Ye, F., Gan, Z., & Zhang, K. Q., (2006). Purification and cloning of a novel serine protease from the nematode-trapping fungus *Dactylellina varietas* and its potential roles in infection against nematodes. *Appl. Microbiol. Biotechnol., 75*, 557–565.

Zasada, I. A., Ingham, R. E., & Phillips, W. S., (2015). Biological insights into *Globodera ellingtonae*. *Asp. Appl. Biol., 130*, 1–8.

Zhang, L., Zhou, Z., Guo, Q., Fokkens, L., Miskei, M., Pócsi, I., Zhang, W., et al., (2016). Insights into adaptations to a near-obligate nematode endoparasitic lifestyle from the finished genome of *Drechmeria coniospora*. *Sci. Rep.*, 1–15.

Zuckerman, B. M., Dicklow, M. B., Coles, G. C., & Jansson, H. B., (1988). Cryopreservation studies on the nematophagous fungus *Drechmeria coniospora*. *Revue Nématol., 11*(3), 327–331.

CHAPTER 8

Role of Fungi in Postharvest Disease Management in Horticultural Crops

RIMA KUMARI[1], PANKAJ KUMAR[1], and ARUN KUMAR[2]

[1]*Department of Botany, Purnea University, Purnia, Bihar, India*

[2]*Department of Agronomy, Bihar Agricultural University, Sabour, Bihar, India*

ABSTRACT

Postharvest diseases are the major factors that cause quality losses in horticultural crops. During the postharvest storage period, horticultural crops suffer from a variety of losses, including rotting, sprouting, physiological weight loss, nutritional loss, and moisture loss. After the crop is harvested, it becomes vulnerable to many postharvest diseases generally caused by bacterial and fungal pathogens. In India, every year, 25–30% of horticulture production and 10–15% of vegetable production is wasted due to lack of postharvest which resulting in massive losses. These postharvest losses are a big issue in various nations. The constant use of fungicides (such as benomyl, streptocyclin, and carbendazim, etc.), not only in the field but also during postharvest stage period has led to microbial resistance cases, which cause the difficulty for controlling many pathogens. In this regard, biological control is the most promising alternative to the use of chemical fungicides. Furthermore, the most effective and an eco-friendly approach for controlling diseases is the use of microbial antagonist fungi like *Aspergillus*, *Trichoderma*, *Gliocladium*, *Penicillium*, *Chaetomium*, *Neurospora*, *Dactylella*, *Glomus*, and *Arthrobotrys*, etc., which have various mechanisms of action to control disease development of horticultural crops.

Applied Mycology for Agriculture and Foods: Industrial Applications. Sanjay K. Singh, Deepak Kumar, Md. Shamim, & Rohit Sharma (Eds.)

8.1 INTRODUCTION

Globally, India is the second largest producer of vegetables and third largest producer of fruits. But the postharvest diseases reveal a major problem worldwide, causing significant losses during a postharvest stage in horticultural crops (Kumar & Iqbal, 2020). The estimated losses are more and reached from 2% to 30%. Horticultural crops are facing maximum trouble during postharvest storage. Tremendous amount of microorganisms has been reported to be responsible for rotting in horticultural crops and among several entities, responsible for postharvest decay, and further among various microorganisms, postharvest fungi is the main causal agent for maximum spoilages of horticultural crops (Currah & Proctor, 1990). More than 40% postharvest losses of horticultural crops have been reported on global level due to different fungi, bacteria, viruses, and protozoa diseases. Fatima et al. (2015) also reported the significant losses in horticultural crops during marketing due to postharvest diseases.

Traditionally, management of postharvest disease in horticultural crops is carried out by the use of many chemical fungicides; however, environmental and health issues as well as microbial resistance play significant role in the development of new approaches to control diseases. Further, biological control approach is a promising alternative to control postharvest disease. But, the applications of different fungi like *Trichoderma*, *Aspergillus*, *Gliocladium*, *Chaetomium*, *Penicillium*, *Neurospora*, *Dactylella*, *Glomus*, and *Arthrobotrys*, etc., are considered as the most promising approach due to their effectiveness for controlling various pathogens in many fruits like banana (*Colletotrichum musae*), strawberry (*Botrytis cinerea*), kiwifruit (*Botrytis cinerea*), citrus (*Penicillium italicum*), guava (*Rhizopus* spp.) and among vegetables. Thus, taking all above into consideration, the present chapter is written to represent the role of fungi in postharvest diseases management in horticultural crops.

8.2 POSTHARVEST DISEASES IN HORTICULTURAL CROPS

India is second in terms of total world fruit and vegetable production, accounting for 10.9% and 11.9% of the total worldwide production. In India, mangoes, banana, and papaya are mostly produced. Additionally, 41.7% of the world's mangoes and 25.7% of the world's bananas are grown in India. In India, grape productivity per unit area is the greatest in the world. The amount of fruits that has been consumed per person each day has risen from

40 grams to 85 grams, while the amount of vegetables has risen from 96 grams to 175 grams in the same decade. Around 5% to 50% or more of the harvest can be lost after harvest. Researchers have estimated that about 30% of fruit and vegetable losses after harvest are caused by postharvest illnesses (Yadav et al., 2013). Horticulture crops may experience a variety of postharvest infections, such as *Alternaria Rot*, *Phytophthora Rot*, *Penicillium Rots*, and *Rhizopus Rot*. Between 20% and 50% of perishable produce is lost because to postharvest illnesses in India.

8.3 POSTHARVEST LOSSES: CAUSES AND CONSEQUENCES

According to a recent study, over 1,300 million tons of food is lost globally each year. Postharvest processing can reduce the amount of produce harvested by 25% to 60%. Diseases and mechanical damage produced by microorganisms that play a key part are just a few of the many elements involved in this only (Estrada et al., 2018). Modern approaches, such as the use of chemical fungicides, are commonly employed for postharvest disease management; nevertheless, they are also harmful to the environment because of their toxicity. Additionally, it has been reported that resistance against microorganisms has increased, decreasing the antifungal impact and complicating control; hence, new, secure, and effective alternatives must be found. Food spoilage due to fungi and bacteria is most commonly seen during handling of produce in the field, after harvest, during shipping and storage, and finally as a result of product deterioration, with reduced supplies and financial losses (Estrada et al., 2018).

8.4 PREHARVEST FACTORS THAT INFLUENCE POSTHARVEST DISEASES

8.4.1 WEATHER

Weather is a significant component when it comes to crop diseases since it influences inoculum and pesticide residue, both of which are linked to plant illnesses. By the time the food is harvested, sufficient inoculum and favorable conditions for infection frequently result in excessive infection. In this case, we have the fungus known as bulls-eye rot, which is rain-dispersed from cankers to infect the bark and cause rotting fruit in storage if rainfall is sustained when fruit is being harvested (Singh et al., 2017).

8.4.2 PHYSIOLOGICAL CONDITION

How long the crop can be preserved depends on the physiological condition of the crop during harvesting. When doing so, pick fruit that is slightly immature in order to ensure that it can be stored for several months. In many fruits, ripening and senescence encourage pathogen growth. Conversely, crop nutrition can be used to help control fruit rot.

8.4.3 FUNGICIDE SPRAYS

Pre-harvest fungicide applications in this situation are unlikely to have a significant impact on storage deterioration. Many research studies have found that pre-harvest ziram fungicide application is quite successful in preventing decay in pome fruit, with an average reduction in decay of 25% to 50% (Coates et al., 1995). Pre-harvest spray application of the organic compound Iprodione has been in use for many years to help keep stone fruit safe from infection by the fungus *Monilinia* spp. Also, since wax and/or oil are utilized in combination with it, the control range for decay is enhanced, and postharvest fungi such as *Alternaria* and *Rhizopus* are controlled. Postharvest control of various diseases is expected with the new class of strobilurin fungicides. This particular method is extremely useful in the fight against apple scab and is useful in preventing the appearance of pin-point scab in storage.

8.5 USE OF SYNTHETIC FUNGICIDE FOR POSTHARVEST DISEASE CONTROL IN HORTICULTURAL CROP

Horticultural crops are mostly subject to disease and decay due to fungus and bacteria. Inhibition of postharvest diseases in horticulture crops is aided by the application of different synthetic fungicide. Not only do agricultural pesticides such as benomyl, streptocyclin, and carbendazim, among others, remain in the field after application, but they are also used throughout the postharvest period, and so promote the development of microbial resistance. There is a substantial alternative to the use of chemical fungicide called biological control. However, using several fungi offers the best hope for managing a variety of pathogenic microbes in horticulture crops.

The team of Futane et al. (2018) studied the antifungal properties of fungi *in vitro*. According to the results, the active ingredients carbendazim (94.55%), safflor (94.55%), and mancozeb (90.98%) exhibited more fungicidal activity than carbendazim (94.55%), safflor (94.55%), and mancozeb (90.98%) with higher mycelial growth inhibitions. With considerably higher mycelial growth suppression of 91.74%, 91.38%, 90.81%, and 88.42%, respectively, SAFF, carbendazim, hexaconazole, and mancozeb were shown to be the most effective against *Fusarium oxysporum*. According to the researchers from Prajapati et al. (2016), both non-systemic fungicides and systemic fungicides showed similar efficiency when used *in vitro* against *Aspergillus niger* utilizing the "Poisoned food methodology method" with adjusted applications (500, 1,000, 1,500, and 2,000 ppm). During the seven days following incubation, the inhibition percentage was observed. *In vitro* and *in vivo* research was undertaken for reducing black mold rot of onion using synthetic fungicides. Among the fungicides tested, the following ingredients ranked highest: Carbendazim (18.2% w/w), Azoxystrobin (11.4% w/w) + Difenconazole (11.4% w/w), Carbendazim (12% w/w) + Mancozeb (63% w/w), Mancozeb (50% w/w) + Carbendazim (25% w/w), Trifloxystrobin (25% w/w) + Tebuconazole (50% w/w) in test tubes at 500 and 100 ppm. Mycelial growth was studied in the lab to identify different sodium salts that would have differing influences on fungal growth. This study found a difference in the sodium salts' impacts on mycelial growth at the p 0.05 level, and 2% (w/v) concentrations of sodium metabisulfite and sodium fluoride completely inhibited fungus growth, while other salts did not. Turkkan & Erper (2014) found that various sodium salts were equally effective in controlling *Fusarium oxysporum* basal rot in onions as synthetic fungicides. According to a 2012 study published in the Journal of Food Science, if you want to inhibit the *A. niger* pathogen, choose Carbendazim 50% WP, which is more effective than the popular Bavistin (Carbendazim 50% WP, 2.0%). However, in contrast, Thiram (80% WP, 2.5%) was more effective at controlling the fungus, while Captan (50% WP, 2.5%) was much less effective (75% WP, 2.5%). In 2007, Dugan et al. isolated the fungal pathogen *Aureobasidium niger*, the fungal pathogen *Aureobasidium ochraceum*, *Fusarium oxysporum* f. sp. Cepae, *Embellisia allii*, *Beauveria bismarkii*, *Pityrosporum hirsutum*, and *Fusarium proliferatum* from diseased garlic. Of the species evaluated for pre-planting and postharvest dips, *F. oxysporum* f. sp. *cepae* and *F. proliferatum* were among the species studied. For preplant dip tests, the pesticides fludioxonil, thiophanatemethyl dimethyl, and benomyl were employed, whereas for postharvest dip tests, the pesticide benomyl was utilized. These synthetic fungicides were used to protect hosts against disease before major contamination. While the efficacy of control was low when postharvest illness was severe or when garlic was already infected, a lower efficacy of control was observed when postharvest disease was severe or when garlic was already infected (2007). According to Srinivasan & Shanmugam (2006), *A. niger* was primarily responsible for black mold rot in onions during storage, and that acetic acid and sulfur oxides were found to be the most effective fungicides when used as foliar sprays, and carbandizim at a concentration of 0.1% was found to be the most effective when used as a postharvest dip. Fungal pathogen disease management via the use of synthetic pesticides for postharvest diseases in vegetable crops has been extensively explored (Jaime et al., 2001; McDonald et al., 2004; Raju & Naik, 2006). A combined treatment of thiabendazole (TBZ), imazalil (IMZ), and iprodione decreased the risk of *Aspergillus* and Botrytis species growing. Padule & colleagues (1996) found that postharvest carbendazim (0.1%), mancozeb (0.25%), captain (0.2%), streptocycline (0.5%), and fumigation with sulfur fumigation (0.4%) were better at inhibiting fungal deterioration than were several treatments that required additional applications.

8.6 ROLE OF FUNGI FOR POSTHARVEST DISEASES MANAGEMENT IN HORTICULTURAL CROPS

The important genera of fungi such as *Trichoderma*, *Aspergillus*, *Gliocladium*, *Penicillium*, *Neurospora*, *Dactylella*, *Chaetomium*, *Glomus*, and *Arthrobotrys*, etc., are used as biocontrol agents (BCAs) in horticultural crops. According to Harman (2010), biological control is a critically needed component for disease management in several horticultural crops. Biocontrol agent is known as antagonist. The most significant, well-studied antagonists against various horticultural crop pathogens are fungi like *Ampelomyces* sp., *Aspergillus* spp. (particularly *A. terreus* and *A. niger*), *Coniothyrium minitans*, *Fusarium* sp., *Chaetomium globosum*, *Gliocladium virens*, *Peniophora gigantea*, *Penicillium citrinum*, *Trichoderma* spp. (particularly *T. viride* and *T. harzianum*) and *Sporodesmium* sp.

8.7 USE OF *TRICHODERMA* SPP. FOR POSTHARVEST DISEASE CONTROL IN HORTICULTURAL CROPS

Trichoderma fungi are saprophytic and free-living, yet they possess a strong affinity for soil, roots, and foliar environments. Fungi from the genus *Trichoderma* were discovered in the 1940s to be hostile. Another substantial advancement was made in the previous few years to increase the effectiveness of *Trichoderma* sp. as a biological control agent. BCAs and commercial utilization of *Trichoderma* products were discussed by Tripathi et al. (2010) (Table 8.1). *Trichoderma* is a widely distributed genus that can be found in many different types of antifungal fungal antagonists. The prevalence of fungi (as represented by its total percentage) in forests is 3% of the total, while the prevalence of fungi in other soils is 1.5% of the whole. Additionally, the pathogen competes with other plant pathogenic fungi for nutrients and room with the other fungi's exudates that spur the development of their spores in soil, as well as with other soil microbes for elements such as nitrogen and phosphate. The use of *Trichoderma* spp. in the fight against a wide range of pathogen-borne, airborne, and soil-borne plant diseases. In practice, these practices have been used on various fruit trees, which include plum, peach, and nectarine trees. Additionally, these practices have been applied to a variety of trees, which include Dutch elm disease on elms, honey fungus (*A. mellea*) on a variety of species, and fungal diseases caused by *Fusarium*, *Pythium*, *Rhizoctonia*, and *Sclerotium* on a variety of crops. According to Lacicowa & Pieta (1994), *Trichoderma* and *Gliocladium* species, both of which belong to the *Basidiomycota*, were shown to be better at controlling soil-borne pathogenic fungi of peas than were chemical pesticides. *Trichoderma* sp. described the mode of action of plant infections to Spiers et al. (2004). The production of mycoparasitism related genes (MRGs), antibiosis, the importance of MRGs in biological control, and *Trichoderma* spp. as a source of genes in crop improvement are now under discussion in Herrera-Estrella & Chet (2004). Because degrading pectinases and other enzymes are essential for pathogenic fungi to produce phytopathology in plants, biocontrol measures are essentially based on these fundamental qualities. While previously noted to be the basis for *Trichoderma* sp. actions on plant growth and development, their direct effects on other fungi are surprising. *Trichoderma* has been reported to affect plant pathogens in many different ways. All of these practices are done in the fight against the destruction of the natural world, including competition, mycoparasitism, antibiosis, siderophore production, systemic resistance induction, growth promotion, and others (Dennis & Webster, 1971; Upadhyay & Mukhopadhyay, 1986; Chet, 1987).

For the past decade, fungal inhibition has been used by numerous researchers. *Trichoderma* species have a strong influence on different storage fungi. *Trichoderma* species from Cherkupally et al. (2017) shown an antagonistic response against the pathogenic fungus, e.g., *Fusarium oxysporum in vitro*. *Trichoderma* species' antagonistic activity was selected

TABLE 8.1 Commercial Use of *Trichoderma* Products in Horticultural Crops

Product	Biocontrol Agent	Effective Against
Antifungus	*Trichoderma* sp.	Various fungi
Basderma	*T. viride* Basarass	Various fungi
Binab T	*T. polysporum* (ATCC 20475) and *T. harzianum* (ATCC 20476)	Control of wound decay and wood rot
Bioderma	*T. harzianum/T. viride*	Various fungi
Biofungus	*Trichoderma* sp.	*Phytophthora*, *Sclerotinia*, *Pythium* spp., *Fusarium*, *Verticillium*, *Rhizoctonia solani*,
Bio-trek 22G	*T. harzianum*	Various fungi
Ecofit	*T. viride*	Various fungi
Root pro and Root care	*T. harzianum*	*Pythium* spp., *R. solani*, *Sclerotium rolfsii*, and *Fusarium* spp.
Root shield, Plant shield	*T. harzianum*	*R. solani* and *Pythium* spp.
Root shield, T-22	*T. harzianum*	*Pythium* spp., *R. solani*, and *Fusarium* spp.
Planter Bo and plant shield	*Rifai* strain	
	KRL-AG (T-22)	
SoilGard	*Trichoderma* sp.	Damping-off diseases caused by *Rhizoctonia* spp. and *Pythium*
Supresivit	*T. harzianum*	Various fungi
T-22 HB and T-22 G	*T. harzianum* strain	Various fungi
	KRL-AG2	
Trichoderma 2000	*Trichoderma* spp.	*S. rolfsii*, *R. solani*, *Fusarium* spp., *Pythium* spp.
Trichophel and Trichodex	*T. harzianum*	Botrytis of grapevines and vegetables
Trichophel, Trichoject, Trichodowels, Trichoseal	*T. viride* and *T. harzianum*	*Botryosphaeria*, *Chondrosternum*, *Nectria*, *Fusarium*, *Rhizoctonia*, *Pythium*, *Phytophthora*

TABLE 8.1 *(Continued)*

Product	Biocontrol Agent	Effective Against
Tri-control	*Trichoderma* sp.	Various fungi
Trieco	*T. viride*	*Pythium* spp., *Rhizoctonia* spp., *Fusarium* spp., root rot, seedling rot, collar rot, red rot and damping-off Fusarium wilt
TY	*Trichoderma* spp.	Various fungi
Tusal	*Trichoderma* spp.	Damping-off diseases caused by *Phoma*, *Pythium*, and *Rhizoctonia* species

Source: Modified from: Tripathi et al. (2010).

in the lab. Fungal pathogens dropped dramatically in all BCAs. For non-volatile chemicals, *T. harzianum* was found to be the most inhibited with an 81.11% inhibition rate, followed by *T. koningii* at 80%, *T. viride* and *T. pseudokoningii* at 78.88%, and *T. atroviride*, *T. virens*, and *T. reesei* at 77.77%. The highest antagonistic potential was discovered in *T. harzianum*, with an effectiveness of 28.88%. In order to find out which biocontrol agent is most effective at preventing fungi like *P. oryzae*, Triveni et al. (2012) conducted various *in vitro* experiments, including tests on *T. pseudokoningii*, *Trichoderma harzianum*, *T. polysporum*, *Paecilomyces variotii, Gliocladium virens*, and *P. lilacinus*. They discovered that *P. lilacinus* was most effective at inhibiting fungus *P. oryzae*. *In vitro* tests were conducted by Reddy et al. (2014) to find Trichoderma strains with strong ability to inhibit a variety of plant diseases, including *Aspergillus niger*, *Alternaria solani*, *Macrophomina phaseolina*, and *Fusarium oxysporum*. Kumar et al. (2012) discovered that 12 filamentous fungi were capable of parasitizing several plant pathogenic fungi, with filamentous fungi, such as *Trichoderma*, from various regions. In the growth medium, all the *Trichoderma* isolates exhibited considerable chitinase and -1,3-glucanase activity. When examining the chitinase and -1,3-glucanase activities, it was found that *T. viride* contained the greatest chitinase activity, with further investigation revealing that chitinase and -1,3-glucanase enzyme activities were increased when the carbon source at a 1% concentration was substituted.

8.8 CONCLUSION

A large number of bacterial, fungal, and virus pathogens cause many postharvest diseases in horticultural crops. Some pathogens infect produce before harvest and after harvesting these pathogens remain quiescent until conditions are more favorable to develop disease. Many other pathogens infect produce through surface injuries during and after harvest. In the development of approaches for postharvest disease control, it is imperative to take a step back and consider the production and postharvest handling systems in their entirety. Several pre-harvest factors directly and indirectly have an effect on the development of postharvest diseases. Traditionally, fungicide has played the major role in postharvest diseases management. However, trends towards less chemical use in horticulture are forcing the development of new approaches. Thus, the use of different fungi is considered as the most promising approach due to their effectiveness in controlling various pathogens in many horticultural crops. On the basis of above-represented information, it may be concluded that significant amounts of harvests are lost due to postharvest diseases in the horticultural crops but, it can be decreased through their management which should be at right place in right time for more profit to the farmers and further share to the national economy.

KEYWORDS

- **diseases management**
- **fungi**
- **horticultural crops**
- **microorganisms**
- **mycoparasitism related genes**
- **postharvest disease**

REFERENCES

Cherkupally, R., Hindumathi, A., & Reddy, B. N., (2017). *In vitro* antagonistic activity of *Trichoderma* species against *Fusarium oxysporum* f. sp. *Melongenae*. *International Journal of Applied Agricultural Research, 12*, 87–95.

Chet, I., (1987). *Trichoderma – Application, Mode of Action, and Potential as a Biocontrol Agent of Soil-Borne Plant Pathogenic Fungi: Innovative Approaches to Plant Disease Control* (pp. 137–160). Wiley Inter-Science, New York.

Coates, L., Cooke, A., Parsley, D., Beattie, B., Wade, N., & Ridgeway, R., (1995). Postharvest diseases of horticultural produce. *Tropical Fruit, 2*. DPI, Queensland.

Currah, L., & Proctor, F. J. (1990). Onions in Tropical Regions. *Natural Resources Institute Bulletin,* 35. https://www.scirp.org/(S(i43dyn45teexjx455qlt3d2q))/reference/References Papers.aspx?ReferenceID=1706688 (accessed on 22 February 2023).

Dennis, C., & Webster, J., (1971). Antagonistic properties of species group of *Trichoderma* I. Production of non-volatile antibiotics. *Transactions of the British Mycological Society, 57*, 25–29.

Dugan, F. M., (2007). Diseases and disease management in seed garlic: Problems and prospects. *American Journal of Plant Sciences and Biotechnology, 1*, 47–51.

Estrada, R. G., Benítez, F. B., Leyva, B. M., Hernández, C. M., Islas, L. D. C. R., Islas, J. R., Peña, R. A., et al., (2018). A review study on the postharvest decay control of fruit by *Trichoderma*. *Trichoderma–The Most Widely Used Fungicide*. doi: http://dx.doi.org/10.5772/intechopen.82784 (accessed on 22 February 2023).

Fatima, A., Abid, S., & Naheed, S., (2015). Trends in wholesale prices of onion and potato in major markets of pakistan: A time series analysis. *Pakistan Journal of Agricultural Research, 28*.

Futane, A. S., Dandnaik, B. P., Salunkhe, P. P., & Magar, S. J., (2018). Management of storage diseases of onion by using different fungicides and antibiotics. *International Journal of Current Microbiology and Applied Sciences, 7,* 1149–1158.

Grinstein, A., Elad, Y., Temkin, G. N., Rivan, Y., & Frankel, H., (1992). Reduced volume application of fungicides for the control of onion rots. *Phytoparasitica., 20*, 293–300.

Gupta, R., Khokhar, M. K., & Lal, R., (2012). Management of the black mold disease of onion. *Journal of Plant Pathology and Microbiology, 3,* 133.

Harman, G., & Mastouri, F., (2010). The role of *Trichoderma* in crop management systems. *Phytopathology, 100*, 16.

Herrera-Estrella, A., & Chet, I., (2004). The biological control agent *Trichoderma* from fundamentals to applications. In: Arora, D. K., (ed.), *Fungal Biotechnology in Agricultural, Food, and Environmental Applications* (pp. 147–156). Marcel Dekker, Inc, New York.

Jaime, L., Martínez, F., Martín, C. M. A., Mollá, E., López-Andréu, F. J., Waldron, K. W., & Esteban, R. M., (2001). Study of total fructan and fructooligosaccharide content in different onion tissues. *Journal of Science of Food and Agriculture, 81*, 177–182.

Kumar, R., Maurya, S., Kumari, A., Choudhary, J., Das, B., Naik, S. K., & Kumar, S., (2012). Bio control potentials of *Trichoderma harzianum* against sclerotial fungi. *The Bioscan., 7*, 521–525.

Kumar, V., & Iqbal, N., (2020). Postharvest pathogens and disease management of horticultural crop: A brief review. *Plant Archives, 20*, 2054–2058.

Lacicowa, B., & Pieta, D., (1994). Protective effect of microbiological dressing of pea seeds (*Pisum sativum* L.) against pathogenic fungi living in soil. *Annals Universitatis Mariae Curie-Sklodowska, Sectio EEE, Horticultura, 2,* 165–171.

McDonald, M. R., De Los, A. J. M., & Hovius, M. H., (2004). Management of diseases of onions and garlic. *In Diseases of Fruits and Vegetables, 2,* 149–200.

Padule, D. N., Lohate, S. R., & Kotecha, P. M., (1996). Control of spoilage of onion bulbs by postharvest fungicidal treatments during storage. *Onion Newsletter for the Tropics, 7*, 44–48.

Prajapati, B. K., Patil, R. K., & Alka, (2016). Bio-efficacy of fungicides in management of black mould rot (*Aspergillus Niger* L.) of onion. *International Journal of Agricultural Science and Research, 6*, 155–160.

Raju, K., & Naik, M. K., (2006). Effect of pre-harvest spray of fungicides and botanicals on storage diseases of onion. *Indian Phytopathology, 59*, 133.

Reddy, B. N., Saritha, K. V., & Hindumathi, A., (2014). *In vitro* screening for antagonistic potential of seven species of *Trichoderma* against different plant pathogenic fungi. *Research Journal of Biology, 2*, 29–36.

Singh, B. K., Yadav, K. S., & Verma, A., (2017). Impact of postharvest diseases and their management in fruit crops: An overview. *J. Bio. Innov., 6*, 749–760.

Spiers, M., Hill, R., & Fullerton, B., (2004). Trials using *Trichoderma* in greenhouse vegetable crops. *Grower*, 37–39.

Srinivasan, R., & Shanmugam, V., (2006). Postharvest management of black mould of onion. *Indian Phytopathology, 59*, 333–339.

Tripathi, A., Sharma, N., & Tripathi, N., (2010). Management of fungal plant pathogens. *Biological Control of Plant Diseases: An Overview and the Trichoderma System as Biocontrol Agents* (pp. 121–137). CAB International, UK.

Triveni, S., Prasanna, R., & Saxena, A. K., (2012). Optimization of conditions for *in vitro* development of *Trichoderma viride* based biofilms as potential inoculants. *Folia Microbiologica, 57*, 431–437.

Türkkan, M. & Erper, I. (2014). Evaluation of antifungal activity of sodium salts against onion basal rot caused by *Fusarium oxysporum* f.sp. *cepae*. *Plant Protect. Sci. 50*(1), 19–25.

Upadhyay, J. P., & Mukhopadhyay, A. N., (1986). Biological control of *Sclerotium rolfsii* by *Trichoderma harzianum* in sugarbeet. *Tropical Pest Management, 32*, 215–220.

Weindling, R., (1934). Studies on a lethal principle effective in the parasitic action of *Trichoderma lignorum* on *Rhizoctonia solani* and other soil fungi. *Phytopathology, 24*, 1153–1179.

Yadav, S. M., Patil, R. K., Balai, L. P., & Niwas, R., (2013). Postharvest diseases of horticultural crops and their management. *Pop. Kheti., 1*(3), 20–25.

CHAPTER 9

Fungal Biofertilizers and Biopesticides and Their Roles in Sustainable Agriculture

DEEPAK KUMAR[1], P. N. SINGH[2], HELGA WILLER[3], SUSHIL K. SHARMA[4], U. B. SINGH[5], AJAY C. LAGASHETTI[2], and RAJA HUSAIN[6]

[1]*R&D Division, Nextnode Bioscience Pvt. Ltd., Kadi, Gujarat, India*

[2]*National Fungal Culture Collection of India (NFCCI), Biodiversity and Paleobiology Group, Agharkar Research Institute, Pune, Maharashtra, India*

[3]*Research Institute of Organic Agriculture FiBL, Frick, Switzerland*

[4]*ICAR–National Institute of Biotic Stress Management, Baronda, Raipur, Chhattisgarh, India*

[5]*ICAR–National Bureau of Agriculturally Important Microorganisms (NBAIM), Maunath Bhanjan, Uttar Pradesh, India*

[6]*Department of Agriculture, Himalayan University, Itanagar, Arunachal Pradesh, India*

ABSTRACT

Biopesticides and biofertilizers have materialized as attractive, environmentally friendly bio-inputs that are augmented for the management of disease and pest infestations and optimum plant growth. Their ability to satisfy plant nutrient needs without the application of artificial fertilizers is enormous. These bio-inputs (as microbial inoculants, which are supplied as macro- and micronutrients in available form to crops) help them develop

Applied Mycology for Agriculture and Foods: Industrial Applications. Sanjay K. Singh, Deepak Kumar, Md. Shamim, & Rohit Sharma (Eds.)

and produce more compared to other production technologies. Entomopathogenic fungi (EPF), particularly insect-pathogenic species, are found in the genera *Metarhizium*, *Lecanicillium*, *Beauveria*, and *Paecilomyces*. In most cases, these fungal applications are used in the form of various formulations with different delivery systems, which entails exposing certain fungal spores to unfavorable temperature, solar radiation, and humidity conditions to control insects and pests. The demand for organically grown foods has increased worldwide because of health, safety, and environmental issues. As a result, demand for bio-inputs (biofertilizers and biopesticides) has also rapidly increased. The efficacy of any fungal-based inoculant is determined by a variety of factors such as the source of origin, formulation, delivery system, toxicity to the target insect pest, and impact on non-target organisms, including animals and humans.

9.1 INTRODUCTION

Abiotic and biotic stresses include nutrient deficiency, insect pests, disease, and poor growing conditions, which contribute to significant crop yield losses (up to 26%; Culliney, 2014). Chemical insecticides and inorganic fertilizers are used by farmers to ensure that their crops produce at their full potential, combating these problems but causing deleterious effects on non-target species, users, and the environment (Bamisile et al., 2021; Fadiji & Babalola, 2020). Using fungal biofertilizers and entomopathogenic fungi (EPF) as biocontrol agents (BCAs) against herbivores has the potential to be a more environmentally friendly and sustainable alternative to traditional nutrient and insect pest management approaches (West & Gwinn, 1993). EPF are well-known for their propensity to infect insects and cause disease when exposed to the appropriate conditions. They do this by piercing the cuticles of insects and colonizing the pests' bodies directly (Kumar et al., 2019). The mineral solubilizing fungi and insect-pathogenic fungi are comprised of more than 700 species from around 90 different genera that have been identified. For example, the most widely used strains of *Trichoderma*, *Aspergillus*, *Penicillium*, *Chaetomium*, *Beauveria*, *Lecanicillium*, *Metarhizium*, *Paecilomyces*, and *Hirsutella* are included in this group (Khachatourians & Qazi, 2008; Bamisile et al., 2021).

As a result of the various characteristics of fungi, they can be used as biostimulant microbes for plant growth promoting activities in addition to their typical employment as pest killers, and they have demonstrated promising importance as biofertilizers so far. These fungal species are regarded as good options to systemic fertilizers and are a more efficient and environmentally friendly approach to food security (Glick, 2014). Moreover, among other benefits, it is becoming more common in organic agriculture systems to use fungi to safeguard the plants from insects and pests and improve yields (Bamisile et al., 2021; Shrivastava et al., 2010). Diverse fungal species have been singled out for consideration because of their prospective as biological control agents indirectly in huge farming applications (Lacey & Neven, 2006). Consequently, the use of biotechnology for agricultural enhancement, such as inoculation of crops with modified fungal strains, would reduce the toxicity of the organisms that are consumed, such as animals, humans, and the environment. Fungal genes could be genetically manipulated by either the elimination of harmful genes or by the introduction of new helpful genes into their genomes (Adeleke & Babalola, 2021). This chapter covers their source, production system, type of formulation, delivery system, regulatory, and legal framework for commercialization and marketing, as well as their significance in sustainable agriculture and its constraints, as part of its assessment of the current phase of fungal biofertilizers and biopesticides. It also gathers together the many forms of fungal biofertilizers and biopesticides as well as proof of their effectiveness in nutrient supply and significant against pests in a variety of crops.

9.2 FUNGAL BIOFERTILIZERS AND BIOFUNGICIDES

Biofertilizers are a group of microbial inoculants (bacteria, fungi, and algae) that have the ability to supply nutrients to soil, fix nitrogen, and solubilize minerals for use by plants in the form of nutrient solutions. Chemical fertilizers have been replaced by organic fertilizers because they are eco-friendlier, cost-efficient, and renewable sources of plant nutrition (Odoh et al., 2020; Kour et al., 2020; Verma et al., 2017b). Fungal biofertilizers, which include the use of fungal inoculants (such as *Trichoderma* sp., *Mycorrhiza* sp., *Gliocladium* sp., and *Chaetomium* sp.), have been developed to deliver nourishment to the host plant while also protecting crops from infections such as fungi and bacteria. When working together as a microbial consortium, they stimulate plant growth, decrease abiotic stress conditions, and have an impact on a variety of biochemical changes and activities (Odoh et al., 2020; Kaewchai et al., 2009). With the long history of the use of fungal biofertilizers, there has been a growing demand in recent years to better understand their application and role as biofertilizers and BCAs in agriculture. According to Ajmal et al. (2018), this capability is attributed to the fungal specie's innate ability to promote crop development while decreasing dependency on synthetic chemical fertilizers and pesticides.

9.2.1 MYCORRHIZAE

Among all the fungal biofertilizers, the Mycorrhizae group is a very prominent group of obligate fungi that is commercially available as a fungal biofertilizer in markets. Mycorrhizas forms mutually beneficial symbiotic interactions with plant roots. In their interactions with these plants, they increase the uptake of phosphorus (P), potash (K), nitrogen (N), iron (Fe), zinc (Zn), sulfur (S), copper (Cu), and boron (B) as well as other trace elements. Many important plant groups (such as trees, shrubs, herbs, epiphytes, helophytes, and xerophytes) interact with *Mycorrhiza* sp. in a mutualistic manner (Rai et al., 2013; Kaewchai et al., 2009). Arbuscular mycorrhizae (AM) and ectomycorrhizae (ECM) are two of the most studied and important mycorrhizae. In recent years, there has been an increased demand for mycorrhizae-based inoculants as well as genetic engineering of the dominant mycorrhiza in order to increase yield and promote stress tolerance in the plant population. Additionally, they protect plant roots against biologically associated stresses by reducing the quantity of pathogens in the soil and moderating the fluctuations of soil pH (Odoh et al., 2019a, b, 2020). Furthermore, as essential members of the microbial population, arbuscular mycorrhizal fungi (AMF) play a critical role in the encouragement of plant growth, the protection of plants, and the enhancement of soil quality. Aside from the fact that they are widely used in cropping systems, they have also gained widespread acceptability as organic farming manure, particularly in areas where the pursuit of sustainable crop cultivation is a top concern.

Considering the significance of fungal mycorrhizal symbioses in agriculture, they are possibly the most important symbioses. A wide variety of fungi are used as biofertilizers in this process, which is a relatively new development (Odoh et al., 2020). In practice, these plant species develop mycorrhizal associations by serving as a vital link between the soil and the roots of the plants they grow with. Plant-fungal interactions are typically marked by the transfer of fungal-acquired nutrients to plants, as well as the transfer of plant-produced carbon to fungi (Johri et al., 2015). Nutrients are absorbed from the soil by the fungal mycelium, which grows from the root system into the soil matrix. The fungal hyphae's narrow width increases the surface area accessible for absorption, hence increasing the plant's capacity to acquire and absorb nutrients.

9.2.1.1 ENDOMYCORRHIZAE

Most agricultural soils are affected by this group of fungi, which provide physiological immunity against soil-borne illnesses and are associated with most agricultural and horticultural soils. Most ecosystems on the planet support their existence, and they can be found in a wide variety of key cereal crops (maize, wheat, rice) as well as horticultural plants, including roses and petunias (Frąc et al., 2018). AMF are the most common of these organisms. Bagyaraj & Ashwin (2017) noted substantial increases in crop yield using the inoculation of AMF, which aids in boosting key effects like root growth, soil structure improvement, nutrient absorption, and ion movement. AMF is a fungal inoculant that has been shown to increase crop yield. Endomycorrhizae are beneficial to plants in a variety of ways, including increasing their resistance to stress and improving their overall health.

9.2.1.2 ECTOMYCORRHIZAE (ECM)

The formation of a broad mantle structure inside the intercellular gaps of the root cortex distinguishes the fungi belonging to this group. They also operate as a connection for nutrient availability by forming a sheath all around the feeding root (Bücking et al., 2012). In the host roots, they are encircled by live cells, which results in the development of a large network known as the Hartig net. It is this Hartig net that serves as both a phosphorus storage and phosphorus transportation organ. Most significantly, ECM are found in abundance among the plant families of Fagaceae, Salicaceae, Pinaceae, Betulaceae, and Myrtaceae (Table 9.1).

9.2.2 TRICHODERMA

Trichoderma species are widespread in soil and root environments, and they are ubiquitous saprobes that can be easily isolated from soil, decomposing wood, or other organic materials (Rosa et al., 2012; Zeilinger & Omann,

TABLE 9.1 Some of the Selected Mycofertilizers and Mycofungicide Available in Global Market

Fungal Inoculant	Species Involved	Product Name	Producing Company	Country of Origin	Mode of Action	References
Trichoderma sp.	*Trichoderma harzanium/Trichoderma viride*	Neemoderma®	Shri Ram Solvent Extractions Pvt. Ltd.	India	Enhanced decomposition, mycoparasitic	https://neemplus.com/
	Trichoderma harzanium/Trichoderma viride	Bioharz™	IPL Biologicals Ltd.	India	Mycoparasitic	https://www.iplbiologicals.com/
	Trichoderma harzianum/Trichoderma hamatum	Bioorganic Plus	Nova Science Co. Ltd.	Thailand	Decomposer, mycoparasitic	http://www.nova-science.com/
	Trichoderma harzanium	RootShield®	Arbico Organics Inc	United States	Mycoparasitic	https://www.arbico-organics.com/
	Trichoderma harzanium	Trianum-G	Koppert Biological Systems	United Kingdom	Mycoparasitic	https://www.koppert.co.uk/
	Trichoderma harzanium	*Trichoderma harzanium* Products	Lifeasible	United States	Mycoparasitic, mycofertilizer, decomposer	https://www.lifeasible.com/
Mycorrhiza sp.	*Vesicular Arbuscular Mycorrhiza*	Vamlet Vam Shakti	IPL Biologicals Ltd.	India	Macro- and micronutrient mobilizer, growth promoting activities	https://www.iplbiologicals.com/
	Vesicular Arbuscular Mycorrhiza	Mycopower™ Mycozone	Agriland Biotech Ltd.	India	Macro- and micronutrient mobilizer, growth promoting activities	https://www.agrilandbiotech.com/
	Combination of *Mycorrhiza* sp.+ Rhizobacteria and other beneficial fungi	Micro-Myco	Arbico Organics Inc.	United States	Macro- and micronutrient mobilizer, growth promoting activities	https://www.arbico-organics.com/

TABLE 9.1 *(Continued)*

Fungal Inoculant	Species Involved	Product Name	Producing Company	Country of Origin	Mode of Action	References
	Vesicular Arbuscular Mycorrhiza	Mycorrhiza Ultrafine Endo	Ferti-Organic, Inc.	United States	Macro- and micronutrient mobilizer, growth promoting activities	https://www.ferti-organic.com/
	Vesicular Arbuscular Mycorrhiza	Premium Mycorrhizal Inoculant	Wallace Organic Wonder	United States	Macro- and micronutrient mobilizing, growth promoting activities, stress reducing	https://wallacewow.com/
	Vesicular Arbuscular Mycorrhiza (*Rhizophagus irregularis*)	MycroRhiza®	Agrilife	India	Helpful in water and nutrient absorption	http://agrilife.in/
	Vesicular Arbuscular Mycorrhiza (*Glomus intraradices*)	Josh Super	CADAgro	India	Increase nutrient supply, growth promoting activities	https://cadagro.co.in/
Aspergillus sp.	*Aspergillus niger* (AN-27)	Kalisena FS Kalisena FA	CADAgro	India	Mycoparasitic and mycofertilizer	https://cadagro.co.in/
Chaetomium sp.	*Chaetomium globosum*	Alglow	Amruth Organic Fertilizers	India	Increase nutrient supply, mycoparasitic	https://amruthgroups.com/
Penicillium sp.	*Penicillium citrinum*	Mn-Sol	Farmers bio-fertilizers and organics	India	Manganese (Mn) solubilizer, nitrate assimilation	https://www.farmers biofertilizers.com/
	Penicillium citrinum	Mn Sol-B®	Agrilife	India	Manganese (Mn) solubilizer, nitrate assimilation	http://agrilife.in/

TABLE 9.1 *(Continued)*

Fungal Inoculant	Species Involved	Product Name	Producing Company	Country of Origin	Mode of Action	References
Ampelomyces sp.	*Ampelomyces quisqualis*	Amices	Amruth Organic Fertilizers	India	Mycoparasitic, highly effective in powdery mildew	https://amruthgroups.com/
	Ampelomyces quisqualis	Ampelo	Anand Agro Care	India	Mycoparasitic	https://anandagrocare.com/
	Ampelomyces quisqualis	Milgo	IPL Biologicals Ltd.	India	Mycoparasitic	https://www.iplbiologicals.com/
	Ampelomyces quisqualis	Katyayani *Ampelomyces quisqualis*	Katyayani organics	India	Mycoparasitic	https://www.katyayaniorganics.com/product
	Ampelomyces quisqualis	Powderycare™	Agrilife	India	Mycoparasitic, highly effective in powdery mildew	http://agrilife.in/
	Ampelomyces quisqualis	Ampeiforte	Vijya Agro Industries	India	Mycoparasitic	https://www.vijayaagro.com/
	Ampelomyces quisqualis	Rom No-Mildew	Rom Vijay Biotech Pvt. Ltd.	India	Mycoparasitic	https://www.indiamart.com/romvijaybiootech/
	Ampelomyces quisqualis	Ampelino	Krishidoot Bio-herbals	India	Mycoparasitic	https://www.krishidoots.com/
	Consortia of *Ampelomyces quisqualis*, *Trichoderma harzanium*, *T. viride*	FungFree	Utkarsh Agrochem Pvt. Ltd.	India	Mycoparasitic, highly effective in powdery mildew	https://www.utkarshagro.com/

TABLE 9.1 *(Continued)*

Fungal Inoculant	Species Involved	Product Name	Producing Company	Country of Origin	Mode of Action	References
Fungal biosynthesized Nanofertilizers	Nanofertilizers	IFFCO Nano-Zinc IFFCO Nano-Copper IFFCO Nano-Nitrogen	IFFCO	India	Bioavailable source of nutrients and disease management	https://www.iffco.in/

2007). Several studies have demonstrated that many plant diseases have been successfully managed by their pretty efficient mycoparasitic and antagonistic activities (Sharma et al., 2019; Kour et al., 2019). Similarly, a number of *Trichoderma* strains interact with their plant hosts, enhancing their direct growth capacity and imposing resistance to diseases and abiotic stressors (Rosa et al., 2012). According to Chandrashekrappa & Basalingappa (2018), *Trichoderma* sp. are supposedly a great contender within the rhizosphere since they exhibit tolerance to soil adverse conditions as well as high performance in utilizing soil macro- and micronutrients while also being averse to phytopathogens. It has been proven over time that *Trichoderma* spp. can be modified to provide successful crop support for many cereals and horticultural crops (such as maize, wheat, barley, potato, soybean, cotton, tomato, and potato) (Sneha et al., 2018). *Trichoderma harzianum* variant T-22 was created through protoplast fusion of *Trichoderma harzianum* T-95 and T-12, and this variant was synthesized as a granular called RootShield® and a powder called PlantShield® by Biworks, Geneva, New York, to treat a variety of plant diseases. *Trichoderma harzianum* T-22 has been shown to be effective against a wide range of phytopathogenic fungi, including *Fusarium*, *Rhizoctonia*, *Botrytis cinerea*, and *Pythium* in several crops (Kaewchai et al., 2009; Paulitz & Belanger, 2001). According to Etebarian et al. (2000), *Trichoderma harzianum* strain T-39 is commercially available under the trade name Trichodex, 20P from Makhteshim Ltd. for the managing of stem rot and pink rot of tomato induced by the pathogen *Phytophthora erythroseptica*. The biocontrol system in *Trichoderma* is a set of mechanisms that work together to maintain balance. Mycoparasitism and antibiosis are the primary mechanisms of action. Mycoparasitism is based on the identification, adhesion, and proteolytic breakdown of the host fungus' cell wall, among other factors (Kumar et al., 2019; Woo & Lorito, 2007). The rapid expansion, high reproductive capabilities, inhibition of a wide range of fungal diseases, variety of control mechanisms, excellent competition in the rhizosphere, ability to modify the microenvironment of the rhizosphere, tolerance or inertia to soil fungicides, ability to sustain in unfavorable environmental conditions, and optimization of resources are all reasons why *Trichoderma* species have been used as mycofungicides with great success (Odoh et al., 2020; Kumar et al., 2019; Kaewchai et al., 2009). The ability of *Trichoderma* species to colonize and continue to expand in interaction with plant root systems is referred to as rhizosphere competence.

9.2.3 CHAETOMIUM

There are a number of *Chaetomium* species that can be found in soil or decomposed matter. Preventive and curative properties are the primary uses of this mycofungicide inoculant. Compost enrichment may be boosted with both *Ch. olivaceum* and *Ch. bostrychodes* (Dhingra et al., 2003). Products of *Chaetomium* species have recently been employed as fungal biofertilizers, which is a fascinating development. As an illustration, consider Ketomium[R], a plant growth biostimulant bioproduct derived from *Ch. cupreum* and *Ch. globoglus* (Jehangir et al., 2017). Raspberry spur blight induced by *Didymella applanata* was shown to be well suppressed by the mycofungicide Ketomium, which also reduced *R. solani* infestation as a potato disease and increased yields of potatoes (Kumar et al., 2019; Shternshis et al., 2005). Similarly, *Chaetomium globosum* isolate CgA-1 is another species that can also work as a biofungicide and reduce soybean stem canker disease induced by *Diaporthe phaseolorum* f. sp. *meridionalis* (Kaewchai et al., 2009; Dhingra et al., 2003). Furthermore, a novel isolate of *Ch. cupreum*, RY202, has been shown to have antagonism against *Rigidoporus microporus*, which promotes white root disease in a certain variety, namely RRIM600, of rubber trees, according to preliminary results (Kaewchai et al., 2009). *Chaetomium* spp. have gained an inhibitory effect on the growth of phytopathogens (fungi and bacteria) as a result of mycoparasitism, competition, and antibiosis, which is a promising approach (Odoh et al., 2020; Marwah et al., 2007).

9.2.4 PENICILLIUM

Penicillium species are phosphorus-solubilizing mycoflora that also function as mycofertilizers by supplementing compost and soil conditions (Sharma et al., 2013). The important species of *Penicillium* are *P. oxalicum*, *P. italicum*, *P. simplicissimum*, *Penicillium bilaji*, *P. rubrum*, and *P. frequentans*, which are some of the notable species in this genus (Yadav et al., 2018b). *P. bilaji* is a genus that contains a number of noteworthy species. In addition to acting as zinc solubilizers, some species of *Penicillium*, including *Penicillium* sp. and *Penicillium simplicissimum,* are utilized as phosphate mobilizing bio-fertilizers in association with *Aspergillus awamori*. According to Mishra et al. (2013), the consortia of *Trichoderma* and *Penicillium* promote the growth of soybean and wheat in greenhouse conditions. Several commercial mycofertilizer products made with *Penicillium* spp. (such as *Penicillium bilaiae*) have been shown to boost available phosphorus (P) to plants and increase yield in *Brassica napus* (canola) (Chandrashekrappa & Basalingappa, 2018; Chang & Yang, 2009). It has been reported that two species of fungi (*P. italicum* and *P. radicum*) are phosphate-solubilizing mycoflora, in which *Penicillium radicum* provided an effective plant growth stimulator in wheat roots, whereas phosphate-soluble solubilizer mycoflora, *P. italicum,* showed the ability to solubilize tricalcium phosphate (TCP) in available phosphorus and facilitate effective soybean growth (Ram et al., 2015). Moreover, Sumita et al. (2015) also reported *Penicillium* sp. as a potent plant growth stimulator and phosphate-solubilizing microbial inoculant in their findings.

9.2.5 *GLIOCLADIUM*

Gliocladium sp. is a genus of asexual fungus belonging to the family Hypocreaceae. *Gliocladium virens* and a few additional species, have lately been moved to the genus *Trichoderma* (https://www.mycobank.org/). In addition to being ubiquitous soil saprobes, some *Gliocladium* species have been documented to be parasites of a wide range of plant pathogens, including *Gliocladium catenulatum* parasites of *Fusarium* sp. and *Sporidesmium sclerotiorum* (Viterbo et al., 2007). It affects the mycofloral host through direct hyphal interaction, resulting in the formation of pseudoappressoria. Moreover, it has also been utilized as a commercially available wettable powder (WP) formulation (a genetically modified strain of *Gliocladium catenulatum* Strain J446) sprayed on foliage, soils, and roots to minimize the prevalence of damping-off disease in greenhouse conditions (Youssef & Eissa, 2014; Viterbo et al., 2007). *Gliocladium virens* has been utilized as a biocontrol agent against a broad range of soil-borne phytopathogens, including *Rhizoctonia* and *Pythium*, both in open field and greenhouse conditions, with promising results. Potent antimicrobial metabolites like gliotoxin, which possess antifungal, antibacterial, antiviral, and antitumor properties, are also produced by this mycoflora (Odoh et al., 2020; Nissipaul et al., 2017).

9.2.6 *ASPERGILLUS*

Aspergillus sp. assists in the breakdown of phosphate aggregates in the soil. In order to achieve this, an organic acid is produced by this fungus, which dissolves unavailable phosphate and makes it available in the soil for plant uptake (Zarafi & Dauda, 2019). In addition to this, several fungal species, such as *A. niger*, *A. flavus*, *A. awamori*, *A. tubingensis*, *A. fumigates*, *A. melleus*, and *A. terreus*, have the ability to dissolve inorganic phosphate into available phosphorus through the formation of citric acid, succinic acid, gluconic acid, and oxalic acid glycolic acid. One of the most common mycoflora identified from decomposed matter, *Aspergillus fumigatus*, has also been found to release potassium in the soil (Kumar et al., 2019; Sharma et al., 2013; Lian et al., 2008; Akintokun et al., 2007). Similarly, Jehangir et al. (2017) also reported *Aspergillus* sp. as microphos mycofertilizers, which are classified as such because of their ability to release phosphate from their bound and insoluble conditions. Moreover, *Aspergillus niger* promotes plant development by encouraging the creation of roots (biopromoter). It enhances soil health and can be used to generate zinc-solubilizing mycofertilizers, which are beneficial to soil health. Ambiphos is one of the mycofertilizers generated by *Aspergillus niger*, and it excretes organic acids that aid in the dissolution of inaccessible phosphate into dissolved form and the availability of that phosphate for plant usage. Apart from that, Ligno Biocompost-culture is an excellent mycofertilizer derived from *Aspergillus awamori*, which has the additional benefit of acting as an enhancer for compost breakdown (Kumar et al., 2019; Sumita et al., 2015). Kalisena™ based on *Aspergillus nige*r isolate (AN-27) was developed by IARI New Delhi and commercialized by CADAgro Company. It is used as a mycofungicide and mycofertilizer, and it improves germination and growth by creating growth-promoting compounds, functioning as a phosphate solubilizer, and inhibiting the growth of key soil-borne fungal infections such as *Macrophomina*, *Rhizoctonia*, *Fusarium*, and *Pythium* (https://cadagro.co.in/bio-products.html).

9.2.7 AMPELOMYCES

Powdery mildews are controlled by *Ampelomyces quisqualis*, a mycoparasitic anamorphic ascomycete that inhibits the growth of powdery mildews and reduces their population in the environment. Because of antibiosis and parasitism, it can have an effect on the pathogen (Kaewchai et al., 2009; Viterbo et al., 2007). For the first time, *A. quisqualis* was identified as a powdery mildew hyperparasite. This fungus is commonly detected in the presence of powdery mildew colonies. Powdery mildew hyphae are penetrated by *Ampelomyces* hyphae, which develop internally and then kill all of the parasitized cells in the hyphae (Kumar et al., 2019; Kiss et al., 2003). The company, Ecogen, Inc. in the United States, has produced a biofungicide based on *Ampelomyces quisqualis* isolate M-10, which is known as AQ10 Biofungicide. Conidia of *A. quisqualis* are used in the formulation of these mycofungicides, which are available as water-dispersible formulations for the treatment of powdery mildew on fruits and vegetables (Kaewchai et al., 2009; Viterbo et al., 2007; Kiss et al., 2003). The mycelia of six different powdery mildew isolates that subsequently infested respective hosts in Japan resulted in a total of 26 *Ampelomyces* strains. The mycoparasitic relationship between eight *Ampelomyces* strains and the tomato powdery mildew phytopathogen (*Pseudoidium neolycopersici* strain KTP-03) was investigated. According to the findings, mycoparasite spores developed on tomato leaves, and their hyphae pierced *Pseudoidium neolycopersici* hyphae. *Ampelomyces* hyphae continued to proliferate inside, causing the powdery mildew conidiophores to atrophy and the parasitized conidiophores to collapse seven days after inoculation. The created model system will benefit this *Ampelomyces*-based formulation of mycoparasites as a biocontrol agent in the future (Nemeth et al., 2021). Furthermore, Thosar et al. (2020) used a bio-intensive strategy for the administration of powdery mildew on grapes, which included the use of chitosan formulations and the inclusion of *Ampelomyces quisqualis*. Similarly, Dominic & Marthamakobe (2016) also used *Ampelomyces quisqualis* Ces to biological control of cashew powdery mildew.

9.3 FUNGAL BIOFERTILIZERS' ROLES AND BENEFITS IN SUSTAINABLE AGRICULTURE

A variety of fungal biofertilizers have been developed from cultures of fungi that have been shown to have the ability to improve soil fertility while simultaneously increasing crop output. They do this through the use of mycoflora species (such as *Aspergillus*, *Penicillium*, *Piriformospora*, and *Curvularia*), which mineralize and solubilize the inorganic phosphorus in the soil (Odoh et al., 2020; Mehta et al., 2019). Mycorrhizal fungi, which are among the most important for sustainable farming system, affect the accessibility of nutrient levels through their hyphae network that allows for more efficacious nutrient absorption by plants. According to Pal et al. (2014) investigation, when mycorrhizae are utilized as biofertilizers, they increase the uptake of minerals like zinc, iron, phosphate, and water, resulting in increased resilience and consistent expansion of transplanted stock. These endophytic fungi transform unavailable soil phosphate into soluble forms through the release of organic acids, leading to changes in the structure, texture, and water retention capacity of the soil as a result of the transformation (Kumar et al., 2019; Tripathi et al., 2017). It has been previously established that for plants to survive in abiotic and biotic conditions, they rely completely on connections in their root zone rhizosphere and associated mycobiota. In addition to the conversion of plant nutrients such as macro- (nitrogen, potash, and phosphorus) and micronutrients (zinc, copper, iron, etc.), these fungal consortiums are also responsible for the encouragement of crop productivity, among other things (Odoh et al., 2020). Fungal biofertilizers have many roles in plant production systems in many ways, which are briefly discussed as follows.

9.3.1 MINERAL SOLUBILIZATION AND TRANSPORT

Strengthen and accelerate mineral uptake and delivery, which are the key objectives of AMF treatment. It has been demonstrated that, upon treatment, the expression patterns of plants and fungal absorption transporters in the mycorrhizal junction contribute to the formation of a significant gradient of concentration across the mycorrhizal interaction. Another example is the expression of high affinity phosphate (P) and nitrogen (N) transporters from AM and ECM fungi in the ericoid mycorrhiza (ERM), while their expression is weaker than expected in the intraradical mycelium (IRM) (Odoh et al., 2020; Kumar et al., 2019). Clark (1997) found that adding AMF to the soil improved plant uptake of mineral elements. Soil structure and porosity are key factors in soil water retention throughout seasonal variations (the dry season). Because of this, "AMF" modulates hormone flow and information exchange between plant shoots and roots, as well as affects the reaction of stomata to variations in soil water potential (Odoh et al., 2019a, 2020). According to the findings of Jagnaseni et al. (2016), mycorrhizal connections boost nutrient uptake in plants during times of water stress by increasing hydraulic conductivity in the root zone of the plants. These fungal mantles act as a substantial apoplastic barrier, resulting in the formation of a shuttered interfacial apoplast that allows for nutrient transport (Bücking et al., 2012). Overall, AM and ECM fungi modulate nutrient delivery in host cells by deposition of polyphosphate while simultaneously increasing carbon supply in the host plant, a mechanism that also results in polyp hydrolysis in the host plant (Bücking & Kafle, 2015; Cruz et al., 2007).

9.3.2 PHOSPHATE SOLUBILIZATION

Phosphorus (P) is a highly stationary element that can be found in soils and groundwater. It is extremely important for the growth of plants. Vascular-arbuscular (VA) fungi play an important role in the provision of phosphorus to plant roots, which is accomplished through the use of phosphate exchangers found in the hyphal layer (Kumar et al., 2019; Odoh et al., 2020). This allows the fungi to deliver phosphorus to the plants in the form of a polyphosphate pool. The frameworks of filamentous, extraradical mycelia of AM fungi, which are found outside the root of the host plant and are termed the "networks of flagellated, extraradical hyphae of AM fungus," aid in the absorption of phosphates that are readily available for plant uptake. In most cases, the mycelium stretches beyond the host root, allowing for greater soil property contact and phosphate uptake. Amphibole fungi (AM fungi) are responsible for the hydrolysis of organic phosphates into soluble forms that can be used by plants. This absorption is accomplished through the action of the Pht1 phosphate transporter family (Odoh et al., 2020). As per findings by Vergara et al. (2019), mycoflora has two different systems for phosphorus uptake: (a) a potent affinity mechanism that operates against this electro-chemical potential gradient; and (b) a weak affinity system that acts in tandem with the potent-affinity system. As a result of protons' cotransport and a weak affinity mechanism, Pi is more easily transported across the mycoflora plasma membrane. Many of these transporters are expressed in response to changes in P concentrations in the environment, as well as fungus's P requirements (Kumar et al., 2019; Johri et al., 2015).

9.3.3 NITROGEN FIXATION

Nitrogen in the soil is regarded as a vital element for plant development. It is also one of the highly important macronutrients for food production, ranking second only to phosphorus and found in the forms of amino acids, ammonium, and nitrate (Odoh et al., 2020). In their investigation, Bücking & Kafle (2015) discovered that the ERM of arbuscular mycorrhizal (AM) and ECM fungi is capable of absorbing inorganic nitrogen substrates (ammonium [NH_4^+] or nitrate [NO_3^+]) from the surrounding soil. Nitrogen is absorbed by plants in the form of ammonium, which is transported via a protein transporter known as AMT1. Glutamate and ammonium combine to make glutamine in the extraradical mycelium during the formation of glutamine synthetase, which is produced by the enzyme. It is worth noting that NH_4^+

has been identified as the preferred nitrogen source for mycorrhizal fungi, presumably because its absorption is more energy efficient than that of NO_3^-. AMT1 and AMT2 are NH_4^+ transporters that are transcribed in *Hebeloma cylindrosporum* and are controlled by the presence of extracellular NH_4^+. The transcription of both transporters is increased in the presence of a limited NH_4^+ influx, although it is downregulated in the presence of an external NH_4^+ supply. The weak affinity NH_4^+ transporter (AMT3), which is particularly significant, allows the mycoflora to retain a basal level of N absorption and assimilation even when exposed to substantial foreign N supplies (Wipf et al., 2019; Bücking et al., 2012).

9.3.4 POTENT BIOCONTROL AGENT

In the case of plant diseases, BCAs are microbes that inhibit the illness. For lack of a better description, they are defined as the use of invasive or indigenous living lifeforms to minimize the pressure and impacts imposed by phytopathogens while also reducing the multiplication of their type (Odoh et al., 2020; Yadav et al., 2020; Kumar et al., 2019). *Trichoderma harzianum*, a species with biocontrol capability against *Fusarium*, *Pythium*, *Botrytis cinerea*, and *Rhizoctonia*, is one example of a fungal biofertilizer (Redda et al., 2018). Similarly, *Coniothyrium minitans*, a mycoparasite of *Fusarium oxysporum* and *Sclerotinia* (Smolińska & Kowalska, 2018); moreover, *Chaetomium cupreum* and *C. globosum*, which have biocontrol behavior against root rot diseases spread by *Phytophthora*, *Fusarium*, and *Pythium* (Aggarwal, 2015). Mycorrhizal connections not only promote biocontrol, but they also safeguard the plant cells against the toxicities of heavy metals (such as zinc, copper, iron, cadmium, manganese, nickel, and so on) that are present in the environment. Ectomycorrhizal fungi provide protection to plants by collecting and immobilizing harmful elements in the mycorrhizal mantle of the tree's root system (Odoh et al., 2020; Odoh, 2017). In addition, fungal biofertilizers can operate as BCAs by contending for resources and space or by creating compounds that inhibit fungal spore germination, antibiosis (killing the pathogenic cells), or change the rhizospheric microbiota (acidifying the soil) immediately to protect pathogen spread (Khokhar et al., 2012). Moreover, for biological control in mycoparasitism, the phytopathogen and the biological control agent interact directly, relying on identification, adherence, and enzymatic destruction of the host fungal cell wall.

9.3.5 IN GREEN SYNTHESIS OF NANOMATERIALS

There has recently been an increase in research on green synthesis approaches for the production of metal nanoparticles (MNPs). Because fungi rely on metallic materials in some way for proliferation, metabolism, and specialization, they have long been recognized as useful for MNP synthesis. This gives fungi a significant advantage over other species in this regard. They have similar efficient processes of endurance to those found in high concentrations of metals, and they are believed to provide an essential supply of molecules capable of converting metallic ions into MNPs. Fungal mycelia synthesize MNPs in two ways: (i) within the cell and (ii) outside the cell. Extracellular synthesis is by far the most common method since fungi secrete large numbers and a diverse range of enzymes that make the process of synthesizing MNPs sustainable, dependable, adaptable, and scalable. Likewise, the fungi that synthesize MNPs can employ a variety of metals, including copper, iron, silver, nickel, gold, silver, copper, phosphorus, potash, and others. Fungi-mediated MNP production is dependent on numerous factors which include biological inoculant (e.g., species and/or type of strains; method or sample preparation and cultivation); process conditions (e.g., type of metal species and its concentration; temperature, pH; and incubation time) which necessitate the development of new approaches to enhance the reproducibility and techno-commercial viability of the processes (Silva et al., 2016; Dorcheh & Vahabi, 2016). Fungal species have a great deal of potential for producing a wide range of bioactive compounds that can be employed in a variety of different applications. The filamentous mycoflora (ascomycetes) and certain other mycofloral species have been shown to synthesize around 6,400 bioactive compounds, the majority of which have antimicrobial and growth-promoting activities (Bérdy et al., 2005; Guilger-Casagrande & de Lima, 2019). According to Shahzad et al. (2019), silver nanoparticles were synthesized using the *Aspergillus fumigatus* BTCB10, with a diameter of 322.8 nm at 25°C and expanding rapidly when the temperature was elevated to a maximum of 1073.45 nm at 55°C. Similarly, Fayaz et al. (2009) synthesized silver nanoparticles by *Trichpoderma viride* by elevating the temperature. Moreover, Ghareib et al. (2018) biosynthesized copper oxide nanoparticles of size 1 nm through *Aspergillus fumigatus* at pH 6, incubation for 60 hours at 30°C under submerged fermentation (SmF). The world's largest fertilizer cooperative, Indian Farmers Fertilizer Cooperative Limited (IFFCO), has started field trials of its most comprehensive green biosynthesized "nanotechnology-based spectrum of nanofertilizers" (Nanozinc, Nanocopper, and

Nannitrogen) in 2019. According to IFFCO, these environmentally friendly solutions have been launched for the first time in India, and they have the potential to reduce the use of conventional chemical fertilizers by 50% while simultaneously increasing crop output by 15–30% (The Economic Times, 2019; https://economictimes.indiatimes.com/).

9.4 FUNGAL BIOPESTICIDES

EPF are another type of pest management pathogen that can be found across both environments (terrestrial and aquatic) and are related to insects. These pests are either obligate or facultative parasites, mutualistic or symbionts, depending on the species. They penetrate and kill sucking pests such as thrips, mealy bugs, aphids, whiteflies, mosquitoes, scale insects, and all kinds of mites when they come into contact with them (Bamisile et al., 2021; Kaul, 2011). EPF are pledging mycopesticides because they have a diverse range of pathogenesis pathways that can be used to control pest insects. As the name implies, these fungi enter the host's epidermis or gut epithelium and colonize the joint surfaces and inner linings. After infecting and killing their hosts, some fungal species such as *Metarhizium anisopilae*, *Beauveria bassiana*, and *Lecanicillium lecanii* cause muscardin disease in insects, which results in the corpses becoming fossilized or encased in mycelial growth (Bamisile et al., 2021; Kumar et al., 2019; Kaul, 2011). They are classified into 12 classes within 6 phyla and are split into 4 major groups: The Zygomycetes, Laboulbeniales, Hyphomycetes, and Pyrenomycetes. *Metarhizium anisopilae*, *Beauveria bassiana*, *Lecanicillium lecanii*, *Nomuraea rileyi*, and *Paecilomyces farinosus* are just a few of the species that are commonly used in the industry. A number of fungi, primarily streptomycetes, are also capable of producing toxins that are toxic to insects. Approximately 50 such bioactive compounds have been identified to be effective against a variety of insects belonging to the orders Homoptera, Coleoptera, Lepidoptera, Orthoptera, and Mites. Most toxins are cycloheximide, novobiocin, and actinomycin A, which are all highly toxic (Koul, 2011; Dowd, 2002; Cole & Rolinson, 1972). A large number of them have been commercialized on a global scale (Kumar et al., 2019; Senthil-Nathan, 2015; Koul, 2011). Actinomycete *Saccharopolyspora spinosa* produces bioactive compound Spinosyns, commercially available biopesticides that are effective against a wide range of insects, including siphonaterans, dipterans, thysanopterans, and hymenopterans, but less effective against nematodes and aphids (Kumar et al., 2019; Koul, 2011; Sparks et al., 1999). Prominent EPF that have been successfully commercialized around the world are briefly discussed as in subsections (Table 9.2).

9.4.1 *METARHIZIUM ANISOPLIAE*

Metarhizium anisopliae Sorokin var. *anisopliae* is an entomopathogenic mycoflora that is crucial for the management of insects. It reproduces in soils all throughout the world, displaying a diverse spectrum of insect hosts. Metschnikoff initially reported this species as *Entomophthora anisopliae* in 1879 as a pathogen of the wheat cockchafer, but it was subsequently changed to *Metarhizium anisopliae* by Sorokin in 1883 (Tulloch, 1976). *M. anisopliae* is a hyphomycete EPF that is the most frequently employed for insect pest management in the world. This fungal species contains a large number of distinct variants and strains from a wide range of diverse origins and from a wide variety of different insect hosts (Senthil-Nathan, 2015; Roberts & St. Leger, 2004). *Metarhizium* is commonly present in soil beneath natural settings, where the wet conditions promote mycelium growth and the formation of contagious spores known as conidia, which infect soil-dwelling bugs when they come into contact with this fungus. In addition to having the ability to be exploited as a biocontrol inoculant, particularly in the context of malaria vector species, *Metarhizium anisopliae* is a good option for additional research and development programs (Mnyone et al., 2010). *Metarhizium anisopliae* is an entomopathogenic mycoflora that develops on the cuticle of a host species and reproduces. According to preliminary findings, the primary enzyme is a subtilisin-like serine protease that is implicated in this growth that triggers protein breakdown (Woessner, 2013). During invasion, this is the primary protein that is injected into the cuticle of the insect host. Afterwards, the work of exopeptidases (such as carboxypeptidases), is triggered, allowing specific amino acids from the host to be released and utilized for nutritional purposes. Despite the fact that fungi normally include serine carboxypeptidase functionality, St. Leger et al. (1992) found a new carboxypeptidase in the fungus *Metarhizium anisopliae* that was suppressed by 1,10-phenanthroline and showed promise as a potential insecticidal activity (Woessner, 2013). *Metarhizium anisopliae* has been approved as a biological control agent worldwide and has been successfully formulated for commercial application against a variety of insect hosts (Kumar et al., 2019; Senthil-Nathan, 2015).

TABLE 9.2 Some of the Selected Mycopesticides have been Commercially Registered and are Available in the Global Market

Fungal Inoculant	Species Involved	Product Name	Producing Company	Country of Origin	Target Insect Pests	References
Metarhizium sp.	*Metarhizium anisopliae*	Jas Meta®	Shri Ram Solvent Extraction Pvt Ltd.	India	*Hemiptera* (Cercopidae), *Coleoptera*	https://neemplus.com/
	Metarhizium anisopliae	Pacer™	Agrilife	India	Root grubs, ants, termites, green hoppers, locusts	http://agrilife.in/
	Metarhizium anisopliae	Kalichakra®	IPL Biologicals	India	Control eggs, larvae, and adult stages of white grubs, *Semiloopers*, termite, mealy bugs, etc.	https://www. iplbiologicals.com/
	Metarhizium anisopliae	Meta Power	K.N. Biosciences (India) Pvt. Ltd.	India	White grubs, termite	http://www. knbiosciences.com/
	Metarhizium anisopliae	Samridhi™	Jaipur Biofertilizers	India	White grubs, termite	https://www.jaipur biofertilizers.com/
	Metarhizium anisopliae	Metacide	Tari Biotech	India	Paddy hoopers, termites, mites, coconut *Rhinoceros* beetles	http://taribiotech. co.in/
	Metarhizium anisopliae	Bio Bullet	Universal Bio-con Pvt. Ltd.	India	White grubs, termite	https://universalbiocon. com/
	Metarhizium anisopliae	Metamole	Micro-Organism Technology (India)	India	White grubs, termite	https://www.indiamart. com/microorganism technologyindia/
	Metarhizium anisopliae	Achieve	Real IPM	Mozambique	*Tetranynchus urticae*	Bamisile et al. (2021); Akutse et al. (2020)
	Metarhizium anisopliae	Bio-Green Granules	Becker-Underwood	Australia	*Coleoptera*	Bamisile et al. (2021); de Faria & Wraight (2007)

TABLE 9.2 *(Continued)*

Fungal Inoculant	Species Involved	Product Name	Producing Company	Country of Origin	Target Insect Pests	References
	Metarhizium anisopliae	BioBlast	EcoScience	United States	*Isoptera* (Termopsidae, Kalotermitidae, Rhinotermitidae)	Bamisile et al. (2021); de Faria & Wraight (2007)
	Metarhizium anisopliae	BioPath	EcoScience	United States	*Blattodea* (Blattidae, Blattellidae)	Bamisile et al. (2021); Zimmermann (2007b)
	Metarhizium anisopliae	Cobican	Probioagro	Venezuala	*Hemiptera*, *Coleoptera*	Bamisile et al. (2021); de Faria & Wraight (2007); Zimmermann (2007b)
	Metarhizium anisopliae	Campaign	Real IPM	Ghana, Uganda	Trips, mealybugs, fruit flies	Akutse et al. (2020)
	Metarhizium anisopliae	Green Muscle	CABI Bioscience/NPP	France/United Kingdom	Grasshopper, locust	Shah & Pell (2003)
	Metarhizium anisopliae	Green Guard SC/ULV	Becker-Underwood	Australia	*Orthoptera* (Acrididae)	Zimmermann (2007b); Bamisile et al. (2021)
	Metarhizium anisopliae	Mazao achieve	Real IPM	Kenya	*T. urticae*, mealybugs	Akutse et al. (2020)
	Metarhizium anisopliae	Real Metarhizium 69	Real IPM	South Africa, Zambia	Mealybugs, leaf miners, thrips, fruit flies	Akutse et al. (2020)
	Metarhizium anisopliae	Real Metarhizium 78	Real IPM	Canada	*T. urticae*, Plant growth regulator	Akutse et al. (2020)
	Metarhizium anisopliae	TAE-001 Technical Bioinsecticide	Novazymes Biologicals	Unites States	Coleopterans, Curculionidae, Elateridae	Zimmermann (2007b); Bamisile et al. (2021)
	Metarhizium anisopliae	Tick-EX EC/G	Novazymes Biologicals	Unites States	*Coleoptera* (Scarabaeidae), Acari (Ixodidae),	Zimmermann (2007b); Bamisile et al. (2021)

TABLE 9.2 *(Continued)*

Fungal Inoculant	Species Involved	Product Name	Producing Company	Country of Origin	Target Insect Pests	References
	Metarhizium anisopliae	Metarhizium Schweizer	Lbu (Eric Schweizer Seeds)	Switzerland	*Coleoptera* (Scarabaeidae)	Zimmermann (2007); Bamisile et al. (2021)
	Metarhizium anisopliae	Metathripol	ICIPE	Kenya	Thrips	https://sitem.herts.ac.uk/aeru/bpdb/Reports/1980.htm
	Metarhizium anisopliae	Bio-Magic	Stanes	India	*Hemiptera*, *Coleoptera*	Bamisile et al. (2021)
		Gran Met-P	Kwizda/Agrifutur	Australia/Italy	*Coleoptera* (Nitidulidae, Scarabaeidae, Curculionidae)	Bamisile et al. (2021); Zimmermann (2007b); de Faria & Wraight (2007)
Beauveria sp.	*Beauveria bassiana*	Jas Bassi	Shri Ram Solvent Extractions Pvt. Ltd	India	Cut worms, root grubs, stem borers, aphids, mealybugs, thrips	https://neemplus.com/
	Beauveria bassiana	Biosoft®	Agriland Biotech Limited	India	White fly, mites, aphids, jassids, mealybugs, hoppers, army worm, etc.	https://www.agrilandbiotech.com/
	Beauveria bassiana	Racer	Agrilife	India	Fruit borers, stem borers, leaf folder, *Spodoptera*, *Helicoverpa*	https://www.agrilife.org.in/
	Beauveria bassiana	Daman	IPL Biologicals Ltd	India	Hairy caterpillars, borers, *Spodoptera*, *Helicoverpa*	https://www.iplbiologicals.com/
	Beauveria bassiana	Bio-Power	T-Stanes	India	Hairy caterpillars, borers, *Spodoptera*, *Helicoverpa*	https://tstanes.com/

TABLE 9.2 *(Continued)*

Fungal Inoculant	Species Involved	Product Name	Producing Company	Country of Origin	Target Insect Pests	References
	Beauveria bassiana	Wah	Amit Biotech Pvt. Ltd.	India	Caterpillars, *Helopeltis theivora*, weevils, berry borers, grubs, leaf feeding insects	http://abplindia.com/
	Beauveria bassiana	Beveroz	Utkarsh Agrochem Pvt. Ltd.	India	Weevils, berry borers, caterpillars, fruit borers, stem borers	https://www.utkarshagro.com/
	Beauveria bassiana	Bio Dawn	Jaipur Biofertilizers	India	Hairy caterpillars, aphids, mealybugs, thrips, *Spodoptera*, *Helicoverpa*	https://www.jaipurbiofertilizers.com/
	Beauveria bassiana	Bauveril	Laverlam S.A., Colombia	Colombia, Dominican Republic	*Coleoptera* (Scarabaeidae, Curculionidae), *Lepidoptera*	Bamisile et al. (2021); de Faria & Wraight (2007)
	Beauveria bassiana	Boverin	Biodron	Russia	*Cydia pomonella*, Colorado potato beetle	Bamisile et al. (2021); Zimmermann (2007a)
	Beauveria bassiana	Naturalis	Intrachem	Italy	*Hymenoptera*, Hemiptera, *Diptera*, *Coleoptera*	Bamisile et al. (2021); Zimmermann (2007a)
	Beauveria bassiana	Conidia	Hoechst Schering AgrEvo	Columbia	*Coleoptera*	Bamisile et al. (2021); Zimmermann (2007a); de Faria & Wraight (2007)
	Beauveria bassiana	Boverol	Fytovita	Czech Republic	*Cleoptera*	Zimmermann (2007a); de Faria & Wraight (2007)
	Beauveria bassiana	Ostrinil	Arysta	France	Crambidae (*Lepidoptera*)	de Faria & Wraight (2007); Zimmermann (2007a); Bamisile et al. (2021)

TABLE 9.2 *(Continued)*

Fungal Inoculant	Species Involved	Product Name	Producing Company	Country of Origin	Target Insect Pests	References
	Beauveria bassiana	Trichobass-L Trichobass-P	AMC Chemical/ Trichodex	Spain	Scarabaeidae, Curulionidae (*Coleoptera*), *Lepidoptera*, *Thysanoptera*, *Hemiptera*	Bamisile et al. (2021); Zimmermann (2007a); de Faria & Wraight (2007)
	Beauveria bassiana	Proecol	Probioagro S.A.	Venezuela	Noctuidae (*Lepidoptera*)	Bamisile et al. (2021); Zimmermann (2007a); de Faria & Wraight (2007)
	Beauveria bassiana	BotaniGard	Loverlam International	United States	Aphids, whiteflies, grasshoppers, and many other insects	Bamisile et al. (2021); de Faria & Wraight (2007)
	Beauveria bassiana	Naturalis-L	Andermatt Biocontrol Troy Biosciences Inc.	United States, Switzerland	Thrips, aphids, whiteflies, grasshoppers, and other insects	Bamisile et al. (2021); Zimmermann (2007a); de Faria & Wraight (2007)
	Beauveria bassiana	Mycotrol	Laverlam International	United States	Whiteflies, aphids. thrips, and other insects	Bamisile et al. (2021); Shan & Pell (2003)
	Beauveria brongniartii	Betel	Arysta	France	Scarabaeidae (Coleoptera)	Zimmermann (2007a); de Faria & Wraight (2007)
	Beauveria brongniartii	Biolisa-Kamikiri	Nitto Denko	Japan	Cerambycidae (*Coleoptera*)	Bamisile et al. (2021); Zimmermann (2007a)
	Beauveria brongniartii	Beauveria Schweizer	Lbu	Switzerland	Scarab beetles, European cockchafer	Bamisile et al. (2021); Shah & Pell (2003)
Lecanicillium (=*Verticillium*) sp.	*Lecanicillium lecanii*	Jas Verti®	Shri Ram Solvent Extractions Pvt Ltd.	India	Whiteflies, thrips	https://neemplus.com/

TABLE 9.2 *(Continued)*

Fungal Inoculant	Species Involved	Product Name	Producing Company	Country of Origin	Target Insect Pests	References
	Lecanicillium lecanii	Mealikil	Agrilife	India	Mealybugs, mites, whitefly, thrips, aphids, jassids, mites	https://www.agrilife.org.in/
	Lecanicillium lecanii	Alestra	Amruth Organic Fertilizers	India	Whiteflies, thrips	https://amruthgroups.com/
	Lecanicillium lecanii	Mycotal	Koppert Biological Systems	Netherlands	Thrips, whiteflies	Bamisile et al. (2021); Shah & Pell (2003)
	Lecanicillium lecanii	Verticoz-P	Utkarsh Agro-chem Pvt. Ltd.	India	Mealybugs, mites, whitefly, thrips, aphids, jassids, mites	https://www.utkarshagro.com/
	Lecanicillium lecanii	Vertalec	Koppert Bio-logical Systems	Netherlands	Aphids	Bamisile et al. (2021); Shah & Pell (2003)
Paecilomyces. (=*Isaria*) sp.	*Paecilomyces lilacinus*	Mysis	Varsha Bioscience and Biotechnology	India	Root knot nematodes, cysts nematodes, burrowing nematodes	http://www.varshabioscience.com/
	Paecilomyces lilacinus	*Criyagen*	Criyagen Agri and Biotech Pvt. Ltd.	India	Root knot nematodes, cysts nematodes, burrowing nematodes	https://criyagen.com/
	Paecilomyces lilacinus	Paecilo®	Agrilife	India	Root knot nematodes, cysts nematodes	http://agrilife.in/
	Isaria farinosa	Melocont-Pilzgerste Paecilomin	Agrifutur-Kwizda	Russia, Italy, Austria	Apple moth, larch caterpillar, Siberian pine caterpillar	Weng et al. (2019); Bamisile et al. (2021)
	Isaria fumosorosea	NoFly™	Natural industries Inc	United States	Mealybugs, thrips, aphids, whiteflies, psyllids, fungus gnats	Weng et al. (2019); Bamisile et al. (2021)

9.4.2 BEAUVERIA BASSIANA

According to Agostino Bassi, who first discovered *Beauveria bassiana* in the cadavers of silkworms in the 19th century, the parasite can infest more than 200 insect species across six orders and 15 families (Nakahara et al., 2009). It grows rapidly, producing a wide range of toxins that cause exogenous infections (Naqqash et al., 2016). For a long time, *B. bassiana* was cultured for its ability to generate conidia on the upper portion of leaves where pests usually sit, allowing an insect to indicate the presence of specific kinds of metabolic compounds by ingesting the leaves. Following that, if the bioactive compounds are detected, the insects may avoid using these crops for oviposition and feeding purposes. As with the pest, *B. bassiana* began its action on the insect by creating appressoria and secreting enzymes and specific proteins after attaching to the insect and penetrating the insect's cuticle. The majority of the time, *B. bassiana* destroys the immune system of an insect, causing it to die (Zhang et al., 2011; Lewis et al., 2009). On the surface of the fungus, proteins having cysteine-rich hydrophobins (Hyd1 and Hyd2) emerge, and they act as virulence agents, allowing fungi to adhere to insects' cuticular surfaces on the outside of the cuticula. After insect cells have been attached to a surface, proteins like cytochrome P450 and enzymes like esterases, aldehyde dehydrogenases, catalases, and long-chain alcohols, break the lipid bilayer and integral proteins of the cell. A fungal infection results in a decreased proclivity for plant feeding as well as a decrease in insect numbers, both of which reduce an insect's capacity to convey harm and infestations to plants (Das et al., 2021; Zhang et al., 2011). There have been a number of studies suggesting the fungus *Beauveria bassiana* is indeed an effective mycoinsecticide for controlling a variety of pests. *Frankliniella occidentalis* (Thysanoptera: Thripidae), a western flower thrips, was tested against the fungus *B. bassiana* strain RSB, which caused 69–96% death rates after 10 days of inoculation of first instar larvae with a concentration of 1×10^7 conidia/mL, indicating that the fungus is effective against this species (Gao et al., 2012). When applied to cucumber plants grown in a greenhouse, a single application of the predatory mite *Neoseiulus barkeri* or *B. bassiana* fungus resulted in a significant reduction in both populations (larval and adult) of the pest *F. occidentalis* (Wu et al., 2016).

9.4.3 LECANICILLIUM LECANII

Lecanicillium lecanii was previously recognized as *Verticillium lecanii* (Zimmerman) Viegas and is now understood to be an anamorphic type of the family of Clavicipitaceae and Cordyceps group of genera (Zare & Gams, 2001). Moreover, *Lecanicillium lecanii* (Zimm.) is a fungus that is usually referred to as "white halo fungus" and is responsible for the transmission of mycosisin to insects and pests (Subramaniam et al., 2021). It is primarily used for the control of thrips and aphids in greenhouses as well as open fields. However, products for the control of homopteran, dipteran, and lepidopteran pests of flowers, fruits, vegetables, and other crops are available in parts of the globe. The multiple products contain blastospores of *Lecanicillium lecanii* produced by submerged culture or conidia formed by solid-state fermentation, depending on the method of production. It was initially believed that the requirement for extremely high humidity was a major barrier to the widespread adoption of this mycoflora for application in greenhouses and polyhouses, where high moisture content also promotes the incidence of plant pathogens. However, the advancement of a better formulation that reduces the moisture content prerequisites of the fungus has significantly reduced the severity of this problem (Goettel & Glare, 2005; Subramaniam et al., 2021). Saruhan (2018) used *V. lecanii* as a mycoinsecticide to control *Aphis fabae* (Hemiptera: Aphididae) in controlled conditions. In a similar vein, the coupled use of *L. lecanii* with other BCAs, like *Encarsia formosa* (whitefly parasites), *Amblyseius* sp., as well as other pest antagonists, has already been evaluated in some cases. The findings revealed that the consortium or mixed application of *L. lecanii* with other mycoinsecticides or insect parasites-based bioagents is indeed a viable option (Alavo, 2015). A composition of *V. lecanii* is now available on the market for purchase under the brand name Mycotal® and is intended exclusively for use against whitefly larvae. When such a product is combined with an emulsifiable vegetable oil as an adjuvant-based formulation, its efficacy is increased (Koppert, 2015; Alavo, 2015).

9.4.4 PAECILOMYCES SP.

The solitary species of *Paecilomyces*, *P. variotii* Bainier, was reported in 1907 as a genus closely connected to *Penicillium* (Bainier, 1907). In the genus *Paecilomyces*, there are species that are thermotolerant, thermophilic, and mesophilic, with yellow-brown colonies that indicate teleomorphic regions correlating to the genera *Talaromyces*, *Thermoascus*, and *Byssochlamys*; and in the genus *Isarioidea*, there are species that are mesophilic and have pink, green, purple, or yellow colonies that show teleomorphic regions. The entomopathogenic or nematophagous species, such as *Paecilomyces lilacinus* or *P. fumosoroseus*, are included in the first section (Moreno-Gavira et al., 2020; Samson, 1974; Ibarra et al., 2006). Several species of the genus *Paecilomyces* exist, both virulent and saprophytic, and can be located in various environments and habitats, comprising soil, insects, compost, nematodes, decomposing plant matter or food, pasteurized food stuffs, marine sediments, and the rhizosphere of a variety of plants, to name a few (Moreno-Gavira et al., 2020; Mohammadi et al., 2016; Barra et al., 2013; Aminuzzamana et al., 2013). In a number of studies, *Paecilomyces* has been shown to have nematicidal action. *P. lilacinus* is a species of this genus that has the ability to enter both the eggshells and internal structures of distinct species of nematodes during their adult and juvenile stages. This is accomplished by consequent hyphal branching, sporulation, and appressoria development. It has been demonstrated that this species can produce lytic enzymes (such as lipases, proteases, chitinases, and amylases) that have a nematicidal effect (Dong et al., 2007; Khan et al., 2006; Moreno-Gavira et al., 2020). Williams et al. (1999) demonstrated that *P. lilacinus* parasitizes eggs at all phases of development, including unfertilized juveniles. Egg infection develops when mycelia are attached to the egg surface and, as a result, appressoria form. The fungus subsequently multiplies, and conidiophores are produced. Eggs of juvenile *M. hapla* were shown to be significantly more sensitive to serine proteases generated by *P. lilacinus* than eggs comprising more fully mature juveniles. On the contrary, there were no symptoms of harm in the larvae. *P. lilacinus* is able to infect cysts of *Globodera* sp., *Heterodera* sp., and female *Meloidogyne* sp. (Jatala et al., 1979). In these instances, hyphae penetrated the nematode body through natural holes (Jatala, 1986). Evidence suggests that cuticle penetration of nematodes and subsequent cell disintegration are mediated by a variety of hydrolyzing proteins and enzymes, including proteases (mostly serine proteases), chitinases, and collagenases (Huang et al., 2004). *Paecilomyces* species have been found to have an inhibitory activity on fungi that cause aerial plant diseases. Investigations have also shown that *P. lilacinus* and *P. variotii* are highly effective, but these were conducted *in vitro*. Most of the antagonistic effects of *Paecilomyces* species can be attributed to the availability of nutrients, competition, and space. Moreover, other secondary metabolite-related pathways have been observed, including plasmolysis of spore germ tubes in *Pyrenophora triticirepentis*, mycoparasitism of *F. oxysporum*, mycelium lysis in *Moniliophthora roreri*, and antibiosis against *R. solani* caused by *P. lilacinus* and *P. variotii*, among others (Subramaniam et al., 2021; Larran et al., 2016; Jacobs et al., 2003).

There are several species of *Paecilomyces* that can form stable propagule substrates like blastospores or conidia, making them good candidates for use in developing mycoinsecticides. Over the course of more than three decades, the EPF *Isaria fumosorosea* and *Isaria farinosa* were known by the names *Paecilomyces fumosoroseus* and *Paecilomyces farinosus*. Both fungi have a global distribution and a relatively broad spectrum of host range. However, whilst *I. farinosa* is generally considered to be of limited value in both study and as a biological control, *I. fumosorosea* is considered to be a species complex, and multiple strains have been effectively employed for biological control of several insect pests, especially whiteflies (Zimmermann, 2008). According to Ruiu (2018), commercial pest management bioformulations have been developed with *P. lilacinus* and *P. fumosoroseus* and have been successfully commercialized worldwide (Jackson et al., 2006; Moreno-Gavira et al., 2020). *Paecilomyces* has been shown to manage pests by either reducing its growth or simply killing them owing to mycosis. The pesticide *P. fumosoroseus (=Isaria fumosorosea*) has also been demonstrated to be more effective against *Frankliniella occidentalis* than conventional insecticides such like fipronil (Jessica et al., 2019; Hunter et al., 2011; Dunlap et al., 2007).

9.5 ALTERNATIVE SAFE SOLUTIONS TO CHEMICAL PESTICIDES AS BIOPESTICIDES

Among the most pressing public issues concerning the applicability of EPF for biocontrol of insects and pests and plant pathogens are the safety of consumers, animals, plants, pollinators, natural enemies, and the overall ecology. Several research has been carried out in accordance with ecotoxicological evaluations of different EPF, and the results have been published (Bamisile et al., 2021; Vestergaard et al., 2003). Roberts (1977) performed the first investigation into the adverse effects of *M. anisopliae* on fish, which was the first of its kind. According to the author, the treatment of spores in waterbodies was shown to have no substantial impact on the fatality of the various fish studied. Similar findings were made in experiments done to determine the deleterious effects of *M. anisopliae* also on *Rana pipiens* Schreber, the African clawed frog, the northern leopard frog, and *Xenopus laevis* Daudin, both of which were conducted in the United States (Peveling & Demba, 2003). The fringe-toed lizard, *Acanthodactylus dumerili*

Milne-Edwards, was also evaluated against another variant of *M. anisopliae* var. acridum, produced as a bioinsecticide widely used for controlling the desert locust. The conidia inhalation and oral exposure of *M. anisopliae* had positive effects against *S. gregaria* locusts and had no deleterious effect on the reptiles. However, the chemical insecticide fipronil has been discovered to have a devastating effect on *A. dumerili* (Peveling & Demba, 2003; Bamisile et al., 2021). Additionally, EPF variants have been tested on a variety of birds, either by feeding them entomopathogenic-affected insects or by feeding them fungal spores that have been deposited in their meal. As an example, *B. bassiana* conidial spores were fed to ring-necked pheasants, *B. brongniartii*-infected white grub was given to hens, and *B. bassiana* conidial spores were provided to the American sparrowhawks, *Falco sparverius* Linnaeus (Johnson et al., 2002; Althouse et al., 1997). *B. bassiana* was shown to be non-toxic to rats and other vertebrates after conducting toxicity testing on the fungus. It was found that the fungus *B. bassiana* only could live in mice for three days after intramuscular injection (Semalulu et al., 1992; Bamisile et al., 2021). Furthermore, as per Weng et al. (2019), there seems to be no published record of the detrimental impact of entomopathogenic fungus from water and the environment on human health up to this point. The fungal spores are unable to persist or grow in the environment for an extended period of time, which is the cause of this phenomenon (Shah & Pell, 2003).

As a result, Milner & colleagues (2002) came to the conclusion that the likelihood of *Metarhizium*-based BCAs having a detrimental impact on aquatic living species is fairly low. Most mycotoxins that are generally called environmental or food web contaminants are already known to be generated as a consequence of plant infestation by mycophytopathogens, instead of as an outcome of fungal entomopathogens colonizing the plant through fungi colonization (Weng et al., 2019; Oyedele et al., 2017). The mycotoxins released by *Aspergillus* sp., *Fusarium* sp., as well as other mycophytopathogens, for example, have been reported to pollute the ecosystem through the plants and goods that they infect (Mallebrera et al., 2018; Oyedele et al., 2017). As a mycoinsecticide, any EPF must be thoroughly tested for its ability to harm important non-target species during its preparation and approval process. As a result, BCAs tend to be safer than synthetic goods in general. Biopesticides are safer and more effective than their chemical counterparts, so there is a great need to raise awareness and inform policymakers and regulators (Bamisile et al., 2021).

9.6 FUNGAL BIOPESTICIDES AND BIOFERTILIZERS' FORMULATIONS AND DELIVERY SYSTEMS

The use of biological products is highly desirable, but it is extremely difficult to create formulations that are acceptable. In addition to the physical properties and ease of use that are required, the formulated product should also maintain the biological agent's functionality while in storage and during application (Bharti & Ibrahim, 2020; Woods, 2003). It is important that commercial fungal biopesticides and biofertilizers are cost-effective to produce, have prolonged storage stability, significant activity, and are easy to handle for mixing and application. Bio-inputs need new formulations to address issues with potency and degradation, as well as convenience in handling and application (Gašić & Tanović, 2013). There are many bio-input formulations and delivery systems available on the market, which are briefly discussed as in subsections (Figure 9.1).

FIGURE 9.1 Some of the fungal-based commercial bioinputs (mycofertilizers, mycopesticides) available in the market.

9.6.1 TYPES OF FORMULATIONS

In most cases, microbial bio-inputs (biopesticides and biofertilizers) active ingredients are developed in the same way as chemical inputs. To make things easier for users, they use the same equipment for multiple procedures. Biopesticides and biofertilizers are derived from living microorganisms. The survival of these microorganisms must be preserved to reasonable standards during the preparation and storage process. The microorganisms must come out of their slumber at the time of application in order to function. Biopesticide formulation issues are significant, and a comprehensive understanding of the mechanisms that lead to the loss of effectiveness is essential for further advancement (Bharti & Ibrahim, 2020; Gašić & Tanović, 2013; Boyetchko et al., 1998).

9.6.1.1 DUST PARTICLES (DP)

Delicately ground, solid carrier powders (clay and talc powder) having particle sizes of 50 to 100 μm are used in the synthesis of dust particles (DPs) type formulations by adsorbate of an active substance. Both manual and mechanical application methods can be used to apply dust. Moreover, ultraviolet (UV) protectants, adhesive materials, and anticaking agents help with adsorption, so they're used as inactive ingredients in this simulation. The typical dust composition of an active ingredient (organism) is 10%. The inhalation dangers of these substances outweigh any benefits they may provide in certain conditions. This is an aged type of formulation that was used for several more years before granules (GRs) were established, and they were restricted because of their negative health effects on the consumers of the medication (Gašić & Tanović, 2013).

9.6.1.2 GRANULES (GRS)

Granules (GRs) are equivalent to dust formulations in size and weight, but they are much bigger and heavier than DPs (granule size: 100–1,000 μm and micro-GRs: 100–600 μm). GRs are produced from complex materials like kaolin, silica, polymers, dry organic fertilizers, decomposed plant residues, vermiculite, etc. (Tadros, 2005). The concentration of microorganism as an active ingredient in GRs varies from 5 to 20%. Whether the active ingredient is coated on GRs or absorbed into the carrier in this case. To make granular products, there is a requirement for a small amount of water to make slurry and mix with the active ingredient that is extruded and dried if needed. A granular formulation of biopesticides and biofertilizers is applied to soil to suppress nematodes, insects, and weeds that live in the soil, and it is also used to increase nutrient uptake by the plant roots (Lyn et al., 2010).

9.6.1.3 WETTABLE POWDERS (WPS)

Wettable powders (WPs) are precisely crushed, dry powders that can be dispersed in water and applied to crops. This is followed by a combination of an active ingredient with the wetting, surfactant, dispersion agents, and inert materials (about 5 microns) required for WPs. Producers face major concerns about safety and health issues due to the products' dustiness, which can trigger problems of inhalation and eye and skin irritation if proper precautionary measures are not taken. For these and other reasons, WPs are rapidly being replaced by water dispersible GRs (WDG) and suspension concentrates (SCs) that are most commonly used globally in biopesticide and biofertilizer formulations (Gašić & Tanović, 2013; Knowles, 2005).

9.6.1.4 WATER DISPERSIBLE GRANULES (WDG)

Water dispersible granules (WDG) that disperse in water have been developed to solve the dustiness issue with powder formulations. Unlike WP, which dissolves in water and forms a slurry, water dispersible grains are made to disperse in water. Aside from being largely dust-free, these WGs have a high degree of stability when it comes to storage. Extrusion granulation, spray drying, fluid bed granulation, and other processes can be used to produce water-dispersible granules. Dispersion agent and wetting agent are present in the products. However, the dispersing agent is normally present in a larger proportion. WDG is generally more expensive than previous forms of formulations (WP, dust), yet they are nevertheless preferred by many consumers because of their health, safety, and ease of application (Bharti & Ibrahim, 2020; Knowles, 2008).

9.6.1.5 SUSPENSION CONCENTRATE (SC)

A suspension concentrate (SC) is a mixture of a finely powdered, solid active component distributed in a liquid phase, typically water, which is used in the production of pharmaceuticals. Because the solid particles do not dissolve inside the liquid state, the mixture must be stirred prior to application in order to keep the particles evenly dispersed throughout. In order to maintain the required consistency of the SC, it is composed of a complicated mixture of different additives like thickening agents, antifoaming agents, wetting/dispersing agents, and other components. Their particle size distribution ranges from 1 to 10 microns, and they are generated using a wet grinding process. The presence of inert substances deposited onto the surfaces of the particles during the milling process prevents small particles from reaggregating. The larger surface area of these tiny particles allows the active ingredient to be more easily accessed by plant tissues, resulting in improved bioefficacy. Because they are water-based, they have several advantages, including easy pouring and measurement, safety for the user as well as environmental friendliness, as well as cost-efficiency (Gašić & Tanović, 2013; Knowles, 2005).

9.6.1.6 CAPSULE SUSPENSION (CS)

Microencapsulated active ingredient (MEA) in an aqueous continuous liquid phase (CS) is designed to be diluted with water prior to its application. To protect the active substance in the bioactive agent, the capsules are coated with gelatin and certain other polymers. As a result, the bioagent is shielded against exposure to harsh environmental factors such as rain, variations in temperature, and UV-ray exposure, and its latent stability is improved. Most commonly, the concept of copolymerization is employed in the encapsulation process. Substantial use of microcapsule encapsulation has been made to reduce the size of mycobiopesticide and mycofertilizer formulations while increasing their efficacy (Gašić & Tanović, 2013; Brar et al., 2006).

9.6.1.7 ULTRA-LOW VOLUME LIQUIDS (UL)

"Ultra-low volumes" (ULs) that contain high concentrations of active ingredients that are exceptionally soluble in crop-compatible fluids. UL formulations are not meant to be mixed with water prior to use and quite often contain surface-active chemicals and drift-control components. Low-volatile liquids can be transported and used easily. A dispersed biological control agent can be used as an active component in UL liquid biopesticides in a similar manner (Gašić & Tanović, 2013; Woods, 2003).

9.6.2 DELIVERY SYSTEMS

Product delivery must be simple, cost-effective, and quick, as well as to the proper action site, and must be compatible with existing agriculture practices, including equipment. Various delivery systems of formulated microbes can be used and applied to soil, tubers, seed pieces, cuttings, transplants, or mature plants (Bharti & Ibrahim, 2020).

9.6.2.1 SEED TREATMENT

In contrast to being percoated or encapsulated onto seeds, BCAs can be blended with seeds at the time of sowing, administered in-furrow, or integrated into the seed bed (Thomashow & Weller, 1990). It is common practice to coat seeds with dormant microorganisms that are capable of sustaining a period of dehydration in the form of liquid, oil-based, or dry powder formulations. In order to improve the adherence of the microbial inoculant to the seeds, additives like gum Arabic and *xanthum* gum are frequently used. Using a specific seed coating procedure known as seed encapsulation, it is possible to extend the survival of microbial inoculants on seeds by encapsulating the seed, the microbe, and maybe other ingredients such as micronutrients or pesticides, in a polymer gel matrix gelatinous. The alginate hydrogel formulation, GEL-COAT, is a type of seed encapsulation bioinput, which was developed as a delivery system for controlling entomopathogenic nematodes and has been patented. Because the active components are adequately encapsulated until they are discharged during germinating seeds, the method of seed encapsulation offers the particular advantage of increased health and safety purposes and reduced environmental hazard. Seed treatment may not be as effective if it is not applied appropriately by the farmer, and the additional workload for producers is a drawback (Bharti & Ibrahim, 2020; Boyetchko, 1996).

9.6.2.2 SOIL TREATMENT

Soil treatment methods have been observed to be more effective than others after fumigation or at the time of sowing or planting (Bharti & Ibrahim, 2020). A large population of biological control agents can limit the colonization of diseases in sterile soils or growth mixtures. "Suppressive soil" is created, making it harder for other organisms to colonize the area. The broadcasting of powder, dust, and granular formulations into the soil is possible, while WP, oil or liquid carrier, and water-dispersible granular formulations can be given in furrows (Lewis, 1991). Soil application can be a beneficial approach for suppressing disease pathogens that overwinter in soil. *Trichoderma harzanium* based product NEEMODERMA, for example, can be applied to the soil to lower the number of pathogenic *Fusarium* species (Kumar et al., 2019).

9.6.2.3 TREATMENT OF PLANTS

Plant roots, foliage, and wounds, all can be treated using BCAs by spraying, drenching, or dipping. Formulated mycoflora can be sprayed directly to the shoots or roots as drench or a dip to promote shoot and root development (Bharti & Ibrahim, 2020; Funk et al., 1997). Freshly cut pine stumps was dusted with biofungicide *Phelbia gigantea* spores in an aqueous dispersion to inhibit the invasion of the pathogenic fungus *Heterobasidion annosum*, thereby preventing open wounds (Bharti & Ibrahim, 2020; Rishbeth, 1975). Depending on the crop, pest, and delivery method, the mycoflora used in foliar sprays can be formulated in a variety of ways. Foliar sprays can be applied with either slurries or liquids; slurries are typically made by reconstituting dry or wet carrier compositions into a reconstituted state. To help the microbes adhere to and disseminate on plant surfaces, as well as safeguard them from adverse climatic conditions like dehydration, unfriendly pH, and UV irradiation, stickers, spreaders, emulsifiers, and other additives and adjuvants are used (Bharti & Ibrahim, 2020; Shieh, 1995). Cory & Bishop (1995) demonstrated that the application of *Verticillium lecanii*, an entomopathogenic mycoflora, to suppress the aphid *Aphis gossypii* is accomplished with a low-volume electrical rotary atomizer and was found effective with foliar spray.

9.7 REGULATORY FRAMEWORK FOR COMMERCIAL PRODUCTION OF FUNGAL-BASED BIOINPUTS

Governments worldwide are working on reducing the use of chemical fertilizers and agrochemicals due to the negative effects they have on the ecosystem, and in order to accomplish this, they have postulated some strategies to decrease the consumption of chemical inputs. At the moment, different countries have different rules and legislation in place to encourage the use of biological fertilizers and biopesticides. Globally, people are becoming more concerned about environmentally friendly and safer food systems after the pandemic COVID-19 incidence, and nations have enacted stringent regulations governing the use of artificial chemicals in food products and other related products (Bamisile et al., 2021; Singh & Arora, 2016; Arora et al., 2016). There are numerous policies and laws in place around the world that not only encourage the use of bio-inputs (fungal biofertilzers and biopesticides) but also regulate their use. Several countries are also attempting to overcome obstacles and create a favorable business environment for the commercialization of bio-inputs, which includes the United States, the United Kingdom, India, and others.

The absence of appropriate governmental guidelines in certain nations, as a result of a lack of an appropriate regulatory framework, creates difficulties in the launch of innovative biological products into the marketplace (Arora et al., 2016; Ochieng, 2015). The product registration process is the most difficult of the main challenges faced by bio-input production companies. It has been discovered that inadequate assessment methods and disproportionally required information frequently result in a long and drawn-out registration system that is unnecessarily costly and complex to complete (AGBR, 2015; Arora et al., 2016). In particular, many small and medium-sized businesses are discouraged from pursuing such time-consuming and expensive registration procedures (AGBR, 2015).

The United States accounts for a significant share of the global biopesticide market. In the United States, the Environmental Protection Agency (EPA) has established an important and sophisticated regulatory regime for the regulation and supervision of biopesticides (USEPA, 2010), and this framework necessarily involves the development of special characteristics for registration that are not required by other regulatory frameworks (Arora et al., 2016; Chandler et al., 2011). The Office of Chemical Safety and Pollution Prevention (OCSPP) and the Office of Pesticide Programs (OPP) are responsible for regulation of BCAs (Matthews, 2014). In the European Union (EU), microbial biological control agents (MBCAs) that contain living microorganisms (fungi, bacteria, viruses, etc.), are regulated at both the European Union (EU) and Member State (MS) level. The European Union (EU) assessment procedure was established first time in Regulations 91/414/EEC, in an effort to harmonize the disparate national registration schemes that existed inside the EU up until 1993. The Regulation No. 1107/2009, which came into effect in 2011, revoked this directive. The amendment was aimed at creating regulatory situations that are more conducive to the particular requirements of MBCAs, which were previously unmet. Comparing the execution of Regulation No. 1107/2009 to the previous Regulation 91/414/EEC, only 26% of registered active ingredients and Plant Protection Products (PPP) were found to have forwarded the review under the latter directive (Frederiksa & Wesseler, 2018). The variations in regulatory requirements between the United States and the European Union may cause a challenge for the latter. Regulatory discrepancies, which are similar to non-tariff trade barriers, put a substantial burden on international trade and commerce (Felbermayr et al., 2013; Frederiksa & Wesseler, 2018).

Rules and regulations are in place in India to encourage large-scale production and licensing of biopesticides. Both the National Agricultural Technology Project (NATP) and the National Farmer Policy (2007) pushed the use of biopesticides in agriculture during the period from 1998 to 2005. The 1968 Insecticide Act simplified the registration and regulating process for biopesticides, resulting in increased manufacture and use. The Central Insecticides Board & Registration Committee (CIB & RC) are powerful organizations for biopesticide regulation under this legislation (http://ppqs.gov.in/divisions/central-insecticides-board-registration-committee), and they worked together under this act (Arora et al., 2016). Experts from all relevant disciplines/fields are on the Apex Advisory Committee (CIB). To comply with the OECD's recommendations, the CIB has reduced both the registration guidelines and data requirements for biopesticides, as well as the minimal infrastructure facilities (NAAS, 2013). The RC verifies and scrutinizes claims of bio-efficacy and safety to humans and animals before approving registrations.

Marketable products, such as biofertilizers, must meet certain legal requirements in order to be considered marketable. There are currently no legal definitions for the term "biofertilizer" in either the United States or the European Union, nor are there any particular legal regulations defining their characteristic features. The organic production system in the European Union (EU) is permitted by the EU Commission Regulation No. 889/2008 on the Production System of Organic Agriculture, which specifically allows for the use of microorganisms for biological control of pests and diseases. As a result, they are classified as biological control agents within the legal framework governing plant protection products. In a similar vein, the National Organic Program (NOP) of the United States envisions only the prospect of using biological entities for plant protection, not their use (Malusá & Vassilev, 2014).

In India, for biofertilizers, the Ministry of Agriculture, Government of India enacted a regulation law on The Fertilizer (Control) Order, 1985, with amendments from time to time (last amended on February 23, 2021, with a new category of biostimulant as schedule VI), which prescribes production and marketing standards for the various types of microorganisms that make up the biofertilizer. The following quality parameters are specified in the criterion: the physical form, the level of contamination, the lowest possible count of viable cells (CFU), particle size in the solid carrier-based formulation, moisture content (w/w) by weight of solid carrier products, and the efficiency character (Fertilizer Control Order 1985, amended up to 2021, https://fertilizerindia.com/). The phosphate solubilizing fungal biofertilizer are registered Schedule-III [(clause 2(h) and (q)] of The Fertilizer Control Order, 1985 has specifications as given in Table 9.3.

TABLE 9.3 Specification of Fungal-based Phosphate Solubilizing Biofertilizer for Production and Commercialization in India*

SL. No.	Components	Specifications
1.	Base material	Carrier based in the form of dry/moist powder or granules, or liquid based
2.	Moisture percentage by weight, maximum (carrier based)	10.0%
3.	pH	Solid carrier: 6.0 to 7.0 Liquid carrier: 3.5 to 5.5
4.	Fungal spore count (per gram or ml)	Minimum 1×10^6 spores/g (in case of solid carrier) Minimum 1×10^7 spores/ml (in case of liquid carrier)
5.	Contamination	1×10^3 cells/g (in case of solid carrier) Nil for liquid carrier
6.	Efficiency character	When tested spectrophotometrically, the strain should have a phosphate solubilization capacity in the range of 30% to 40%. In terms of zone formation, a minimum 10 mm solubilization zone in the prescribed media having at least 3 mm thickness.

*The Fertilizer (Control) Order, 1985, Amended up to February 2019; https://www.faidelhi.org/.

9.8 CONCLUSION AND FUTURE PROSPECTS

The most recent advancements in microbiological studies have contributed to the recognition of the significance of microorganisms in agriculture, industry, and medicine. In order to fully utilize the ecological functions that these beneficial microorganisms provide and maximize their incorporation and use in agriculture, it is important to gain a clear understanding of the functions that these organisms play. Eco-friendly nutrient and pest management technologies have been implemented to address concerns about the deleterious effects of chemical fertilizers and pesticides. All of the multiple species of entomopathogenic and rhizospheric mycoflora, as well as associated microorganisms that might be used as fungal biofertilizers or BCAs, are now widely accepted as sustainable nutrients in soils and pest management alternatives to chemical fertilizers and pesticides. Mycopesticides show great promise as chemical pesticide alternatives, but more research is needed to determine their full potential. Most fungal biopesticides are thought to be safe for use in the agriculture system based on the evidence currently available and could effectively reduce synthetic pesticide abuse. If new fungal strains are to be registered for mineral solubilization, toxicity, and pathogenicity, they must first be tested in non-target species, including vertebrates. To avoid hazardous effects, it has been advised that current safety precautions should be followed throughout development, processing, and application. As a result of their numerous benefits, fungal-based biofertilizers and BCAs merit promotion. Aside from laboratory and field experiments, policies, and legislative changes will be required to address the underlying issues. The major applications of these microorganisms must be properly explored from a scientific perspective as well as economically, socially, and politically.

Moreover, it is encouraged that additional research is conducted to bridge the gaps in the formulations of biofertilizers and biopesticides. The ability of bio-inputs to remain constant during storage and field circumstances will serve as an assurance of their complete efficacy in nutrient and crop pest management. Researchers should collaborate with technologists in government and business, including with farmers, in order to develop stable, long-lasting formulations and delivery systems for biofertilizers and biopesticides for use in agriculture (Figures 9.2 and 9.3).

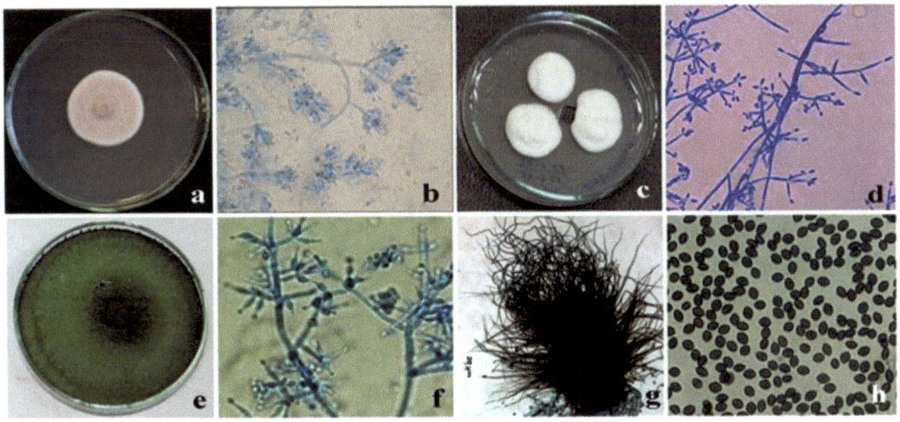

FIGURE 9.2 (a) Colony morphology of *Paecilomyces lilacinum* (a nematophagous bioagent); (b) branched conidiophores with phialides and conidia of *P. lilacinum*; (c) colony morphology of *Lecanicillium lecanii* on potato carrot agar; (d) conidiophores with phialides and conidia of *L. lecanii*; (e) dark green colony of *Trichoderma harzianum* (biocontrol and biofertilizer agent) on PDA; (f) branched hyaline conidiophores phialides and conidia of *T. harzianum*; (g) an ascomata with ascomatal hairs of *Chaetomium* sp. (biocontrol and biofertilizer agent); and (h) numerous fusoid ascospores of *Chaetomium* sp. [*Source:* National Fungal Culture Collection of India (NFCCI), Biodiversity and Paleobiology Group Laboratory, DST-Agharkar Research Institute, Pune, India (Original figure)].

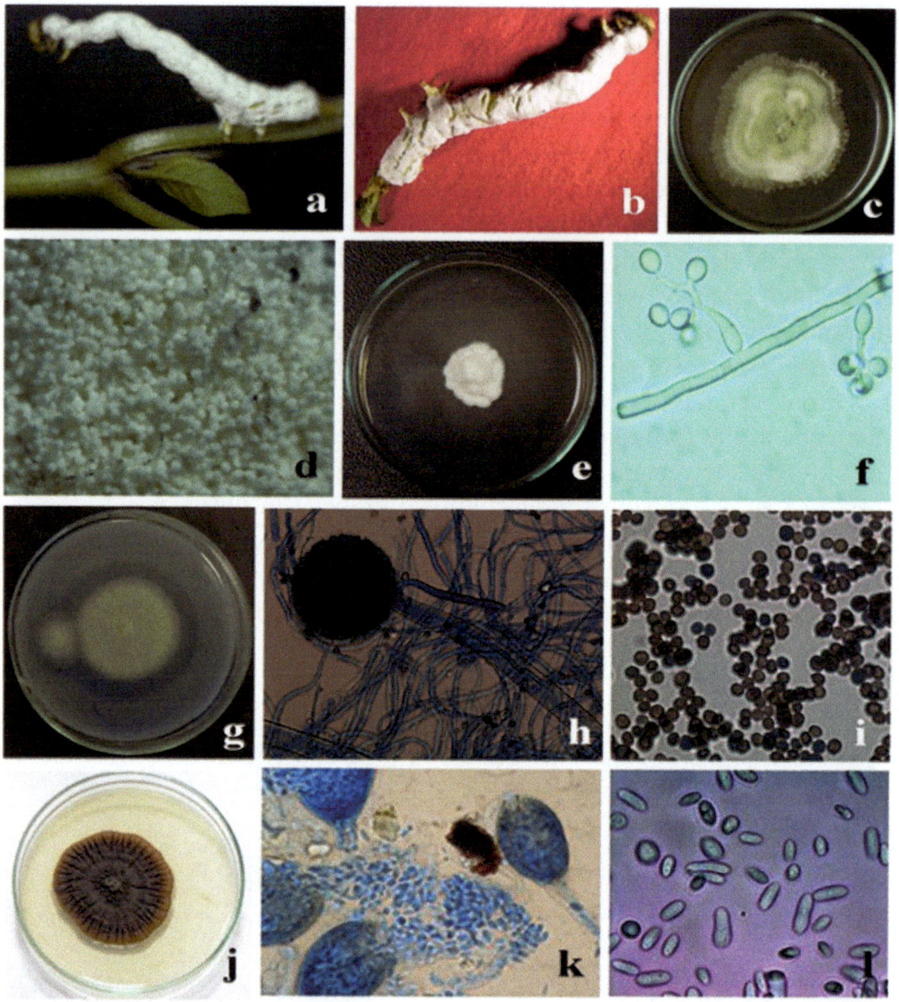

FIGURE 9.3 (a) An infected caterpillar (semilooper) of an entomopathogenic fungus *Nomuraea* sp.; (b) an infected caterpillar of (a natural epizootics) an entomopathogenic fungus *Nomuraea* sp.; (c) coloy morphology of *Nomuraea* sp. on MGYP; (d) stereoscopic view of sporodochia of entomopathogenic fungus *Beauveria bassiana* colonized on unidentified dead insect; (e) *in-vitro* culture of *B. bassiana* on PCA; (f) slide culture of *B. bassiana* showing conidiophore with phialides and conidia; (g) phosphate solubilizing fungus *Aspergillus niger* showing zone of clearance on Pikovaskiyas medium; (h) hyphae and conidiophore of *A. niger* with vesicle along with sterigmata; (i) numerous rough walled conidia of *A. niger*; (j) colony of a biocontrol fungus *Ampelomyces quisqualis* on PCA; (k) conidia of *A. quisqualis* liberating from vesiculate conidiomata; and (l) numerous hyaline conidia of *A. quisqualis*. [*Source:* National Fungal Culture Collection of India (NFCCI), Biodiversity and Paleobiology Group Laboratory, DST-Agharkar Research Institute, Pune, India (Original figure)].

KEYWORDS

- **arbuscular mycorrhizal fungi**
- **ectomycorrhizae**
- **entomopathogenic fungi**
- **ericoid mycorrhiza**
- **formulations**
- **fungal bio-inputs**
- **regulatory framework**
- **sustainable agriculture**

REFERENCES

Adeleke, B. S., & Babalola, O. O., (2021). Biotechnological overview of agriculturally important endophytic fungi. *Hortic. Environ. Biotechnol., 62*, 507–520. doi: 10.1007/s13580-021-00334-1.

AGBR, (2015). *Agrow Global Biopesticide Regulations*. Commodity analysis. Available at www.agra-net.com (accessed on 13 January 2023).

Aggarwal, R., (2015). *Chaetomium globosum*: A potential biocontrol agent and its mechanism of action. *Indian Phytopathol., 68*, 8–24.

Ajmal, M., Ali, H. I., Saeed, R., Akhtar, A., Tahir, M., Mehboob, M. Z., & Ayub, A., (2018). Biofertilizer as an alternative for chemical fertilizers. *J Agri Sci., 7*, 1–7.

Akintokun, A. K., Akande, G. A., Akintokun, P. O., Popoola, T. O., & Babalola, A. O., (2007). Solubilization on insoluble phosphate by organic acid-producing fungi isolated from Nigerian soil. *Int J Soil Sci., 2*, 301–307.

Akutse, K., Subramanian, S., Maniania, N., Dubois, T., & Ekesi, S., (2020). Biopesticide research and product development in Africa for sustainable agriculture and food security – experiences from the international center of insect physiology and ecology (icipe). *Front. Sustainable Food Syst., 4*, 152. doi: 10.3389/fsufs.2020.563016.

Alavo, T. B. C., (2015). The insect pathogenic fungus *Verticillium lecanii* (Zimm.) viegas and its use for pests' control: A review. *Journal of Experimental Biology and Agricultural Sciences*. http://dx.doi.org/10.18006/2015.3(4).337.345.

Althouse, C., Petersen, B., & McEwen, L., (1997). Effects on young American kestrels (*Falco sparverius*) exposed to *Beauveria bassiana* bioinsecticide. *Bull. Environ. Contam. Toxicol., 59*, 507–512. doi: 10.1007/s001289900507.

Aminuzzamana, F. M., Xieb, H. Y., Duanc, W. J., Sundand, B. D., & Liu, X. Z., (2013). Isolation of nematophagous fungi from eggs and females of *Meloidogyne* spp. and evaluation of their biological control potential. *Biocontrol Sci. Technol., 23*, 170–182.

Arora, N. K., Verma, M., Prakash, J., & Mishra, J., (2016). Regulation of biopesticides: Global concerns and policies. In: Arora, N., Mehnaz, S., & Balestrini, R., (eds.), *Bioformulations: For Sustainable Agriculture*. Springer, New Delhi. https://doi.org/10.1007/978-81-322-2779-3_16.

Bagyaraj, D. J., & Ashwin, R., (2017). Soil biodiversity: Role in sustainable horticulture. *Biodivers Hortic. Crops, 5*, 1–18.

Bainier, G., (1907). Mycothe 'que de l'e'cole de pharmacie. XI *paecilomyces*, genre nouveau de muce'dine'es. *Bull. Soc. Mycol. Fr., 23*, 26–27.

Bamisile, B. S., Akutse, K. S., Siddiqui, J. A., & Xu, Y., (2021). Model application of entomopathogenic fungi as alternatives to chemical pesticides: Prospects, challenges, and insights for next-generation sustainable agriculture. *Frontiers in Plant Science*, 741B04. doi: 10.3389/fpls.2021.741804.

Barra, P., Rosso, L., Nesci, A., & Etcheverry, M., (2013). Isolation and identification of entomopathogenic fungi and their evaluation against *Tribolium confusum*, *Sitophilus zeamais*, and *Rhyzopertha dominica* in stored maize. *J. Pest Sci., 86*, 217–226.

Bérdy, J., (2005). Bioactive microbial metabolites. *J. Antibiot., 58*, 1–26. doi: 10.1038/ja.2005.1.

Bharti, V., & Ibrahim, S., (2020). Biopesticides: Production, formulation and application systems. *Int. J. Curr. Microbiol. App. Sci., 9*(10), 3931–3946.

Boyetchko, S. M., (1996). Impact of soil microorganisms on weed biology and ecology. *Phytoprotection, 77*, 41–56.

Boyetchko, S., Pedersen, E., Punja, Z., & Reddy, M., (1998). Formulation of biopesticides. In: Hall, F. R., & Menn, J. J., (eds.), *Biopesticides: Use and Delivery Methods in Biotechnology* (p. 487–508). Humana Press, Totowa, NJ.

Brar, S. K., Verma, M., Tyagi, R. D., & Valero, J. R., (2006). Recent advances in downstream processing and formulations of *Bacillus thuringiensis* based biopesticides. *Process Biochemistry, 41*(2), 323–342.

Bücking, H., & Kafle, A., (2015). Role of arbuscular mycorrhizal fungi in the nitrogen uptake of plants: Current knowledge and research gaps. *Agronomy, 5*, 587–612.

Bücking, H., Liepold, E., & Ambilwade, P., (2012). The role of the mycorrhizal symbiosis in nutrient uptake of plants and the regulatory mechanisms underlying these transport processes. *Plant Sci., 4*, 108–132.

Chandler, D., Bailey, A. S., Tatchell, G. M., Davidson, G., Greaves, J., & Grant, W. P., (2011). The development, regulation and use of biopesticides for integrated pest management. *Philos. Trans. R Soc. Lond. B: Biol. Sci., 366*(1573), 1987–1998.

Chandrashekrappa, G. K., & Basalingappa, K. M., (2018). Biofertilizer for crop production and soil fertility. *Acad. J. Agric. Res., 6*, 299–306.

Chang, C. H., & Yang, S., (2009). Thermo-tolerant phosphate-solubilizing microbes for multi-functional biofertilizer preparation. *Bioresour. Technol., 100*, 1648–1658.

Clark, R. & Soil, Appalachian. (1997). Arbuscular mycorrhizal adaptation, spore germination, root colonization, and host plant growth and mineral acquisition at low pH. Plant and Soil - PLANT SOIL. 192. 15–22. 10.1023/A:1004218915413.

Cole, M., & Robinson, G. N., (1972). Microbial metabolites with insecticidal properties. *Applied Microbiology, 24*, 660–662.

Cory, J. S., & Bishop, D. H. L., (1995). Use of baculoviruses as biological insecticides, In: Richardson, C. D., (ed.), *Methods in Molecular Biology, Baculovirus Expression Protocols,* (Vol. 39, pp. 277–294). Humana, Totowa, NJ.

Cruz, C., Egsgaard, H., Trujillo, C., Ambus, P., Requena, N., Martins-Loucao, M. A., & Jakobsen, I., (2007). Enzymatic evidence for the key role of arginine in nitrogen translocation by arbuscular mycorrhizal fungi. *Plant Physiol., 144*, 782–792.

Culliney, T. W., (2014). Crop losses to arthropods. In: Pimentel, D., & Peshin, R., (eds.), *Integrated Pest Management* (pp. 201–225). Springer, Dordrecht.

Das, T., Tudu, C. K., Nandy, S., Pandey, D. K., & Dey, A., (2021). Role of fungal metabolites as biopesticides: An emerging trend in sustainable agriculture. In: Kumar, A., Singh, J., &

Samuel, J., (eds.), *Volatiles and Metabolites of Microbes* (pp. 385–407). Academic Press. https://doi.org/10.1016/B978-0-12-824523-1.00014-6.

De Faria, M. R., & Wraight, S. P., (2007). Mycoinsecticides and myco*acari*cides: A comprehensive list with worldwide coverage and international classification of formulation types. *Biol. Control., 43*, 237–256. doi: 10.1016/j.biocontrol.2007.08.001.

Dhingra, O. D., Mizubuti, E. S. G., & Santana, F. M., (2003). *Chaetomium globosum* for reducing primary inoculum of *Diaporthe phaseolorum* sp. *meridionalis* in soil-surface soybean stable in field condition. *Biol. Control., 26*, 302–310.

Dominic, M., & Marthamakobe, (2016). Biological control of cashew powdery mildew using *Ampelomyces quisqualis* Ces. *Journal of Biological Control, 30*(4), 226–235. doi: 10.18311/jbc/2016/15591.

Dong, L. Q., Yang, J. K., & Zhang, K. Q., (2007). Cloning and phylogenetic analysis of the chitinase gene from the facultative pathogen *Paecilomyces lilacinus*. *J. Appl. Microbiol., 103*, 2476–2488.

Dorcheh, S. K., & Vahabi, K., (2016). Biosynthesis of nanoparticles by fungi: Large-scale production. In: Mérillon, J. M., & Ramawat, K., (eds.), *Fungal Metabolites. Reference Series in Phytochemistry*. Springer, Cham. https://doi.org/10.1007/978-3-319-19456-1_8-1.

Dowd, P. F., (2002). Anti-insectan compounds derived from microorganisms. In: Koul, O., & Dhaliwal, G. S., (eds.), *Microbial Biopesticides* (pp. 113–116). Taylor & Francis, London.

Dunlap, C. A., Jackson, M. A., & Wright, M. S., (2007). A foam formulation of *Paecilomyces fumosoroseus*, an entomopathogenic biocontrol agent. *Biocontrol Sci. Technol., 17*, 513–523.

Fadiji, A. E., & Babalola, O. O., (2020). Exploring the potentialities of beneficial endophytes for improved plant growth. Saudi *J. Biol. Sci., 27*, 3622–3633. doi: 10.1016/j.sjbs.2020.08.002.

Fayaz, A. M., Balaji, K., Kalaichelvan, P. T., & Venkatesan, R., (2009). Fungal based synthesis of silver nanoparticles—An effect of temperature on the size of particles. *Colloids Surf. B Biointerfaces, 74*, 123–126. doi: 10.1016/j.colsurfb.2009.07.002.

Felbermayr, G., Larch, M., Krüger, F., Flach, L., Yalcin, E., & Benz, S., (2013). In: Felbermayr, G., Larch, M., Krüger, F., Flach, L., Yalcin, E., & Benz, S., (eds.), *Dimensions and Implications of a Free Trade Agreement between the EU and the US* (pp. 21–56). Ifo Institute, Munich.

Frąc, M., Hannula, S. E., Bełka, M., & Jędryczka, M., (2018). Fungal biodiversity and their role in soil health. *Front Microbiol., 9*, 707.

Frederiksa, C., & Wesseler, H. H., (2018). *A Comparison of the EU and US Regulatory Frameworks for the Active Substance Registration of Microbial Biological Control Agents*. SCI, wileyonlinelibrary.com. doi: 10.1002/ps.5133.

Funk, L. M., He, D. N., Pedersen, E. A., & Reddy, M. S., (1997). Optimization of product delivery for a microbial inoculant, *Burkholderia cepacia*, for commercial use in the forest industry. *Journal of Plant Pathology, 19*, 108.

Gao, Y., Reitz, S. R., Wang, J., et al., (2012). Potential of a strain of the entomopathogenic fungus *Beauveria bassiana* (*Hypocreales*: *Cordycipitaceae*) as a biological control agent against western flower thrips, *Frankliniella occidentalis* (*Thysanoptera*: *Thripidae*). *Biocontrol Sci Tech., 22*(4), 491–495. https://doi.org/10.1080/09583157.2012.662478.

Gašić, S., & Tanović, B., (2013). Biopesticide formulations, possibility of application and future trends. *Pestic. Phytomed. (Belgrade), 28*(2), 97–102. doi: 10.2298/PIF1302097G.

Ghareib, M., Tahon, M. A., Abdallah, W. E., & Tallima, A., (2018). Green biosynthesis of copper oxide nanoparticles using some isolated from the Egyptian soil. *International Journal of Research in Pharmaceutical and Nano Sciences, 7*(4), 119–128.

Glick, B. R., (2014). Bacteria with ACC deaminase can promote plant growth and help to feed the world. *Microbiol. Res., 169*, 30–39. doi: 10.1016/j.micres.2013.09.009.

Goettel, M. S., & Glare, E. T., Entomopathogenic fungi and their role in regulation of insect populations. In: Gilbert, L. I., (ed.), *Comprehensive Molecular Insect Science* (Vol. 6, pp. 361–405). https://doi.org/10.1016/B0-44-451924-6/00088-0.

Guilger-Casagrande, M., & De Lima, R., (2019). Synthesis of silver nanoparticles mediated by fungi: A review. *Front. Bioeng. Biotechnol.* https://doi.org/10.3389/fbioe.2019.00287.

Huang, X. W., Zhao, N. H., & Zhang, K. Q., (2004). Extracellular enzymes serving as virulence factors in nematophagous fungi involved in infection of the host. *Res. Microbiol., 115*, 811–816.

Hunter, W. B., Avery, P. B., Pick, D., & Powell, C. A., (2011). Broad spectrum potential of *Isaria fumosorosea* against insect pests of citrus. *Fla. Entomol., 94*, 1051–1054.

Ibarra, J. E., Del Rincón, C. M. C., Galindo, E., Patiño, M., Serrano, L., García, R., Carrillo, J. A., Pereyra, A. B., Alcázar, P. A., Luna, O. H., et al., (2006). Los microorganismos en el control biológico de insectos y fitopatógenos. *Rev. Latinoam. Microbiol., 48*, 113–120.

Inglis, P. W., & Tigano, M. S., (2006). Identification and taxonomy of some entomopathogenic *Paecilomyces* spp. (*Ascomycota*) isolates using rDNA-ITS Sequences. *Genet. Mol. Biol., 29*, 132–136.

Jackson, M. A., Erhan, S., & Poprawski, T. J., (2006). Influence of formulation additives on the desiccation tolerance and storage stability of blastospores of the entomopathogenic fungus *Paecilomyces fumosoroseus* (*Deuteromycotina*: *Hyphomycetes*). *Biocontrol. Sci. Technol., 16*, 61–75.

Jacobs, H., Gray, S. N., & Crump, D. H., (2003). Interactions between nematophagous fungi and consequences for their potential as biological agents for the control of potato cyst nematodes. *Mycol. Res., 107*, 47–56.

Jagnaseni, B., Aveek, S., Babita, S., & Siraj, D., (2016). Mycorrhiza: The oldest association between plant and fungi. *Resonance, 21*, 1093–1104.

Jatala, P., (1986). Biological control of plant-parasitic nematodes. *Ann. Rev. Phytopathol., 24*, 453–489.

Jatala, P., Kaltenback, R., & Bocangel, M., (1979). Biological control of Meloidogyne incognita acrita and *Globodera pallida* on potatoes. *J. Nematol., 11*, 303.

Jehangir, I. A., Mir, M. A., Bhat, M. A., & Ahangar, M. A., (2017). Biofertilizers an approach to sustainability in agriculture: A review. *Int. J. Pure Appl. Biosci., 5*, 327–334.

Jessica, J. J., Peng, T. L., Sajap, A. S., Lee, S. H., & Syazwan, S. A., (2019). Evaluation of the virulence of entomopathogenic fungus, *Isaria fumosorosea* isolates against subterranean termites *Coptotermes* spp. (*Isoptera*: *Rhinotermitidae*). *J. For. Res., 30*, 213–218.

Johnson, D. L., Smits, J. E., Jaronski, S. T., & Weaver, D. K., (2002). Assessment of health and growth of ring-necked pheasants following consumption of infected insects or conidia of entomopathogenic fungi, *Metarhizium anisopliae* var. acridum and *Beauveria bassiana*, from Madagascar and North America. *J. Toxicol. Environ. Health, 65*, 2145–2162. doi: 10.1080/00984100290071847.

Johri, A. K., Oelmüller, R., Dua, M., Yadav, V., Kumar, M., Tuteja, N., & Stroud, R. M., (2015). Fungal association and utilization of phosphate by plants: Success, limitations, and future prospects. *Front Microbiol., 6*, 984.

Kaewchai, S., Soytong, K., & Hyde, K. D., (2009). Mycofungicides and fungal biofertilizers. *Fungal Diversity, 38*, 25–50.

Khachatourians, G. G., & Qazi, S. S., (2008). Entomopathogenic fungi: Biochemistry and molecular biology. In: Brakhage, A. A., & Zipfel, P. F., (eds.), *Human and Animal Relationships* (pp. 33–61). Springer, Berlin, Heidelberg.

Khan, A., Williams, K. L., & Nevalainen, H. K., (2006). Infection of plant-parasitic nematodes by *Paecilomyces lilacinus* and *Monacrosporium lysipagum. BioControl, 51*, 659–678.

Khokhar, M., Renu, G., & Sharma, R., (2012). Biological control of plant pathogens using biotechnological aspects: A review. *Epidemiology, 1*, 277–289.

Kiss, L., (2003). A review of fungal antagonists of powdery mildews and their potential as biocontrol agents. *Pest Management Science, 59*, 475–483.

Knowles, A., (2005). *New Developments in Crop Protection Product Formulation* (pp. 53–156). Agrow Reports UK: T and F Informa UK Ltd.

Knowles, A., (2008). Recent developments of safer formulations of agrochemicals. *Environmentalist, 28*(1), 35–44. doi: 10.1007/s10669-007-9045-4.

Koppert, B. V., (2015). *Mycotal*: *Verticillium Lecanii*. http://www.koppert.com (accessed on 13 January 2023).

Koul, O., (2011). Microbial biopesticides: Opportunities and challenges. *CAB Reviews: Perspectives in Agriculture, Veterinary Science, Nutrition and Natural Resources, 6*, 056.

Kour, D., Rana, K. L., Yadav, A. N., Yadav, N., Kumar, M., Kumar, V., Vyas, P., et al., (2020). Microbial biofertilizers: Bioresources and eco-friendly technologies for agricultural and environmental sustainability. *Biocatal. Agric. Biotechnol., 23*, 101487. https://doi.org/10.1016/j.bcab.2019.101487.

Kour, D., Rana, K. L., Yadav, N., Yadav, A. N., Singh, J., Rastegari, A. A., & Saxena, A. K., (2019). Agriculturally and industrially important fungi: Current developments and potential biotechnological applications. In: Yadav, A. N., Singh, S., Mishra, S., & Gupta, A., (eds.), *Recent Advancement in White Biotechnology Through Fungi: Perspective for Value-Added Products and Environments* (Vol. 2, pp. 1–64). Springer International Publishing, Cham. https://doi.org/10.1007/978-3-030-14846-1_1.

Kumar, D., Singh, M. K., Singh, H. K., & Singh, K. N., (2019). Fungal biopesticides and their uses for control of insect pest and diseases. In: Kaushik, B. D., Kumar, D., & Shamim, Md., (eds.), *Biofertilizers and Biopesticides in Sustainable Agriculture* (pp. 43–70). Apple Academic Press, USA.

Larran, S., Simon, M. R., Moreno, M. V., Siurana, M. S., & Perelló, A., (2016). Endophytes from wheat as biocontrol agents against tan spot disease. *Biol. Control., 92*, 17–23.

Lewis, J. A., (1991). Formulation and delivery systems of biocontrol agents with emphasis scent on fungi. In: Keisterand, D. L., & Cregan, P. B., (eds.), *The Rhizosphere and Plant Growth* (pp. 279–287). Kluwer, Dordrecht, The Netherlands.

Lewis, M. W., Robalino, I. V., & Keyhani, N. O., (2009). Uptake of the fluorescent probe FM4-64 by hyphae and haemolymph-derived *in vivo* hyphal bodies of the entomopathogenic fungus *Beauveria bassiana. Microbiology, 155*, 3110–3120.

Lian, B., Wang, B., Pan, M., Liu, C., & Teng, H., (2008). Microbial release of potassium from K-bearing minerals by thermophilic fungus *Aspergillus fumigatus. Geo et Cosmo Acta, 72*, 87–98.

Lyn, M. E., Burnett, D., Garcia, A. R., & Gray, R., (2010). Interaction of water with three granular biopesticide formulations. *Journal of Agricultural and Food Chemistry, 58*(1), 1804–1814.

Mallebrera, B., Prosperini, A., Font, G., & Ruiz, M. J., (2018). *In vitro* mechanisms of beauvericin toxicity: A review. *Food Chem. Toxicol., 111*, 537–545. doi: 10.1016/j.fct.2017.11.019.

Malusá, E., & Vassilev, N., (2014). A contribution to set a legal framework for biofertilizers. *Appl Microbiol Biotechnol., 98*, 6599–6607. Doi: 10.1007/s00253–014–5828-y.

Marwah, R. G., Fatope, M. O., Deadman, M. L., Al-Maqbali, Y. M., & Husband, J., (2007). Musanahol: A new aureonitol-related metabolite from a *Chaetomium* sp. *Tetrahedron, 63*, 8174–8180.

Matthews, K. A., (2014). Regulation of biopesticides by the environmental protection agency. In: Coats, J. R., (ed.), *Biopesticides: State of the Art and Future Opportunities, Symposium* (Vol. 18, Series 1172, pp. 267–279). Sidley Austin, Washington, DC.

Mehta, P., Sharma, R., Putatunda, C., & Walia, A., (2019). Endophytic fungi: Role in phosphate solubilization. In: Singh, B., (ed.), *Advances in Endophytic Fungal Research, Fungal Biology* (pp. 183–209). Springer, Cham.

Milner, R. J., Lim, R. P., & Hunter, D. M., (2002). Risks to the aquatic ecosystem from the application of *Metarhizium anisopliae* for locust control in Australia. *Pest Manag. Sci., 58*, 718–723. doi: 10.1002/ps.517.

Mishra, D. J., Rajivir, S., Mishrra, U. K., & Kumar, S. S., (2013). Role of biofertilizers in organic agriculture: A review. *Res J Rec Sci., 2*, 39–41.

Mnyone, L. L., Koenraadt, C. J. M., Lyimo, I. N., Mpingwa, M. W., Takken, W., & Russell, T. L., (2010). Anopheline and culicine mosquitoes are not repelled by surfaces treated with the entomopathogenic fungi *Metarhizium anisopliae* and *Beauveria bassiana. Parasite Vectors, 3*, 80.

Mohammadi, S., Soltani, J., & Piri, K., (2016). Soilborne and invertebrate pathogenic *Paecilomyces* species show activity against pathogenic fungi and bacteria. *J. Crop Prot., 5*, 377–387.

Moreno-Gavira, A., Huertas, V., Diánez, F., Sánchez-Montesinos, B., & Santos, M., (2020). *Paecilomyces* and its importance in the biological control of agricultural pests and diseases. *Plants, 9*, 1746. doi: 10.3390/plants9121746.

NAAS, (2013). National academy of agricultural sciences. *Biopesticides – Quality Assurance*. http://naasindia.org/Policy%20Papers/Policy%2062.pdf (accessed on 13 January 2023).

Nakahara, Y., Shimura, S., Ueno, C., Kanamori, Y., Mita, K., Kiuchi, M., et al., (2009). Purification and characterization of silkworm hemocytes by flow cytometry. *Dev. Comp. Immunol., 33*, 439–448. doi: 10.1016/j.dci.2008.09.005.

Naqqash, M. N., Gökçe, A., Bakhsh, A., & Salim, M., (2016). Insecticide resistance and its molecular basis in urban insect pests. *Parasitol. Res., 115*, 1363–1373. doi: 10.1007/s00436-015-4898-9.

Ne´meth, M. Z., Mizuno, Y., Kobayashi, H., Seress, D., Shishido, N., Kimura, Y., et al., (2021). *Ampelomyces* strains isolated from diverse powdery mildew hosts in Japan: Their phylogeny and mycoparasitic activity, including timing and quantifying mycoparasitism of *Pseudoidium neolycopersici* on tomato. *PLoS One, 16*(5), e0251444. https://doi.org/10.1371/journal.pone.0251444.

Nissipaul, M., Sodimalla, T., Subhashreddy, R., & Suman, B., (2017). Novel biofilm biofertilizers for nutrient management and *Fusarium* wilt control in chickpea. *Int. J. Curr. Microbiol. App. Sci., 6*, 1846–1852.

Ochieng, J. R. A., (2015). *Towards a Regulatory Framework for Increased and Sustainable Use of Bio-fertilizers in Kenya*. MA Thesis, Centre for Advanced Studies in Environmental Law and Policy (CASELAP), University of Nairobi.

Odoh, C. K., (2017). Plant growth promoting rhizobacterial (PGPR): A bioprotectant bioinoculant for sustainable agrobiology. *Int J Adv Res Biol Sci., 4*, 123–142.

Odoh, C. K., Eze, C. N., Akpi, U. K., & Unah, V. U., (2019a). Plant growth promoting rhizobacteria (PGPR): A novel agent for sustainable food production. *Am J Agri Biol Sci., 14*, 35–54.

Odoh, C. K., Eze, C. N., Obi, C. J., Anyah, F., Egbe, K., Unah, U. V., Akpi, U. K., & Adobu, U. S., (2020). Fungal biofertilizers for sustainable agricultural productivity. In: Yadav, A., Mishra, S., Kour, D., Yadav, N., & Kumar, A., (eds.), *Agriculturally Important Fungi for Sustainable Agriculture, Fungal Biology*. Springer, Cham, https://doi.org/10.1007/978-3-030-45971-0_9.

Odoh, C. K., Zabbey, N., Sam, K., & Eze, C. N., (2019b). Status, progress and challenges of phytoremediation – An African scenario. *J. Environ. Manag., 237*, 365–378.

Oyedele, O. A., Ezekiel, C. N., Sulyok, M., Adetunji, M. C., Warth, B., Atanda, O. O., et al., (2017). Mycotoxin risk assessment for consumers of groundnut in domestic markets in Nigeria. *Int. J. Food Microbiol., 251*, 24–32. doi: 10.1016/j.ijfoodmicro.2017.03.020.

Pal, S., Singh, H. B., & Rakshit, A., (2014). The arbuscular mycorrhizal symbiosis: An underground world wide web. In: Singh, D. P., & Singh, H. B., (eds.), *Microbial Communities for Sustainable Soil Health and Ecosystem Productivity* (pp. 219–253). Studium Press LLC, Houston.

Paulitz, T. C., & Belanger, R. R., (2001). Biological control in greenhouse system. *Annual Reviews Phytopathology, 39*, 103–133.

Peveling, R., & Demba, S. A., (2003). Toxicity and pathogenicity of *Metarhizium anisopliae* var. acridum (Deuteromycotina, *Hyphomycetes*) and fipronil to the fringe-toed lizard *Acanthodactylus dumerili* (*Squamata*: *Lacertidae*). *Environ. Toxicol. Chem., 22*, 1437–1447. doi: 10.1002/etc.5620220704.

Rai, A., Rai, S., & Rakshit, A., (2013). Mycorrhiza-mediated phosphorus use efficiency in plants. *Environ. Exp. Biol., 11*, 107–117.

Ram, H., Malik, S. S., Dhaliwal, S. S., Kumar, B., & Singh, Y., (2015). Growth and productivity of wheat affected by phosphorus-solubilizing fungi and phosphorus levels. *Plant Soil Environ., 61*, 122–126.

Redda, E. T., Ma, J., Mei, J., Li, M., & Wu, B., (2018). Antagonistic potential of different isolates of *Trichoderma* against *Fusarium oxysporum*, *Rhizoctonia solani*, and *Botrytis cinerea*. *Eur. Exp. Biol., 8*, 12.

Rishbeth, J., (1975). Stump inoculation: A biological control of domes annosus. In: Bruehl, G. W., (ed.), *Biology and Control of Soil-Borne Plant Patholgy* (pp. 158–162). American Phytopathological Society.

Roberts, D. W., & St. Leger, R. J., (2004). *Metarhizium* spp., cosmopolitan insect-pathogenic fungi: Mycological aspects. *Adv. Appl. Microbiol., 54*, 1–70.

Roberts, D. W., (1977). Isolation and development of fungus pathogens of vectors. In: Briggs, J. E., (ed.), *Biological Regulation of Vectors* (pp. 85–93). CABI, Wallingford, UK.

Rosa, H., Ada, V., Ilan, C., & Enrique, M., (2012). Plant-beneficial effects of *Trichoderma* and of its genes. *J. Microbiol., 158*, 17–25.

Ruiu, L., (2018). Microbial biopesticides in agroecosystems. *Agronomy, 8*, 235.

Samson, R. A., (1974). *Paecilomyces* and some allied hyphomycetes. *Stud. Mycol., 6*, 1–119.

Saruhan, I., (2018). Efficacy of some entomopathogenic fungi against *Aphis fabae* Scopoli (*Hemiptera*: *Aphididae*). *Egypt. J. Biol. Pest Control, 28*, 89.

Semalulu, S., MacPherson, J., Schiefer, H., & Khachatourians, G., (1992). Pathogenicity of *Beauveria bassiana* in mice. *J. Veterinary Med. Ser., B, 39*, 81–90. doi: 10.1111/j.1439-0450.1992.tb01141.x.

Senthil-Nathan, S. (2015). A review of biopesticides and their mode of action against insect pests. In: Thangavel, P., Sridevi, G. (eds.) Environmental Sustainability. Springer, New Delhi. pp. 49–63. https://doi.org/10.1007/978-81-322-2056-5_3.

Shah, P., & Pell, J., (2003). Entomopathogenic fungi as biological control agents. *Appl. Microbiol. Biotechnol., 61*, 413–423. doi: 10.1007/s00253-003-1240-8.

Shahzad, A., Saeed, H., Iqtedar, M., Hussain, S. Z., Kaleem, A., & Abdullah, R., (2019). Size-controlled production of silver nanoparticles by *Aspergillus fumigatus* BTCB10: Likely antibacterial and cytotoxic effects. *J. Nanomater, 2019*, 5168698. doi: 10.1155/2019/5168698.

Sharma, S. B., Sayyed, R. Z., & Trivedi, M. H., (2013). *Phosphate Solubilizing Microbes: Sustainable Approach for Managing Phosphorus Deficiency in Agricultural Soils, 2*, 587. Springer Plus.

Sharma, S., Kour, D., Rana, K. L., Dhiman, A., Thakur, S., Thakur, P., Thakur, S., et al., (2019). *Trichoderma*: Biodiversity, ecological significances, and industrial applications. In: Yadav, A. N., Mishra, S., Singh, S., & Gupta, A., (eds.), *Recent Advancement in White Biotechnology Through Fungi, Diversity and Enzymes Perspectives* (Vol. 1, pp. 85–120). Springer International Publishing, Cham. https://doi.org/10.1007/978-3-030-10480-1_3.

Shieh, T. R., (1995). Biopesticide formulations and their applications, In: Ragsdale, N. N., Kearney, P. C., & Plimmer, J. R., (eds.), *Proceedings of American Chemical Society* (pp. 104–114). Eighth International Congress of Pesticide Chemistry options 2000, Washington, DC.

Shrivastava, G., Rogers, M., Wszelaki, A., Panthee, D. R., & Chen, F., (2010). Plant volatiles-based insect pest management in organic farming. *Crit. Rev. Plant Sci., 29*, 123–133. doi: 10.1080/07352681003617483.

Shternshis, M., Tomilova, O., Shpatova, T., & Soytong, K., (2005). Evaluation of Ketomium mycofungicide on Siberian isolates of phytopathogenic fungi. *Journal of Agricultural Technology, 1*, 247–253.

Silva, L. P., Bonatto, C. C., & Polez, V. L. P., (2016). Green synthesis of metal nanoparticles by fungi: Current trends and challenges. In: Prasad, R., (ed.), *Advances and Applications Through Fungal Nanobiotechnology, Fungal Biology.* Springer, Cham, https://doi.org/10.1007/978-3-319-42990-8_4.

Singh, R., & Arora, N. K., (2016). Bacterial formulations and delivery systems against pests in sustainable agro-food production. *Module in Food Sciences,* 1–11. Elsevier.

Smolińska, U., & Kowalska, B., (2018). Biological control of the soil-borne fungal pathogen *Sclerotinia sclerotiorum*: A review. *J. Plant Pathol., 100*, 1–12.

Sneha, S., Anitha, B., Sahair, R. A., Raghu, N., Gopenath, T. S., Chandrashekrappa, G. K., & Basalingappa, K. M., (2018). Biofertilizer for crop production and soil fertility. *Acad. J. Agric. Res., 6,* 299–306.

Sparks, T. C., Thompson, G. D., Kirst, H. A., Hertlein, M. B., Mynderse, J. S., Turner, J. R., et al., (1999). In: Hall, F. R., & Menn, J. J., (eds.), *Biopesticides: Use and Delivery* (pp. 171–188). Humana Press, Totowa, NJ.

St. Leger, R. J., Frank, D. C., Roberts, D. W., & Staples, R. C., (1992). Molecular cloning and regulatory analysis of the cuticle-degrading-protease structural gene from the entornopathogenic fungus *Metarhizium anisopliae*. *Eur. J. Biochem., 204*, 991–1001.

Subramaniam, M. S. R., Babu, A., & Deka, B., (2021). *Lecanicillium lecanii* (zimmermann) Zare & gams, as an efficient biocontrol agent of tea thrips, *Scirtothrips bispinosus* Bagnall (*Thysanoptera:Thripidae*).*EgyptJ. Biol. PestControl, 31*,38.https://doi.org/10.1186/s41938-021-00380-y.

Sumita, P., Singh, H. B., Alvina, F., & Amitava, R., (2015). Fungal biofertilizers in Indian agriculture: Perception, demand and promotion. *J. Eco. Agric, 10,* 101–113.

Tadros, F., (2005). *Applied Surfactants, Principles and Applications* (pp. 187–256). Wiley-VCH Verlag GmbH and Co. KGaA.

The Economic Times, (2019). *IFFCO Launches Nano-Tech Based Fertilizers for on-Field Trials.* https://economictimes.indiatimes.com/news/economy/agriculture/iffco-launches-nano-tech-based-fertilisers-for-on-field-trials/articleshow/71878538.cms?from=mdr (accessed on 13 January 2023).

The Fertilizer (Control) Order, (1985). Amended up to February 2019. https://www.faidelhi.org/general/FCO-July-21-content.pdf (accessed on 13 January 2023).

Thomashow, L. S., & Weller, D. M., (1990). Application of fluorescent pseudomonads to control root diseases of wheat and some mechanisms of disease suppression. In: Hornby, D., (ed.), *Biological Control of Soil-Borne Plant Pathogens* (pp. 109–122). Redwood, Melkshan Wiltshire, UK.

Thosar, R. U., Sawant, I., Chavan, V. M., Sawant, S. D., Saha, S., & Suthakar, A. V., (2020). Generation of a bio-intensive strategy using chitosan formulations and *Ampelomyces quisqualis* for the management of powdery mildew of grapes. *International Journal of Current Microbiology and Applied Sciences, 9*(10), 1190–1201. https://doi.org/10.20546/ijcmas.2020.910.143.

Tripathi, S., Mishra, S. K., & Varma, A., (2017). Mycorrhizal fungi as control agents against plant pathogens. In: Varma, A., Prasad, R., & Tuteja, N., (eds.), *Mycorrhiza – Nutrient Uptake, Biocontrol* (pp. 161–178). *Ecorestoration,* Springer, Cham.

USEPA, (2010). United States Environmental Protection Agency. Regulating biopesticides. http://npic.orst.edu/reg/risk.html (accessed on 13 January 2023).

Vergara, C., Araujo, K., Emanuelle, C. S., Sônia, R., Nivaldo, S. J., Orivaldo, J. S., Marcus, V. L., & Zilli, J. É., (2019). Plant-mycorrhizal fungi interaction and response to inoculation with different growth promoting fungi. *Pesqui Agropecu Bras, 54,* e25140.

Verma, P., Yadav, A. N., Kumar, V., Singh, D. P., & Saxena, A. K., (2017b). Beneficial plant-microbes' interactions: Biodiversity of microbes from diverse extreme environments and its impact for crop improvement. In: Singh, D. P., Singh, H. B., & Prabha, R., (eds.), *Plant-Microbe Interactions in Agroecological Perspectives, Microbial Interactions and Agro-Ecological Impacts* (Vol. 2, pp. 543–580). Springer, Singapore. https://doi.org/10.1007/978-981-10-6593-4_22.

Vestergaard, S., Cherry, A., Keller, S., & Goettel, M., (2003). Safety of hyphomycete fungi as microbial control agents. In: Hokkanen, H. M. T., & Hajek, A. E., (eds.), *Environmental Impacts of Microbial Insecticides* (pp. 35–62). Springer, Dordrecht.

Viterbo, A., Inbar, J., Hadar, Y., & Chet, I., (2007). Plant disease biocontrol and induced resistance via fungal mycoparasites. In: Kubicek, C. P., & Druzhinina, I. S., (eds.), *Environmental and Microbial Relationships: The Mycota* (2nd edn., Vol. 4, pp. 127–146). Springer-Verlag, Berlin, Heidelberg.

Weng, Q., Zhang, X., Chen, W., & Hu, Q., (2019). Secondary metabolites and the risks of *Isaria fumosorosea* and *Isaria farinosa. Molecules, 24,* 664. doi: 10.3390/molecules24040664.

West, C., & Gwinn, K., (1993). Role of acremonium in drought, pest, and disease tolerances of grasses. In: Hume, D. E., Latch, G. C. M., & Easton, H. S., (eds.), *Proceedings of the Second International Symposium on Acremonium/Grass Interactions* (pp. 131–140). Plenary papers, Palmerston North, New Zealand.

Williams, K., Khan, A., & Holland, R., (1999). Infection of *Meloidogyne javanica* by *Paecilomyces lilacinus. Nematology, 1,* 131–139.

Wipf, D., Krajinski, F., Tuinen, D., Recorbet, G., & Courty, P., (2019). Trading on the arbuscular mycorrhiza market: From arbuscules to common mycorrhizal networks. *New Phytol., 223,* 1127–1142.

Woessner, J. F., (2013). Carboxypeptidase MeCPA. In: Rawlings, N. D., & Salvesen, G., (eds.), *Handbook of Proteolytic Enzyme* (Vol. 1, pp. 1329–1331).

Woo, L. S., & Lorito, M., (2007). Exploiting the interactions between fungal antagonists, pathogens and the plant for biocontrol. In: Vurro, M., & Gressel, J., (eds.), *Novel Biotechnologies for Biocontrol Agent Enhancement and Management* (pp. 107–130). Springer.

Woods, T. S., (2003). Pesticide formulations. In: *AGR 185 in Encyclopedia of Agrochemicals* (pp. 1–11). Wiley & Sons, New York.

Wu, S., Gao, Y., Smagghe, G., et al., (2016). Interactions between the entomopathogenic fungus *Beauveria bassiana* and the predatory mite *Neoseiulus barkeri* and biological control of their shared prey /host *Frankliniella occidentalis. Biol Control, 98*, 43–51. https://doi.org/10.1016/j.biocontrol.2016.04.001.

Yadav, A. N., Rastegari, A. A., Yadav, N., & Kour, D., (2020). *Advances in Plant Microbiome and Sustainable Agriculture: Diversity and Biotechnological Applications.* Springer, Singapore.

Yadav, A. N., Verma, P., Kumar, V., Sangwan, P., Mishra, S., Panjiar, N., Gupta, V. K., & Saxena, A. K., (2018b). Biodiversity of the genus *Penicillium* in different habitats. In: Gupta, V. K., & Rodriguez-Couto, S., (eds.), *New and Future Developments in Microbial Biotechnology and Bioengineering: Penicillium System Properties and Applications* (pp. 3–18). Elsevier, Amsterdam. https://doi.org/10.1016/B978-0-444-63501-3.00001-6.

Youssef, M. A., & Eissa, F. M., (2014). Biofertilizers and their role in management of plant parasitic nematodes: A review. *Biotechnol. Pharma. Resour., 5*, 1–6.

Zarafi, A. B., & Dauda, W. P., (2019). Exploring the importance of fungi in agricultural biotechnology. *Int. J. Agric. Sci. Vet. Med., 7,* 1–12.

Zare, R., & Gams, W., (2001). A revision of *Verticillium* sect. *Prostrata*. III. Generic classification. *Nova Hedwigia, 72*, 329–337.

Zeilinger, S., & Omann, M., (2007). *Trichoderma* biocontrol: Signal transduction pathways involved in host sensing and mycoparasitism. *Gene Regulation and Systems Biology, 1*, 227–234.

Zhang, S., Xia, Y. X., Kim, B., & Keyhani, N. O., (2011). Two hydrophobins are involved in fungal spore coat rodlet layer assembly and each play distinct roles in surface interactions, development and pathogenesis in the entomopathogenic fungus, *Beauveria bassiana. Molecular Microbiology, 80*(3), 811–826.

Zimmermann, G., (2007a). *Review on Safety of the Entomopathogenic Fungi Beauveria bassiana and Beauveria Brongniartii, 17*, 553–596. doi: 10.1080/09583150701309006.

Zimmermann, G., (2007b). Review on safety of the entomopathogenic fungus *Metarhizium anisopliae. Biocontrol Sci. Tech., 17*, 879–920. doi: 10.1080/09583150701593963.

Zimmermann, G., (2008). The entomopathogenic fungi *Isaria farinosa* (formerly *Paecilomyces farinosus*) and the *Isaria fumosorosea* species complex (formerly *Paecilomyces fumosoroseus*): Biology, ecology and use in biological control. *Biocontrol Science and Technology, 18*(9), 865–901. doi: 10.1080/09583150802471812.

CHAPTER 10

Pre-Harvest Management of Aflatoxin Contamination in Groundnut Through Biocontrol Products

DEEPAK KUMAR[1], L. J. DESAI[2], CHANDRA BHANU[3], SANJAY K. SINGH[4], K. P. SINGH[5], N. BALASUBRAMANI[6], and A. SADALAXMI[6]

[1]*R&D Division, Nextnode Bioscience Pvt. Ltd., Kadi, Gujarat, India*

[2]*Center for Research on Integrated Farming System, S.D. Agricultural University, Sardarkrushinagar, Banaskantha, Gujarat, India*

[3]*ICAR–Indian Institute of Farming Systems Research, Modipuram, Meerut, Uttar Pradesh, India*

[4]*Biodiversity and Paleobiology, Agharkar Research Institute, Pune, Maharashtra, India*

[5]*Vice Chancellor, M.J.P. Rohilkahand University, Bareilly, Uttar Pradesh, India*

[6]*Coordinator, CFA Program, MANAGE, Rajendranagar, Hyderabad, Telangana, India*

ABSTRACT

Aflatoxin poisoning of food and feed crops is common in tropical and subtropical areas. The main contaminants are *Aspergillus flavus* and *Aspergillus parasiticus*. Aflatoxin in food and feed causes harm directly or indirectly to human health as well as trade agricultural sectors. Aflatoxin levels may be reduced to acceptable levels through various pre- and post-harvest measures, but this is not always the case. Using atoxigenic (non-toxin producing) *A. flavus* and other biocontrol isolates to remove

Applied Mycology for Agriculture and Foods: Industrial Applications. Sanjay K. Singh, Deepak Kumar, Md. Shamim, & Rohit Sharma (Eds.)

aflatoxin producers reduces aflatoxin contamination in groundnuts from field to plate. Toxicological isolates as active ingredients are approved in many African countries, the United States and soon in Italy. This chapter provides an overview of biocontrol technology, explains how active ingredients such as atoxigenic isolates of *Aspergillus* sp. and other biological inoculants are selected, how compositions are designed and tested, and how biocontrol products are used effectively.

10.1 INTRODUCTION

Groundnut (*Arachis hypogaea* L.) is one of the world's most economically significant oilseed and feed crops, accounting for about a quarter of global production. The colonization of *Aspergillus flavus* and *A. parasiticus* reduces groundnut quality and yield globally during cultivation, harvesting, processing, and storage. The best growth environments for these infections include ambient temperatures ranging between 82°C and 98°C, soil pH values ranging from 3.5 to 8, and humid conditions (relative humidity of more than 85%) (Wheeler et al., 1991). When groundnut kernels and plants come into direct touch with these pathogenic fungi (soil inhabitants), they become infected. In addition to reducing yield, the causing fungi (*A. flavus* and *A. parasiticus*) have the potential to infect the crop with aflatoxin. Mycotoxin contamination is common in agro-ecological zones with high temperatures and humidity (Marin et al., 2013).

Aflatoxins are toxic compounds generated by *A. flavus* and *A. parasiticus*, which are harmful to humans and animals. In the food and feed business, these fatal poisons are a serious hazard to public health. Suitable food substrate and favorable environmental conditions are required for the growth of these fungi's spores. Aflatoxins are divided into several classes based on the amount of fluorescence they produce when exposed to ultraviolet (UV) light. Aflatoxins that exhibit blue fluorescence when exposed to UV light are called B1 and B2, whereas those that emit green fluorescence are labeled G1, and those that emit green-blue fluorescence are labeled G2. Each *Aspergillus* sp. does not synthesize all types of aflatoxins found in the cropping system (Alshannaq & Yu, 2017; Abbas et al., 2006). Aflatoxins AFB1, AFB2, AFG1, AFG2, and aflatoxins AFM1 are among the 18 various types of aflatoxins that have been reported, with the principal poisons. According to the International Agency for Research on Cancer (IARC), both AFB1 and AFM1 are classified as group-1 carcinogens (FSSAI, 2019). Carcinogens with the highest toxicity levels include aflatoxins, which have been linked to liver cancer in both humans and animals (Wogan et al., 2004). *Aspergillus* spp. infect and contaminate a wide range of crops, including groundnuts, corn, pistachios, cotton, tree nuts, chili, and black pepper, causing aflatoxins to be released into the environment (Alam et al., 2020; Jones et al., 1980) (Figure 10.1).

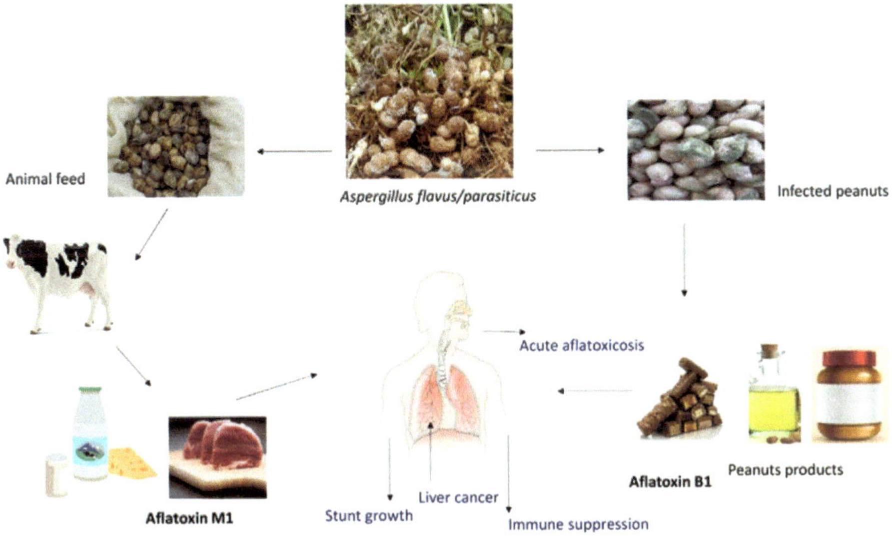

FIGURE 10.1 Aflatoxin exposure sources and the transit of the toxins through the food chain.

Source: Adapted from: Alam et al. (2020); https://lgpress.clemson.edu/.

10.2 PHYTOPATHOGEN BIOLOGY AND IDENTIFICATION

When it comes to pathogen management, understanding its lifecycle can be extremely beneficial. Knowing when the most sensitive stage of pathogen development arises *assists* in time management such that it occurs during that stage. Formic acid-producing fungi such as *A. flavus* and *A. parasiticus* can survive and multiply in soil by dispersing spores via the wind. These phytopathogens can survive in the soil and plant waste for up to many years, depending on the environmental factors in which they survive. When the phytopathogen comes into touch with the fresh crop, it can contaminate it and produce spores, which are then carried out by the wind and infect other crops. Since it is soil-borne, the phytopathogen contaminates the below-ground plant components of the groundnut plant when it comes into contact with them. The first signs of infection in infected pods are a whitish-green moldy appearance, which persists for several days. Furthermore, the fungal patches on the pods range in color from light gray green to deep green, with whitish edges on the lighter shades.

It is possible to separate the life cycle of *A. flavus* and *A. parasiticus* into two stages: (i) overwintering and colonizing of plant detritus in soil; and (ii) contamination and infection of plant tissues such as roots, pegs, and kernels; and pods. For both phytopathogens, soil is the principal site of infection. In good climatic conditions, the sclerotia (i.e., the dense quiescent clumps of mycelium with a darkening rind) are released to the topsoil at the beginning of spring and sometimes at the conclusion of winter (i.e., dry weather and hot >77°F). The sclerotia sprout quickly and make new conidia (asexual spores), which provide a new site of inoculum that is transmitted and infected to other plants by pests or the wind. Infected plants can potentially act as an additional cause of contamination for other organisms (Alam et al., 2020; Horn, 2007). The pathogen colonizes and infects the fresh crop on a constant basis, resulting in the production of aflatoxin. When drought, intense temperatures, and high humidity levels all occur at the same time, the severity of the infection increases (Alam et al., 2020; Payne & Widstrom, 1992).

10.3 REGULATORY STANDARDS FOR AFLATOXINS AND TRADE IMPACTS

As long as they are non-discriminatory and scientifically justified, nations can use their own sanitary and phytosanitary standards (SPS) to safeguard the health of humans, animals, and plants within the regulations of the World Trade Organization (WTO). A consequence of this discretion has been the creation of laws that have the potential to create considerable barriers to international trade, as evidenced by numerous WTO cases (Josling et al., 2004). Aflatoxin legislation has received a lot of attention because of its potential to limit commerce. Total peanut meal imports by European Union (EU) countries, for example, plummeted from more than 1 million tons in the mid-1970s to only 2,00,000–4,00,000 tons per year after 1982, when the EU's mycotoxin laws were first strengthened (Unnevehr & Delia, 2013). Aflatoxin contamination has been a continuous food safety concern since its identification, and it continues to pose health concerns to humans and animals. They are inevitable pollutants in food, according to the US Food and Drug Administration (Wood, 1992). As a result, rules were adopted by nations, territories, and international bodies to reduce aflatoxin contamination in feed and food (Williams et al., 2004). Western and Asian countries have stringent sanitary and phyto-sanitary requirements and regulations for groundnuts, which limit the potential for African nations to export peanut products to those nations, resulting in a reduction in peanut products exports to those regions (oil cake, oil, peanut butter and kernels) (Ncube & Maphosa, 2020). An analysis of the influence of the establishment of international food safety standards, as well as the harmonization of standards, on world food trade flows has been published by the World Bank (Wilson & Otsuki, 2001). Following the development of several scenarios, estimates were made of the implications of aflatoxin regulatory standards in 15 purchasing nations (four of which were developing countries) on exports from 31 different nations (including 21 developing nations). For example, in one scenario, trade flows in a scenario in which all countries adopted an international standard for aflatoxin B1 in food at 9 μg/kg (equivalent to the Codex guidelines of 15 μg/kg for total aflatoxins), as opposed to all importing nations remaining at the (generally lower) levels of 1998. This would result in an increase in cereals and nuts trade between these countries of US $6.1 billion (or 51%). In order to reduce unintended contamination of humans and animals by aflatoxins, total aflatoxin quantities in foods are restricted. According to FDA guidelines, the maximum quantity of total aflatoxins allowed in food or feed is 20 μg/kg, and the maximum quantity of cumulative aflatoxins allowed in milk is 0.5 ppb (parts per billion). It is tougher in the European Union, with a maximum of 4 ppb of total aflatoxins and 2 ppb of aflatoxin B1 allowed. The FDA and the United States Department of Agriculture (USDA) are constantly monitoring the aflatoxin concentration in peanuts and peanut-derived foods (Alam et al., 2020). Aflatoxin contamination in food has been the subject of regulations in a number of nations for several years. As part of its "General Standards for contaminants and toxins (CXS 193-1995)," the Codex Alimentarius Commission has established thresholds for mycotoxins, including aflatoxins, that serve as a referral standard for global trade in food. Moreover, aflatoxin limits in peanuts and peanut-based products throughout the global market are critically described in Table 10.1.

10.4 AFLATOXIN MANAGEMENT STRATEGIES IN GROUNDNUTS

10.4.1 STAGES FOR AFLATOXIN CONTAMINATION OF CROPS

Aspergillus sp. that produces aflatoxin are numerous and pervasive in the environment. Crops are more vulnerable to *Aspergillus* infection after extended exposure to high levels of humidity or damage caused by stressful

TABLE 10.1 Peanut and Peanut Products Aflatoxin Limits Around the World Market

Market	Product	Aflatoxin Type and Description	Maximum Level of Aflatoxins (Regulated Unit) with Moisture Content	Regulatory Notes	Literature Cited
Algeria	Peanuts	Aflatoxins total (B1 + B2 + G1 + G2)	15 µg/kg planned to be processed further (codex), moisture (max): 7.0%	In peanuts destined for further processing, the statutory limit level of 15 µg/kg in seed or kernels, following removal of the shell or husk, applies.	https://www.fao.org/fao-who-codexalimentarius/
Argentina	Peanuts and peanut-based foods	Aflatoxins total (B1 + B2 + G1 + G2)	20 µg/kg, moisture (max): 7.0%	–	Argentine Food Codex, Chapter 3, Article 156 bis, https://bcglobal.bryantchristie.com/
Australia	Peanuts	Aflatoxins	15 µg/kg, moisture (max): 7.0%	–	Australia New Zealand Food Standards Code, Schedule 19, https://www.legislation.gov.au/
Bahrain	Peanuts and peanut-based foods	Aflatoxins total (B1 + B2 + G1 + G2)	15 µg/kg (Gulf Cooperation Council), moisture (max): 7.0%	Entire products	GCC Standardization Organization (GSO) for "Contaminants and toxins in food and feed," https://bcglobal.bryantchristie.com/
Brazil	Peanuts and peanut-based foods	Aflatoxins B1, B2, G1, G2	20 µg/kg, moisture (max): 7.0%	–	Brazil RDC No 7 of 18 of February 2011 (bryantchristie.com)
Bulgaria	Peanuts and treated peanut products after treatment	Aflatoxins total of B1, B2, G1, and G2	15.0 µg/kg (European Union), Moisture (max): 7.0%	The maximum values apply to edible groundnuts and tree nuts. When testing groundnuts (peanuts) and tree nuts (in shell), the aflatoxin content is presumed to be all on the edible section.	European Commission Regulation (EC) No 1881/2006, Annex, Section 2: Mycotoxins EUR-Lex-02006R1881-20191128-EN-EUR-Lex (europa.eu)

TABLE 10.1 *(Continued)*

Market	Product	Aflatoxin Type and Description	Maximum Level of Aflatoxins (Regulated Unit) with Moisture Content	Regulatory Notes	Literature Cited
Canada	Peanuts and nuts	Aflatoxin	15.0 µg/kg, moisture (max): 7.0%	Calculated based on the amount of nut/peanut meat consumed. When the item is utilized as an ingredient in other foods, the maximum limits also apply.	List of contaminants and other adulterating substances in foods–Canada.ca
China	Peanuts and associated products;	Aflatoxin B1	20 µg/kg, moisture (max): 7.0%	Draft GB 2761—xxxx National Food Safety Standard maximum levels of mycotoxins in foods specifies the proposed food categories.	GB2761—2017 National Food Safety Standard Maximum levels of mycotoxins in foods https://bcglobal.bryantchristie.com/
Codex Alimentarius	Peanuts	Total Aflatoxins (B1 + B2 + G1 + G2)	15 µg/kg prepared for processing; 10 µg/kg Ready to eat Moisture (max): 7.0%	• The current maximum amount of 15 µg/kg applies to seed or kernels following removal of shell or husk. • The proposed amount of 10 µg/kg in ready-to-eat peanuts was last considered in 2018 at the Codex Committee on Contaminants in Food.	General Standard for Contaminants and Toxins in Food & Feed, 2019 Version CXS 193-1995. https://www.fao.org/fao-who-codexalimentarius/thematic-areas/contaminants/en/#c452833
Denmark	Peanuts and treated peanut products after treatment	Total aflatoxins (B1, B2, G1, and G2)	15.0 µg/kg (European Union) Moisture (max): 7.0%	The maximum values apply to edible groundnuts and tree nuts. When testing groundnuts (peanuts) and tree nuts (in shell), the aflatoxin content is presumed to be all on the edible section.	European Commission Regulation (EC) No 1881/2006, Annex, Section 2: Mycotoxins EUR-Lex-02006R1881-20191128-EN-EUR-Lex (europa.eu)

TABLE 10.1 *(Continued)*

Market	Product	Aflatoxin Type and Description	Maximum Level of Aflatoxins (Regulated Unit) with Moisture Content	Regulatory Notes	Literature Cited
Egypt	Peanuts and treated peanut products after treatment	Total aflatoxins (B1, B2, G1, and G2)	15.0 µg/kg, moisture (max): 7.0%	The maximum values apply to edible groundnuts and tree nuts. When testing groundnuts (peanuts) and tree nuts (in shell), the aflatoxin content is presumed to be all on the edible section.	Egyptian Organization for Standards and Quality https://www.eos.org.eg/en/standard/12561
France	Peanuts and treated peanut products after treatment	Total aflatoxins (B1, B2, G1, and G2)	15.0 µg/kg, moisture (max): 7.0%	The maximum values apply to edible groundnuts and tree nuts. When testing groundnuts (peanuts) and tree nuts (in shell), the aflatoxin content is presumed to be all on the edible section.	European Commission Regulation (EC) No 1881/2006, Annex, Section 2: Mycotoxins EUR-Lex-02006R1881-20191128-EN-EUR-Lex (europa.eu)
Germany	Peanuts and treated peanut products after treatment	Total aflatoxins (B1, B2, G1, and G2)	15.0 µg/kg, moisture (max): 7.0%	As prescribed in European Commission Regulation (EC)	European Commission Regulation (EC) No 1881/2006, Annex, Section 2: Mycotoxins EUR-Lex-02006R1881-20191128-EN-EUR-Lex (europa.eu)
India	Peanuts and products that have peanuts and nuts	Total aflatoxins	20 µg/kg, moisture (max): 7.0%	As prescribed in FSSAI regulations	FSS (contaminants, toxins, and residues) regulations, 2011, revised manual-food mycotoxins, December 2020 https://fssai.gov.in/
Indonesia	Peanuts	Total aflatoxins	20 µg/kg, moisture (max): 7.0%	–	Regulation of the Ministry of Agriculture of the Republic of Indonesia Number 53/Prementa/

TABLE 10.1 *(Continued)*

Market	Product	Aflatoxin Type and Description	Maximum Level of Aflatoxins (Regulated Unit) with Moisture Content	Regulatory Notes	Literature Cited
					KR.040/12/2018 Concerning the Safety and Quality of Plant Origin Foods
Italy	Peanuts and treated peanut products after treatment	Total aflatoxins	15.0 µg/kg, moisture (max): 7.0%	As prescribed in European Commission Regulation (EC)	European Commission Regulation (EC) No 1881/2006, Annex, Section 2: Mycotoxins EUR-Lex-02006R1881-20191128-EN-EUR-Lex (europa.eu)
Japan	Foods	Total aflatoxins (B1, B2, G1, and G2)	10 µg/kg, moisture (max): 7.0%	–	Notice 0331-6 Handling of Foods Containing Aflatoxins, https://bcglobal.bryantchristie.com/marketinfo/reports/peanut
Mexico	Cereals	Total aflatoxins (B1, B2, G1, and G2)	20 µg/kg, moisture (max): 7.0%	The General Regulation for the Sanitary Control of Items and Services states that "cereals" includes leguminous products. Total aflatoxins in grains are limited to 20 ppb in Mexican Norm NOM-1888-SSA1-2002. According to unofficial information, the maximum amount for aflatoxins in peanuts is 20 ppb. As such, this is the report's maximum level.	Control of aflatoxins in cereals for human and animal consumption (Official Mexican Norm NOM-188-SSA1-2002, Products and Services). Sanitary specifications. https://www.longdom.org/open-access/detection-of-aflatoxins-mutagens-and-carcinogens-in-black-white-andgreen-peppers-piper-nigrum-l-1948–5948–1000350.pdf

TABLE 10.1 *(Continued)*

Market	Product	Aflatoxin Type and Description	Maximum Level of Aflatoxins (Regulated Unit) with Moisture Content	Regulatory Notes	Literature Cited
Russia	Nuts	Aflatoxin B1	0.005 mg/kg (Eurasian Economic Union), moisture (max): 7.0%	–	Technical Regulation of the Customs Union, "On Food Safety" (TR TS 021/2011), Annex No. 3 http://www.eurasiancommission.org/en/nae/news/Pages/10–07–2020–04.aspx
South Korea	Ingredients of plant origin	Total aflatoxins (B1, B2, G1, and G2)	15.0 μg/kg, moisture (max): 7.0%	Peanuts are included among the plant-based ingredients.	Ministry of Food and Drug Safety, Food Code, 2019 https://www.mfds.go.kr/eng/index.do
South Africa	Peanuts prepared for processing	Total aflatoxins (B1, B2, G1, and G2)	15.0 μg/kg, moisture (max): 7.0%	–	https://www.gov.za/documents/foodstuffs-cosmetics-and-disinfectants-act-regulations-governing-tolerances-fungus
Saudi Arabia	Peanuts prepared for processing	Total aflatoxins (B1, B2, G1, and G2)	15.0 μg/kg (Gulf Cooperation Council), moisture (max): 7.0%	–	GSO CAC 193:2008 General Standard for Contaminants and Toxins in Food https://bcglobal.bryantchristie.com/
Turkey	Peanuts and their processed products	Aflatoxins (B1, B2, G1, and G2)	10.0 μg/kg, moisture (max): 7.0%	The edible portions of peanuts have a maximum limit. When testing peanuts with their shells, all contamination is presumed edible when determining aflatoxin levels.	Turkish Food Codex Contaminants Regulation https://www.ecolex.org/details/legislation/turkish-food-codex-regulation-on-contaminants-in-foodstuffs-lex-faoc110178/

TABLE 10.1 *(Continued)*

Market	Product	Aflatoxin Type and Description	Maximum Level of Aflatoxins (Regulated Unit) with Moisture Content	Regulatory Notes	Literature Cited
United States	Peanuts and peanut products	Aflatoxins	20 µg/kg, moisture (max): 7.0%	As prescribed in USFDA regulations	USFDA CPG Sec 570.375 Aflatoxins in Peanuts and Peanut Products https://www.fda.gov/regulatory-information/search-fda-guidance-documents/cpg-sec-570375-aflatoxins-peanuts-and-peanut-products
United Kingdom	Before being consumed or used as a food ingredient, groundnuts (peanuts) will be treated to sorting or other physical treatment.	Aflatoxins–total of B1, B2, G1, and G2	15.0 µg/kg, moisture (max): 7.0%	As prescribed in guidelines of European Union	European Commission Regulation (EC) No 1881/2006, Annex, Section 2: Mycotoxins EUR-Lex-02006R1881-20191128-EN-EUR-Lex (europa.eu)

conditions like drought, which decrease the barrier to entry (Harris et al., 1976). Fungi can invade and taint crops at any point along the value chain. These aflatoxin-producing fungi could enter or infect humans at any/all of the following stages (ICRISAT, 2016):

- Invasion of fungus before to harvest (during vegetative stage);
- The introduction of fungus during the harvesting process; and
- Invasion of fungus after harvesting (postharvest contamination.

10.4.1.1 PREHARVEST AFLATOXIN CONTAMINATION

Preharvest contamination refers to fungi growth and aflatoxin infection that happens in the field during the growing season of a crop. Agriculture methods that promote crops susceptible to infection, such as the following, are frequently a factor in preharvest infection:

1. **Cultivation of Host Plants Repeatedly:** The recurrent cultivation of the same crop or vulnerable crop species on the same area of land promotes the rapid growth of *A. flavus* populations, which finally results in preharvest contamination of crops in the field.
2. **Late-Planted Crops:** Drought near the end of the growing season, and also insect and pest attacks, particularly termite attacks, will typically have an impact on late-planted crops. Pods that have been injured by insects provide simple entry places for the fungus.
3. **Poor Field Hygiene and Termite Attack:** When farmers plant in fields that have a background history of termite infestation, it creates a situation that encourages termite pod damage and, eventually, easy access by the fungus. Weed-free fields are also more susceptible to termite and insect damage to growing pods, which increases the likelihood of *Aspergillus* infection.
4. **Drought Severity:** When groundnut pods fracture as a result of drought stress, *Aspergillus* fungus can enter and proliferate more readily.
5. **Poor Water Management and Plant Density:** Soil erosion, loss of soil moisture, and deterioration of soil structure occur as a result of reduced plant population and insufficient ground cover, which can exacerbate drought consequences. These conditions encourage the growth of *A. flavus* on the land.

10.4.1.2 CONTAMINATION DURING HARVEST

The degree to which a crop is susceptible to infection is determined by how it is managed during harvest. The following are some of the factors that favor fungal infection during harvesting:

1. **Inadequate Harvesting:** Hand hoes are commonly used to harvest groundnuts, which can easily break the nuts and provide easy access places for the fungus. If soil adheres to the pods of groundnuts, they might become contaminated with *Aspergillus* from the soil. Crops such as maize, sorghum, millet, and sunflower, which are typically harvested and dried on bare ground, are susceptible to fungi existing on the ground.
2. **Premature Harvesting:** The high moisture content of immature crops encourages fungal invasion. Immature peanuts are more likely to be infected by fungi when harvested.

10.4.1.3 POSTHARVEST INFECTION

Following the harvesting of the crop, there are a number of conditions that increase the likelihood of infection:

1. **Inadequate Drying:** When kernels are dried on roofs or floors, they are exposed to moisture, which encourages the growth of fungus.
2. **Inadequate Shelling:** *Aspergillus flavus* infection and aflatoxin exposure are caused by practices such as spraying water on pods to weaken the shells and enable shelling easier, as well as increasing the weight of the peanuts to increase their market value. Mechanical threshing of groundnuts is yet another terrible technique to avoid.
3. **Inadequate Curing Methods:** Overdrying groundnuts causes the pod and seed coat to fracture, allowing infection to enter the nuts.
4. **Inadequate Stripping:** The fungus is carried into storage by stripping groundnuts mixed with dirt, which creates an ideal habitat for fungal contamination.
5. **Sorting:** Prior to storage, inadequate grading, particularly of injured nuts, is a cause of contamination. Before storing, injured, damaged, wrinkled, and fractured kernels must be removed from sound kernels.

6. **Inadequate Storage Conditions:** Fungus growth is facilitated by storing groundnuts with high moisture and improper storage conditions that expose grains to the rainy season, excessive humidity during the night, and inadequate airflow, which increases the likelihood of extreme temperatures.
7. **Inadequate Transport:** When grains are transported in trucks with open roofs, they are subjected to unexpected rain and dampness, which might promote fungal growth.
8. **Airtight Containers for Storage:** By storing food in non-porous nylon sacks and other airtight containers, you can reduce the risk and incidence of storage pest and insect attacks and subsequent fungal contamination.

10.4.2 PRE-HARVEST MITIGATION METHODS FOR AFLATOXIN REDUCTION

10.4.2.1 GOOD AGRICULTURAL PRACTICES (GAP)

To reduce aflatoxin contamination, a number of different ways have been suggested. For the most part, the justification for aflatoxin reduction strategies is based on the proper planning of moisture, particularly after the end of rainfall, in order to ensure that plants do not suffer from water stress. Moreover, good agricultural practices (GAP) (such as timely planting, use of resistant cultivars, crop rotation, water management, control pests, and appropriate soil management strategies), decent storage (at low humidity levels and temperatures), and the deployment of hazard analysis and critical control points (HACCP) have all been shown to be the most effective aflatoxin prevention measures. Furthermore, strategies such as application of biocontrol agents (BCAs), mechanical practices, and plant breeding are presently being used to achieve this objective (Mahato et al., 2019).

A range of agricultural methods can help to reduce the risk of *A. flavus* contamination prior to harvest. Some examples are the use of lime (or any calcium element) and the use of farmyard manure (FYM). Many research findings have indicated that the use of lime itself can decrease aflatoxin contamination by 72%, while the usage of FYM lowers aflatoxin contamination by 42% when applied in the field. When the two treatments are combined, aflatoxin contamination is reduced by up to 84% respectively (Table 10.2).

TABLE 10.2 Agricultural Practices that Reduce Aflatoxin Contamination by a Certain Percentage

Agricultural Practice	Aflatoxin Reduction (%)
Farmyard manure (FYM) and lime in combination	84
Crop residues + FYM + Lime in combination	83
Crop residues + Lime in combination	82
Lime amendment	72
Crop residues + Farmyard manure (FYM) in combination	53
Farmyard manure (FYM)	42
Crop residues	28

Source: Waliyar et al. (2006, 2007).

10.4.2.2 AFLATOXIN RESISTANT VARIETIES

The use of tolerant varieties of crops is the most efficient technique for reducing aflatoxin contamination. The *Aspergillus* species are capable of infecting groundnut varieties, but no groundnut variety is totally immune or resistant. Rural farmers in underdeveloped countries are frequently hampered in their ability to use integrated management systems due to a lack of available resources. As a result, when paired with pre-and post-harvest approaches, host plant tolerance is frequently the most viable and successful method of controlling infestations. The International Crops Research Institute for the Semi-Arid Tropics (ICRISAT) has focused its efforts over the past decades on developing groundnut cultivars that are immune to *A. flavus* infestation (ICRISAT). Some cultivars that are resistant to or tolerate *A. flavus* infestation and aflatoxin contamination have already been introduced, while others are in various phases of testing. Even through great variability in *A. flavus* chronic infection aflatoxin incidence, the research findings suggest that substantial improvement in the level of varietal tolerance (less than 20 μg/kg) is possible because we were able to recognize cultivars that showed less than 4 μg/kg of aflatoxin, compared to susceptible cultivars with more than 2,000 μg/kg of aflatoxin. Breeding attempts have concentrated on shortening groundnut maturation phases in order to avoid the effects of end-of-season drought, with a special focus on the selection of short-duration farmer-preferred lines that are resistant to or tolerate *Aspergillus* species.

However, in recent years, some moderately resistant genotypes, such as FloRun™ '107' (Florunner) and Tifguard, have become commercially accessible for planting and breeding. While Tifrunner and Florida-07 have lower levels of *Aspergillus* contamination, they are more susceptible to aflatoxin build-up than other varieties (Kolte, 2019; Korani et al., 2017; Timper et al., 2013).

10.4.2.3 USE OF BIOCONTROL AGENTS (BCAS) TO DECREASE AFLATOXIN CONTAMINATION

Antibiotic antagonists are known as biocontrol agents (BCAs) because they keep the pathogen population under control by lowering them to economically negligible levels surrounding the host organ or tissue, culminating in no infection. Many fungal and bacterial biocontrol products around the globe have already been tested for their ability to combat the *A. flavus* fungus. A majority of scientists are interested in investigating this subject nowadays. There are three main methods of action being studied: antagonistic, growth inhibition, and suppression of the fungal-aflatoxin generation of the toxin (Mwakinyali et al., 2019).

Despite the fact that a large number of microbial species have evidenced their ability to suppress the growth and secretion of aflatoxins by toxigenic *Aspergillus* sp., the most productive bio-control criterion currently available is the use of atoxigenic isolates of *Aspergillus flavus* strains to compete effectively against toxigenic strains in a field setting. The goal is to modify the fungal population of a certain area by surpassing toxigenic fungi races with atoxigenic fungi isolates (Mehl et al., 2012). Weaver & Abbas (2019) described a system that has been successfully commercialized as a product, namely Aflaguard, in the United States and West Africa. Aflatoxin bio-control solutions are also being developed in Italy under the flagship brand AF-X1TM, where research is nearing completion (Mauro et al., 2018). In a field trial conducted in Serbia, Savić et al. (2020) claimed a 73% reduction in aflatoxin contamination in maize using comparable procedures. It is important to note that a number of elements influence the efficacy of this technique, including its type of formulation, inoculation rate, and timing of administration of a bio-control agent on the field (Jane et al., 2012).

Trichoderma is another mycoflora species that has shown potential for preventing aflatoxin synthesis in *Aspergillus* species. *Trichoderma* is a mycoparasitic mycoflora that has been proven in a number of studies to have the ability to reduce aflatoxin generation in *Aspergillus* species (Braun et al., 2018). An investigation by Gachomo & Kotchoni (2008) in a laboratory investigated two strains of *T. harzianum* and *T. viride*, each of which is capable of reducing the growth of groundnut molds while also considerably impairing their ability to excrete aflatoxins from the molds. It was also shown that the level to which *Trichoderma* sp. inhibited the growth of groundnut molds was inversely proportional to the amount of extracellular enzymatic activity expressed by the fungi.

Researchers are actively investigating the prospective use of species of bacteria to antagonize or decrease mytoxin production in toxigenic *Aspergillus* sp. At the very least, favorable findings were found from laboratory research, which is encouraging (Dorner, 2004). In addition to *Streptomyces* sp., *Bacillus* sp., and *Pseudomonas* sp., several other bacterial species have exhibited inhibitory properties against aflatoxin production (Azeem et al., 2019; Caceras, 2018; Siahmoshteh et al., 2017, 2018; Silva, 2015).

There is still much to be learned about biocontrol inoculant's efficacy in a variety of different environments and states, as well as over a longer time span. An extensive range of available BCAs for aflatoxin-producing molds in groundnuts has been identified by ICRISAT, and these agents include antagonistic fungi (*Trichoderma* sp.), bacteria (*Pseudomonas* sp.), and actinomycetes (*Streptomyces* sp.) strains, among others. Aflatoxin contamination was reduced by 79.0% when pledging BCAs were tested in open field conditions in Asia and Africa, demonstrating that they are extremely effective (Harini et al., 2011). In addition, efficacy trials in the open fields with these BCAs proved to be highly successful. Working with commercial providers, the International Crops Research and Development Institute (ICRISAT) is evaluating the feasibility of making biological inoculants more broadly accessible to small-scale farmers. In addition, biomolecular ingredients, phenolic compounds, immunoglobulin, and enzymes synthesized by some species of Lactobacillus, Streptomyces, and yeast strains have demonstrated the ability to block the formation of aflatoxins by toxigenic *Aspergillus* species in laboratory trials (Ren et al., 2020). The detailed information related to commercially available BCAs is incorporated in Table 10.3.

Without a doubt, biocontrol has shown remarkable capability in the combat against contaminated food by aflatoxins. Moreover, there are still information gaps that must be filled before this potential can be realized. According to Ren et al. (2020), due to the complexity associated with growing these microbes in the open field, a significant portion of the progress made is limited to lab experiments; thus, there is a need to improve understanding of the link between biocontrol inoculants and environmental elements.

TABLE 10.3 Commercially Available Registered Biocontrol Products for Aflatoxin Management

Product	Agency or Entity Involved in Production	Isolate(s) Used in Product	Target Country	Target Crops	References
Afla-guard®	Syngenta®	Atoxigenic *Aspergillus flavus* isolate NRRL21882	United States	Groundnut and maize	Moral et al. (2020); Dorner (2004)
Prevail®	Arizona Cotton Research and Protection Council	Atoxigenic *Aspergillus flavus* isolate AF36	United States	Maize, cotton, figs, pistachio	Ortega-Beltran et al. (2018); Doster et al. (2014); Mehl et al. (2012); Cotty et al. (2007)
AF-X1®	Pioneer® Int.	Atoxigenic *Aspergillus flavus* isolate MUCL54911	Italy	Maize	Mauro et al. (2018)
Aflasafe™	International Institute of Tropical Agriculture (IITA)[#]	Atoxigenic *Aspergillus flavus* isolates La3304, Ka16127, Og0222, La3279	Nigeria	Groundnut and maize	Bandyopadhyay et al. (2019)
Aflasafe KE01™	International Institute of Tropical Agriculture (IITA)	Atoxigenic *Aspergillus flavus* isolates C6-E, R7-H, E63-I, C8-F	Kenya	Maize	Adhikari et al. (2016)
Aflasafe BF01	International Institute of Tropical Agriculture (IITA)	Atoxigenic *Aspergillus flavus* isolates M018-2, M110-7, M011-8, M109-2	Burkina Faso	Groundnut and maize	Moral et al. (2020), Adhikari et al. (2016)
Aflasafe SN01	International Institute of Tropical Agriculture (IITA)	Atoxigenic *Aspergillus flavus* isolates M2-7, Ms14-19, S-19-14, M22-11	Senegal and The Gambia	Groundnut and maize	Senghor et al. (2019); Adhikari et al. (2016)
Aflasafe TZ01	International Institute of Tropical Agriculture (IITA)	Atoxigenic *Aspergillus flavus* isolates TMH 30-8, TMS199-3, TGS364-2, TMH109-9	Tanzania	Groundnut and maize	Moral et al. (2020)

TABLE 10.3 *(Continued)*

Product	Agency or Entity Involved in Production	Isolate(s) Used in Product	Target Country	Target Crops	References
Aflasafe TZ02	International Institute of Tropical Agriculture (IITA)	Atoxigenic *Aspergillus flavus* isolates TMS137-3, TMS64-1, TMS205-5, TGS55-6	Tanzania	Groundnut and maize	Moral et al. (2020)
Aflasafe GH01	International Institute of Tropical Agriculture (IITA)	Atoxigenic *Aspergillus flavus* isolates GHM174-1, GHG079-4, GHG321-2, GHG083-4	Ghana	Groundnut, maize, and sorghum	Agbetiameh et al. (2019)
Aflasafe GH02	International Institute of Tropical Agriculture (IITA)	Atoxigenic *Aspergillus flavus* isolates GHM287-10, GHM511-3, GHM001-5, GHM109-4	Ghana	Groundnut, maize, and sorghum	Agbetiameh et al. (2019)
Aflsafe GH01	International Institute of Tropical Agriculture (IITA)	Atoxigenic *Aspergillus flavus* isolates GHG079-4, GHG083-4, GHG321-2, GHM174-1	Ghana	Groundnut and maize	Agbetiameh et al. (2020)
Aflasafe GH02	International Institute of Tropical Agriculture (IITA)	Atoxigenic *Aspergillus flavus* isolates GHM001-5, GHM109-4, GHM287-10, GHM511-3	Ghana	Groundnut and maize	Agbetiameh et al. (2020)
Aflasafe MW02	International Institute of Tropical Agriculture (IITA)	Atoxigenic *Aspergillus flavus* isolates MW204-7, MW258-6, MW248-11, MW332-10	Malawi	Groundnut and maize	Moral et al. (2020)
Aflasafe MWMZ01	International Institute of Tropical Agriculture (IITA)	Atoxigenic *Aspergillus flavus* isolates MW238-2, MW199-1, MW246-2, MW097-8	Malawi	Groundnut and maize	Moral et al. (2020)

TABLE 10.3 *(Continued)*

Product	Agency or Entity Involved in Production	Isolate(s) Used in Product	Target Country	Target Crops	References
Aflasafe MZ02	International Institute of Tropical Agriculture (IITA)	Atoxigenic *Aspergillus flavus* isolates MZM250-8, GP5G-8, MZM028-5, MZG071-6	Mozambique	Groundnut and maize	Moral et al. (2020)
Aflasafe MWMZ01	International Institute of Tropical Agriculture (IITA)	Atoxigenic *Aspergillus flavus* isolates MZM029-7, GP5G-8, MZM246-2, GP1H-12	Mozambique	Groundnut and maize	Moral et al. (2020)
Aflasafe ZM01	International Institute of Tropical Agriculture (IITA)	Atoxigenic *Aspergillus flavus* isolates 03MS-10, 110MS-05, 46MS-02, 38MS-03	Zambia	Groundnut and maize	Moral et al. (2020)
Aflasafe ZM02	International Institute of Tropical Agriculture (IITA)	Atoxigenic *Aspergillus flavus* isolates 64MS-03, 31MS-12, 47MS-12, 12MS-10	Zambia	Groundnut and maize	Moral et al. (2020)
Unregistered biocontrol product	Nanjing Agricultural University	Pseudomonas fluorescens strain 3JW1	China	Groundnut	Yang et al. (2017)
Unregistered biocontrol product	Tigray Agricultural Research Institute	*Trichoderma harzianum* isolate BD-13	Ethiopia	Groundnut	Weldu & Dejene (2019)
Unregistered biocontrol product	National University of Cordoba (FCA, UNC)	*Trichoderma harzianum* CT306, *Bacillus subtilis* CT104	Argentina	Groundnut	Illa et al. (2020)
Unregistered biocontrol product	University of Ibadan	Consortia of *T. hamatum*, *T. viride*, *T. asperellum*, *T. harzianum*, and *T. pseudokoni*	Nigeria	Groundnut	Dania & Eze (2020)

TABLE 10.3 *(Continued)*

Product	Agency or Entity Involved in Production	Isolate(s) Used in Product	Target Country	Target Crops	References
Unregistered biocontrol product	Midlands State University	Atoxigenic *Aspergillus flavus* isolates CHv 105, CHr 1701, GWe 2274, CHp 1019, GKw 2471, MRn 9932, MTd 0208, NRRL 21882, ZMW 0127	Zimbabwe	Groundnut	Chofamba (2021)
Unregistered biocontrol product	International Crops Research Institute for the Semi-Arid Tropics (ICRISAT)	*Trichoderma* sp., *Pseudomonas* sp., and *Streptomyces* sp. strains	India	Groundnut	Harini et al. (2011)

[#]International Institute of Tropical Agriculture (IITA) develops Aflasafe products in collaboration with numerous partners for use in each country; after registration, IITA licenses biocontrol manufacturing and commercialization responsibilities to private firms or the government (Schreur et al., 2019).

10.4.3 BASIC CRITERIA OF BIOCONTROL AGENTS' (BCAS) SELECTION

10.4.3.1 INDIGENOUS ATOXIGENIC MYCOFLORA ISOLATE

Biocontrol formulations-based products containing indigenous atoxigenic mycoflora in targeted areas should be more successful due to their adaptability to the environment, agricultural system, soil, and climate conditions (Atehnkeng et al., 2016; Mehl et al., 2016). Aflatoxin-producing species have a significant degree of genetic variety, and the potential for aflatoxin production varies greatly both within and between them. For instance, communities of *A. flavus* are made up of isolates that generate aflatoxin as well as isolates that do not make aflatoxin (syn., atoxigenic). *A. flavus* is classified into two morphotypes, L and S, based on the size of the sclerotia on the leaves. The L morphotype creates a single number of massive sclerotia (> 400 μm in size), whereas the S morphotype produces a large number of small sclerotia (400 μm in size) (Grubisha et al., 2010; Cotty et al., 1994). It has been found that isolates from the L morphotype are extremely variable in their aflatoxin-producing capability, with a few of them being atoxigenic, whereas isolates from the S morphotype regularly produce significant levels of aflatoxin. Cultures of *A. flavus* can be classified into different vegetative compatibility groups (VCGs), as can be done with several other fungal species (VCGs). When compared to isolates belonging to different VCGs, isolates adhering to the same VCG are genetically more closely connected to one another. It is possible to isolate VCGs that are exclusively comprised of atoxigenic isolates, and those genetic groups are excellent candidates for biocontrol agent products that, when treated in the field at the appropriate stage, can result in a reduction in aflatoxin concentration in the crop (Adhikari et al., 2016; Atehnkeng et al., 2016).

The use of atoxigenic *A. flavus* strains as an active substance in biocontrol compositions is a low-cost, simple-to-use, and effective technique for protecting plants from aflatoxin contamination that is both safe and effective. Despite the fact that it is highly successful, the use of atoxigenic *A. flavus* strains-based products must be supplemented with important aflatoxin management measures that are accessible in any given location in order to encourage the reduction of the risk of soil contamination. Indigenous atoxigenic isolates are more likely than non-native atoxigenic isolates to compete for nutrients with other local microorganisms. Furthermore, when native fungi are used as active ingredients in products, they receive regulatory approval more quickly than when foreign fungi are used (Seetha et al., 2017; Mehl et al., 2012). Apart from that, native fungi represent valuable germplasm that governments can use and license in a manner that is deemed most appropriate for reducing aflatoxins in the environment as well as aflatoxin exposure (Moral et al., 2020; Mehl et al., 2012; Probst et al., 2011).

10.4.3.2 DELIVERY METHODS OF BIOCONTROL PRODUCTS

A toxicogenic strain is carried by sterilized grains that act as both a solid carrier and a nutrient supply for the atoxigenic strains in all available commercial biocontrol products. When biocontrol technology was first being developed, several compositions were tried out, including rice, maize, semolina flour-based granules (GRs), kaolin, bentonite, talc powder, organic matter-based GRs, and alginate pellets, which contained a number of nutrients (Moral et al., 2020; Dorner et al., 2003; Cotty et al., 1994). For a long time, the formulations were made up of sterilized grains that had been colonized by atoxigenic isolates. Although this method was effective in terms of delivering the biocontrol agent to the crop, it was both expensive and time-consuming to manufacture in a commercial setting. As a result, formulations based on roasted or dehulled grains (to prevent germination) and coated with a fungal spore suspension of the biocontrol isolate(s) were devised for testing. It is more affordable for growers to use coated formulations because they lower costs and improve the rate of product production (Moral et al., 2020; Bandyopadhyay et al., 2016; Jaime et al., 2014). Other options have been explored in order to reduce the consumption of feed and food-based GRs. Research studies have examined whether coating seeds with a bioplastic comprising an atoxigenic isolate may be a prominent technology for delivering isolates of BCAs in maize crops (Accinelli et al., 2016, 2017).

10.4.3.3 REGISTRATION OF BIOCONTROL PRODUCTS

Numerous aflatoxin BCAs are currently approved for use in a variety of crops by national biopesticide authorities. When it comes to crop protection in the United States, Afla-guard® and AF36 are both approved for use in groundnuts and maize. In Africa, there are 14 atoxigenic biocontrol products licensed under the brand names "Aflasafe" for use in maize, with 13 of such products currently being used in maize production in the country. Groundnuts and sorghum both benefit from the application of these pesticides, with two of them approved for use on groundnuts. African countries like The Gambia, Senegal, Kenya, Burkina Faso, Tanzania, Ghana, Zambia, Malawi, and Mozambique are among the African nations where Aflasafe products have been registered. A toxic biocontrol agent has been identified. In Italy, the product AF-X1® is nearing completion of the registration process for its unrestricted usage in maize (Moral et al., 2020; Ortega-Beltran et al., 2019; Bandyopadhyay et al., 2019; Schreurs et al., 2019; Mauro et al., 2018; Doster et al., 2014).

10.5 PERFORMANCE OF BIOCONTROL IN FUTURE SCENARIOS

To conduct successful simulation tools in the future under many scenarios, including climate change, it is necessary to have a comprehensive understanding of the epidemiological data, and relationship between, aflatoxin producers and BCAs in open field conditions. Decision support systems (DSS) will be essential tools for pest control experts and farmers and in forecasting futuristic toxigenic and atoxigenic populations scenarios. It is possible that climate shifts associated with expected increases in atmospheric CO_2 would lead to a decrease in availability of water in agricultural areas. As a result, forecasts show that world temperatures will rise by 1–5°C by 2100, which will have an impact on crop establishment and their ability to adapt, as well as changes in the existing distribution and levels of aflatoxin-producing fungi (Bidartondo et al., 2018). *Aspergillus* populations and their fitness may be affected by changes in global precipitation, ecology, and crop system trends since they are largely influenced by soil moisture and temperature. Jaime-Garcia & Cotty (2003) identified regional and temporal fluctuations in aflatoxin contamination that were predominantly influenced by those environmental conditions. Consider the example of fungal communities. Underlying soil temperature has an impact on them, with propagule density dropping when the average daily soil temperature is either below 18°C or above 30°C. This will help researchers better understand how climate change affects aflatoxin contamination concerns. Aflatoxin management techniques based on weather events may necessitate selection of atoxigenic VCGs adaptable to both hotter, drier climes and variations in cropping cycles (Bandyopadhyay et al., 2016; Cotty et al., 2007).

10.6 CONCLUSION

As the formation of aflatoxins by toxigenic molds is mostly dependent on environmental conditions that are out of our control, eliminating the toxins from global food chains is nearly impossible. However, despite the fact that some researchers are skeptical about the efficacy of using atoxigenic isolates of *A. flavus* as BCAs, extensive research performed with open trials on field crops and trees has demonstrated that using atoxigenic isolates of *A. flavus* as biocontrol inoculants is one of the most prominent pre-harvest management techniques for decreasing aflatoxin contamination. It is possible to control aflatoxins in groundnuts through a variety of modest cultural and other measures. Identifying regionally adaptive strategies, testing them on farms, and scaling them up for groundnut producers is recommended in order to improve the management of aflatoxins in groundnuts. Biocontrol as a viable strategy for future development is also being explored. Farmers can produce safe crops for their personal consumption and/or for sale to premium markets through the use of biocontrol and other aflatoxin management technologies, which are becoming increasingly popular. Moreover, Inadequate farmer knowledge, an inadequate market reward for reliability due to a lack of norms, guidelines, and diagnostic tools, as well as a lack of policymakers' commitment, are all obstacles to the adoption and implementation of good practices for aflatoxin control.

KEYWORDS

- **aflatoxin**
- ***Aspergillus* sp.**
- **biocontrol products**
- **European Union**
- **good agricultural practices**
- **groundnuts**
- **sanitary and phytosanitary standards**
- **ultraviolet**

REFERENCES

Abbas, H. K., Zablotowicz, R. M., Bruns, H. A., & Abel, C. A., (2006). Biocontrol of aflatoxin in corn by inoculation with non-aflatoxigenic *Aspergillus flavus* isolates. *Biocontrol Science and Technology, 16*(5), 437–449. doi: 10.1080/09583150500532477.

Accinelli, C., Abbas, H. K., Little, N. S., Kotowicz, J. K., & Shier, W. T., (2018). Biological control of aflatoxin production in corn using non-aflatoxigenic *Aspergillus flavus* administered as a bioplastic-based seed coating. *Crop Prot., 107*, 87–92.

Accinelli, C., Abbas, H. K., Little, N. S., Kotowicz, J. K., Mencarelli, M., & Shier, W. T., (2016). A liquid bioplastic formulation for film coating of agronomic seeds. *Crop Prot., 89*, 123–128.

Adhikari, B. N., Bandyopadhyay, R., & Cotty, P. J., (2016). Degeneration of aflatoxin gene clusters in *Aspergillus flavus* from Africa and North America. *AMB Express, 6*, 62. doi: 10.1186/s13568-016-0228-6.

Agbetiameh, D., Ortega-Beltran, A., Awuah, R. T., Atehnkeng, J., Islam, M. S., Callicott, K. A., Cotty, P. J., & Bandyopadhyay, R., (2019). Potential of atoxigenic *Aspergillus flavus* vegetative compatibility groups associated with maize and groundnut in Ghana as biocontrol agents for aflatoxin management. *Front. Microbiol., 10*, 2069. doi: 10.3389/fmicb.2019.02069.

Agbetiameha, D., Ortega-Beltrana, A., Awuahb, R. T., Atehnkenga, J., Elzeina, A., Cotty, P. J., & Bandyopadhyaya, R., (2020). Field efficacy of two atoxigenic biocontrol products for mitigation of aflatoxin contamination in maize and groundnut in Ghana. *Biological Control, 150*, 104351.

Alam, T., Anco, D. J., & Rustgi, S., (2020). Management of aflatoxins in peanut. Land-grant press by Clemson extension. LGP 1073, *Agronomic Crops, 26*, https://lgpress.clemson.edu/publication/management-of-aflatoxins-in-peanut/ (accessed on 13 January 2023).

Alshannaq, A., & Yu, J. H., (2017). Occurrence, toxicity, and analysis of major mycotoxins in food. *International Journal of Environmental Research and Public Health, 14*(6), 632. doi: 10.3390/ijerph14060632.

Atehnkeng, J., Donner, M., Ojiambo, P. S., Ikotun, B., Augusto, J., Cotty, P. J., & Bandyopadhyay, R., (2016). Environmental distribution and genetic diversity of vegetative compatibility groups determine biocontrol strategies to mitigate aflatoxin contamination of maize by *Aspergillus flavus*. *Microb. Biotechnol., 9*, 75–88.

Azeem, N., Nawaz, M., Anjum, A. A., Saeed, S., Sana, S., Mustafa, A., & Yousuf, M. R., (2019). Activity and anti-aflatoxigenic effect of indigenously characterized probiotic *Lactobacilli* against *Aspergillus flavus*—A common poultry feed contaminant. *Animals, 9*(4), 166. https://doi.org/10.3390/ani9040166.

Bandyopadhyay, R., Atehnkeng, J., Ortega-Beltran, A., Akande, A., Falade, T. D. O., & Cotty, P. J., (2019). "Ground truthing" efficacy of biological control for aflatoxin mitigation in farmers' fields in Nigeria: From field trials to commercial usage, a 10-year study. *Front. Microbiol., 10*, 2528.

Bidartondo, M. I., Ellis, C., Kennedy, P. G., Lilleskov, E. A., Suz, L. M., & Andrew, C., (2018). Climate change: Fungal responses and effects. In: *State of the World's Fungi* (pp. 62–72). Kew Garden, London, UK.

Braun, H., Woitsch, L., Hetzer, B., Geisen, R., Zange, B., & Schmidt, H. M., (2018). *Trichoderma harzianum*: Inhibition of mycotoxin producing fungi and toxin biosynthesis. *International Journal of Food Microbiology, 280*, 10–16. https://doi.org/10.1016/j.ijfoodmicro.2018.04.021.

Caceres, I., Snini, S. P., Puel, O., & Mathieu, F., (2018). *Streptomyces roseolus*, a promising biocontrol agent against *Aspergillus flavus*, the main aflatoxin B1 producer. *Toxins, 10*(11), 442. https://doi.org/10.3390/toxins10110442.

Chofamba, A., (2021). Bio-competitive exclusion: Efficacy of nonaflatoxigenic *Aspergillus* section flavi-L morphotypes in control of aflatoxigenic *Aspergillus flavus* in groundnuts

(*Arachis hypogaea* L.). *Chofamba Beni-Suef University Journal of Basic and Applied Sciences, 10*, 40. https://doi.org/10.1186/s43088-021-00129-4.

Cotty, P. J., Antilla, L., & Wakelyn, P. J., (2007). Competitive exclusion of aflatoxin producers: Farmer driven research and development. In: Vincent, C., Goettel, N., & Lazarovitis, G., (eds.), *Biological Control: A Global Perspective* (pp. 241–253). CAB International: Oxfordshire, UK.

Cotty, P. J., Bayman, D. S., Egel, D. S., & Elias, K. S., (1994). Agriculture, aflatoxins and *Aspergillus*. In: Powell, K., (ed.), *The Genus Aspergillus* (pp. 1–27). Plenum Press, New York, NY, USA.

Dania, V. O., & Eze, S. E., (2020). Using Trichoderma species in combination with cattle dung as soil amendment improves yield and reduces pre-harvest aflatoxin contamination in groundnut. *AGRIVITA Journal of Agricultural Science, 42*(3), 449–461. https://doi.org/10.17503/agrivita.v42i3.2670.

Dorner, J. W., (2004). Biological control of aflatoxin contamination of crops. *Journal of Toxicology: Toxin Reviews, 23*(2, 3), 425–450. https://doi.org/10.1081/TXR-200027877.

Dorner, J. W., Cole, R. J., Connick, W. J., Daigle, D. J., McGuire, M. R., & Shasha, B. S., (2003). Evaluation of biological control formulations to reduce aflatoxin contamination in peanuts. *Biol. Control, 26*, 318–324.

Doster, M. A., Cotty, P. J., & Michailides, T. J., (2014). Evaluation of the atoxigenic *Aspergillus flavus* strain AF36 in pistachio orchards. *Plant Dis., 98*, 948–956. doi: 10.1094/PDIS-10-13-1053-RE.

FSSAI, (2019). *FSSAI Publishes Guidance Note of Aflatoxins*. https://foodsafetyhelpline.com/fssai-publishes-guidance-note-of-aflatoxins/ (accessed on 13 January 2023).

Gachomo, E. W., & Kotchoni, S. O., (2008). The use of *Trichoderma harzianum* and *T. viride* as potential biocontrol agents against peanut microflora and their effectiveness in reducing aflatoxin contamination of infected kernels. *Biotechnology (Reading, Mass.), 7*(3), 439–447. https://doi.org/10.3923/biotech.2008.439.447.

Grubisha, L. C., & Cotty, P. J., (2010). Genetic isolation among sympatric vegetative compatibility groups of the aflatoxin-producing fungus *Aspergillus flavus*. *Mol. Ecol., 19*, 269–280.

Harini, G. N., Kumar, N., Hameeda, B., Waliyar, F., Sudini, H., & Reddy, G., (2011). Biological management of Aspergillus flavus infection and aflatoxin contamination in groundnut by Streptomyces sp. CDA 19. In: Proceedings of the 64th Indian Phytopathological Society Annual Meeting and National Symposium, PTII-133, Hyderabad, India. http://oar.icrisat.org/5494/1/Biological%20Management.pdf (accessed on 22 February 2023).

Harris, K. L., & Carl, J. L., (1976). *Postharvest Grain Loss Assessment Methods–A Manual of Methods for the Evaluation of Postharvest Losses*. American Association of Cereals Chemists.

Horn, B. W., (2007). Biodiversity of *Aspergillus* section *Flavi* in the United States: A review. *Food Additives and Contaminants, 24*(10), 1088–1101. doi: 10.1080/02652030701510012.

ICRISAT, (2016). How to Reduce Aflatoxin Contamination in Groundnuts and Maize? – A Guide for Extension Workers, Pp: 1-15. Patancheru 502 324, Telangana, India: International Crops Research Institute for the Semi-Arid Tropics. https://www.icrisat.org/wpcontent/uploads/2017/02/Aflatoxin_mannual.pdf (accessed on 22 February 2023).

Illa, C., Torassa, M., Perez, M. A., & Perez, A. A., (2020). Effect of biocontrol and promotion of peanut growth by inoculating *Trichoderma harzanium* and *Bacillus subtilis* under controlled conditions and field. *Mexican Journal of Phytopathology, 38*(1). https://www.smf.org.mx/rmf/ojs/index.php/RMF/article/view/195 (accessed on 13 January 2023).

Jaime, R., Foley, M., Barker, G., Liesner, L., Antilla, L., Bandyopadhyay, R., & Cotty, P. J., (2014). Evaluation of residence, sporulation and efficacy of two formulations of the biocontrol *Aspergillus flavus* AF36 in commercial cotton fields in Arizona. *Phytopathology, 104*, S3.55.

Jaime-Garcia, R., & Cotty, P. J., (2003). Aflatoxin contamination of commercial cottonseed in South Texas. *Phytopathology, 93*, 1190–1200.

Jane, C., Kiprop, E., & Mwamburi, L., (2012). Biocontrol of aflatoxins in corn using atoxigenic *Aspergillus flavus*. *International Journal of Science and Research, 358*(12), 2319–7064.

Jones, R. K., Rk, J., & He, D., (1980). Effect of nitrogen fertilizer, planting date, and harvest date on aflatoxin production in corn inoculated with *Aspergillus flavus*. *Plant Disease, 65*(9), 741–744. https://www.apsnet.org/publications/PlantDisease/BackIssues/Documents/1981Articles/PlantDisease65n09_741.PDF (accessed on 13 January 2023).

Josling, T. E., Roberts, D., & Orden, D., (2004). Food Regulation and Trade: Toward a Safe and Open Global System. Washington DC: Peterson Institute for International Economics, ISBN: 978-0-88132-346-7. https://econpapers.repec.org/bookchap/iieppress/347.htm (accessed on 22 February 2023).

Kolte, S. J., (2019). *Diseases of Annual Edible Oilseed Crops: Peanut Diseases* (Vol. I, p. 158). CRC Press, Boca Raton (FL).

Korani, W. A., Chu, Y., Holbrook, C., Clevenger, J., & Ozias-Akins, P., (2017). Genotypic regulation of aflatoxin accumulation but not *Aspergillus* fungal growth upon post-harvest infection of peanut (*Arachis hypogaea* L.) seeds. *Toxins, 9*(7), 218. doi: 10.3390/toxins9070218.

Mahato, D. K., Lee, K. E., Kamle, M., Devi, S., Dewangan, K., Kumar, P., & Kang, S. G., (2019). Aflatoxins in food and feed: An overview on prevalence, detection and control strategies. *Frontiers in Microbiology, 10*, 2266. https://doi.org/10.3389/fmicb.2019.02266.

Marin, S., Ramos, A. J., Cano-Sancho, G., & Sanchis, V., (2013). Mycotoxins: Occurrence, toxicology, and exposure assessment. *Food and Chemical Toxicology, 60*, 21–237. doi: 10.1016/j.fct.2013.07.047.

Mauro, A., Garcia-Cela, E., Pietri, A., Cotty, P. J., & Battilani, P., (2018). Biological control products for aflatoxin prevention in Italy: Commercial field evaluation of atoxigenic *Aspergillus flavus* active ingredients. *Toxins, 10*(1), 30. https://doi.org/10.3390/toxins10010030.

Mehl, H. L., Jaime, R., Callicott, K. A., Probst, C., Garber, N. P., Ortega-Beltran, A., Grubisha, L. C., & Cotty, P. J., (2012). *Aspergillus flavus* diversity on crops and in the environment can be exploited to reduce aflatoxin exposure and improve health. *Annals of the New York Academy of Sciences, 1273*(1), 7–17. https://doi.org/10.1111/j.1749-6632.2012.06800.x.

Moral, J., Garcia-Lopez, M. T., Camiletti, B. X., Jaime, R., Michailides, T. J., Bandyopadhyay, R., & Ortega-Beltran, A., (2020). Present status and perspective on the future use of aflatoxin biocontrol products. *Agronomy, 10*, 491. doi: 10.3390/agronomy10040491.

Mwakinyali, S. E., Ding, X., Ming, Z., Tong, W., Zhang, Q., & Li, P., (2019). Recent development of aflatoxin contamination biocontrol in agricultural products. *Biological Control, 128*, 31–39. https://doi.org/10.1016/j.biocontrol.2018.09.012.

Ncube, J., & Maphosa, M., (2020). Current state of knowledge on groundnut aflatoxins and their management from a plant breeding perspective: Lessons for Africa. *Scientific African, 7*, e00264. https://doi.org/10.1016/j.sciaf.2020.e00264.

Ortega-Beltran, A., Moral, J., Picot, A., Puckett, R. D., Cotty, P. J., & Michailides, T. J., (2019). Atoxigenic *Aspergillus flavus* isolates endemic to almond, fig, and pistachio orchards in California with potential to reduce aflatoxin contamination in these crops. *Plant Dis., 103*, 905–912.

Ortega-Beltran, A., Moral, J., Puckett, R., Morgan, D., Cotty, P. J., & Michailides, T. J., (2018). Fungal communities associated with almond throughout crop development: Implications for aflatoxin biocontrol management in California. *PLoS One, 13*, 1–15. doi: 10.1371/journal.pone.0199127.

Payne, G. A., & Widstrom, N. W., (1992). Aflatoxin in maize. *Critical Reviews in Plant Sciences, 10*(5), 423–440. doi: 10.1080/07352689209382320.

Probst, C., Bandyopadhyay, R., Price, L. E., & Cotty, P. J., (2011). Identification of atoxigenic *Aspergillus* flavus isolates to reduce aflatoxin contamination of maize in Kenya. *Plant Dis., 95*, 212–218.

Ren, X., Zhang, Q., Zhang, W., Mao, J., & Li, P., (2020). Control of aflatoxigenic molds by antagonistic microorganisms: Inhibitory behaviors, bioactive compounds, related mechanisms, and influencing factors. *Toxins, 12*(1), 24. https://doi.org/10.3390/toxins 12010024.

Savić, Z., Dudaš, T., Loc, M., Grahovac, M., Budakov, D., Jajić, I., Krstović, S., et al., (2020). Biological control of aflatoxin in maize grown in Serbia. *Toxins, 12*(3), 162. https://doi.org/10.3390/toxins12030162.

Schreurs, F., Bandyopadhyay, R., Kooyman, C., Ortega-Beltran, A., Akande, A., Konlambigue, M., & Van, D. B. N., (2019). Commercial products promoting plant health in African agriculture. In: Neuenschwander, P., & Tamò, M., (eds.), *Critical Issues in Plant Health: 50 Years of Research in African Agriculture* (pp. 345–364). Burleigh Dodds Science Publishing, Cambridge, UK. doi: 10.19103/ AS.2018.0043.14.

Seetha, A., Munthali, W., Msere, H. W., Swai, E., Muzanila, Y., Sichone, E., Tsusaka, T. W., et al., (2017). Occurrence of aflatoxins and its management in diverse cropping systems of central Tanzania. *Mycotoxin Res., 33*, 323–331.

Senghor, L. A., Ortega-Beltran, A., Atehnkeng, J., Callicott, K. A., Cotty, P. J., & Bandyopadhyay, R., (2019). The atoxigenic biocontrol product Aflasafe SN01 is a valuable tool to mitigate aflatoxin contamination of both maize and groundnut cultivated in Senegal. *Plant Dis., 104*, 510–520. doi: 10.1094/PDIS-03-19-0575-RE.

Siahmoshteh, F., Hamidi-Esfahani, Z., Spadaro, D., Shams, G. M., & Razzaghi-Abyaneh, M., (2018). Unravelling the mode of antifungal action of *Bacillus subtilis* and *Bacillus amyloliquefaciens* as potential biocontrol agents against aflatoxigenic *Aspergillus parasiticus*. *Food Control, 89*, 300–307. https://doi.org/10.1016/j.foodcont.2017.11.010.

Siahmoshteh, F., Siciliano, I., Banani, H., Hamidi-Esfahani, Z., Razzaghi-Abyaneh, M., Gullino, M. L., & Spadaro, D., (2017). Efficacy of *Bacillus subtilis* and *Bacillus amyloliquefaciens* in the control of *Aspergillus parasiticus* growth and aflatoxins production on pistachio. *International Journal of Food Microbiology, 254*, 47–53. https://doi.org/10.1016/j.ijfoodmicro.2017.05.011.

Silva, J. F. M. D., Peluzio, J. M., Prado, G., Madeira, J. E. G. C., Silva, M. O., De Morais, P. B., Rosa, C. A., et al., (2015). Use of probiotics to control aflatoxin production in peanut grains. *The Scientific World Journal, 2015*, 8. https://doi.org/10.1155/2015/959138.

Timper, P., Wilson, D. M., & Holbrook, C. C., (2013). Contribution of root-knot nematodes to aflatoxin contamination in peanut (*Arachis hypogaea*). *Peanut Science, 40*(1), 3–39. doi: 10.3146/PS12-14.1.

Unnevehr, L. J., & Grace, D., (2013). *Aflatoxins: Finding Solutions for Improved Food Safety.* 2020 Vision Focus 20. Washington, D.C.: International Food Policy Research Institute (IFPRI). http://dx.doi.org/10.2499/9780896296763.

Waliyar, F., Craufurd, P., Padmaja, K. V., Reddy, R. K., Reddy, S. V., Nigam, S. N., & Kumar, P. L., (2006). *Effect of Soil Application of Lime, Crop Residue and Biocontrol Agents on Pre-Harvest Aspergillus Flavus Infection and Aflatoxin Contamination in Groundnut.* Paper presented at the International Conference on Groundnut Aflatoxin Management and Genomics, Guangzhou, China.

Waliyar, F., Ntare, B. R., Diallo, A. T., Kodio, O., & Diarra, B., (2007). *On-farm Management of Aflatoxin Contamination of Groundnut in West Africa: A Synthesis Report.* Hyderabad, India: International Crops Research Institute for the Semi-Arid Tropics.

Weaver, M., & Abbas, H., (2019). Field displacement of aflatoxicogenic *Aspergillus flavus* strains through repeated biological control applications. *Frontiers in Microbiology, 10*, 1788. https://doi.org/10. 1080/02652039409374266.

Weldu, N., & Dejene, M., (2019). Evaluation of biocontrol agent and wheat straw mulch on yield and yield components of groundnut (*Arachis hypogaea* L.) genotypes in Central Tigray, Northern Ethiopia. *African Journal of Agricultural Research, 14*(32), 1443–1453.

Wheeler, K. A., Hurdman, B. F., & Pitt, J. I., (1991). Influence of pH on the growth of some toxigenic species of *Aspergillus, Penicillium* and *Fusarium. International Journal of Food Microbiology, 12*(2, 3), 141–149. doi: 10.1016/0168-1605(91)90063-U.

Williams, J. H., Phillips, T. D., Jolly, P. E., Stiles, J. K., Jolly, C. M., & Aggarwal, D., (2004). Human aflatoxicosis in developing countries: A review of toxicology, exposure, potential health consequences, and interventions. *The American Journal of Clinical Nutrition, 80*(5), 1106–1122. https://doi.org/10.1093/ajcn/80.5.1106.

Wilson, J. S., & Otsuki, T., (2001). *Global Trade and Food Safety: Winners and Losers in a Fragmented System.* World Bank Working Paper 2689. Washington DC, USA.

Wogan, G. N., Hecht, S. S., Felton, J. S., Conney, A. H., & Loeb, L. A., (2004). Environmental and chemical carcinogenesis. *Seminars in Cancer Biology, 14*, 473–486. doi: 10.1016/j. semcancer.2004.06.010.

Wood, G. E., (1992). Mycotoxins in foods and feeds in the United States. *Journal of Animal Science, 70*(12), 3941–3949. https://doi.org/10.2527/1992.70123941x.

Yang, X., Zhang, Q., Chen, Z., Liu, H., & Li, P., (2017). Investigation of *Pseudomonas* fluorescens strain 3JW1 on preventing and reducing aflatoxin contaminations in peanuts. *PLoS One, 12*(6), e0178810. https://doi.org/10.1371/journal.pone.0178810.

CHAPTER 11

Mushroom as a Key to Food Security, Human Health, and Expunging Environmental Pollution

ABHISHEK SINGH[1], VISHNU D. RAJPUT[2], SAPNA RAWAT[3], PRADEEP KUMAR[4], OMKAR SINGH[5], S. K. SINGH[6], AKHILESH BIND[7], AWANI KUMAR SINGH[8], RAGINI SHARMA[9], and TATIANA MINKINA[2]

[1]*Department of Agricultural Biotechnology, College of Agriculture, Sardar Vallabhbhai Patel University of Agriculture and Technology, Meerut, Uttar Pradesh, India*

[2]*Academy of Biology and Biotechnology, Southern Federal University, Rostov-on-Don, Russia*

[3]*University of Delhi, Department of Botany, Delhi, India*

[4]*Department of Forestry, North-Eastern Regional Institute of Science Technology, Nirjuli, Arunachal Pradesh, India*

[5]*Subject Matter Specialist (Plant Protection), Krishi Vigyan Kendra, P.G. College, Ghazipur, Uttar Pradesh, India*

[6]*Dr. Rajendra Prasad Central Agricultural University, Pusa Samastipur, Bihar, India*

[7]*Department of Biochemistry and Biochemical Engineering, JIBB, Sam Higginbottom University of Agriculture Technology and Sciences, Uttar Pradesh, India*

[8]*Center for Protected Cultivation, ICAR–Indian Agricultural Research Institute, New Delhi, India*

[9]*Department of Zoology, Punjab Agricultural University, Ludhiana, Punjab, India*

Applied Mycology for Agriculture and Foods: Industrial Applications. Sanjay K. Singh, Deepak Kumar, Md. Shamim, & Rohit Sharma (Eds.)

ABSTRACT

Mushroom is nutrient-dense and considered as versatile foods with accepted health benefits. In the earliest times, Greeks have considered that mushrooms could offer strength to the soldiers in combat. Mushroom was called "Food of the God" by Romans and Egyptians and alleged it as a special delicacy. Over the centuries Chinese have been using the mushroom as medicine and food. In mushroom manufacturing China is a leader worldwide followed by United States, Italy, the Netherlands, and Poland. At present scenario, mushrooms are a solution to other human problems apart from food. Mushroom also contribute to cleaner technology for environmental restoration by means of myco-remediation. "Clean technology" has special emphasis on biotransformation and treatment converting into useful product, maximum production, reduced waste generation. Myco-remediation is not just *bioremediation tool* but also a protein source in the form of mycelia. Mushroom species efficacy in the production of food protein (biomass/fruiting bodies) from waste can be attributed to its ability in secreting various kind of hydrolyzing and oxidizing enzymes that has drawn research attention in mushroom cultivation and waste remediation area. Amongst all, three main problems that is being addressed by mushroom are: food security, human health, and expunging environmental pollution. Because of all these qualities of mushrooms, it seems imperative to pool in resources for more intensive study by numerous worldwide technological and scientific institutes. The mushroom marketplace is estimated to grow due to its nutritional and potential health benefits, containing cognition, oral health, weight management, and can reduce risk of cancer.

11.1 INTRODUCTION

Mushroom (toadstool) are macrofungus that have fresh fruit edible bodies. It is originated from fungus with no tissue structure, which can be unicellular or multicellular organism. Although mushroom is a type of fungi, do not contain cellulose and chlorophyll, unable to synthesize their own food, therefore, they must have to live on plant or animal to absorb nutrition from them. Mushrooms consist of protein, vitamins (good source of B vitamins), minerals, and antioxidants. In the food world, it is one of the highest sources of selenium as well as Vitamin D. Antioxidant like selenium, vitamin C and choline enriches the mushroom and provide nutraceutical benefits to humans. It is a small product of forest which rises and develops on cellulose which is found plenteous in the biosphere. In the current scenario, mushrooms are microfungi with unique fruiting bodies that can be epigeous or hypogeous and are picked manually because their bodies are big enough that are visible to the naked eyes (Chang & Miles, 1992). Mushrooms exploitation as food is because of distinct taste and nutritional supplemented with bioremediation of waste throughout the world. Today more than 2,000 mushroom species exist in nature, but only 25 amongst the available species can be used in diet and have commercial cultivation.

Due to high nutritive and functional value, mushrooms are called nutraceutical foods and also considered as 'delicacy.' Due to their merits in medicine, organoleptic, and economic importance, it has received significant interest (Chang & Miles, 2008; Ergonul et al., 2013). There is no difference between medical and edible mushroom as there are edible mushrooms with therapeutic value while numerous are used in medicines which are also edible (Guillamon et al., 2010). *Agaricus bisphorus, Lentinus edodes, Pleurotus* spp., and *Flammulina velutipes* are most cultivated mushrooms worldwide. Mushroom production continuously attaining increment and China being the largest producer in the world (Change & Miles, 2008; Aida et al., 2008; Patel et al., 2012). Though, wild mushroom is getting more importance because of its sensory, nutritive, and pharmacological features (Ergonul et al., 2013).

Mushroom has antimicrobial properties and also act as a source of secondary metabolites like sesquiterpenes including other anthraquinones, steroids, terpenes, quinolones, and derivates of benzoic acid. It is also a source of primary metabolites such as peptide, oxalic acid and protein. *Lentinus edode* is a highly studied mushroom species that have broad antimicrobial properties against gram-positive as well as gram-negative bacteria. Antimicrobial property of one another mushroom species *Pseudoplectania nigrelle* is highest against gram-positive bacteria, while against gram-negative bacteria, *Leucopaxilus albissimus* produces 2-aminoquinoline, which is most potent with largest antimicrobial property (Ergonul et al., 2013).

Fungi can act as perfect class for the purpose of remediation of harmful contaminants due to their vast growth, huge network of hyphae, formation of lignoytic enzymes, larger area of surface, adaptable nature toward fluctuating parameters like temperature, pH, heavy metals with the presence of metal-binding proteins (Khan et al., 2019). Numerous harmful pollutants that are released by several industries, like pharmaceuticals, dyes, and herbicides can be done using fungi. As a myco-remediation agent fungi can be grown in bioreactors under controlled physiological conditions that can promote their growth (Aragao et al., 2020). Bioreactors projected for waste treatment from several industries including pharmaceutical can be designed with suitable modifications (Morals et al., 2012) that can accelerate degradation of pollutants in the bioreactor with regulated metabolite use and fungal biomass in a controlled manner (Couto et al., 2006). Harmful chemicals such as pesticides, herbicides, polycyclic aromatic hydrocarbons (PAH), explosive compounds and chlorinated solvents are present in soil can be remediated ex-situ (Tekere et al., 2019).

Myco-remediation can combat major issues of water and soil contamination, which has reasonable cost and sustainable strategy. Myco-remediation tool refers to degrade numerous kinds of substrate such as converting wastes from agro-industries or persistent pollutants of the environment into useful substances, using enzymes produced by mushrooms. Mushrooms as myco-remediating agents can sought to be considered as solution of 3P issues, i.e., Problem of human health, Problem of food security and Problem of environmental pollution (Figure 11.1).

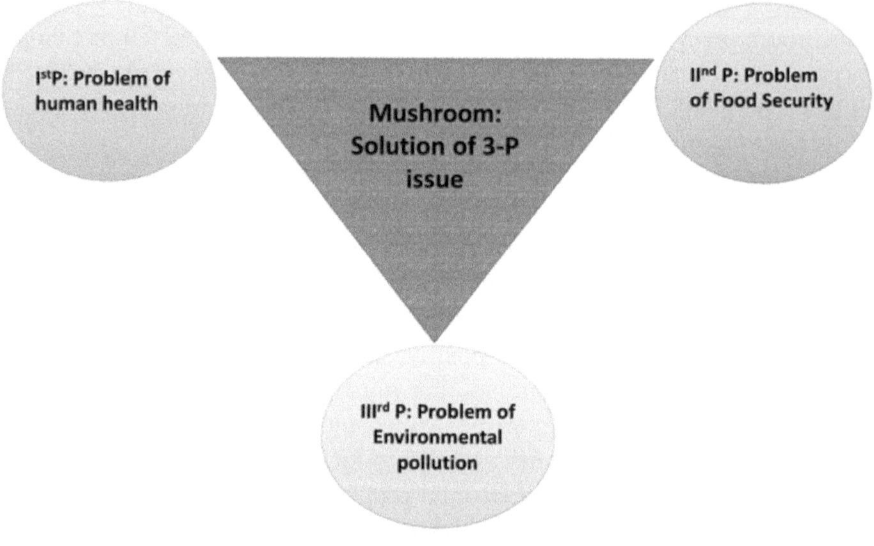

FIGURE 11.1 Diagrammatic representation of mushroom base 3-P issue solution.

11.2 FIRST P: PROBLEM OF HUMAN HEALTH

Indian human diet is generally centered around cereals like maize, wheat, and rice that are not very rich in protein. Daily usage of Mushrooms as supplementary food in the human diet will fulfill the deficiency of proteins and enhance good health of social and economic communities of poor/

developing countries like India, Pakistan, Bangladesh, Bhutan, etc. Mushrooms in the past were thought to be highly priced vegetable, and rich people favored them for special food items. Presently, ordinary people also started considering mushrooms as quality food as a complete diet food for all ages group, due to their healthy nutritional goods. The nutritional content of mushroom can be affected by various aspects like stage of development, species, and environmental circumstances. It is a major source of proteins, carbohydrate fiber, minerals, and vitamins.

11.2.1 CARBOHYDRATE

Carbohydrate is a biomolecule having carbon, hydrogen, and oxygen atom play numerous roles in a living organism, serves for the energy storage (e.g., starch in the plant, glycogen in human) and structural component such as cellulose in plant and chitin in arthropods. The whole carbohydrate content in mushroom contrast from 50–65% on the basis of dry weight of mushrooms of various species. Polysaccharides obtained from mushrooms have anti-tumor and immunomodulation properties. Numerous types of sugar found in various different species of mushrooms are: fructose, mannitol, sucrose, and trehalose. The mannitol is also called mushroom sugar because mannitol is constituted about 80% of the total free sugar in mushrooms (Wannet et al., 2000) (Tables 11.1–11.3).

TABLE 11.1 Biochemical Composition of *Agaricus bisporus*

Carbohydrate	Value (%)
Mannitol	0.9
Reducing sugar	0.28
Glycogen	0.59
Hemicellulose	0.91

Source: Singh & Singh (2002).

The most interesting fact of common mushroom containing polysaccharides that is peculiar to the animal kingdom, such as glycogen and does not contain either starch or cellulose, which are characteristic of the plant. The glycemic index acts as an indicator of the ability of a different type of food that contains carbohydrates that boost the blood glucose level. Mushrooms have complex carbohydrates and fiber that are very low in glycemic index food (IG =10). Breakdown of these complex carbohydrates are slower and immediately after meal; it does not accelerate the amount of sugar in the blood. Foods with a low glycemic index are highly recommended for the people suffering from diabetes, and that's the reason for mushroom can be good edible food for diabetic people.

TABLE 11.2 Composition of Sugar (g/100 g Dry Matter) in Selected Mushroom Species (Dry Weight)

Species	Common Name	Mannitol	Trehalose	Maltose	Total Sugar
Agaricus bisporus	Button mushroom	19.6	0.8	ND	20.9
Lentinula edodes	Shiitake	10.01	3.38	0.69	14.03
Agaricus silvaticus	Pinewood mushroom	2.7	0.3	0.4	3.7
Boletus edulis	Hog mushroom	3.5	9.7	ND	13.5
Calocybe gambosa	St George's mushroom	0.3	8.0	ND	9.1
Cantharellus cibarius	Chanterelle	8.3	6.1	ND	14.5

ND: Not detected.

Source: Barros et al. (2008).

TABLE 11.3 Glycemic Index of Different Vegetable

Vegetable	GI (Glycemic Index) Value
Bell peppers	10
Broccoli	10
Cabbage	10
Lettuce	10
Mushroom	10
Onion	10
Asparagus	14
Green beans	14
Artichoke	15
Spinach	15
Tomato	15
Chickpeas	33

11.2.2 PROTEIN

The protein composition varies for dry weight between 15% and 35% that depends on the stage of development and species of fruiting body

(Manzi et al., 2004; Diez & Alvarez, 2001). *P. ostreatus* and *L. edodes* have good protein digestibility varies from 73.5% and 76.3% respectively (Adewusi et al., 1993; Dabbour & Takruri, 2002). Mushrooms have comparable content of proteins (Wong & Cheung, 1998) with legumes but lower than animal proteins with high digestibility (Mc Donough et al., 1990) and also rich in essential amino acids like Met, Phe, Ile, Lys, Val, Trp, Leu, Thr, Val. Edible mushrooms are rich source of glutamic acid, aspartic acid, and arginine according to the data of WHO (World Health Organization). Glutamic acid plays a key role in amino acid cellular metabolism, aspartic acid in the urea cycle, and arginine is a precursor for the biosynthesis of nitric acid that is a signaling molecule. In common mushrooms, methionine and cysteine amino acids are available in the greatest amount in common mushrooms (Manzi et al., 1999). Methionine amino acid a precursor for various biosynthesis pathways like ethylene synthesis and is an important amino acid that is needed for the development of the human body. Cysteine is a semi-essential amino acid, and cysteine has antioxidant properties.

It also helps in reducing the toxic effect of alcohol reduced the risk of liver damage. In mushrooms, the amino acid has a significant role in aroma and flavor, such as glutamic acid is principally responsible for umami taste. As compared to most of the vegetables, mushroom's quality of proteins has improved (FAO, 1981). Mushroom is a source of animal amino acid for vegetarian people.

11.2.3 DIETARY FIBER

Dietary fibers are components of plant products that are not broken by the digestive enzymes of human. Human body's enzymes can't digest dietary fibers that are present in cereals, fruits, vegetables, dried peas, nuts, lentils, mushrooms, and grain. Fibers are also called insoluble and resistant starch. Dietary fiber keeps the gut healthy and also avoid the threat of different kinds of diseases like cancer, diabetes, and heart diseases. Various food soluble fibers and resistant starch also function as prebiotics and support the probiotics. Mushroom has dietary fiber in high content.

Mushroom contains a very higher amount of insoluble fiber (2.28–8.99/100 g edible part) as compared to soluble fiber (0.32–2.20 g/100 g edible part) (Mani et al., 2004). Common mushroom fiber is beta-glucan (4–13% of the total dietary fiber) followed by chitin (Guillamon et al., 2010). The fruiting body of an edible mushroom, *Lentinulla edodes* like lentinan is used to separate beta-glucans. This lentinan help in stimulating the immune system of human also have anticancer property. Beta-glucan is being used as immunomodulatory in cancer treatment some success has also been achieved from treatment (Reshetnikov et al., 2001) (Table 11.4).

TABLE 11.4 Fiber Components of Various Food Products

Food	Total Dietary Fiber (gram)
Baked beans	6.6
Untoasted muesli	2.7
Green pea	3.4
Almond	4.0
Apple	2.3
Carrot	4.0
Potato	2.0
Multigrain bread	3.1
Mushroom	1.0

Source: Nestle; https://www.nestle.com.

11.2.4 LIPID

A lipid is a heterogenous group of compounds related by chemical properties than physical that consist of fats, oils, steroids, waxes, and related compounds. Lipids are biomolecules that have some common property like they are soluble in a nonpolar solvent like alcohol, ether, chloroform, but not soluble in polar solvent, for example, water. Lipid molecules participates in various biological functions directly or indirectly like storing energy, signaling, and acting as structural and functional components of biological membrane. Mushroom is low in fat, approximately less than 5% in dry weight. Fat contents and its concentration in mushrooms are affected by various environmental conditions like growing conditions, nutritional factors, oxygen availability, temperature, and nutrition of the substrate (Pedneault et al., 2007).

The edible mushrooms are rich in unsaturated fatty acid. An omega-3 fatty acid, Linolenic acid and an omega-6 fatty acid, linoleic fatty acid are essential fatty acids for human health, and both are polyunsaturated fatty acids (PUFAs). A comparative study was done by Reis et al. (2012) on different varieties of edible and cultivable mushroom, they concluded that the Shiitake is the one with the highest level of polyunsaturated fatty acid and a lower amount of saturated than the rest of the mushroom studied.

11.2.5 VITAMINS

Vitamins are a group of organic nutrients which are needed in less amount for a number of biochemical activities, are not synthesized by the human body therefore, must be taken in diet. Mushrooms contain high content of vitamin B like riboflavin (B_2), niacin (B_3), pantothenic acid (B_5), folate (B_9) while, thiamine (B_1), cobalamin (B_{12}), and ascorbic acid (Vitamin C) are present.

In small amount (Mattila et al., 2002). Vitamin B helps the body to get energy from food, they help in red blood cells formation and is also important for the growth and development of the brain. Doctors advise pregnant women to take folic acid (B_9) during pregnancy for the growth and development of the fetus.

Mushroom are also a good source of vitamin D, due to its role in recent years vitamin D remains a hotspot for researchers. Vitamin D has an important role in muscular function, cardiac immunology, cardiovascular disease, and cancer, etc. (Phillips et al., 2012). Recent studies show that mushroom produces more vitamin D than daily requirement, when exposed to UV light under certain conditions (Rathore et al., 2017) (Table 11.5).

TABLE 11.5 Different Types of Vitamins and Their Function and Deficiency Disease

Vitamin	Functional Product	Function in Metabolism	Deficiency Disease
Thiamine	Thiamine pyrophosphate	Transfer of aldehyde group	Beriberi
Riboflavin	FMN and FAD	Redox reaction involving one or two-electron transfers	Rare-inflamed tongue, mouth lesions
Niacin	NAD^+ and $NADP^+$	Redox reactions involving one or two-electron transfers	Pellagra
Pantothenic acid	Pantothenic acid	Synthesis of coenzyme-transfer of acyl group and play an important role in metabolism	Rare-impaired energy production, fatigue neurological symptoms
Folic acid	Tetrahydrofolate	Nucleotide synthesis: provides methyl group for thymine of DNA	Megaloblastic anemia: neural tube defects causing spina bifida if a severe deficiency in pregnancy
Ergocalciferol	1,25-dihydroxycholecalciferol (a steroid hormone)	Promotes absorption of dietary calcium synthesized in the skin on exposure to UV light so light treatment of children necessary above 'arctic circle'	Rickets

11.2.6 MINERALS

Minerals are inorganic compounds required for the human body in a minimum quantity for different metabolic functions. Minerals help in the formation of bones and teeth of the organism. They are the vital composition of body fluid, tissue, enzyme system and also involved in nerve system functions. The human body requires different minerals in small or large amounts for the growth and development. Some minerals are required for humans in greater than 100 mg per day, i.e., sodium, potassium, calcium, phosphorus, and other minerals needed in lesser quantity, i.e., iron, fluoride, iodine, copper, zinc, selenium, etc. Mineral content in mushrooms differs between 6% and 11% in the dry matter according to different species (Table 11.6).

TABLE 11.6 Vitamin and Minerals Components in Mushroom

Major Vitamin and Minerals	Daily Requirement	Mushroom Content
Thiamine (vitamin B1)	1.4 mg	4.8–8.9 mg
Riboflavin (vitamin B2)	1.5 mg	3.7–4.7 mg
Niacin	18.2 mg	42–108 mg
Phosphorus	450 mg	708–1,348 mg
Iron	9 mg	15–17 mg
Calcium	450 mg	33–199 mg
Copper	2 mg	12–22 mg

Mushroom has a reasonable amount of minerals compared to vegetables (Manzi et al., 1999). Some edible mushroom *Aagaricus bisporus* (Button mushroom), *Pleurotus ostreatus* (oyster mushroom), and *Lentinula edodes* (Shitake) they are rich in potassium (2,670–4,730 mg/100 g in dry matter), phosphorus (493–1,390 mg/100 g dry matter), zinc (470–920 mg/100 g in dry matter), and copper (0.52–350 mg/100 g in dry matter) (Cheung, 2008). Cultivated mushroom species of *Boletus* genus is an important source of selenium (Cocchi et al., 2006). Selenium has antioxidant property that helps to neutralize free radicals, induces apoptosis, stimulate the immune system and intervenes in the various function of the thyroid gland. Selenium is compound that form various selenoprotein which act as anticancer molecules against cancer (Lu & Holmgren, 2009).

11.3 SECOND P: PROBLEM OF FOOD SECURITY

At present, the world population is 7.8 billion which can increase to 9.7 billion by the year 2050. According to population explosion, agricultural lands are getting limited that directly affect crop production and food supply. This is a very alarming situation and threatens to food security to feed millions of people worldwide (Singh et al., 2021). For a higher amount of crop production in a limited agricultural area, the area required hodiernal technology that helps farmers to produce large quantity of crops without compromising quality and production rate (Singh et al., 2021). Along with increased population, food supply is also affected by climate change like abiotic stresses salinization, floods, heat, etc. Plant breeders dependent on conventional breeding methods along with mutations, either natural or artificial, that help in crop improvement and food security. But conventional plant breeding technology is a time taking process with uncertainty of stable characteristics due to which the new crop varieties cannot be developed in a short period (Sedeek et al., 2019).

In today's modern era, mushrooms are being seen as an attractive food item due to their low calories, rich in carbohydrates, fat, sodium, and are free from cholesterol. Apart from this, mushroom also provides important nutrient containing selenium, riboflavin, potassium, vitamin D, niacin, protein, and fiber. Together with all these facts, it can be said that mushrooms have been used primarily as food traditional medicines since ancient times. It has been reported that mushrooms beneficial effect on the health and can help in the treatment of various diseases. Several nutraceutical properties are defined in mushroom which is useful for the treatment of Parkinson, Hypertension, Alzheimer, and heart disease. Mushroom is also utilized to reduce metastasis and cancer invasion because of antitumoral attributes. Mushrooms help lose body fat, boost immune response, act as antibacterial, balance sugar level prevent the allergic effect, lower blood lipid promote metabolism. Some mushroom extracts are used for health benefits and used in dietary supplements due to its unique properties.

11.4 THIRD P: PROBLEM OF ENVIRONMENTAL POLLUTION

Synthetic organic compounds causing the environmental pollution, is a major issue worldwide (El-Ramady et al., 2021). Many of the substances which do not have natural occurrence in the biosphere and do not get degraded by indigenous microfauna and microflora easily are xenobiotic compounds (Sullia, 2004). United States Environmental Agency (USEPA) have categorized various chemicals as major pollutants because of their toxic impacts on health and environment. Including polychlorinated biphenyls (PCBs), PAH, pentachlorophenol, 1,1,1-trichloro-2,2-bis (4-chlorophenyl) ethane, toluene, benzene, trinitrotoluene, and, ethylbenzene xylene. PAH are recalcitrant environmental contaminants which are produced from coal mining, drilling for oil, burning of wood and fossil fuels (Verdin et al., 2004). Nowadays, a very effective and common remediation is incineration but is not cost-effective considering money. These chemical compounds create a serious problem for human health and liveliness in the world (Hamman, 2004). For the development of sustainable society with minor impact on environment is the removal of these wide range of pollutants from nature is important.

For cleaning up, a rapid, cost-effective, and ecologically responsible process is desirable due to the extent of this problem and the absence of the lucid solution (Hamman, 2004). PH residues (PAH) are stated as major soil contaminants which are released from the processing of coal, tar, oil, and similar substances (Loske et al., 1990). Similarly, in transformers, PCBs are utilized as cooling agent and fly ashes from combustion process and chemical manufacturing Digoxin is released as byproduct. The method of enhancing live organisms of soil like bacteria, green plants, and fungi to decompose organic contaminants and hydrocarbons is called bioremediation (Atlas & Bartha, 1992). For this purpose, contaminated soil is treated with the of organism and nutrients to enhance the process of biodegradation. Natural biodegradation can be enhanced by natural organisms called intrinsic bioaugmentation.

11.4.1 MUSHROOM-BASED MYCOREMEDIATION

Myco-remediation is the bioremediation technology in which fungus is for the removal of toxic compounds. In myco-remediation, presence of filamentous fungi like molds as well as macro-fungi such as mushrooms (Bosco et al., 2019) (Figure 11.2). Some examples of pollutants which degraded by mushroom through the myco-remediation process are discussed in subsections.

11.4.1.1 ORGANIC WASTE REMEDIATION

Due to the processes in food, forests, and agricultural industries, organic solid wastes are produced (Rajput et al., 2017). They mainly consist of 3 components are cellulose, hemicellulose, and lignin. In plant fiber, the building block is lignocellulose (Chang, 1987, 1989), they consist of organic compounds made up of hydrogen and carbon long chains that are insoluble and complex and have similarity with various organic pollutants. These may be processed by different chemical treatments but that are not cost effective and have adverse effect on environment. Mushrooms and other various fungi release oxidative and hydrolytic enzymes which can convert complex organic matter into soluble that mushrooms absorb for their nutrition (Chang & Miles, 2004). On the basis of particular enzyme secreted by individual mushrooms, they have different ability to utilize these substrates.

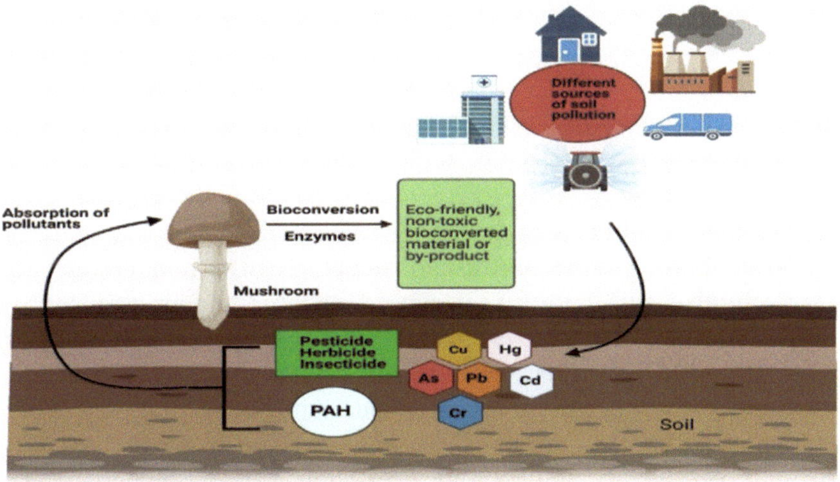

FIGURE 11.2 Diagrammatic representation of mushroom-based myco-remediation for soil pollutants.

After examining the profiles of lignocellulolytic enzyme from three main mushroom varieties which are cultivated commercially displayed variable potential to use lignocellulose as a growth substrate (Buswell and Chang, 1994; Buswell et al., 1996). (i) Two extracellular enzymes, i.e., manganese peroxidase and laccase are present in *Lentinula edodes* which have function in depolymerization of lignin, due to which they are cultivated on sawdust (SD), wood, and other highly lignified matter. (ii) *Volvariella volvacea* prefers paddy cotton wastes due to their high cellulose, less content containing lignin. Though no one is considered as degrading enzyme for lignin, a family of cellulolytic enzymes is produced by it, containing two beta glucosidases, five cellobihydrolases, and five endoglucanases. (iii) Similarly, an oyster mushroom, *Pleurotus pulmonaris var. stechangii* is highly adaptable species among these three. It can grow extensively on several wastes of agriculture with different composition such as polysaccharide/lignin ratio, because it can release cellulose as well as lignin degrading enzymes. A notable effect on pollution at regional and national levels can be seen due to bioconversion of biomass containing lignocellulose into useful food and products (Chang, 1984; Chang & Buswell, 2003; Koutrotsios et al., 2014).

In the bioremediation process, mushroom mycelia is used to degrade and remove contaminants and absorption of pollutants through the process of biosorption, thus mushrooms have an influential role in the ecosystem (Dai, 2016; Miller, 2013). Therefore, fungi and mushrooms are considered as agents of environmental bioremediation utilizing green technology.

11.4.1.2 MYCOREMEDIATION AND POLYCYCLIC AROMATIC HYDROCARBONS (PAH)

One of the major pollutants named PAH is released by incomplete combustion of wood, coal, and petroleum (Sushkova et al., 2018). Therefore, due to the human activities, such harmful substances can enter into air, soil, and water and pollute them (Pozdnyakova et al., 2012 and Samanta et al., 2002). These organic molecules have benzene rings fused with them, such as chrysene, acenaphthene, pyrene, 2-methyl naphthalene, anthracene. Such contaminants that are ubiquitous in nature, are classified into three main classes including pyrogenic, petrogenic, and biological (Abdel-Shafy et al., 2016). These pollutants can have harmful effect on humans and animals. Such chemicals when move into the surroundings, will stay for longer time periods and impose serious harm to the environment. So, contaminated water and soil requires immediate action for removing such harmful ubiquitous, and persistent compounds (Abdel-Shafy et al., 2016). Soil adsorption, chemical oxidation, bioaccumulation, biodegradation, photodegradation, chemical oxidation, and leaching are the methods used for removal of these harmful compounds. Amongst these methods, biodegradation using micro-organisms has been extensively researched (Bhattacharya et al., 2014).

Ligninolytic enzymes that are released by various fungi seems to be promising, economical, and eco-friendly strategy in PAH removal from contaminated sites (Rodarte-Morales et al., 2012). Numerous extracellular enzymes for lignolysis like manganese peroxidase, lignin peroxidase, and laccase are released by fungi (Pozdnyakova et al., 2012). PAH is converted into quinones via oxidation that are further degraded by ring fission and this process is catalyzed by extracellular ligninolytic enzymes (Pozdnyakova et al., 2012; Hammel et al., 1991). *Pleurotus pulmonarius* (oyster mushroom) spent can use a broad spectrum of waste substrates (Chiu et al., 1998, 2000). The third most cultivated mushroom is *Pleurotus ostreatus* that is used for food purposes. PAHs, atrazine, organophosphorus, and waste waters can be degraded, and toxic chemicals can be mineralized by *P. ostreatus* (Pointing, 2001).

With respect to natural attenuation (SM) and adjacent to a shooting range, PAH are found for bioremediation of polluted soil in *Agaricus bisporus* substrate (SAS) spent application method. Thus, in multi-polluted soil, *Agaricus bisporus* spent can be feasibly utilized for PAHs biodegradation. SAS enhanced the PAH-degrading bacterial population, however fungal activity is lowered. Due to high levels of ligninolytic activity and increased PAH-degrading bacteria population, *A. bisporus* is used as a noble inoculum carrier (Delgado et al., 2015).

11.4.1.3 MYCOREMEDIATION AND HEAVY METAL POLLUTION

For human health and biodiversity, heavy metal deposition can cause a higher risk in the environment. At the site of contamination, various heavy metals are found like mercury, arsenic, cadmium, silver, copper, lead, chromium, and nickel (Bind et al., 2019; Rajput et al., 2019). Through various natural activities such as earth's crust weathering, volcanic eruption, and soil erosion and anthropogenic means such as combustion of fossil fuels, insecticides, fungicides, fertilizers, discharge from factories like textile, paint, metal parts, electronic, mining, leaded petrol, these heavy metals can enter into water and soil (Morals et al., 2012). They may can have deleterious effect on the human body when present in higher concentration (Rajput et al., 2020).

Various reports have suggested that bioremediation of waste using mushrooms can be done by three biological processes such as bioconversion, bio-absorption, and biodegradation (Akinyele et al., 2012; Kulshreshtha et al., 2013; Bind et al., 2018; Kumhomkul & Panich-Pat, 2013; Lamrood & Ralegankar, 2013). The process of bioconversion by mushroom can be used in which they can use lignocellulosic waste that are released via industries for its cultivation, which can be further used as a product. For cultivation of *Pleurotus citrinopileatus*, the sludge of industrial waste of cardboard and handmade paper can be used. These kinds of industrial waste along with the wheat straw (WS) are used for the purpose of mushroom cultivation

(Kulshreshtha et al., 2013). In the same way, *Lentinus connotus and Pleurotus citrinopileatus* use banana pseudostem, sorghum stalk, and paddy straw waste for bioconversion with great biological efficiency (Rani et al., 2008).

Second important process is *biosorption* used to remove pollutants, mainly metals from the environment. This can be considered as an alternate for the remediation of effluents from industries and recovery of metals present in them. This process is based upon the marked tolerance of living and dried biomass to metals and other adverse conditions which helps them in sorption of xenobiotic compounds, pollutants, and metallic ions from waste (Gavrilescu, 2004). Spent mushroom concept and mushroom mycelium are used to prepare biosorbents. *Agaricus bisporus* and *Lactarius piperatus* have shown high removal capacity for Cadmium (II) ions (Nagy et al., 2013). Mushrooms such as *Agaricus bisporus, Calocybe indica, and Pleurotus platypus* are efficient biosorbent for removing Zinc, Copper, Cadmium, Nickle, Lead, Iron from aqueous solution (Lamrood & Ralegankar, 2013).

The third major process is biodegradation which is utilized to eventually degrade and recycle larger molecules molecule to complete mineralization to the simpler components like CO_2, H_2O, NO_3, and some more inorganic molecules. Degradation of various wastes and factors through this process have been published in various reports. One of the examples is degrading Oxo-Biodegradable Plastic by *Pleurotus ostreatus* (da Luz et al., 2013), and similarly using Lentinula edodes (shiitake mushroom), 2,4-dichlorophenol is degraded where activator is vanillin (Tsujiyama et al., 2013).

11.4.1.4 MYCOREMEDIATION AND AGRICULTURAL EFFLUENTS

In agricultural practices, pesticides and herbicides are used extensively. Leaching of minerals such as sulfur, nitrogen, and phosphorus can affect soil, ground water and nearby water bodies that can cause eutrophication, which can decrease dissolved oxygen and affect aquatic flora and fauna. Harmful toxic chemicals can accumulate in fruits and vegetables, and thus, they can enter into the food chain and affect the health of humans. These herbicides and pesticides have aided farmers in combating a variety of plant and insect diseases while also increasing agricultural productivity. On the other hand, such chemicals are also damaging the ecosystem. *Pleurotus ostreatus* (Oyster mushroom) is able to degrade its major metabolite, aldrin, and dieldrin.

Epoxidation and hydroxylation reactions by *P. ostreatus* leads to biotransformation of dieldrin and aldrin (Purnomo et al., 2017). Various studies have shown participation of specific enzymes for ligninolysis in degrading several insecticides and herbicides but crucial function of these enzymes in degradation of xenobiotic compounds is still not recognized. Bio-mixture was also evaluated with bioaugmentation using WRF Lentinula edodes inoculation into cork. Study results clearly indicated the need of employing natural biosorbents such as cork residues to improve capability of pesticides (diflufenican, pendimethalin, terbuthylazine, and difenoconazole, as well as creation of bio-beds via dissipation of biomixtures (Pinto et al., 2016).

11.4.1.5 MYCOREMEDIATION OF PHARMACEUTICALS

Environmental conditions that are already severe, it is getting more worse with pharmaceutical compounds that are released drug manufacturing industries. These harmful compounds are severely affecting the health of humans and animals and also disturbing the environmental balance and decreasing available drinking water level. For treating wastes from pharmaceuticals, various mushrooms can be a possible contestant. *Pleurotus ostreatus*, is a lignolytic fungi can absorb antibiotic and oxytetracycline from the media of liquid culture in two weeks duration (Migliore et al., 2012). Similarly, edible mushroom *Lentinula edodes'* mycelia is known to absorb and eliminate antifungal drugs such as bifonazole and clotrimazole (Kryczyk-Poprawa et al., 2019).

11.5 FUTURE PERSPECTIVE

Using the enzymes and biomass of various fungi, mycoremediation of environmental contaminants from disturbed soil based on mushroom may be accomplished. That is a green or ecological technology that has more potential as compared to other conventional approaches. However, the newly added contaminants can be biodegraded via enzymatic activities of newly isolated mushrooms. Using the current biotechnological techniques such as genetic engineering of newly isolated mushrooms by utilizing whole mushroom cell and their enzymatic function for the process of mycoremediation will create a path for the future. It can be used as a tonic, a meal, or as medicine. They have high nutrition value as they have differential value in medicine, and rich source of vitamins, proteins, and crude fiber, and low in calories and fat. Due to easy availability of its raw materials and low price, they have plentiful scope in India as well as in some developing countries. This technology is sustainable and labor intensive. Subsequently, can generate employment, predominantly for rural women and youths thus can elevate their status in the society. During winter months, when farmers are less busy, it can provide supplementary work for them. Tropical mushrooms like paddy straw (*Volvaricella volvacea*), milky mushroom (*Calocybe indica),* and oyster (*Pleurotus spp.)* production can utilize nearby obtainable wastes of agriculture like lignocellulosic wastes, cotton wastes, chickpea, mustard, Lathyrus, soybean, wheat, and paddy straw. Self-help women groups, either in small or larger scale can do farming of oyster mushroom, which can establish an important income source for them. In Orissa, commonly grown is *Volvariella volvacea* (paddy straw). In India, cultivation of milky mushrooms is done in the entire plain whole year. It is estimated that the integrated rural development program will include employment by mushroom cultivation and farming, and it will turn out to be very essential small-scale industry.

11.6 CONCLUSION

On earth, mushroom seems like an all-purpose product that uses for a potential solution "3P Problems," i.e., the problem of human health, problem of food security, and problem of environmental pollution. Mushrooms contain high-value carbohydrate, protein, fiber, fate, lipids, and vitamins that help in improving human health, especially in the area where nutritional food grain availability is minimum. On the other hand, the mushroom also plays a potential role to overcome the problem related to food security because of climate change food production and protection both are major problems that reduced crop production the born the problem related to food availability for a population of developed, developing, and poor country. For this problem, the mushroom is a prompt solution because it's produced in less area at a low cost. As edible mushrooms are proficient in converting wastes from industries and agriculture into feeds and foods, their cultivation can be a valuable process. Thus, mushroom has the efficiency to convert the various type of waste by their enzymatic activity through mycoremediation process that is a revolutionary solution for decomposition of wastage. Hence, more research is needed to exploit and understand the potential of mushrooms as a mycoremediation agent along with its safety aspects to be consumed as a product.

ACKNOWLEDGMENT

This work is supported by the Ministry of Science and Higher Education of the Russian Federation within the framework of the state task in the field of scientific activity (No. 0852-2020-0029).

KEYWORDS

- **dietary fibers**
- **environmental pollution**
- **food security**
- **macrofungus**
- **mushroom**
- **mycoremediation**
- **polycyclic aromatic hydrocarbons**

REFERENCES

Abdel-Shafy, H. I., & Mansour, M. S. M., (2016). A review on polycyclic aromatic hydrocarbons: Source, environmental impact, effect on human health and remediation. *Egypt. J. Pet., 25*, 107–123.

Adewusi, S. R. A., Alofe, F. V., Odeyemi, O., Afolabi, O. A., & Oke, O. L., (1993). Studies on some edible wild mushrooms from Nigeria: 1. Nutritional, teratogenic and toxic considerations. *Plant Foods for Human Nutrition, 43*, 115–121.

Akinyele, J. B., Fakoya, S., & Adetuyi, C. F., (2012). Anti-growth factors associated with *Pleurotus ostreatus* in a submerged liquid fermentation. *Malaysian J. Microbiol., 8*, 135–140.

Alves, M., Ferreira, I. F. R., Dias, Teixeira, V., Martins, A., & Pintado, M., (2012). A review on antimicrobial activity of mushroom (*Basidiomycetes*) extracts and isolated compounds. *Planta Medica, 78*(16), 1707–1718.

Aragão, M. S., Menezes, D. B., Ramos, L. C., Oliveira, H. S., Bharagava, R. N., Romanholo, F. L. F., Teixeira, J. A., et al., (2020). Mycoremediation of vinasse by surface response methodology and preliminary studies in air-lift bioreactors. *Chemosphere, 244*, 125432.

Atlas, R. M., & Bartha, R., (1992). Hydrocarbon biodegradation and oil spill bioremediation. *Adv. Microbiol. Ecol., 12*, 287–338.

Barros, L., Cruz, T., Baptista, P., Estevinho, L. M., & Ferreira, I. C., (2008). Wild and commercial mushrooms as source of nutrients and nutraceuticals. *Food Chem. Toxicol., 46*(8), 2742–2747.

Bhattacharya, S., Das, A., Prashanthi, K., Palaniswamy, M., & Angayarkanni, J., (2014). Mycoremediation of benzo[a]pyrene by *Pleurotus ostreatus* in the presence of heavy metals and mediators. *3 Biotech, 4*, 205–211.

Bosco, F., & Mollea, C., (2019). Mycoremediation in soil. *Environmental Chemistry and Recent Pollution Control Approaches*, 1–16.

Buswell, J. A., & Chang, S. T., (1994). Biomass and extracellular hydrolytic enzyme production by six mushroom species grown on soybean waste. *Biotechnology Letters, 16*, 1317–1322.

Buswell, J. A., Cai, Y. J., Chang, S. T., Peberdy, J. F., Fu, S. T., & Yu, H. S., (1996). Lignocellulytic enzyme profiles of edible mushroom fungi, *World Journal of Microbiology & Biotechnology, 12*, 537–542.

Chang, S. T., & Buswell, J. A., (2003). Bioconversion technology: A tool for economic and technological development in island communities. *International Journal of Island Affairs*, 17–20.

Chang, S. T., & Miles, P. G., (1992). Mushroom biology – A new discipline. *Mycologist, 6*, 64–65.

Chang, S. T., & Miles, P. G., (2004). *Mushroom: Cultivation, Nutritional Value, Medicinal Effect, and Environmental Impact* (2nd edn.). Boca Raton, FL: CRC Press.

Chang, S. T., & Miles, P. G., (2008). *Mushrooms: Cultivation, Nutritional Value, Medicinal Effect, and Environmental Impact*. CRC Press (2), Boca Raton, Fla, USA.

Chang, S. T., (1984). Conversion of agricultural and industrial wastes into fungal protein. *Conservation and Recycling, 7*, 175–180.

Chang, S. T., (1987). Microbial biotechnology: Integrated studies on utilization of solid organic wastes. *Resources and Conservation, 13*, 75–82.

Chang, S. T., (1989). Recycling of solid wastes with emphasis on organic wastes. *Green Productivity*, 23–26.

Cheung, P. C. K., (2008). *Mushrooms as Functional Foods*. Editorial John Wiley & Sons. Wiley: Hoboken, NJ.

Chiu, S. W., Chan, Y. H., Law, S. C., Cheung, K. T., & Moore, D., (1998). Cadmium and manganese in contrast to calcium reduce yield and nutritional value of the edible mushroom *Pleurotus pulmonarius*, *Mycological Research, 102*, 449–457.

Chiu, S. W., Law, S. C., Ching, M. L., Cheung, K. W., & Chen, M. J., (2000). Themes for mushroom exploitation in the 21st century: Sustainability, waste management and conservation, *Journal of General and Applied Microbiology, 46*(2000), 269–282.

Cocchi, L., Vescovi, L., Petrini, L. E., & Petrini, O., (2006). Heavy metals in edible mushrooms in Italy. *Food Chemistry, 98*, 277–284.

Couto, S. R., Rodríguez, A., Paterson, R. R. M., Lima, N., & Teixeira, J. A., (2006). Laccase activity from the fungus *Trametes hirsuta* using an air-lift bioreactor. *Lett. Appl. Microbiol., 42*, 612–616.

Da Luz, J. M. R., Paes, S. A., Nunes, M. D., Da Silva, M. C. S., & Kasuya, M. C. M., (2013). Degradation of oxo-biodegradable plastic by *Pleurotus ostreatus*. *PLoS One, 8*(8), 69386.

Dabbour, I., & Takruri, H. R., (2002). Protein quality of four types of edible mushrooms found in Jordan. *Plan. Foods Hum. Nutri., 57*, 1–11.

Dai, E., (2016). *Mushroom as Pollution Mask. China Daily.*
Delgado, C. G., Yunta, F., & Eymar, E., (2015). Bioremediation of multi-polluted soil by spent mushroom (*Agaricus bisporus)* substrate: Polycyclic aromatic hydrocarbons degradation and Pb availability. *Journal of Hazardous Materials, 300*, 281–288.
Diez, V. A., & Alvarez, A., (2001). Compositional and nutritional studies on two wild edible mushrooms from northwest Spain. *Food Chem., 75*, 417–422.
El-Ramady, H., Singh, A., Rajput, V., Amer, M., Omara, A. E. D., Elsakhawy, T., Elbehiry, F., et al., (2021). Environment, biodiversity and soil security: A new dimension in the era of COVID-19. *Environment, Biodiversity and Soil Security*. 5, 1–14.
Ergonul, P. G., Akata, I., Kalyoncu, & Ergonul, B., (2013). Fatty acid compositions of six wild edible mushroom species. *The Sci. World J., 163964*, 4.
Guillamón, E., García-Lafuente, A., Lozano, M., D'Arrigo, M., Rostagno, M.A., Villares, A., & Martínez, J.A., (2010). Edible mushrooms: Role in the prevention of cardiovascular diseases. *Fitoterapia. 81*, 715–723.
Hamman, S., (2004). Bioremediation capabilities of white-rot fungi, *Biodegradation, 52*, 1–5.
Hammel, K. E., Green, B., & Gai, W. Z., (1991). Ring fission of anthracene by a eukaryote, *Proc. Natl. Acad. Sci. U. S. A, 88*, 10605–10608.
Khan, I., Aftab, M., Shakir, S., Ali, M., Qayyum, S., Rehman, M. U., Haleem, K. S., & Touseef, I., (2019). Mycoremediation of heavy metal (Cd and Cr)–polluted soil through indigenous metallotolerant fungal isolates. *Environ. Monit. Assess,* 191.
Koutrotsios, G., Mountzouris, K. C., Chatzipavlidis, L., & Zervakis, G., (2014). Bioconversion of lignocellulosic residues by *Agrocybe cylindracea* and *Pleurotus ostreatus* mushroom fungi: Assessment of their effect on the final product and spent substrate properties, *Food Chemistry, 161*, 127–136.
Kryczyk-Poprawa, A., Zmudzki, P., Maslanka, A., Piotrowska, J., Opoka, W., & Muszynska, B., (2019). Mycoremediation of azole antifungal agents using *in vitro* cultures of *Lentinula* edodes. *3 Biotech,* 9.
Kulshreshtha, S., Mathur, N., Bhatnagar, P., & Kulshreshtha, S., (2013). Cultivation of *Pleurotus citrinopileatus* on handmade paper and cardboard industrial wastes. *Ind. Crop Prod., 41*, 340–346.
Kumhomkul, T., & Panich-pat, T., (2013). Lead accumulation in the straw mushroom, *Volvariella volvacea*, from lead contaminated rice straw and stubble. *Bull. Environ. Contam. Toxicol., 91*, 231–234.
Lamrood, P. Y., & Ralegankar, S. D., (2013). Biosorption of Cu, Zn, Fe, Cd, Pb and Ni by non-treated biomass of some edible mushrooms. *Asian J. Exp. Biol. Sci., 4*, 190–195.
Loske, D., Huttermann, A., Majerczk, A., Zadrazil, F., Lorsen, H., & Waldinger, P., (1990). Use of white rot fungi for the clean-up of contaminated sites. In: Coughlan, M. P., & Collaco, (eds.), *Advances in Biological Treatment of Lignocellulosic Materials* (pp. 311–321). Elsevier.
Lu, J., & Holmgren, A., (2009). Selenoproteins. *Journal of Biological Chemistry, 284*, 723–727.
Manzi, P., Grambelli, L., Marconi, S., Vivanti, V., & Pizzoferrato, L., (1999). Study of the embryofoetotoxicity of alpha-terpinene in the rat. *Food Chemistry, 65*, 477–482.
Manzi, P., Marconi, S., Guzzi, A., & Pizzoferrato, L., (2004). Commercial mushrooms: Nutritional quality and effect of cooking. *Food Chemistry, 84*, 201–206.
Mattila, P., Lampi, A. M., Ronkainen, R., Toivo, J., & Piironen, V., (2002). Sterol and vitamin D2 contents in some wild and cultivated mushrooms. *Food Chemistry, 76*, 293–298.

McDonough, F. E., Steinke, F. H., Sarwar, G., Eggum, B. O., Bressani, R., Huth, P. J., Barbeau, W. E., et al., (1990). *In vivo* rat assay for true protein digestibility: Collaborative study. *Journal of the Association of Official Analytical Chemists, 73*, 801–805.

Migliore, L., Fiori, M., Spadoni, A., & Galli, E., (2012). Biodegradation of oxytetracycline by *Pleurotusostreatus* mycelium: A mycoremediation technique. *J. Hazard. Mater*, 215, 216.

Miller, K., (2013). *How mushrooms Can Save The World. Discover*. https://www.discover-magazine.com/environment/how-mushrooms-can-save-the-world (accessed on 13 January 2023).

Morais, S., Costa, F. G., & Lourdes, P. M. D., (2012). Heavy metals and human health. *Environ. Heal.–Emerg. Issues Pract.* InTech.

Nagy, B., Măicăneanu, A., Indolean, C., Mânzatu, C., & Silaghi-Dumitrescu, M. C., (2013). Comparative study of Cd(II) biosorption on cultivated *Agaricus bisporus* and wild *Lactarius piperatus* based biocomposites, Linear and nonlinear equilibrium modelling and kinetics. *J. Taiwan Inst. Chem. E.*

Patel, S., & Goyal, A., (2012). Recent developments in mushrooms as anticancer therapeutics: A review. *3 Biotech, 2*(1), 1–15.

Pedneault, K., Angers, P., Avis, T. J., Gosselin, A., & Tweddell, R. J., (2007). Fatty acid profiles of polar and non-polar lipids of *Pleurotus ostreatus* and *P. cornucopiae* var. 'citrino-pileatus' grown at different temperatures. *Mycological Research, 111*, 1228–1234.

Phillips, K. M., Horst, R. L., Koszewski, N. J., & Simon, R. R., (2012). Vitamin D4 in mushrooms. *PLoS One, 7*(8), 40702.

Pinto, A. P., Rodrigues, S. C., Caldeira, A. T., & Teixeira, D. M., (2016). Exploring the potential of novel biomixtures and *Lentinula edodes* fungus for the degradation of selected pesticides. Evaluation for use in biobed systems. *Sci. Total Environ., 541*, 1372–1381.

Pointing, S. B., (2001). Feasibility of bioremediation by white rot fungi. *Appl. Microbiol. Biotechnol., 57*, 20–33.

Pozdnyakova, N. N., (2012). Involvement of the ligninolytic system of white-rot and litter-decomposing fungi in the degradation of polycyclic aromatic hydrocarbons. *Biotechnol. Res. Int.*, 1–20.

Purnomo, A. S., Nawfa, R., Martak, F., Shimizu, K., & Kamei, I., (2017). Biodegradation of Aldrin and Dieldrin by the white-rot fungus *Pleurotus ostreatus. Curr. Microbiol., 74*, 320–324.

Rajput, V. D., Minkina, T., Sushkova, S., Tsitsuashvili, V., Mandzhieva, S., Gorovtsov, A., Nevidomskyaya, D., & Gromakova, N., (2017). Effect of nanoparticles on crops and soil microbial communities. *Journal of Soils and Sediments, 18*(6), 2179–2187.

Rajput, V., Minkina, T., Ahmed, B., Sushkova, S., Singh, R., Soldatov, M., Laratte, B., et al., (2019). Interaction of copper-based nanoparticles to soil, terrestrial, and aquatic systems: Critical review of the state of the science and future perspectives. *Reviews of Environmental Contamination and Toxicology*, 51–96.

Rajput, V., Minkina, T., Mazarji, M., Shende, S., Sushkova, S., Mandzhieva, S., Burachevskaya, M., et al., (2020). Accumulation of nanoparticles in the soil-plant systems and their effects on human health, *Annals of Agricultural Sciences, 65*(2), 137–143.

Rani, P., Kalyani, N., & Prathiba, K., (2008). Evaluation of lignocellulosic wastes for production of edible mushrooms. *Appl. Biochem. Biotechnol., 151*, 151–159.

Reis, F. S., Barros, L., Martins, A., & Ferreira, I., (2012a). Chemical composition and nutritional value of the most widely appreciated cultivated mushrooms: An inter-species comparative study. *Food and Chemical Toxicology, 50*, 191–197.

Reis, F. S., Martins, A., Barros, L., & Ferreira, I. C. F. R., (2012b). Antioxidant properties and phenolic profile of the most widely appreciated cultivated mushrooms: A comparative study between *in vivo* and *in vitro* samples, *Food and Chemical Toxicology, 50*, 1201–1207.

Reshetnikov, S. V., Wasser, S. P., & Tan, K. K., (2001). Higher basidiomycetes as a source of antitumor and immunostimulating polysaccharides (review). *International Journal of Medicinal Mushroom, 3*, 361–394.

Rodarte-Morales, A. I., Feijoo, G., Moreira, M. T., & Lema, J. M., (2012). Biotransformation of three pharmaceutical active compounds by the fungus *Phanerochaete chrysosporium* in a fed batch stirred reactor under air and oxygen supply. *Biodegradation, 23*, 145–156.

Samanta, S. K., Singh, O. V., & Jain, R. K., (2002). Polycyclic aromatic hydrocarbons: Environmental pollution and bioremediation. *Trends Biotechnol., 20*, 243–248.

Singh, A., Rajput, V., Mehrotra, R., Pal, N., Singh, V. K., Chokheli, V. A., & Singh, R. K., (2021). *Sustainable Soil Fertility Management*. pp. 11788, Nova Science Publishers, Inc. 415 Oser Avenue, Suite N. Hauppauge, NY.

Singh, A., Rajput, V., Singh, A. K., Sengar, R. S., Singh, R. K., & Minkina, T., (2021). Transformation techniques and their role in crop improvements: A global scenario of GM crops. *Policy Issues in Genetically Modified Crops*, 515–542.

Singh, N. B., & Singh, P., (2002). Biochemical composition of *Agaricus bisporus*. *J. Indian Bot. Soc., 81*, 235–237.

Sullia, S. B., (2004). Environmental applications of biotechnology. *Asian J. Microbiol. Biotechnol. Environ. Sci., 4*, 65–68.

Sushkova, S., Minkina, T., Deryabkina, I., Rajput, V., Antonenko, E., Nazarenko, O., Yadav, B. K., et al., (2019). Environmental pollution of soil with PAHs in energy producing plants zone. *Science of The Total Environment, 655*, 232–241.

Tekere, M., (2019). Microbial bioremediation and different bioreactors designs. *Applied Biotechnol. Bioeng.* IntechOpen.

Tsujiyama, S., Muraoka, T., & Takada, N., (2013). Biodegradation of 2,4-dichlorophenol by shiitake mushroom (*Lentinula edodes*) using vanillin as an activator. *Biotechnol Lett, 35*, 1079–1083.

Verdin, A. A., Sahraoui, L. H. R., & Durand, R., (2004). Degradation of benzo(a)pyrene by mitosporic fungi and extracellular oxidative enzymes. *Int. Biodeterior. Biodegr., 53*, 65–70.

Wannet, W. J. B., Hermans, J. H. M., Vander, D. C., & Op Den, C. H. J. M., (2000). HPCL detection of soluble carbohydrates involved in mannitol and trehalose metabolism in the edible mushroom, *Agaricus bisporus*. *J. Agric. Food Chem., 48*(2), 287–291.

CHAPTER 12

The Emergence of Mushrooms as Novel Resources of Potential Prebiotics: An Updated View

SOMANJANA KHATUA[1,2], SOUMI BOSE[1], LUCIMARA M. C. CORDEIRO[3], and KRISHNENDU ACHARYA[1]

[1]*Molecular and Applied Mycology and Plant Pathology Laboratory, Center of Advanced Study, Department of Botany, University of Calcutta, Kolkata, West Bengal, India*

[2]*Department of Botany, Faculty of Science, University of Allahabad, Prayagraj 211002, Uttar Pradesh, India*

[3]*Department of Biochemistry and Molecular Biology, Federal University of Paraná, Curitiba, Paraná, Brazil*

ABSTRACT

Prebiotics are non-digestible functional ingredients that can facilitate the growth of helpful bacteria in the gut, responsible for several health benefits. The concept has thus received escalating attention from nutraceutical industries and prompted an interest toward the identification of new molecules that can promote health. Currently, edible mushrooms have emerged as an excellent source of prebiotics as they contain numerous bioactive components of which polysaccharides, β-(1→3)-D-glucans and polysaccharide-peptide/protein complexes are of special significance. The constituents cannot only tolerate gastro-intestinal (GI) conditions, but also be fermented by beneficial members of the colonic population resulting in superior production of short chain fatty acids. The consequence contributes to improved gut barrier integrity, nutrient absorption and gastrointestinal as well as systemic immunity. In this review,

Applied Mycology for Agriculture and Foods: Industrial Applications. Sanjay K. Singh, Deepak Kumar, Md. Shamim, & Rohit Sharma (Eds.)

we aimed to summarize the current knowledge on the effect of macrofungi on improvement of gut dysbiosis and prevention of diabetes, osteoporosis, colitis, and metabolic disorders that may pave the way for translating the health benefits observed during investigation into real-life outcomes.

12.1 INTRODUCTION

Mushrooms are traditionally being considered as a gourmet cuisine across the globe owing to their great taste, subtle flavor and amazing health benefits. To date, around 200 species are popularly recognized as edible where many are enriched in nutrients, such as proteins, essential amino acids, carbohydrates with modest amount of fiber, vitamins, and minerals; although low in fat and calorie (Jayachandran et al., 2017). Along with that, Basidiomycetes are also acknowledged to contain various bioactive substances such as polysaccharides, lectins, peptides, phenolics, ergosterols, lactones, and terpenoids where high molecular weight metabolites have attracted more and more attention (Sawangwan et al., 2018). Indeed, macrofungal polysaccharides, specifically β-glucans, contain diverse medical activities, including anti-tumor, anti-inflammatory, antioxidant, anti-ulcer, antibacterial, hypoglycemic, and immune-enhancing effects (Khatua & Acharya, 2021; Maity et al., 2021; Gong et al., 2020; Khatua & Acharya, 2019; Chatterjee et al., 2013). As a result, mushroom cultivation has witnessed a tremendous growth resulting in a steep escalation in the global market value of fresh fruit bodies which reached 38 billion US dollars in 2018 (Wan Mahari et al., 2020). Despite that, research on the group is scarce as only 10% of total members are presumed to be known indicating that the taxon still represents a vast and untapped area for research (Ruthes et al., 2021).

Currently, prebiotics isolated from natural and novel sources has become an emerging research area. They are essentially recognized as non-digestible food ingredients that fuel the growth and/or activity of beneficial gut bacteria leading to an associated positive consequence on colonic health (Aida et al., 2009). Extensive studies have revealed that the intestinal flora provides nutrients and a protective barrier against pathogens; thus, playing a key role in host health maintenance. Several factors can change the gut microbiota (GM), including host genetic, mode of birth, antibiotics, diet, and age (Hasan & Yang, 2019). Perturbation of the GM population is associated with several chronic diseases such as overweight/obesity, dyslipidemia, hypertension, type-2 diabetes, cardiovascular diseases, osteoporosis, and dental caries (Florowska et al., 2016; Cencic & Chingwaru, 2010). In this context, mushrooms have shown promising effects as they contain chitin, β- and α-glucans, mannans, polysaccharide-peptide/protein complexes, proteoglycans, xylans, galactans, phenolic compounds and terpenes (Khangwal & Shukla, 2019; Aida et al., 2009). These components, specifically mushroom D-glucans can escape digestion in small intestine and are fermented in large intestine stimulating the growth of beneficial microbes and reducing the richness of opportunistic pathogens improving immune response. Further, the fermentation process augments production of short-chain fatty acids (SCFAs) which are quickly absorbed in the colon to offer additional energy to the host (Ruthes et al., 2021). As a result, these prebiotics help in minerals absorption, metabolites production, GM regulation and prevention to curing of various ailments such as diabetes, metabolic disorders, colitis, gastro-intestinal (GI) injury and cancer indicating their high potential of applications in both food and pharmaceutical industries (Yin et al., 2020). Hence, in this review article, we attempt to congregate recent findings on the effects of mushrooms on GM to improve gut dysbiosis assisting disease treatment.

12.2 HUMAN GUT MICROBIOTA (GM)

Human GI or digestive tract is extended from mouth to anus and thus represents one of the largest interfaces among the host, environmental parameters and antigens inside the body. During a normal lifetime, approximately 60 tons of food pass through the alimentary canal, along with plenty of microbes from the milieu that might be hazardous for gut integrity (Thursby & Juge, 2017). Assemblage of these colonizing microbes encompassing bacteria, archaea, and eukarya is termed as gut microbiota (GM) that coevolves within the host (Carlson et al., 2018). Almost 90% of the flora belong to three major phyla, such as Firmicutes (Gram positive), Bacteroidetes (Gram negative) and Actinobacteria (Gram positive); while rest of the microbial phyla are represented by Proteobacteria, Fusobacteria, and Verrucomicrobia (Yin et al., 2020; Rinninella et al., 2019). The Firmicutes phylum is comprised of over 200 different types of genera including *Bacillus*, *Lactobacillus*, *Enterococcus*, *Roseburia, Clostridium* (represent about 95% of the phylum) and *Ruminicoccus*. Bacteroidetes consists mainly of *Bacteroides*, *Prevotella*, *Parabacteroides*, and *Alistipes* where the first two genera are predominant. In contrast, Actinobacteria phylum is comparably less abundant and chiefly represented by the genus *Bifidobacterium* (elegantly reviewed by Rinninella et al., 2019). Overall, Bifidobacteria and lactobacilli are the prime members of GM where Bifidobacteria alone make up around 25% of the total quantity of bacteria present (Azmi et al., 2012). However, such composition of gut flora varies depending on intestine anatomical regions containing different physiology, substrate availability, O_2 tension, digesta flow rates (speedy in mouth to caecum, slow-moving afterward), pH and host secretions. The microbial colonizers face the most challenging environment in the small intestine being characterized by short transit time and high bile concentration. The large intestine, on the other hand, provides neutral to mildly acidic pH, slow flow rates and readily available nutrients harboring the largest microbial community where obligate anaerobic bacteria are prevalent (Rinninella et al., 2019). These anaerobes express carbohydrate-active enzymes and digest substrates like dietary carbohydrates, proteins, lipids, and indigenous secretion including mucin (Sanders et al., 2019). This fermentation process produces some end products of which secondary bile acids (BAs) and SCFAs are the two major groups (Zeng et al., 2019). SCFAs are one to six carbons containing saturated aliphatic organic acids of which acetate (C2), propionate (C3) and butyrate (C4) are typically found in an approximate molar ratio of 3:1:1 in the colon as well as stool (den Besten et al., 2013). SCFAs are water-soluble and readily absorbed into the blood stream. Colonic epithelial cells preferentially take up butyrate to extract around 60–70% of their energy requirements essential for proliferation, even when competing substrates such as glucose and glutamine are available (Tan et al., 2014). The SCFA is thus considered as a key nutrient not only to determine the growth and metabolic activity of colonocytes; but also, for, modulation of growth and variation of intestinal cells, exertion of strong anti-inflammatory potency, stimulation of cell apoptosis and exhibition of anticancer effect (Mitsou et al., 2020). Propionate is usually exploited by intestinal epithelial cells as an energy source; however, it can also be released into portal vein and reach the liver where it is metabolized by hepatocytes. Previous studies indicated that propionic acid exerts cholesterol-lowering, anti-lipogenic, anti-inflammatory, anti-microbial, and anti-carcinogenic activities (Maheshwari et al., 2021). Acetate also reaches liver through portal vein and remains either in the liver where it contributes to lipogenesis or is released systemically to peripheral venous system and can be metabolized in the human kidney, muscle, heart, and brain (Tan et al., 2014). Studies have elucidated that acetic acid could augment $CD103^+$ DCs and T_{REG} cell responses that aid in protection from food allergy (Ding et al., 2017). Further metabolism of SCFAs into the muscle, liver or other peripheral tissues is regarded to contribute with about 7–8% of the host's day-to-day energy requirements (Slavin, 2013). Thus, SCFA signaling through the central nervous system moderates a range of physiological procedures, including energy homeostasis, carbohydrate, and lipid metabolism as well as inflammation (Maheshwari et al., 2021). Other benefits offered by the GM to the host include shaping intestinal epithelium or strengthening gut integrity, nutrient absorption, promotion of angiogenesis, lowering both glycemic level and body weight, modulation of cardiovascular biomarkers as well as increased inhibition of carcinogen toxicity (Yin et al., 2020; de Paulo Farias et al., 2019; Azmi et al., 2012). Production of SCFA also inhibits the growth of pathogens through reducing the luminal and fecal pH. Low pH diminishes peptide degradation, activity of undesirable bacterial enzymes and synthesis of toxic components such as amines, ammonia, and phenolics (Slavin, 2013). In this manner, the essential gut microflora prospers and maintains intestinal cells healthy by delivering them with substances which upsurge blood flow in these cells (Singla & Chakkaravarthi, 2017).

Many factors may affect the normal GM composition. For instance, the mother's vaginal and intestinal microbiota can influence the fetus microbiota. Further, therapeutic treatments, antibiotics, and drug intake, hygiene level, genetic background, climate, illness, stress, poor nutrition, surgery, diarrhea, aging, and lifestyle can also disturb the GM balance (Gagliardi et al., 2018; Thursby & Juge, 2017; Panesar et al., 2012). The dysbiosis has been associated with allergies, inflammatory bowel disease (IBD), irritable bowel syndrome (IBS), ulcerative colitis (UC), Crohn's disease (CD), colorectal cancer, gastric cancer, esophageal cancer, non-intestinal autoimmune diseases, neurological disorders, obesity, diabetes, cardiovascular diseases, decreased bone density, depression, autism, and even multiple organ failure (Kerezoudi et al., 2021; Yin et al., 2020). Besides, alterations in the intestinal function are evident in elderly persons encompassing changes in immune function, increased mucosal membrane permeability and microbial dysbiosis. Previous studies have shown an upsurge in facultative anaerobes, such as enterococci, streptococci, and enterobacteria, in the elderly populations and decrease in intestinal levels of *Bifidobacterium* and *Bacteroides* spp. (Mitsou et al., 2020). Disequilibrium of commensal flora causes reduced expression of tight junction proteins and increased intestinal permeability resulting in metabolic endotoxemia and insulin resistance. Altered microbiota thus become unable to protect the host from pathogenic organisms that in turn induces inflammation and generates genotoxins or carcinogenic metabolites (Gomes et al., 2014). The composition of gut bacteria hence can reflect the risk of diseases in each person (Zhang et al., 2015). In this backdrop, the development of products that can improve gut health has become one of the main targets in recent years for the functional food industry (Hernandez & Pandiella, 2013)

12.3 PROBIOTICS

One approach for positively modulating the GM was the administration or ingestion of live bacteria or probiotics to prevent some diseases (Butel, 2014). Probiotics are currently defined as "live strains of strictly selected microorganisms which when administered in adequate amounts, confer a health effect on the host" (Markowiak & Śliżewska, 2018). Repertoire of the most consumed probiotics consists of lactic acid producing bacteria, chiefly lactobacilli, *Lactococci*, *Bifidobacterium*, and *Streptococci*. Along with them, bacilli, yeast, and some non-pathogenic *Escherichia coli* strains are also used (Lerner et al., 2019). Mechanisms of probiotics include regulation of intestinal microbial communities, stimulation of epithelial cell proliferation, immunomodulation, and differentiation and fortification of the intestinal barrier (Hemarajata & Versalovic, 2013). Besides, they may effectively inhibit development of pathogenic bacteria, such as *Clostridium perfringens*, *Salmonella enteritidis*, *Campylobacter jejuni*, *E. coli*, various species of *Staphylococcus*, *Yersinia*, and *Shigella* thus inhibiting food poisoning. As such, a positive effect of probiotics on dealing with food allergies, digestion processes and dental caries has also been confirmed (Markowiak and Śliżewska, 2017). As a result, probiotics are now widely being marketed constituting a growing multi-billion-dollar trade and one of the most regularly consumed dietary supplements or functional foods worldwide (Hemarajata & Versalovic, 2013). Nowadays, yogurt, ice cream, snacks, infant formulas, breakfast cereals, cheese, and nutrition bars are supplemented with probiotics, so as are cosmetics. Probiotics are also marketed as lyophilized pills (Suez et al., 2019). However, the use of probiotics in many cases becomes difficult owing to low viability of bacteria after pelleting and during storage. In addition, there is a high chance of environmental issues as probiotics may enter into the environment (Lerner et al., 2019).

12.4 PREBIOTICS

As an alternative, the concept of prebiotics has been aroused in an attempt to overcome issues connected to probiotic applications. Instead of introducing living bacteria, the purpose of prebiotics is to stimulate functionality of selected beneficial indigenous microbiota populations (Lauzon et al., 2014). The perception was introduced for the first time in 1995 by Glenn Gibson and Marcel Roberfroid defining prebiotic as "a non-digestible food ingredient that beneficially affects the host by selectively stimulating the growth and/or activity of one or a limited number of bacteria in the colon and thus improves host health" (Davani-Davari et al., 2019). Prebiotics cannot be digested by gastric acidity and α-amylase or other types of hydrolases in the upper gut segment of the intestinal tract and thus must not be absorbed in the small intestine (Azmi et al., 2012). They should be fermented by the GM (large intestine) and selectively enhance the growth as well as activity of a limited number of colonic bacteria (such as *Lactobacillus acidophilus*, *Bifidobacterium bifidum*, *Faecalibacterium prausnitzii,* and *Bifidobacterium adolescentis*), but not pathogens causing GI diseases such as *C. perfringens*; hence can alter the colonic microflora equilibrium towards a healthier composition (Figure 12.1) (Dwivedi et al., 2014).

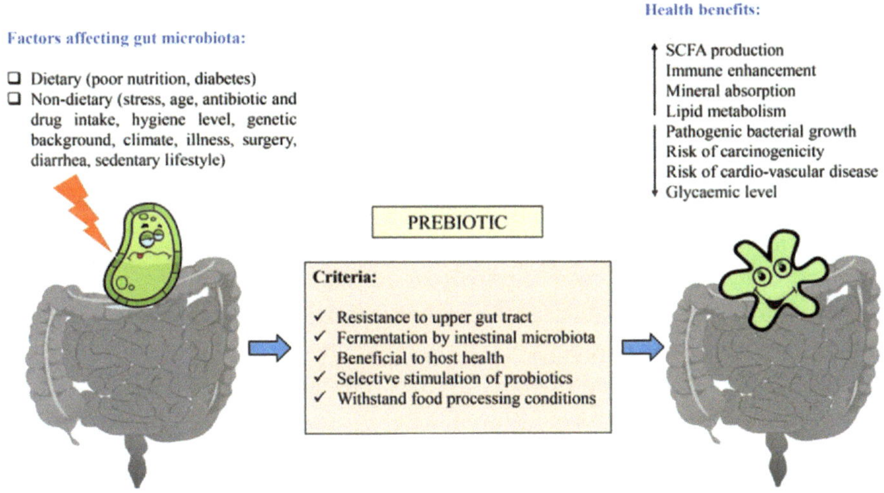

FIGURE 12.1 Overview of a prebiotic.

Currently, the established prebiotics can be categorized into three major groups: (i) polyols (mannitol, lactulose, xylitol); (ii) oligosaccharides such as galacto-oligosaccharide (GOS), fructo-oligosaccharides (FOS), isomalto-oligosaccharides (IMO), xylo-oligosacharides (XOS), raffinose oligosaccharides (RFOs), mannano-oligosaccharides (MOS), arabinoxylan-oligosaccharides (AXOS), isomaltulose (palatinose); and (iii) fibers (dextrins, cellulose, β-glucans, pectins). These are known as "colonic foods" where Inulin and FOS are the most widely researched prebiotics (Bajury et al., 2018). Besides them, peptides, proteins, and lipids are also considered as candidates to prebiotics, however only the carbohydrate molecules have been examined in depth so far (Hernandez & Pandiella, 2013). As such, low molecular weight polysaccharides are efficiently metabolized by microbes such as bifidobacteria. While members of the *Bacteroides* genus and *Ruminococcus* spp. are adept to break down high molecular weight carbohydrates and resistant starch, respectively (Sanders et al., 2019). Many scientists thus have dedicated their research on the prebiotic carbohydrates that can be synthesized through various chemical and biochemical reactions including hydrolysis, isomerization, transgalactosylation, fructosyltransfer, oxidation, and reduction. Parallelly, biotechnological production encompassing microbial fermentation process is regarded as a good initiative to enlarge production scale, quality of the final product and promote a decrease in production cost, making the procedure economically viable (de Paulo Farias et al., 2019).

12.5 MECHANISM OF ACTION OF PREBIOTICS AND HEALTH BENEFITS

An important mode of action for prebiotics lies in the fermentation process in the colon and alterations in the gut microflora (Slavin, 2013). During the digestion process, they resist hydrolytic activity in the upper part of the GI tract due to lack of suitable microflora and thus reach the colon in an intact form. There they are selectively fermented by beneficial microflora such as bifidobacteria and lactobacilli as they contain relatively high amounts of β-fructosidase and glycosyltransferases, respectively (Yang et al., 2020; Singla & Chakkaravarthi, 2017). Besides, many members of the GM encode vast ranges of carbohydrate-active enzymes or CAZymes to forage carbohydrate-enriched diet and host-related glycans present in the gut. The enzymatic saccharification and subsequent fermentation by the GM produce SCFAs which provide myriad of benefits to the host (Briggs et al., 2021). Acetate and lactate, the major metabolic end products of bifidobacterial and lactic acid bacteria, are exploited by other microorganisms to produce butyrate and propionate, respectively (Sanders et al., 2019; Al-Sheraji et al., 2013). About 90–95% of SCFAs produced are absorbed through the intestinal wall, in the caecum and ascending part of the colon (Hernandez & Pandiella, 2013). The acidic environment in the colon in turn suppresses growth of the pathogens. However, bacteria belonging to the genus, *Bifidobacterium*, demonstrate tolerance to the SCFAs and low pH. This mechanism permits prebiotics to manipulate the colonic microbiota composition improving the host health including enhancement of immune function, improve digestion, creating vitamins and digestive enzymes, inducing oral tolerance for dietary allergens and elimination of feces (Singla & Chakkaravarthi, 2017; Panesar et al., 2012; Azmi et al., 2012). Prebiotics are also fermented to some antibacterial constituents, such as bacteriocine against the pathogenic species. These substances may not only benefit the intestinal microbial configuration but also enhance the integrity of intestinal epithelial cells, which further upsurge absorption of nutrients resulting improvement in growth performance of the host (Teng & Kim, 2018). Likewise, prebiotics can augment the uptake of iron, calcium, and zinc. Besides, they possess remarkable potential to diminish colon cancer and levels of triglycerides (TGs) as well as cholesterol (Markowiak & Śliżewska, 2017; Dwivedi et al., 2014).

12.6 SYNBIOTICS

When probiotics and prebiotics are administered simultaneously, the combination is named as synbiotics. As the term implies synergy, it should be used for those products where a prebiotic constituent selectively favors a probiotic microorganism. The chief purpose of this type of mixture is the enhancement of the survival of probiotic microbes in the GI tract. Henceforth, an appropriate amalgamation of both the substances in a single product should confirm a superior outcome in comparison with the action of the individual probiotic or prebiotic (Markowiak & Śliżewska, 2017). The mechanism of synbiotic effects is *via* the development of beneficial microbiota, modulation of metabolic effect in the intestine and the concomitant inhibition of potential pathogens present in the GI tract (Gyawali et al., 2017).

12.7 MUSHROOMS AS POTENTIAL SOURCE OF PREBIOTICS

The increasing demand for prebiotics has paved the way for searching new sources of relatively low price in comparison with commercially available components. Extensive studies are now focusing on the prebiotic potential of metabolites extracted from natural products such as mushrooms (Azmi et al., 2012). Macrofungi have always been cherished for their culinary as well as nutritional values and are nowadays increasingly exploited for numerous important therapeutic properties (Khatua et al., 2019, 2021;

Khatua & Acharya, 2018). The most well-known consumed mushroom species include *Auricularia auricula*, *Flammulina velutipes*, *Ganoderma lucidum*, *Grifola frondosa*, *Hericium erinaceus* and *Lentinula edodes* (Yin et al., 2020). At present, only 60 species are commercially cultivated and amongst them 10 members including black fungus or wood-ear mushrooms (*A. auricula*, *A. polytricha*), white button mushroom (*Agaricus bisporus*), shiitake (*L. edodes*), oyster mushroom (*Pleurotus* spp.) and paddy straw mushroom (*Volvariella* spp.) are produced on an industrial scale (Dudekula et al., 2020; Kapahi, 2018). These cultivatable specimens are the leading component in global mushroom industry accounting for around $34 billion, while medicinal and wild samples made up $24 and $5 billion, respectively, in 2013 (Royse et al., 2017).

Edible mushrooms endow with high quality of proteins (19–35%), including most of the essential amino acids, carbohydrates (50–65%) encompassing dietary fiber, essential fatty acids (2–6%) and trace of vitamins as well as minerals. They could be a good source of numerous different nutraceuticals including polysaccharides, β-glucans, proteins, glycoproteins, phenolic compounds, unsaturated fatty acids, toco-pherols, ergosterols, and lectins (Ma et al., 2018). Taking advantage of the synergistic and additive effects of all these bioactive compounds, Basidiomycetes can execute a range of health benefits, such as antitumor, immunomodulatory, antinociceptive, antimicrobial, and anti-inflammatory properties (Ruthes et al., 2021). They have thus high potential to be used as functional foods in the form of nutraceuticals, mycotherapy products and dietary supplements to maintain a good state of health and cure or prevent ailments (Venturella et al., 2021). Mushrooms contain a plethora of bioactive components, as stated earlier, amongst them polysaccharides are the most potent prebiotic substance being responsible for various physiological activities. They are present predominantly as linear and branched glucans possessing various types of glycosidic linkages, for instance, (1→3)-α-glucans and (1→3),(1→6)-β-glucans; while some are true heteroglycans consisting of mannose, galactose, xylose, glucuronic acid, ribose or arabinose (Mitsou et al., 2020; Synytsya et al., 2009). These components justify their role as prebiotics to improve the growth of intestinal bacteria and promote beneficial effects on the health of the host (Yin et al., 2020). Another important prebiotic component in mushrooms is polysaccharopeptide which is a protein-bound polysaccharide complex. A recent clinical study enumerated that polysaccharopeptide from *Trametes versicolor* might function as a prebiotic through modulation of human intestinal microbiome composition (Pallav et al., 2014).

12.8 *IN VITRO* DIGESTION AND FERMENTATION

To determine the prebiotic effect of an investigated drug, *in vitro* digestion and fermentation methods are widely followed being simply cheaper, faster, and more ethical in comparison with *in vivo* assays. The *in vitro* digestion models simulate gastrointestinal conditions and are thus used to study digestibility, structural alterations and release of food components providing high-accuracy results in a short time (Bajury et al., 2018). Conversely, the rationale of *in vitro* fermentation models is to cultivate a complex intestinal microbiota for carrying out microbial modulation and metabolism studies (Nissen et al., 2020). In this view, human or other animal feces are commonly used as inoculum during fermentation being noninvasive. The most prevalent anaerobic bacteria found in feces are *Lactobacillus*, *Eubacterium* spp., *Bifidobacterium* spp., *Bacteroides* spp., *Clostridia*, and *Fusobacterium* (Bajury et al., 2018). As such, the main beneficial property of prebiotics is their ability to stimulate growth and/or activity of certain groups of bacteria, mainly of lactobacilli and bifidobacteria (strains of *L. paracasei*, *L. plantarum*, *B. bifidium*) (Florowska et al., 2016). Bifidobacteriaceae and Lactobacillaceae possess the antibacterial, anti-tumor, anti-inflammatory, cholesterol lowering, age delaying and laxative properties (Yang et al., 2020). Along with that, prebiotics may inhibit growth of the pathogens (*C. perfringens*., *E. Coli*, *C. jejuni*, *Enterobacterium* spp., *S. enteritidis* or *Salmonella typhimurium*) by lowering the cecal pH and creating an unfavorable environment for their development (Florowska et al., 2016).

To date, many researchers have depicted the prebiotic effects of mushrooms by performing *in vitro* digestion and fermentation assays (Table 12.1). For instance, Rodrigues et al. (2016) described bioactivity of *Pholiota nameko* using cellulase and Flavourzyme digested fractions. In particular, the extract prepared with Flavourzyme was found to suppress growth of *Clostridium histolyticum* and members of *Clostridium cocoides/Eubacterium rectale* group (the major anaerobic population in human gut). Whilst the population of *Bifidobacterium* spp. was enhanced; although *Lactobacillus* spp. remained unaffected. In consistence, increased production of total SCFA and lactic acid was also recorded (Yang et al., 2020). The observation was partially in according to Chou et al. (2013) describing bioactivity of hot water extracted crude polysaccharides from *L. edodes* stipe, *Pleurotus eryngii* base and *F. velutipes* base which are normally discarded. The extracts apparently maintained higher populations of the probiotics such as *Lactobacillus casei*, *L. acidophilus* and *Bifidobacterium longum* subsp. *longum* during cold storage.

Further, tolerance and stabilities of the probiotics supplemented with mushroom polysaccharides were found to improve in simulated gastric juice and bile acid. In another study, carbohydrate-rich water extracts were isolated from *Pleurotus sajor-caju*, *P. florida* and *Lentinus edodes* to determine prebiotic effect. The studied mushrooms, particularly *P. sajor-caju*, executed good prebiotic score as calculated by comparing probiotic growth (*L. acidophilus*) with respect to growth of enteric pathogens (*E. coli*, *Enterobacter cloacae* and *S. typhimurium*) (Mallick & Bhawsar, 2018). Similar trend of action was reported by Mitsou et al. (2020) as well highlighting potential of four edible mushrooms namely *Cyclocybe cylindracea*, *Pleurotus ostreatus*, *P. eryngii* and *H. erinaceus* as prebiotic candidates. Their study indicated the beneficial influence of the studied samples on GM of elderly donors (65 years old). In sum, *P. ostreatus* and *C. cylindracea* augmented bifidogenic effect; while *P. eryngii* induced lactogenic effect increasing molar ratio of propionate and butyrate. Avci et al. (2017) showed that *Lactobacillus rhamnosus* could use aqueous extract from *A. polytricha* as a carbon source. Nowak et al. (2018) performed an extensive study involving polysaccharide extracts from 53 wild mushrooms. Obtained results revealed that all isolated macromolecules stimulated the growth of *Lactobacillus* strains; while some polymers remain undigested by the artificial human gastric juice. Yang et al. (2020) isolated three oligosaccharide fractions from broken and unbroken spores of *G. lucidum*. All the studied preparations, particularly the one isolated by precipitation with 80% ethanol, exhibited effective prebiotic effect which encompassed resistance to simulated human stomach and gastric juice hydrolysis; boosting growth of Ruminococcaceae, Bifidobacteriaceae, and Lactobacillaceae as well as diminishing abundance of Enterobacteriaceae and Lachnospiraceae as observed after 24 h fermentation of fecal samples collected from healthy human. In a separate study, crude polysaccharides derived from fruiting bodies and mycelia of *G. lucidum* and *Poria cocos* exhibited regulatory effect on mice GM structure. A profound decrease in the pro-obesity taxa (*Pseudobutyrivibrio* spp., *Barnesiella* spp. and *Dehalobacterium* spp.) and increase in the anti-obesity taxa (*Bacteroides acidifaciens*, *Akkermansia muciniphila* and *Anaerotruncus* sp.) appeared when treated with the macromolecules from fruiting bodies. Overall, enhancement in SCFA-producing bacteria was recorded in all treated sets compared to the control group (Khan et al., 2018). Su et al. (2019) demonstrated *in vitro* digestion and fermentation ability of *F. velutipes* derived polysaccharides. The fraction, composed of different monosaccharides, modulated GM composition by elevating the amounts of Bifidobacteriaceae and Bacteroidaceae and reducing the numbers of genera under Lachnospiraceae and Enterococcaceae increasing synthesis of SCFAs. The study also revealed that the polymers were able to escape digestion by human saliva and simulated GI conditions. The outcome could be justified by the dominant presence of β-glucan in the isolated extract that may have played a key role behind the beneficial effects in GM and/or SCFAs production. Chaikliang et al. (2015) evaluated the prebiotic effect of β-glucan from *Schizophylum commune* and *Auricularia auricula-judae* by fecal fermentation in batch culture. The polymers isolated from *A. auricula-judae* augmented numbers of bifidobacteria and lacto-bacillus as well as prebiotic index. Acetate was the most prevalent SCFA produced in all treatments followed by propionate, butyrate, and lactate. Similarly, two glucans (Glucan A and Glucan B), extracted from *Cookeina speciosa*, were found to be highly propiogenic and butyrogenic, with low gas formation during *in vitro* fecal fermentation. The observation might be due to specific increase in *Bacteroides uniformis* and genera from *Clostridium* cluster *XIVa*, including butyrogenic *Anaerostipes* and *Roseburia*. Overall, Glucan A presented a faster fermentation profile than that of Glucan B (Cantu-Jungles et al., 2018). A glucan-enriched crude polysaccharide fraction was also isolated from *H. erinaceus* following conventional hot water extraction method. Results indicated that both simulated gastric and small intestinal digesta of the polymers can boost the proliferation of six tested probiotics in comparison with the fraction alone. The prebiotic activity increased, evidenced by the high amount of SCFAs production, when the polymers were treated with simulated gastric and small intestinal juice for 2 and 4 h, respectively, indicating efficient release of monomers throughout the gastrointestinal digestion period (Yang et al., 2018). Hess et al. (2018) revealed that consumption of *A. bisporus* rich diet could affect fecal microbiota composition in healthy adults compared to the meat diet, as evident by the greater abundance of Bacteroidetes and lower abundance of Firmicutes. The ingestion also increased stool weight suggesting that inclusion of the mushroom into the diet may influence laxation in healthy adults.

12.9 PROTECTION AGAINST INTESTINAL INJURY

Together with a mucus layer, the large intestinal layer of the specialized epithelial cells, connected by tight junction proteins (claudins, zonula occludens-1 (ZO-1), and occludin), serve as a barrier. Foremost cells of the intestinal epithelium are enterocytes (control absorption of nutrients) and goblet cells (secrete mucin glycoproteins). The mucus layer is enriched in immunoglobulin A and antimicrobial peptides that help in the protection of the epithelial cell surface from insults and enhance epithelium integrity. However, a large number of opportunistic pathogens in the gut may disrupt this intestinal barrier, causing it leaky (Maheswari et al., 2021). The concept of "leaky gut" is derived from the phenomena in which tight junctions are compromised resulting in increased permeability of the epithelial layer that allows translocation of antigens from the lumen into the blood stream. The phenomenon can trigger low-grade chronic inflammation through activation of toll like receptor (TLR)-4 signaling cascade in various cells resulting in a condition known as metabolic endotoxemia. In case, if interruption of the symbiotic relationship between GM and the defense mechanism persists, it aids in the development of several diseases and disorders such as IBD, IBS, and obesity (Pham et al., 2018; Carlson et al., 2018). In this context, prebiotics can play a key role in the improvement of intestinal barrier function, and the effect occurs mainly through the production of SCFAs (Carlson et al., 2018).

Till date, only a few studies have reported focusing on the ability of mushrooms to improve gut barrier via regulating composition and metabolites of the intestinal microbiota. Wang et al. (2019) used piglet model possessing LPS-induced intestinal injury to determine whether Lentinan (a glucan derived from *L. edodes*) could alleviate the consequence. Results showed that pigs fed with Lentinan had higher claudin-1 protein expression in ileum. Further, the ingestion down-regulated mRNA expression of jejunal TLR-4, nuclear factor (NF)-κB and associated several proteins together with ileal tumor necrosis factor receptor (TNFR)-associated factor 6 (TRAF6) and nucleotide-binding oligomerization domain 1 (NOD1) indicating immune suppressing activity. The outcome might be due to change in GM as the polysaccharide diet lowered Firmicutes and increased Bacteroidetes abundances resulting amplified level of butyrate, propionate, isovalerate, and isobutyrate in cecal digesta. In another study, Kanwal et al. (2018) administered clindamycin and metronidazole (broad-spectrum antibiotics) in the healthy mice to induce gut dysbiosis. To ameliorate the condition, a crude polysaccharidic fraction was isolated from *D. indusiata*. Results showed that the ingestion of the fraction reduced the abundance of pathogenic bacteria and increased beneficial flora such as, Lactobacillaceae and Ruminococaceae. Further, the treatment augmented mucus-producing goblet cells and triggered expression of tight-junction proteins (claudin-1, occludin, and ZO-1) as well as MUC2. In addition, administration of the extract reduced endotoxin/pro-inflammatory cytokine levels in mice with intestinal lesions improving gut epithelial barrier and ameliorating the structure of colon tissue. Polysaccharides from *F. velutipes* have also shown potential in improvement of gut health as evident by higher villus height and villus height/crypt depth value in the fraction treated mice. The polymers altered microbiota composition by increasing the Firmicutes phylum and decreasing Bacteroidetes phylum (Hao et al., 2021).

12.10 EFFECT ON IMMUNE RESPONSE

Prebiotics can act on the local and systemic immune cells as well as gut-associated epithelial cells, mainly through G protein-coupled receptor (GPR) mediated pathways. However, other cascades such as histone deacetylase inhibition and inflammasome pathway are also being evident to regulate the immunomodulatory effect. The prebiotics can stimulate the gut-associated epithelial and innate immune cells through TLRs as well (Pujari & Banerjee, 2021). Prebiotics and the metabolites that are formed by their fermentation can influence T_{REG} cells, effector T cells, natural killer cells, and B cells. The cumulative effect results in preservation of the epithelial barrier integrity and modulation of innate immunity (Carlson et al., 2018).

A β-glucan enriched polysaccharide (M_w 426 kDa) was isolated from *P. eryngii* to investigate *in vivo* fermentation behavior and consequence on defense mechanism in mice. The oral administration induced level of different SCFA, decreased pH value and improved moisture contents of the cecum and colon. The treatment also caused an increase in the abundance of Porphyromonadaceae, Rikenellaceae, Bacteroidaceae, and Lactobacillaceae. Further, induction in the concentrations of tumor necrosis factor (TNF)-α, interferon (IFN)-γ, interleukin (IL)-1, IL-2, IL-6 in serum and SlgA in cecum as well as colon was also noticed indicating that the sample possesses remarkably ability to modulate immune response of the host (Ma et al., 2017). The outcome was in accord to Zhao et al. (2019) describing the effect of a crude polysaccharidic fraction from *F. velutipes* on mice intestinal microbiota and immune response. Results showed that the treatment decreased the ratio of Firmicutes/Bacteroidetes and augmented total SCFAs concentration in each intestinal section. Further, an increase in the levels of Ig-A, Ig-G, and several cytokines (IFN-γ, TNF-α, IL-6, and IL-8) in serum was also noted indicating the immune regulatory effect of the studied mushroom. In another study, purified polysaccharide from hot water extract of *A. auricula* was used to investigate the impact of the sample on a cyclophosphamide (CP) induced immunosuppressive model. The treatment increased the expression of IFN-γ, several ILs and TNF-α indicating immune boosting effect. It was also observed that mRNA levels of tight junction proteins were upregulated, suggesting that administration of the macromolecule could enhance colon intestinal barrier function. Besides, the sample regulated the composition of GM by decreasing the ratio of Firmicutes/Bacteroidetes and enhancing the synthesis of SCFAs in immunosuppressed mice (Kong et al., 2020). The traditional hot water extraction process was also applied to *Cordyceps sinensis* to extract crude polysaccharidic fraction which was then administered in mice possessing CP-induced intestinal mucosal immunosuppression and microbial dysbiosis. The ingestion augmented expressions of TLRs (TLR-2, 4, and 6) as well as NF-κB pathway key proteins (NF-κB p65 and p-IκB-α). As a result, augmentation in transcription factor production (GATA-3, T-bet, Foxp3, RORγt) and cytokines secretion (IL-2, 4, 6, 10, 12, 13, 17, and 21; IFN-γ, TNF-α, TGF-β3) were also evident. Further, the treatment restored gut microbial dysbiosis by modulating microbial community diversity, structure, and composition (Ying et al., 2020). In a separate study, Solano-Aguilar et al. (2018) fed pigs a diet supplemented with *A. bisporus* prebiotics and after six weeks alveolar macrophages (AM) and peripheral blood mononuclear cells (PBMC) were isolated and stimulated with lipopolysaccharide (LPS). Results revealed an anti-inflammatory effect with a noteworthy reduction in IL-1β expression and cytokine production. The mushroom diet affected the composition of fecal and proximal colon microbiota by inducing the abundance of butyrate producing families such as Lachnospiraceae and Ruminococcaceae suggesting a dose dependent increase in beneficial bacterial genera within the order Clostridiales.

12.11 INHIBITION OF IBD

IBD is characterized by immune activation in the GI tract, causing inflammation and damage to mucosa or submucosa resulting in severe watery and bloody diarrhea accompanied by abdominal pain. The disorder affects both the colon and small intestine and includes UC (inflammation and ulcers along both the colon and rectum), CD (inflammation along GI tract) and pouchitis (Rasmussen & Hamaker, 2017). The factors causing IBD involves environmental, genetic, type of intestinal microbes, dysregulation of immune system and oxidative stress (Pandey et al., 2015). Studies have shown that the structure of intestinal microbiota is altered in patients with IBD resulting lesser secretion of antimicrobial peptides from Paneth and Goblet cells and activation of REGIIIγ. IBD patients shows, in particular, reduction of some *Verrucomicrobia* (*A. muciniphila*; a propionate-producer), *Firmicutes* (*Faecalibacterium prausnitzii;* a butyrate-producer) and an increase of *Proteobacteria* causing reduction of SCFA. Besides, loss of the intestinal barrier integrity amplifies bacterial antigen translocation and inflammatory response of intestinal mucosa, executing the core pathological feature of IBD (Meng et al., 2019). In this context, prebiotics have been reported to play a beneficial role in controlling IBD (Pandey et al., 2015). Although their exact mechanism of action is still uncertain; however, it is hypothesized that skill of prebiotics to increase synthesis of SCFAs, modulate cytokines production within gut mucosa and regulate the GM composition might play a significant role (Scaldaferri et al., 2013).

Different fractions (alcoholic fractions, polysaccharide, and whole extracts) were isolated from *H. erinaceus* and were subjected to rats with IBD induced by trinitro-benzene-sulfonic acid (TNBS) enema. The treatments improved histological and clinical changes in IBD rats by reducing expression of myeloperoxidase (MPO) which is synthesized in higher level in patients with severe forms of IBD. Further the extracts, specifically the alcoholic preparation, suppressed immune functionality as evident by stimulation of T-cells as well as decreased activity of NF-κB p65 and TNF-α. Growth of beneficial gut bacteria was also observed indicating the clinical potential of the fractions in relieving IBD by regulating GM and immune system (Diling et al., 2017b). The similar observation has also been reported in case of 50 to 55-kDa single-band protein from *H. erinaceus* exhibiting anti-inflammatory activity in TNBS induced IBD rat model. The treatment augmented Foxp3- and IL-10-positive cells; while diminished TNF-α and NF-κB p65-positive cells significantly. Further, administration of the protein restored all the pro-inflammatory cytokine levels to near normal. Moreover, the protein was also subjected to mice with high-dose CP-induced immunotoxicity to elucidate the immunomodulatory activity in depth. The treatment augmented neutral red engulfment, thymus, and spleen index, splenocyte proliferation and T cell activation, indicating that the protein could reverse the immunotoxicity in mice (Diling et al., 2017a). Similarly, polysaccharide extracted from *A. auricula-judae* showed a surprising preventive effect against colitis. The dextran sulfate sodium (DSS) induced UC mice suffered from weight loss, colon shortening, damage of intestinal barrier, mucosal inflammation and dysbiosis. Pre-administration of the studied fraction relieved these symptoms, reduced D-lactic acid and diamine oxidase levels in the plasma and prevented damage of the intestinal barrier. The outcome might be justified by the ability of the sample to alter GM composition (Zhao et al., 2020). Xie et al. (2019) found polysaccharides from *G. lucidum* not only regulated the GM composition and SCFAs production in DSS-induced UC rats, but also controlled expression of genes involved in inflammation-related KEGG pathways. The studies indicate potential of macrofungi polysaccharides in UC may involve regulation of different pathways; however, further studies are required for better understanding.

12.12 PREVENTION OF OSTEOPENIA

Osteoporosis is designated as a systemic skeletal ailment that is categorized by low bone mass and microarchitectural deterioration of bone tissue resulting increase in bone fragility as well as susceptibility to fracture (Compston et al., 2019). It is the most prevalent bone disease in humans where the risk increases in Caucasians, women, and older people (Sözen et al., 2017). Traditionally, the disorder is being classified into primary and secondary osteoporosis. Primary osteoporosis is usually accompanied with aging and diminished gonadal function, such as low level of estrogen (Lau & Guo, 2011). Whilst, secondary osteoporosis occurs due to a variety of pathological factors, such as type-1 diabetes (T1D), smoking, parathyroid disease, arthritis, IBD, and glucocorticoid therapy. Growing evidence suggests that prebiotics may improve mineral absorption and skeletal health via changes in GM composition, SCFAs synthesis, varying intestinal pH, modification of biomarker and regulation of immune system (Whisner & Castillo, 2018). Gut microbes can increase bone mass and decrease osteoporosis by hindering osteoclast proliferation and differentiation, reducing bone resorption, promoting apoptosis or inducing osteoblast proliferation and maturation (Ding et al., 2020).

Several medicinal mushrooms including genera *Cordyceps/Ophiocordyceps*, *Grifola*, *Ganoderma*, *Phellinus*, *Lentinula*, *Taiwanofungus*, *Pleurotus*, *Wolfiporia*, and *Trametes* are able to enhance bone stability by influencing different phases of bone formation, mineralization or resorption (Lindequist & Haertel, 2021). Even though, research on the beneficial effects of macrofungi derived prebiotics on bone metabolism is still scarce. Recently Kerezoudi et al. (2021) reported their research in this context focusing on two mushrooms namely oyster mushroom (*P. ostreatus*) and reishi mushroom (*G. lucidum*). For that, an *in vitro* fermentation study was accomplished using fresh fecal inocula from osteopenic and healthy women and lyophilized mushrooms (enriched in β-glucans) as substrates. Data suggested fermentation of the mushroom powder increased counts of *Bifidobacterium* spp., particularly in the osteopenic group. Alongside, fermentation of oyster mushroom increased growth of *Faecalibacterium prausnitzii* accompanied by a substantial enhancement in butyrate production. Both mushrooms exhibited lower RANKL (receptor activator of nuclear factor kappa-B ligand) levels which appears to be a promising strategy to treat osteoporosis and related disorders.

12.13 HYPOCHOLESTEROLEMIC AND ANTI-DIABETIC EFFECT

Cholesterol is an indispensable element of eukaryotic cell membranes and a precursor of steroid hormones and BAs. Numerous diseases have been found to be associated with dysregulation of cholesterol metabolism namely neurodegenerative diseases, non-alcoholic hepatitis, atherosclerosis, and cardiovascular ailments and cancers (Le Roy et al., 2019). Recently, the GM has appeared as a crucial factor that impacts cholesterol metabolism. A high ratio of Firmicutes/Bacteroidetes refers to a superior capability to absorb calories from foods by GM and it has been linked to the development of obesity and type-2 diabetes. In contrast, a reduction in the ratio of Firmicutes/Bacteroidetes has been related with a healthy gut and lean body weight (Magne et al., 2020). Some prebiotics in general increase viscosity in the upper intestinal tract and reduce cecal pH decreasing lipid absorption. The hypocholesterolemic property of prebiotics is also derived from metabolic effect as the fermentation by-products such as propionate inhibits cholesterol pathways by down-regulating β-Hydroxy β-methylglutaryl-CoA (HMG-CoA) reductase (Al-Sheraji et al., 2013).

In this context, several mushrooms have shown hypo-lipidemic effect in conjugation with GM alteration (Table 12.1 and Figure 12.2). Anwar et al. (2019) administered a diet encompassing *L. edodes* to hypercholesterolemic rats and results showed prevention of weight gain, induction of the level of high-density lipoprotein-cholesterol (HDL-c) and reduction in the level of TGs, total cholesterol (TC) and low-density lipoprotein-cholesterol (LDL-c). The treatment also augmented abundance of *Clostridium*, *Bacteroides* spp., and *Butyricimonas* illustrating that administration of the mushroom could manage dyslipidemia. Increased production of total SCFAs has also been reported after inclusion of *A. auricula* fruit body to the diet of hyperlipidemic rats. The phenomenon could be due to induction of the abundance of the genera *Bacteroides* and *Paraprevotella* (Zhang et al., 2020). In addition, crude polysaccharide fractions isolated from *Cordyceps sinensis*, *Agaricus blazei*, *Grifola frondosa*, *Pleurotus eryngii* and *A. auricula* have also executed their potential to inhibit obesity and change GM in mice. Supplementation of these macromolecules effectively prevented body weight gain, induced

TABLE 12.1 Prebiotic Effect of Mushrooms

Name of Mushroom	Bioactive Compound	Experimental Model	Biological Property	References
Agaricus bisporus (FB)	Powdered fruit body	C57BL/6 mice	↑ *Helicobacter*, *Lactobacillus*, *Bacilli*, *Coprobacillus*, *Prevotella;* ↓ *Pseudoflavonifractor*, *Dorea*, *Clostridium_XI*, Peptostreptococcaceae; ↑ Cecal propionate, succinate; ↓ Levels of glucose and glycogen in liver	Tian et al. (2018)
Agaricus bisporus (FB)	Fruit body	Healthy adults	↑ Bacteroidetes; ↓ Firmicutes; ↑ Stool weight	Hess et al. (2018)
Agaricus bisporus (FB)	Powdered fruit body	Pig	↑ *Oscillibacter*, *Butyricicoccus*, *Fusicatenibacter*, *Robinsoniella*, *Eisenbergiella*; ↓ LPS-induced inflammatory response in alveolar macrophages	Solano-Aguilar et al. (2018)
Agaricus blazei (FB)	Glucan enriched acidic polysaccharide fraction	HFD induced SD rat	↑ Clostridium_sensu_stricto, Proteobacteria, Peptostreptococcaceae, Allobaculum, Erysipelotrichaceae, Clostridiaceae_1; ↓ Firmicutes, Ruminococcaceae_unclassified, Ruminococcaceae; ↓ Ratio of Firmicutes/Bacteroidetes; ↓ Body weight, TC, TG, LDL-c; ↑ HDL-c;	Li et al. (2020)

TABLE 12.1 *(Continued)*

Name of Mushroom	Bioactive Compound	Experimental Model	Biological Property	References
			↑ Expression of CYP7A1; ↓ Expression of SREBP-1C	
Auricularia auricula (NM)	Water soluble β-glucan	BALB/C mice with DSS induced colitis	↑ *Bacteroidetes*; ↓ *Ruminococcus*, *Firmicutes*, *Deferribacteres*, Actinobacteria; ameliorated colon damage, weight loss, mucosal inflammation; prevented intestinal barrier damage	Zhao et al. (2020)
Auricularia auricula (FB)	Mushroom powder	HFD induced Sprague-Dawley rats	↑ *Lactobacillus*, *Oscillibacter*, *Bacteroides*, *Paraprevotella*; ↑ Acetate, propionate, butyrate; ↓ TC, LDL-c	Zhang et al. (2020)
Auricularia auricula (FB)	Hot water extracted crude polysaccharide	HFD induced Sprague-Dawley rats	↑ *Lactobacillus*, *Oscillibacter*, *Flavonifractor*, *Clostridium IV*; ↑ Acetate, propionate, butyrate; ↓ TC, LDL-c	Zhang et al. (2020)
Auricularia auricula (NM)	Hot water extracted pure polysaccharide	CP induced immunosuppressive BALB/c mice	↑ *Bacteroides*, *Alloprevotella*, *Blautia*, *Lactobacillus*; ↑ Acetate, propionate, butyrate; ↑thymus index, spleen index; ↑ Cytokine synthesis; improved colon intestinal barrier function	Kong et al. (2020)
Auricularia auricula-judae (NM)	β-glucan	*In vitro* batch fecal fermentation	↑ *Bifidobacteria*, *Lactobacillus*; ↓ *Clostridia*; ↑ PI; ↑ Acetate, propionate, butyrate, total SCFAs	Chaikliang et al. (2015)

TABLE 12.1 *(Continued)*

Name of Mushroom	Bioactive Compound	Experimental Model	Biological Property	References
Auricularia auricula-judae (NM)	Oligo-β-D-glucan	*In vitro* batch fecal fermentation	↑ *Bifidobacteria*, *Lactobacillus*, *Bacteroides*, *Clostridia*, *Eubacteria*; ↑ Acetate	Chaikliang et al. (2015)
Cephalosporium sinensis (MY)	Water soluble pure polysaccharide (mannose: glucose: galactose) = 3:2:1, M_w 16.7 kDa	Adenine induced chronic kidney disease male Wistar rats	↑ *Clostridium coccoides* group, *Lactobacillus* group, *Bifidobacterium*; ↓ *Escherichia* subgroup; ↑ Acetate, propionate, butyrate in colon and cecum; ↓ Expression of NF-κB, GPR41; improve renal dysfunction	Zha et al. (2020)
Clitocybe gibba (FB)	Water soluble crude polysaccharides	*In vitro*	↑ *Lactobacillus acidophilus*, *Lactobacillus rhamnosus*; tolerated artificial human gastric juice	Nowak et al. (2018)
Cookeina speciosa (FB)	Insoluble branched (1→ 3), (1→6)-β-D-glucan	*In vitro* batch fecal fermentation	↑ *Bacteroides uniformis*, *Anaerostipes*, *Roseburia*; ↓ *Bacteroides fragilis*, *Bacteroides plebeius*, *Ruminococcus*, *Blautia*; ↑ Acetate, propionate, butyrate, total SCFAs	Cantu-Jungles et al. (2018)
Cookeina speciosa (FB)	Insoluble linear (1→3)-β-D-glucan	*In vitro* batch fecal fermentation	↑ *Bacteroides uniformis*, *Anaerostipes*, *Roseburia*; ↓ *Bacteroides plebeius*, *Bacteroides fragilis*, *Blautia*, *Ruminococcus*; ↑ Acetate, propionate, butyrate, total SCFA, total BCFA	Cantu-Jungles et al. (2018)
Cordyceps sinensis (FB)	Water extracted pure polysaccharide (Mw 6.486×10^4 Da)	HFD-fed C57BL/6 J mice	↓ Bacteroidetes; ↑ Acidobacteria, Actinobacteria, *Olsenella*; reduced body weight, although aggravated liver fibrosis and steatosis	Chen et al. (2020)

TABLE 12.1 *(Continued)*

Name of Mushroom	Bioactive Compound	Experimental Model	Biological Property	References
Cordyceps sinensis (MY)	Hot water extracted crude polysaccharide	CP induced immunosuppressive female BALB/c mice	↑ *Lactobacillus*, *Bifidobacterium*, *Bacteroides*; ↓ *Clostridium*, *Flexispira*; ↑ Acetate, propionate, butyrate, valerate; ↑ Expression of TLRs, NF-κB and cytokines	Ying et al. (2020)
Coriolus versicolor	Hot water extracted crude polysaccharide	HFD induced C57Bl/6J mice	↑ *Akkermansia muciniphila*; ↑ Expression of ZO-1, occludin-1; ↓ Expression of TNF-α, IL (1β, 17A, 6)	Li et al. (2019)
Dictyophora indusiata (FB)	Crude polysaccharide	HFD induced male BALB/c mice	↑ *Lactobacillus*; ↓ Gammaproteobacteria, Bacteroidaceae, Enterobacteriaceae; ↓ Adipocyte size; ↑ Expression of claudin-1, occludin, zonula occluden; ↓ Expression of TNF-α, IL-1β, IL-6, MCP-1	Kanwal et al. (2020)
Dictyophora indusiata (FB)	Hot water extracted crude polysaccharide	Antibiotic induced intestinal dysbiotic male BALB/c mice	↑ Lactobacillaceae, Ruminococaceae; ↓ *Enterococcus*, *Bacteroides*, Proteobacteria; ↑ Body weight; ↓ Expression of IL-6, TNF-α, IL-1β; ↑ Expression of occludin, claudin-1, zonula occludens-1	Kanwal et al. (2018)
Flammulina velutipes (FB)	Hot water extracted crude polysaccharide	*In vitro* digestion and fermentation	↑ Bifidobacteriaceae, Bacteroidaceae; ↓ Lachnospiraceae, Enterococcaceae; ↑ Acetate, propionate, butyrate, valerate, total SCFAs; tolerated human gastrointestinal conditions	Su et al. (2019)

TABLE 12.1 *(Continued)*

Name of Mushroom	Bioactive Compound	Experimental Model	Biological Property	References
Flammulina velutipes (FB)	Hot water extracted crude polysaccharide	Male ICR mice	↓ Firmicutes/Bacteroidetes ratio, Lachnospiraceae, Lactobacillaceae; ↑ Bacteroidaceae, Porphyromonadaceae; ↑ Total SCFAs, acetate, propionate, *n*-butyrate	Zhao et al. (2019)
Flammulina velutipes (FB)	β-glucan enriched crude polysaccharide	DSS-induced colitis C57BL/6J male mice	↓ Ratio of Firmicutes/Bacteroidetes; ↑ Porphyromonadaceae; ↑ Total SCFAs, acetate, propionate, and *n*-butyrate	Zhao et al. (2020)
Flammulina velutipes (FB)	Water soluble crude polysaccharide	Male C57BL/6 mice with scopolamine induced learning and memory damages	↑ Bacteroidia, Erysipelotrichia, Actinobacteria; ↓ Clostridia, Bacilli; ↓ Expression of pro-inflammatory cytokines; prevented learning and memory impairment	Su et al. (2018a)
Flammulina velutipes	Water soluble pure polysaccharide (1→3)-linked-β-D-Gal, (1→6)-linked-β-D-Gal, (1→6)-linked-α-D-Glc and (1→3,6)-linked-α-D-Man, M_w 18.3 kD	Sprague Dawley rats	↑ Lachnospiraceae, Ruminococcaceae; ↓ Ratio of Firmicutes/Bacteroidetes; ↑ Isobutyrate, butyrate	Ye et al. (2020)
Flammulina velutipes (FB)	Hot water extracted polysaccharide	Male C57BL/6 mice	↑ Firmicutes; ↓ Bacteroidetes; improved intestine structure	Hao et al. (2021)

TABLE 12.1 *(Continued)*

Name of Mushroom	Bioactive Compound	Experimental Model	Biological Property	References
Flammulina velutipes (Base)	Hot water extracted crude polysaccharide	*In vitro*	↑ Survival rate of *Bifidobacterium longum* subsp. *Longum*, *Lactobacillus acidophilus*, *L. casei* during cold storage; protected probiotics in simulated gastric and bile juice conditions	Chou et al. (2013)
Ganoderma atrum (FB)	Hot water extracted pure polysaccharide (Mw 198 kDa)	*In vitro* and male Kunming mice	↑ Total SCFAs, acetate, propionate, butyrate *in vitro*; ↑ Total SCFAs *in vivo*	Ding et al. (2017)
Ganoderma atrum (FB)	Pure polysaccharide (glucose: mannose: galactose: galacturonic acid = 4.91: 1: 1.28: 0.71; Mw 1,013 kDa)	HFD induced male Wistar rats	↑ Acetate, propionate, butyrate; ↓ ALT, AST; ↑ Expression of PPAR-γ, GLUT4, PI3K, p-Akt	Zhu et al. (2016)
Ganoderma lucidum (spore)	Three oligosaccharide fractions	*In vitro* fermentation	↑ Lactobacillaceae, Bifidobacteriaceae, Ruminococcaceae; ↓ Enterobacteriaceae, Lachnospiraceae	Yang et al. (2020)
Ganoderma lucidum (FB)	Hot water extracted crude polysaccharide	C57BL/6J mice	↑ *Eubacterium rectale*, *Bifidobacterium choerinum*, *Lactococcus lactis*, *Lactobacillus johnsonii*; ↓ *Barnesiella*, *Tannerella*, *Pseudobutyrivibrio*, *Dehalobacterium*	Khan et al. (2018)
Ganoderma lucidum (MY)	Hot water extracted crude polysaccharide	C57BL/6J mice	↑ Eubacteriaceae, Sutterellaceae	Khan et al. (2018)
Ganoderma lucidum (Sporoderm)	Hot water extracted crude polysaccharide	4T1-breast cancer, xenograft mice	↑ *Firmicutes*, *Proteobacteria*; ↓ Actinobacteria, *Bacteroidetes*;	Su et al. (2018b)

TABLE 12.1 *(Continued)*

Name of Mushroom	Bioactive Compound	Experimental Model	Biological Property	References
			↑ Cytotoxic T-cell population, ratio of T-cell to helper T-cell in peripheral blood; ↓ Two immune checkpoints, programmed cell death protein-1, cytotoxic T-lymphocyte antigen-4	
Ganoderma lucidum (NM)	β-glucan enriched polysaccharide	DSS-induced colitis in male Wistar rats	↑ *Ruminococcus_1*, *Pasteurella*, *Fusicatenibacter*, *Firmicutes*, *Lachnospiraceae*_UCG-006, *Enterorhabdus*, *Marvinbryantia*, *Erysipelatoclostridium*, *Ruminococcaceae*_UCG-008, and *Anaerofilum*, etc.; ↓ *Escherichia-Shigella*, *Proteobacteria*, *Anaerotruncus*, *Barnesiella*, *Tyzzerella*; ↑ Total SCFA, acetate, propionate, and butyrate	Xie et al. (2019)
Ganoderma lucidum (FB)	Polysaccharides	4T1 breast cancer-bearing BALB/C mice	↑ *Bacteroidetes*, *Desulfovibrio*, *Alistipes*, *Prevotellaceae*_UCG-001; ↓ *Firmicutes*	Li et al. (2018)
Ganoderma sinense (FB)	Polysaccharides	4T1 breast cancer-bearing BALB/C mice	↑ *Bacteroidetes*, *Alistipes*, *Prevotellaceae*_UCG-001; ↓ *Firmicutes*	Li et al. (2018)
Grifola frondosa (FB)	95% ethanol extract (flavone, fatty acid)	HFD induced male Wistar rats	↑ *Intestinimonas*, *Butyricimonas*, total SCFAs; ↓ TG, TC, FFA, LDL-c; ↑ HDL-c, total bile acid, GSH-Px, T-SOD; ↑ Expression of AMPK-α, CYP7A1; ↓ Expression of FAS, SREBP-1c, IL-1β, ACC	Pan et al. (2018)

TABLE 12.1 *(Continued)*

Name of Mushroom	Bioactive Compound	Experimental Model	Biological Property	References
Grifola frondosa (FB)	Water extracted crude polysaccharide	HFD induced male Wistar rats	↑ *Barnesiella*, *Helicobater*, *Intestinimonas*, *Ruminococcus*, *Parasutterella*, *Flavonifracter*; ↓ *Clostridium*-XVIII, *Butyricicoccus*, and *Turicibacter*; ↓ TG, TC, free fatty acids, suppressed hepatic lipid accumulation and steatosis; ↑ Expression of CYP7A1, BSEP	Li et al. (2019)
Hericium erinaceus (FB)	Single band protein (M_w 50–55 kDa)	Rats with IBD induced by TNBS	↑ *Bifidobacterium*; ↓ *Bacteroides vulgatus*, *Desulfovibrio*; ↑ Foxp3, IL-10; ↓ TNF-α, NF-κB	Diling et al. (2017a)
Hericium erinaceus (FB)	Alcohol extract	Rats with IBD induced by TNBS	↑ *Bacteroides*, *Bifidobacterium*, *Prevotella*, *Parabacteroides*, *Coprococcus*, *Desulfovibrio*, Lactobacillus; ↓ *Corynebacterium*, *Staphylococcus*, *Ruminococcus*, *Roseburia*, *Dorea*, *Sutterella*; ↓ IL(1α, 8, 12), VEGF, MPO, MIP-α, TNF-α, NF-κB p65; ↑ Foxp3, IL-2, IL-10, IL-11, TNF-γ	Diling et al. (2017b)
Hericium erinaceus (FB)	Gastric and intestinal juice treated crude polysaccharide	*In vitro* fermentation and digestion	↑ *Lactobacillus rhamnosus*, *L. plantarum*, *L. acidophilus*, *L. paracasei*, *Bifidobacterium*, *Streptococcus thermophilus*; ↑ Acetate, isovalerate, lactate, butyrate	Yang et al. (2018)
Hygrophoropsis aurantiaca (FB)	Water soluble crude polysaccharides	*In vitro*	↑ *Lactobacillus acidophilus*, *L. rhamnosus*; tolerated artificial human gastric juice	Nowak et al. (2018)

TABLE 12.1 *(Continued)*

Name of Mushroom	Bioactive Compound	Experimental Model	Biological Property	References
Lactarius aurantiacus (FB)	Water soluble crude polysaccharides	*In vitro*	↑ *Lactobacillus acidophilus*, *L. rhamnosus*; tolerated artificial human gastric juice	Nowak et al. (2018)
Lentinula edodes (Stipe)	Hot water extracted crude polysaccharide	*In vitro*	↑ Survival rate of *Lactobacillus acidophilus*, *Bifidobacterium longum* subsp. *Longum*, *L. casei* during cold storage; protected probiotics in simulated gastric and bile juice conditions	Chou et al. (2013)
Lentinula edodes (FB)	Lentinan (β-(1,3)-glucan with β-(1,6) branches)	Weaned piglets with LPS induced intestinal injury	↑ *Bacteroidetes*, *Faecalibacterium*, *norank_f__ Ruminococcaceae*, *Prevotella_9*; ↓ Firmicutes, *Actinobacillus*, *Coprococcus_3*, *Oscillospira*, *Sutterella*, *Phascolarctobacterium*; ↑ Propionate, butyrate, isobutyrate, and isovalerate; ↑ Expression of claudin-1; ↓ Expression of jejunal MyD88, TLR4, IRAK1, NOD1, TRAF6, NF-κB, RIP2, IL-1β, IL-6; ↓ Ileal TRAF6, NOD1; ↑ Ileal HSP70	Wang et al. (2019)
Lentinus edodes (FB)	Powdered fruit body	Diet induced hyper-cholesterolemic Wistar rats	↑ *Bacteroides*, *Clostridium*, *Dorea*, *Eubacterium*, *Lachnoclostridium*, *Methanobrevibacter*, *Paludibacter*, *Ruminococcus*; ↓ *Lactobacillus*, *Bifidobacterium*; ↑ HDL-c; ↓ LDL-c, TC, TG	Anwar et al. (2019)

TABLE 12.1 *(Continued)*

Name of Mushroom	Bioactive Compound	Experimental Model	Biological Property	References
Macrolepiota procera (FB)	Water soluble crude polysaccharides	*In vitro*	↑ *Lactobacillus acidophilus*, *L. rhamnosus*; tolerated artificial human gastric juice	Nowak et al. (2018)
Piptoporus betulinus (FB)	Water soluble crude polysaccharides	*In vitro*	↑ *Lactobacillus acidophilus*, *L. rhamnosus*; tolerated artificial human gastric juice	Nowak et al. (2018)
Pleurotus eryngii (FB)	Crude polysaccharide	HFD induced obese C57BL/6J mice	↑ *Parabacteroides*, *Lactococcus*; ↓ *Roseburia*; ↑ Expression of SREBP2, LDLR (in liver), GPR43 (in fat)	Nakahara et al. (2020)
Pleurotus eryngii (FB)	β-glucan enriched polysaccharide (M_w 426 kDa)	C57BL/6 male mice	↑ Porphyromonadaceae, Rikenellaceae, Bacteroidaceae, Lactobacillaceae; ↑ Acetate, propionate, *i*-butyrate, *n*-butyrate, *i*-valerate; ↑ Expression of TNF-α, IFN-γ, IL-(1, 2, 6), SIgA	Ma et al. (2017)
Pleurotus eryngii (Base)	Hot water extracted crude polysaccharide	*In vitro*	↑ Survival rate of *Lactobacillus acidophilus*, *Bifidobacterium longum* subsp. *Longum*, *L. casei* during cold storage; protected probiotics in simulated gastric and bile juice conditions	Chou et al. (2013)
Pleurotus eryngii	Water and alkali fractions	*In vitro* fermentation	↑ *Lactobacillus* ssp., *Bifidobacterium* ssp., *Enterococcus faecium*; ↑ SCFA	Synytsya et al. (2009)
Pleurotus ostreatus	Water and alkali fractions	*In vitro* fermentation	↑ *Lactobacillus* ssp., *Bifidobacterium* ssp., *Enterococcus faecium*; ↑ SCFA	Synytsya et al. (2009)

TABLE 12.1 *(Continued)*

Name of Mushroom	Bioactive Compound	Experimental Model	Biological Property	References
Pleurotus sajor-caju (MY)	NM	Zucker rats	↑ *Faecalibaculum*, *Bifidobacterium*, *Roseburia*, *Blautia*; ↓ *Escherichia – Shigella*; ↑ Butyrate, secondary BAs	Maheswari et al. (2021)
Poria cocos (FB)	Hot water extracted crude polysaccharide	C57BL/6J mice	↑ *Akkermansia*, *Lactococcus*; ↓ *Barnesiella*, *Pseudobutyrivibrio*, *Dehalobacterium*	Khan et al. (2018)
Poria cocos (MY)	Hot water extracted crude polysaccharide	C57BL/6J mice	↑ Eubacteriaceae; ↓ Sutterellaceae, ratio of Firmicutes/Bacteroidetes	Khan et al. (2018)
Poria cocos (Sclerotium)	Water insoluble (1-3)-β-D-glucan (M_w 4.486×10^6 Da)	*ob/ob* mice	↑ *Lachnospiracea*, *Clostridium XIVa*, *Clostridium IV*; ↑ Butyrate; ↓ TG, TC, LDL-c, ALT, AST; ↑ Expression of Muc-5, ZO-1, Occludin; improved gut mucosal integrity though intestinal PPAR-γ pathway	Sun et al. (2019)
Pseudoclitocybe cyanthiformis (FB)	Water soluble crude polysaccharides	*In vitro*	↑ *Lactobacillus acidophilus*, *L. rhamnosus*; tolerated artificial human gastric juice	Nowak et al. (2018)
Psilocybe capnoides (FB)	Water soluble crude polysaccharides	*In vitro*	↑ *Lactobacillus acidophilus*, *L. rhamnosus*; tolerated artificial human gastric juice	Nowak et al. (2018)
Psilocybe fascicularis (FB)	Water soluble crude polysaccharides	*In vitro*	↑ *Lactobacillus acidophilus*, *L. rhamnosus*; tolerated artificial human gastric juice	Nowak et al. (2018)
Schizophylum commune (NM)	β-glucan	*In vitro* fecal fermentation	↓ *Bifidobacteria*, *Lactobacillus*, *Bacteroides*, *Clostridia*; ↑ Acetate, butyrate, total SCFA	Chaikliang et al. (2015)

TABLE 12.1 *(Continued)*

Name of Mushroom	Bioactive Compound	Experimental Model	Biological Property	References
Schizophylum sp.	β-glucan	LFD induced gut dysbiosis C57BL/6J male mice	↑ *Lactobacillus*, *Anaerostipes*	Muthura-malingam et al. (2019)
Sparassis crispa (FB)	Water soluble crude polysaccharides	*In vitro*	↑ *Lactobacillus acidophilus*, *L. rhamnosus*; tolerated artificial human gastric juice	Nowak et al. (2018)
Suillus bovinus (FB)	Water soluble crude polysaccharides	*In vitro*	↑ *Lactobacillus acidophilus*, *L. rhamnosus*; tolerated artificial human gastric juice	Nowak et al. (2018)
Trametes versicolor (NM)	Polysaccharide peptide	*In vitro*	↑ *Bifidobacterium* spp., *Lactobacillus* spp.; ↓ *Clostridium* spp., *Staphylococcus* spp., *Enterococcus* spp.; ↑ Acetate, *n*-butyrate, propionate	Yu et al. (2013)

Abbreviations: ACC: Acetyl-CoA carboxylase; AMPK-α: adenosine-activated protein kinase-α; BCFA: branched chain fatty acids; BSEP: bile salt export pump; CP: cyclophosphamide; CYP7A1: cholesterol 7-alpha hydroxylase; DSS: dextran sulfate sodium; FAS: fatty acid synthase; FB: fruit body; FFA: free fatty acid; HFD: high fat diet; GRP: G protein-coupled receptor; GSH-Px: glutathione peroxidase; GLUT: glucose transporter; HDL-c: High-density lipoprotein cholesterol; IBD: inflammatory bowel disease; IL: interleukin; LDL-c: low-density lipoprotein cholesterol; LDLR: low density lipoprotein receptor; LPS: lipopolysaccharide; MY: mycelia; NM: not mentioned; p-Akt: phosphorylated-Akt; PI: prebiotic index; PI3K: phosphoinositide 3-kinase; PPAR-γ: peroxisome proliferator-activated receptor-γ; SCFA: short chain fatty acids; SD: Sprague Dawley; SREBP-1c: sterol regulatory element-binding transcription factor-1c; TC: total cholesterol; TG: triglycerides; TNBS: trinitrobenzene sulfonic acid; T-SOD: total superoxide dismutase; ZO-1: zonula occludens-1.

the level of HDL-c and reduced the levels of serum TG, TC, LDL-c, and fat accumulation (Chen et al., 2020; Zhang et al., 2020; Nakahara et al., 2020; Li et al., 2019, 2020). The outcome was in according to Pan et al. (2018) describing the lipid-lowering effect of a flavon and fatty acid enriched ethanol extract from *G. frondosa.* A similar outcome has also been demonstrated by Li et al. (2019) investigating on possible regulatory mechanism of crude polysaccharidic extract isolated from *G. frondosa* on lipid metabolism in high fat diet (HFD) induced hyperlipidemic rats. The outcome portrayed that ingestion of the fraction diminished free fatty acids (FFA), lipid droplets, fatty degeneration in hepatocytes, levels of malondialdehyde (MDA) of serum as well as liver, aspartate aminotransferase (AST) and alanine aminotransferase (ALT) as well illustrating beneficial effects on the treatment of dyslipidemia. The treatment further stimulated expressions of cholesterol 7α-hydroxylase (CYP7A1) (rate-limiting enzyme for transformation of cholesterol into BA in liver) and bile salt export pump (BSEP) increasing the excretion of fecal BAs. Enhanced expression of CYPA1 together with decreased synthesis of SREBP-1C (the key transcription factor of TG metabolism) have also been reported in the case of *A. blazei.* Additionally, ingestion of the isolated polysaccharides alleviated disordered hepatic lobules and steatosis of hepatocytes in hyperlipidemic rats as well (Li et al., 2020). Sun et al. (2019) reported that alkali soluble polysaccharides from *P. cocos* increased abundance of butyrate-producing bacteria and improved glucose tolerance as well as insulin resistance in *ob/ob* mice. Moreover, reduced the levels of TG, TC, ALT, and AST together with increased concentration of superoxide dismutase (SOD) in obese mice were also recorded. Further, the treatment boosted expression of peroxisome proliferator-activated receptor-γ (PPAR-γ) mucosal integrity proteins (Muc-5) and tight junction proteins in the ileum indicating ability of the fraction in prevention and cure of metabolic diseases. A similar trend of beneficial activity has also been reported in the case of polysaccharides isolated from *Ganoderma atrum* on liver function in type-2 diabetic rats (Zhu et al., 2016). Kanwal et al. (2020) unraveled the anti-obesity effect of *D. indusiata* polysaccharide using HFD-induced mouse model. Findings suggested that the macromolecules were able to change the obesity parameters and inflammatory cascades by reducing the level of pro-inflammatory cytokines and boosting anti-inflammatory mediators. Moreover, the treatment modified the GM community by lowering the Firmicutes to Bacteroidetes ratio modulating intestinal integrity. Li et al. (2019b) highlighted that β-glucan enriched crude polysaccharide fraction from *Coriolus versicolor* can reduce obesity and metabolic inflammation in mice fed with HFD as evident by suppressed expression of pro-inflammatory cytokines. The administration increased expression of the tight junction components namely ZO-1 and occludin-1. Analysis of GM revealed that the fraction markedly elevated the abundance of *A. muciniphila*. Maheswari et al. (2021) observed that supplementation of diets with 5% biotechnologically produced *P. sajor-caju* mycelium altered bacterial population in the colon of lean and obese Zucker rats. As a result, enhanced generation of protective SCFAs, altered level of plasma BAs, diminished liver inflammation and hepatic steatosis were also noted supporting prebiotic effect of the macrofungus.

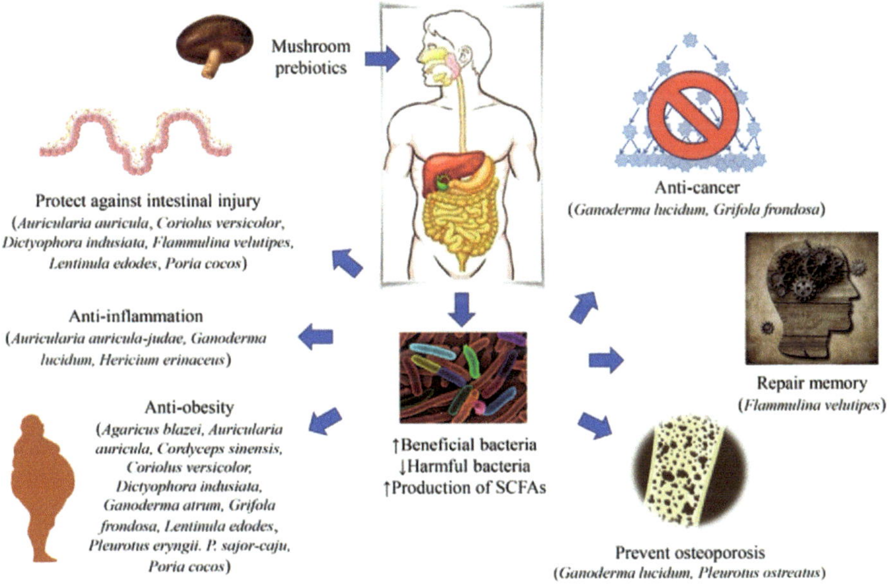

FIGURE 12.2 General scheme of the benefits of mushroom derived prebiotics.

Source: Ma et al. (2021); Davani-Davari et al. (2019); Khangwal & Shukla (2019).

12.14 MUSHROOMS IN SYMBIOTIC FORMULATIONS

Literature survey revealed that mushrooms could act as synbiotics as well, although limited studies have been performed to date (Table 12.2). In a study, rats with hypercholesterolemia were supplemented with probiotics and *A. bisporus* fruit body powder mixture or the probiotics and prebiotics separately to examine their effects on dyslipidemia. Results showed that the treatment decreased TC, LDL-c, TG, total oxidant status (TOS) and increase in HDL-c and total antioxidant status (TAC) where administration of the mushroom alone exhibited the best outcome. In respect to GM, the prebiotic

TABLE 12.2 Synbiotic Properties of Mushroom

Prebiotic Source	Probiotic Source	Experimental Model	Biological Effect	References
Agaricus bisporus (FB)	Commercial probiotic mixture composed of 14 live bacterial strains	Atherogenic diet induced hypercholesteremic Wistar rats	↑ *Ruminococcus*, *Methanobrevibacter*, *Clostridium*, *Dorea*, *Lactobacillus*, *Eubacterium*, *Lachnoclostridium*; ↓ *Bifidobacterium*, *Collinsella*, *Blautia*; ↑ HDL-c, TAC; ↓ LDL-c, TG, TOS	Asad et al. (2020)
Auricularia auricula (FB) (water extract)	Yogurt	*In vitro*	↑ *Lactobacillus acidophilus*, *Bifidobacterium bifidum*; ↑ Phenolic content, antioxidant activity	Faraki et al. (2020)
Pluerotus ostreatus (FB) powder	Yogurt	*In vitro*	↑ Viability of lactic acid bacteria; ↓ pH level	Tupamahu & Budiarso (2017)
Schizophyllum commune (NM) β-(1,3/1,6)-glucan	Concoction made of eight different bacterial strains	CR mice	↑ *Bifidobacterium*, *Streptococcus*, *Lactobacillus*, Porphyromonadaceae; ↓ Lachnospiraceae, Porphyromonadaceae	Singh et al. (2021)

Abbreviations: FB: Fruit body; HDL-c: high-density lipoprotein cholesterol; LDL-c: low-density lipoprotein cholesterol; NM: not mentioned; TAC: total antioxidant status; TG: triglycerides; TOS: total oxidant status.

supplementation increased *Ruminococcus*, *Blautia*, *Methanobrevibacter*, *Lachnoclostridium*, and *Eubacterium* in comparison to that of hypercholesterolemic control group. The researchers suggested that supplementation of the mushroom and probiotics can reduce oxidative stress and dyslipidemia with partial effects on phylogenetic makeup in the GM (Asad et al., 2020). Faraki et al. (2020) isolated an aqueous extract from *A. auricula* and mixed with yogurt in the presence or absence of probiotics to determine the effect of the mixture on the bacteria. Results showed that the mushroom at the level of 0.1% enhanced survival of *L. acidophilus* and *B. bifidum*. The addition of the species to yogurt also increased total phenolic content and antioxidant activity, indicating improved functional properties. However, inclusion of *A. auricula* to the yogurt decreased sensorial acceptance while increased syneresis. In contrast, a separate study showed that addition of 1.5% of oyster mushroom in yogurt improved the quality as evident by the proliferation of lactic acid bacteria, increased lactic acid concentration, reduced acidity. However, the addition of 1% powder produced the most favored yogurt in terms of color, aroma, flavor, and texture (Tupamahu & Budiarso, 2017). Sawangwan et al. (2018) selected seven edible mushrooms (*A. auricula-judae*, *L. edodes*, *Pleurotus citrinopileatus*, *Pleurotus djamor*, *P. ostreatus* and *Pleurotus pulmonarius*) to isolate aqueous-alcohol fractions. After HPLC analysis, the chromatograms showed galactose as the predominant component in all extracts followed by lactulose. Results of bioactivity assays showed that all the investigated fractions, specifically isolated from *L. edodes* and *P. pulmonarius*, stimulated the growth of *L. acidophilus* and *L. plantarum*. Interestingly, the cultivation of *L. acidophilus* using extracts executed a superior inhibitory effect on *Samonella paratyphi*, *Bacillus cereus*, and *Stapphylococcus aureus*. Conversely, *L. plantarum* culture with the extract inhibited *B. cereus* more efficiently. These cultured probiotics also showed GI tolerance suggesting that the fractions could act as efficient substrates for the growth of probiotics under study. In a separate study, Srisuk & Jirasatid (2020) revealed that the fruit body of *D. indusiata* has the capability to be used as a matrix for co-encapsulation of probiotic cells such as *L. acidophilus* against simulated GI condition.

12.15 CONCLUSION AND FUTURE PERSPECTIVE

Mushroom derived prebiotics possess a significant effect on the human health and great possibilities for incorporation into a vast range of common foodstuffs. Their role is mainly played by fermentable carbohydrates, β-glucans in particular, which preferentially stimulate proliferation of probiotic bacteria (bifidobacteria and lactic acid bacteria) and inhibit pernicious bacterial growth with enhancement of the gastrointestinal and immune systems. In addition, macrofungal prebiotics have been shown to influence blood glucose levels, improve plasma lipids and increase absorption of calcium. Also noteworthy is the synbiotic formulation, a combination of prebiotics and probiotics, can be established to improve the therapeutic effect. Further, the ingredients should be utilized for preparing different food products possessing better sensorial and technological features. Thus, more studies of these formulations should be accomplished in the future to provide deep insights and for that long-term clinical trials need to be conducted to confirm the health benefits in humans.

KEYWORDS

- **dysbiosis**
- **human gut microbiota**
- **immune function**
- **lipid metabolism**
- **mechanism of action**
- **mushroom metabolites**
- **synbiotics**

REFERENCES

Aida, F. M. N. A., Shuhaimi, M., Yazid, M., & Maaruf, A. G., (2009). Mushroom as a potential source of prebiotics: A review. *Trends Food Sci. Technol., 20*, 567–575.

Al-Sheraji, S. H., Ismail, A., Manap, M. Y., Mustafa, S., Yusof, R. M., & Hassan, F. A., (2013). Prebiotics as functional foods: A review. *J. Funct. Foods, 5*, 1542–1553. doi: 10.1016/j.jff.2013.08.009.

Anwar, H., Suchodolski, J., Ullah, M. I., Hussain, G., Shabbir, M. Z., Mustafa, I., & Sohail, M. U., (2019). Shiitake culinary-medicinal mushroom, *Lentinus edodes* (*Agaricomycetes*), supplementation alters gut microbiome and corrects dyslipidemia in rats. *Int. J. Med. Mushrooms, 21*, 79–88. doi: 10.1615/IntJMedMushrooms.2018029348.

Asad, F., Anwar, H., Yassine, H. M., Ullah, M. I., Rahman, A., Kamran, Z., & Sohail, M. U., (2020). White button mushroom, *Agaricus bisporus* (*Agaricomycetes*), and a probiotics mixture supplementation correct dyslipidemia without influencing the colon microbiome profile in hypercholesterolemic rats. *Int. J. Med. Mushrooms., 22*(3), 235–244. doi: 10.1615/IntJMedMushrooms.2020033807.

Avci, G. A., Ozcelik, B., & Kesepara, C., (2017). Effect of inulin and *Auricularia polytricha* extract on proliferation of *Lactobacillus rhamnosus*. *Hittite J. Sci. Eng., 4*(1), 17–21. doi: 10.17350/HJSE19030000043.

Azmi, A. F., Mustafa, S., Hashim, D. M., & Manap, Y. A., (2012). Prebiotic activity of polysaccharides extracted from *Gigantochloa levis* (Buluh beting) shoots. *Molecules, 17*(2), 1635–1651. doi: 10.3390/molecules17021635.

Bajury, D. M., Nashri, S. M., Hung, P. K. J., & Sarbini, S. R., (2018). Evaluation of potential prebiotics: A review. *Food Rev. Int., 34*(7), 639–664. doi: 10.1080/87559129.2017.1373287.

Briggs, J. A., Grondin, J. M., & Brumer, H., (2021). Communal living: Glycan utilization by the human gut microbiota. *Environ. Microbiol., 23*(1), 15–35. doi: 10.1111/1462-2920.15317.

Cantu-Jungles, T. M., Ruthes, A. C., El-Hindawy, M., Moreno, R. B., Zhang, X., Cordeiro, L. M. C., Hamaker, B. R., & Iacomini, M., (2018). *In vitro* fermentation of *Cookeina speciosa* glucans stimulates the growth of the butyrogenic *Clostridium* cluster XIVa in a targeted way. *Carbohydr. Polym., 183*, 219–229. doi: 10.1016/j.carbpol.2017.12.020.

Carlson, J. L., Erickson, J. M., Lloyd, B. B., & Slavin, J. L., (2018). Health effects and sources of prebiotic dietary fiber. *Curr. Dev. Nutr., 2*(3), nzy005. doi: 10.1093/cdn/nzy005.

Cencic, A., & Chingwaru, W., (2010). The role of functional foods, nutraceuticals, and food supplements in intestinal health. *Nutrients., 2*, 611–625. doi: 10.3390/nu2060611.

Chaikliang, C., Wichienchot, S., Youravoug, W., & Graidist, P., (2015). Evaluation on prebiotic properties of β-glucan and oligo-β-glucan from mushrooms by human fecal microbiota in fecal batch culture. *Funct. Food Health Dis., 5*(11), 395–405. doi: 10.31989/ffhd.v5i11.209.

Chatterjee, A., Khatua, S., Chatterjee, S., Mukherjee, S., Mukherjee, A., Paloi, S., Acharya, K., & Bandyopadhyay, S. K., (2013). Polysaccharide-rich fraction of *Termitomyces eurhizus* accelerate healing of indomethacin induced gastric ulcer in mice. *Glycoconj. J., 30*(8), 759–768. doi: 10.1007/s10719-013-9479-5.

Chen, L., Zhang, L., Wang, W., Qiu, W., Liu, L., Ning, A., Cao, J., et al., (2020). Polysaccharides isolated from *Cordyceps Sinensis* contribute to the progression of NASH by modifying the gut microbiota in mice fed a high-fat diet. *PLoS One, 15*(6), e0232972. doi: 10.1371/journal.pone.0232972.

Chou, W. T., Sheih, I. C., & Fang, T. J., (2013). The applications of a polysaccharides from various mushroom wastes as prebiotics in different systems. *J. Food Sci., 78*, 1041–1048. doi: 10.1111/1750-3841.12160.

Compston, J. E., McClung, M. R., & Leslie, W. D., (2019). Osteoporosis. *Lancet, 393*(10169), 364–376. doi: 10.1016/S0140-6736(18)32112-3.

Davani-Davari, D., Negahdaripour, M., Karimzadeh, I., Seifan, M., Mohkam, M., Masoumi, S. J., Berenjian, A., & Ghasemi, Y., (2019). Prebiotics: Definition, types, sources, mechanisms, and clinical applications. *Foods, 8*(3), 92. doi: 10.3390/foods8030092.

De Paulo, F. D., De Araújo, F. F., Neri-Numa, I. A., & Pastore, G. M., (2019). Prebiotics: Trends in food, health and technological applications. *Trends Food Sci Technol., 93*, 23–25. doi: 10.1016/j.tifs.2019.09.004.

Den Besten, G., Van, E. K., Groen, A. K., Venema, K., Reijngoud, D. J., & Bakker, B. M., (2013). The role of short-chain fatty acids in the interplay between diet, gut microbiota, and host energy metabolism. *J. Lipid Res., 54*(9), 2325–2340. doi: 10.1194/jlr.R036012.

Diling, C., Chaoqun, Z., Jian, Y., Jian, L., Jiyan, S., Yizhen, X., & Guoxiao, L., (2017). Immunomodulatory activities of a fungal protein extracted from *Hericium erinaceus* through regulating the gut microbiota. *Front. Immunol., 8*, 666. doi: 10.3389/fimmu.2017.00666.

Diling, C., Xin, Y., Chaoqun, Z., Jian, Y., Xiaocui, T., Jun, C., Ou, S., & Yizhen, X., (2017). Extracts from *Hericium erinaceus* relieve inflammatory bowel disease by regulating immunity and gut microbiota. *Oncotarget., 8*(49), 85838–85857. doi: 10.18632/oncotarget.20689.

Ding, K., Hua, F., & Ding, W., (2020). Gut microbiome and osteoporosis. *Aging Dis., 11*(2), 438–447. doi: 10.14336/AD.2019.0523.

Dudekula, U. T., Doriya, K., & Devarai, S. K., (2020). A critical review on submerged production of mushroom and their bioactive metabolites. *3 Biotech., 10*(8), 337. doi: 10.1007/s13205-020-02333-y.

Dwivedi, S., Sahrawat, K., Puppala, N., & Ortiz, R., (2014). Plant prebiotics and human health: Biotechnology to breed prebiotic-rich nutritious food crops. *Electron. J. Biotechnol., 17*(5), 238–245. doi: 10.1016/j.ejbt.2014.07.004.

Faraki, A., Noori, N., Gandomi, H., Banuree, S. A. H., & Rahmani, F., (2020). Effect of *Auricularia auricula* aqueous extract on survival of *Lactobacillus acidophilus* La-5 and *Bifidobacterium bifidum* Bb-12 and on sensorial and functional properties of synbiotic yogurt. *Food Sci. Nutr., 8*(2), 1254–1263. doi: 10.1002/fsn3.1414.

Florowska, A., Krygier, K., Florowski, T., & Dłużewska, E., (2016). Prebiotics as functional food ingredients preventing diet-related diseases. *Food Funct., 7*(5), 2147–2155. doi: 10.1039/c5fo01459j.

Gagliardi, A., Totino, V., Cacciotti, F., Iebba, V., Neroni, B., Bonfiglio, G., Trancassini, M., et al., (2018). Rebuilding the gut microbiota ecosystem. *Int. J. Environ. Res. Public Health, 15*(8), 1679. doi: 10.3390/ijerph15081679.

Gomes, A. C., Bueno, A. A., De Souza, R. G., & Mota, J. F., (2014). Gut microbiota, probiotics, and diabetes. *Nutr. J., 13*, 60. doi: 10.1186/1475-2891-13-60.

Gong, P., Wang, S., Liu, M., Chen, F., Yang, W., Chang, X., Liu, N., et al., (2020). Extraction methods, chemical characterizations and biological activities of mushroom polysaccharides: A mini review. *Carbohydr. Res., 494*, 108037. doi: 10.1016/j.carres.2020.108037.

Gyawali, R., Nwamaioha, N., Fiagbor, R., Zimmerman, T., Newman, R. H., & Ibrahim, S. A., (2019). The role of prebiotics in disease prevention and health promotion. In: *Dietary Interventions in Gastrointestinal Diseases* (pp. 151–167). Academic Press. doi: 10.1016/B978-0-12-814468-8.00012-0.

Hao, Y., Wang, X., Yuan, S., Wang, Y., Liao, X., Zhong, M., He, Q., et al., (2021). *Flammulina velutipes* polysaccharide improves C57BL/6 mice gut health through regulation of intestine microbial metabolic activity. *Int J Biol Macromol., 167*, 1308–1318. doi: 10.1016/j.ijbiomac.2020.11.085.

Hasan, N., & Yang, H., (2019). Factors affecting the composition of the gut microbiota, and its modulation. *PeerJ., 7*, e7502. doi: 10.7717/peerj.7502.

Hemarajata, P., & Versalovic, J., (2013). Effects of probiotics on gut microbiota: Mechanisms of intestinal immunomodulation and neuromodulation. *Therap. Adv. Gastroenterol., 6*(1), 39–51. doi: 10.1177/1756283X12459294.

Hernandez, E., & Pandiella, S.S. (2017). Development of probiotics and prebiotics. In: Teixeira, J. A., & Vicente, A. A., (eds.), Engineering Aspects of Food Biotechnology, pp. 22–49. CRC Press. doi: 10.1201/b15426.

Hess, J., Wang, Q., Gould, T., & Slavin, J., (2018). Impact of *Agaricus bisporus* mushroom consumption on gut health markers in healthy adults. *Nutrients, 10*, 1402. doi: 10.3390/nu10101402.

Jayachandran, M., Xiao, J., & Xu, B., (2017). A critical review on health promoting benefits of edible mushrooms through gut microbiota. *Int. J. Mol. Sci., 18*(9), 1934. doi: 10.3390/ijms18091934.

Kanwal, S., Aliya, S., & Xin, Y., (2020). Anti-obesity effect of *Dictyophora indusiata* mushroom polysaccharide (DIP) in high fat diet-induced obesity via regulating inflammatory cascades and intestinal microbiome. *Front. Endocrinol., 11*, 558874. doi: 10.3389/fendo.2020.558874.

Kanwal, S., Joseph, T. P., Owusu, L., Xiaomeng, R., Meiqi, L., & Yi, X., (2018). A polysaccharide isolated from *Dictyophora indusiata* promotes recovery from antibiotic-driven intestinal dysbiosis and improves gut epithelial barrier function in a mouse model. *Nutrients., 10*(8), 1003. doi: 10.3390/nu10081003.

Kapahi, M., (2018). Recent advances in cultivation of edible mushrooms. In: Singh, B., Lallawmsanga, & Passari, A., (eds.), *Biology of Macrofungi*. Springer. doi: 10.1007/978-3-030-02622-6_13.

Kerezoudi, E. N., Mitsou, E. K., Gioti, K., Terzi, E., Avgousti, I., Panagiotou, A., et al., (2021). Fermentation of *Pleurotus ostreatus* and *Ganoderma lucidum* mushrooms and their extracts by the gut microbiota of healthy and osteopenic women: Potential prebiotic effect and impact of mushroom fermentation products on human osteoblasts. *Food Funct., 12*(4), 1529–1546. doi: 10.1039/d0fo02581j.

Khan, I., Huang, G., Li, X., Leong, W., Xia, W., & Hsiao, W. L. W., (2018). Mushroom polysaccharides from *Ganoderma lucidum* and *Poria cocos* reveal prebiotic functions. *J. Funct. Foods., 41*, 191–201. doi: 10.1016/j.jff.2017.12.046.

Khangwal, I., & Shukla, P., (2019). Potential prebiotics and their transmission mechanisms: Recent approaches. *J. Food Drug Anal., 27*, 649–656.

Khatua, S., & Acharya, K., (2018). Water soluble antioxidative crude polysaccharide from *Russula senecis* elicits TLR modulated NF-κB signaling pathway and pro-inflammatory response in murine macrophages. *Front. Pharmacol., 9*, 985. doi: 10.3389/fphar.2018.00985.

Khatua, S., & Acharya, K., (2019). Alkali treated antioxidative crude polysaccharide from *Russula alatoreticula* potentiates murine macrophages by tunning TLR/NF-κB pathway. *Sci. Rep., 9*, 1713. doi: 10.1038/s41598-018-37998-2.

Khatua, S., & Acharya, K., (2021). Isolation of crude polysaccharides from *Russula senecis* (*Agaricomycetes*): Characterization, antioxidant activity and immune-enhancing property. *Int. J. Med. Mushrooms, 23*(1), 47–57. doi: 10.1615/IntJMedMushrooms.2020037158.

Khatua, S., Chandra, S., & Acharya, K., (2019). Expanding knowledge on *Russula alatoreticula*, a novel mushroom from tribal cuisine, with chemical and pharmaceutical relevance. *Cytotechnology, 71*(1), 245–259. doi: 10.1007/s10616-018-0280-y.

Khatua, S., Paloi, S., & Acharya, K., (2021). An untold story of a new myco-resource from tribal cuisine: An ethno-medicinal, taxonomic, antioxidant and immune-potentiating approach. *Food Funct., 12*, 4679–4695. doi: 10.1039/D1FO00533B.

Kong, X., Duan, W., Li, D., Tang, X., & Duan, Z., (2020). Effects of polysaccharides from *Auricularia auricula* on the immuno-stimulatory activity and gut microbiota in immunosuppressed mice induced by cyclophosphamide. *Front. Immunol., 11*, 595700. doi: 10.3389/fimmu.2020.595700.

Lau, R. Y., & Guo, X., (2011). A review on current osteoporosis research: With special focus on disuse bone loss. *J. Osteoporos., 2011*, 293808. doi: 10.4061/2011/293808.

Lauzon, H. L., Dimitroglou, A., Merrifield, D. L., Ringø, E., & Davies, S. J., (2014). Probiotics and prebiotics: Concepts, definitions and history. In: Merrifield, D., & Ringø, E., (eds.), *Aquaculture Nutrition: Gut Health, Probiotics and Prebiotics* (pp. 169–184). Wiley-Blackwell Publishing, Oxford, UK. doi: 10.1002/9781118897263.ch7.

Le Roy, T., Lécuyer, E., Chassaing, B., Rhimi, M., Lhomme, M., Boudebbouze, S., Ichou, F., et al., (2019). The intestinal microbiota regulates host cholesterol homeostasis. *BMC Biol., 17*(1), 94. doi: 10.1186/s12915-019-0715-8.

Lerner, A., Shoenfeld, Y., & Matthias, T., (2019). Probiotics: If it does not help it does not do any harm. really? *Microorganisms, 7*, 104. doi: 10.3390/microorganisms7040104.

Li, L. F., Liu, H. B., Zhang, Q. W., Li, Z. P., Wong, T. L., Fung, H. Y., Zhang, J. X., et al., (2018). Comprehensive comparison of polysaccharides from *Ganoderma lucidum* and *G. sinense*: Chemical, antitumor, immunomodulating and gut-microbiota modulatory properties. *Sci. Rep., 8*(1), 1–12. doi: 10.1038/s41598-018-22885-7.

Li, L., Guo, W. L., Zhang, W., Xu, J. X., Qian, M., Bai, W. D., Zhang, Y. Y., et al., (2019). *Grifola frondosa* polysaccharides ameliorate lipid metabolic disorders and gut microbiota dysbiosis in high-fat diet fed rats. *Food Funct., 10*(5), 2560–2572. doi: 10.1039/c9fo00075e.

Li, X., Chen, P., Zhang, P., Chang, Y., Cui, M., & Duan, J., (2019). Protein-bound β-glucan from *Coriolus versicolor* has potential for use against obesity. *Mol. Nutr. Food Res., 63*(7), 1801231. doi: 10.1002/mnfr.201801231.

Li, Y., Lu, X., Li, X., Guo, X., Sheng, Y., Li, Y., Xu, G., et al., (2020). Effects of *Agaricus blazei* Murrill polysaccharides on hyperlipidemic rats by regulation of intestinal microflora. *Food Sci. Nutr., 8*(6), 2758–2772. doi: 10.1002/fsn3.1568.

Lindequist, U., & Haertel, B., (2021). Medicinal mushrooms for prevention and therapy of osteoporosis (review). *Int. J. Med. Mushrooms., 23*(4), 13–22. doi: 10.1615/IntJMed Mushrooms.2021038084.

Ma, G., Du, H., Hu, Q., Yang, W., Pei, F., & Xiao, H., (2021). Health benefits of edible mushroom polysaccharides and associated gut microbiota regulation. *Crit. Rev. Food Sci. Nutr.*, 1–18. doi: 10.1080/10408398.2021.

Ma, G., Kimatu, B. M., Zhao, L., Yang, W., Pei, F., & Hu, Q., (2017). *In vivo* fermentation of a *Pleurotus eryngii* polysaccharide and its effects on fecal microbiota composition and immune response. *Food Funct., 8*(5), 1810–1821. doi: 10.1039/c7fo00341b.

Ma, G., Yang, W., Zhao, L., Pei, F., Fang, D., & Hu, Q., (2018). A critical review on the health promoting effects of mushrooms nutraceuticals. *Food Sci. Hum. Wellness, 7*(2), 125–133. doi: 10.1016/j.fshw.2018.05.002.

Magne, F., Gotteland, M., Gauthier, L., Zazueta, A., Pesoa, S., Navarrete, P., & Balamurugan, R., (2020). The firmicutes/Bacteroidetes ratio: A relevant marker of gut dysbiosis in obese patients? *Nutrients., 12*(5), 1474. doi: 10.3390/nu12051474.

Maheshwari, G., Gessner, D. K., Neuhaus, K., Most, E., Zorn, H., Eder, K., & Ringseis, R., (2021). Influence of a biotechnologically produced oyster mushroom (*Pleurotus sajor-caju*) on the gut microbiota and microbial metabolites in obese Zucker rats. *J. Agric. Food Chem., 69*, 1524−1535. doi: 10.1021/acs.jafc.0c06952.

Maity, G. N., Maity, P., Khatua, S., Acharya, K., Dalai, S., & Mondal, S., (2021). Structural features and antioxidant activity of a new galactoglucan from edible mushroom *Pleurotus djamor*. *Int. J. Biol. Macromol., 168*, 743–749. doi: 10.1016/j.ijbiomac.2020.11.131.

Mallick, B. P., & Bhawsar, H., (2018). Evaluation of prebiotic score of edible mushroom extract. *Int. J. Eng. Res. Technol., 7*(10), 122–127.

Markowiak, P., & Śliżewska, K., (2017). Effects of probiotics, prebiotics, and synbiotics on human health. *Nutrients, 9*, 1021. doi: 10.3390/nu9091021.

Meng, X., Zhang, G., Cao, H., Yu, D., Fang, X., De Vos, W. M., & Wu, H., (2020). Gut dysbacteriosis and intestinal disease: Mechanism and treatment. *J. Appl. Microbiol., 129*(4), 787–805. doi: 10.1111/jam.14661.

Mitsou, E. K., Saxami, G., Stamoulou, E., Kerezoudi, E., Terzi, E., Koutrotsios, G., Bekiaris, G., et al., (2020). Effects of rich in B-glucans edible mushrooms on aging gut microbiota characteristics: An *in vitro* study. *Molecules, 25*(12), 2806. doi: 10.3390/molecules25122806.

Muthuramalingam, K., Singh, V., Choi, C., Choi, S. I., Park, S., Kim, Y. M., Unno, T., & Cho, M., (2019). Effect of mushroom (*Schizophyllum* spp.) derived β-glucan on low-fiber diet induced gut dysbiosis. *J. Appl. Biol. Chem., 62*(2), 211–217. doi: 10.3839/jabc.2019.029.

Nakahara, D., Nan, C., Mori, K., Hanayama, M., Kikuchi, H., Hirai, S., & Egashira, Y., (2020). Effect of mushroom polysaccharides from *Pleurotus eryngii* on obesity and gut microbiota in mice fed a high-fat diet. *Eur. J. Nutr., 59*(7), 3231–3244. doi: 10.1007/s00394-019-02162-7.

Nissen, L., Casciano, F., & Gianotti, A., (2020). Intestinal fermentation *in vitro* models to study food-induced gut microbiota shift: An updated review. *FEMS Microbiol. Lett., 367*(12), fnaa097. doi: 10.1093/femsle/fnaa097.

Nowak, R., Nowacka-Jechalke, N., Juda, M., & Malm, A., (2018). The preliminary study of prebiotic potential of polish wild mushroom polysaccharides: The stimulation effect on *Lactobacillus* strains growth. *Eur. J. Nutr., 57*, 1511–1521. doi: 10.1007/s00394-017-1436-9.

Pallav, K., Dowd, S. E., Villafuerte, J., Yang, X., Kabbani, T., Hansen, J., Dennis, M., et al., (2014). Effects of polysaccharopeptide from *Trametes versicolor* and amoxicillin on the gut microbiome of healthy volunteers: A randomized clinical trial. *Gut Microbes, 5*, 458–467. doi: 10.4161/gmic.29558.

Pan, Y. Y., Zeng, F., Guo, W. L., Li, T. T., Jia, R. B., Huang, Z. R., Lv, X. C., Zhang, J., & Liu, B., (2018). Effect of *Grifola frondosa* 95% ethanol extract on lipid metabolism and gut microbiota composition in high-fat diet-fed rats. *Food Funct., 9*(12), 6268–6278. doi: 10.1039/c8fo01116h.

Pandey, K. R., Naik, S. R., & Vakil, B. V., (2015). Probiotics, prebiotics and synbiotics: A review. *J. Food Sci. Technol., 52*(12), 7577–7587. doi: 10.1007/s13197-015-1921–1.

Panesar, P. S., Kumari, S., & Panesar, R., (2013). Biotechnological approaches for the production of prebiotics and their potential applications. *Crit. Rev. Biotechnol., 33*(4), 345–364. doi: 10.3109/07388551.2012.709482.

Pham, V. T., Seifert, N., Richard, N., Raederstorff, D., Steinert, R. E., Prudence, K., & Mohajeri, M. H., (2018). The effects of fermentation products of prebiotic fibres on gut barrier and immune functions *in vitro*. *PeerJ, 6*, e5288. doi: 10.7717/peerj.5288.

Pujari, R., & Banerjee, G., (2021). Impact of prebiotics on immune response: From the bench to the clinic. *Immunol. Cell Biol., 99*(3), 255–273. doi: 10.1111/imcb.12409.

Rasmussen, H. E., & Hamaker, B. R., (2017). Prebiotics and inflammatory bowel disease. *Gastroenterol. Clin. North Am., 46*(4), 783–795. doi: 10.1016/j.gtc.2017.08.004.

Rinninella, E., Raoul, P., Cintoni, M., Franceschi, F., Miggiano, G. A. D., Gasbarrini, A., & Mele, M. C., (2019). What is the healthy gut microbiota composition? a changing ecosystem across age, environment, diet, and diseases. *Microorganisms, 7*(1), 14. doi: 10.3390/microorganisms7010014.

Rodrigues, D., Walton, G., Sousa, S., Rocha-Santos, T. A. P., Duarte, A. C., Freitas, A. C., & Gomes, A. M. P., (2016). *In vitro* fermentation and prebiotic potential of selected extracts from seaweeds and mushrooms. *LWT-Food Sci. Technol., 73*, 131–139. doi: 10.1016/j.lwt.2016.06.004.

Royse, D. J., Baars, J., & Tan, Q., (2017). Current overview of mushroom production in the world. In: Cunha, Z. D., & Pardo-Giménez, A., (eds.), *Edible and Medicinal Mushrooms: Technology and Applications* (pp. 5–14). John Wiley & Sons Ltd.

Ruthes, A. C., Cantu-Jungles, T. M., Cordeiro, L. M. C., & Iacomini, M., (2021). Prebiotic potential of mushroom D-glucans: Implications of physicochemical properties and structural features. *Carbohydr. Polym., 262*, 117940. doi: 10.1016/j.carbpol.2021.117940.

Sanders, M. E., Merenstein, D. J., Reid, G., Gibson, G. R., & Rastall, R. A., (2019). Probiotics and prebiotics in intestinal health and disease: From biology to the clinic. *Nat. Rev. Gastroenterol Hepatol., 16*(10), 605–616. doi: 10.1038/s41575-019-0173-3.

Sawangwan, T., Wansanit, W., Pattani, L., & Noysang, C., (2018). Study of prebiotic properties from edible mushroom extraction. *Agric. Nat. Resour., 52*, 519–524. doi: 10.1016/j.anres.2018.11.020.

Scaldaferri, F., Gerardi, V., Lopetuso, L. R., Del Zompo, F., Mangiola, F., Boškoski, I., Bruno, G., et al., (2013). Gut microbial flora, prebiotics, and probiotics in IBD: Their current usage and utility. *BioMed. Res. Int., 2013*, 435268. doi: 10.1155/2013/435268.

Singh, V., Muthuramalingam, K., Kim, Y. M., Park, S., Kim, S. H., Lee, J., Hyun, C., Unno, T., & Cho, M., (2021). Synbiotic supplementation with prebiotic *Schizophyllum commune* derived β-(1,3/1,6)-glucan and probiotic concoction benefits gut microbiota and its associated metabolic activities. *Appl. Biol. Chem., 64*, 7. doi: 10.1186/s13765-020-00572-4.

Singla, V., & Chakkaravarthi, S., (2017). Applications of prebiotics in food industry: A review. Food *Sci. Technol. Int., 23*(8), 649–667. doi: 10.1177/1082013217721769.

Slavin, J., (2013). Fiber and prebiotics: Mechanisms and health benefits. *Nutrients, 5*(4), 1417–1435. doi: 10.3390/nu5041417.

Solano-Aguilar, G. I., Jang, S., Lakshman, S., Gupta, R., Beshah, E., Sikaroodi, M., Vinyard, B., et al., (2018). The effect of dietary mushroom *Agaricus bisporus* on intestinal microbiota composition and host immunological function. *Nutrients, 10*(11), 1721. doi: 10.3390/nu10111721.

Sözen, T., Özışık, L., & Başaran, N. Ç., (2017). An overview and management of osteoporosis. *Eur. J. Rheumatol., 4*(1), 46–56. doi: 10.5152/eurjrheum.2016.048.

Srisuk, N., & Jirasatid, S., (2020). Characteristics co-encapsulation of *Lactobacillus acidophilus* with *Dictyophora indusiata. Curr. Res. Nutr. Food Sci., 8*(3), 1013–1024. doi: 10.12944/CRNFSJ.8.3.28.

Su, A. X., Ma, G. X., Xie, M. H., Ji, Y., Li, X. F., Zhao, L. Y., & Hu, Q., (2019). Characteristic of polysaccharides from *Flammulina velutipes in vitro* digestion under salivary, simulated gastric and small intestinal conditions and fermentation by human gut microbiota. *Int. J. Food Sci. Technol., 54*, 2277–2287. doi: 10.1111/ijfs.14142.

Su, A., Yang, W., Zhao, L., Pei, F., Yuan, B., Zhong, L., Ma, G., & Hu, Q., (2018). *Flammulina velutipes* polysaccharides improve scopolamine-induced learning and memory impairment in mice by modulating gut microbiota composition. *Food Funct., 9*(3), 1424–1432. doi: 10.1039/c7fo01991b.

Su, J., Su, L., Li, D., Shuai, O., Zhang, Y., Liang, H., Jiao, C., et al., (2018). Antitumor activity of extract from the sporoderm-breaking spore of *Ganoderma lucidum*: Restoration on exhausted cytotoxic T cell with gut microbiota remodeling. *Front Immunol., 9*, 1765. doi: 10.3389/fimmu.2018.01765.

Suez, J., Zmora, N., Segal, E., & Elinav, E., (2019). The pros, cons, and many unknowns of probiotics. *Nat. Med., 25*, 716–729. doi: 10.1038/s41591-019-0439-x.

Sun, S. S., Wang, K., Ma, K., Bao, L., & Liu, H. W., (2019). An insoluble polysaccharide from the sclerotium of *Poria cocos* improves hyperglycemia, hyperlipidemia and hepatic steatosis in ob/ob mice via modulation of gut microbiota. *Chin. J. Nat. Med., 17*(1), 3–14. doi: 10.1016/S1875-5364(19)30003-2.

Synytsya, A., Mícková, K., Synytsya, A., Jablonsky, I., Spevácek, J., Erban, V., Kováríková, E., & Copíková, J., (2009). Glucans from fruit bodies of cultivated mushrooms *Pleurotus ostreatus* and *Pleurotus eryngii*: Structure and potential prebiotic activity. *Carbohydr. Polym., 76*, 548–556. doi: 10.1016/j.carbpol.2008.11.021.

Tan, J., McKenzie, C., Potamitis, M., Thorburn, A. N., Mackay, C. R., & Macia, L., (2014). The role of short-chain fatty acids in health and disease. *Adv. Immunol., 121*, 91–119. doi: 10.1016/B978-0-12-800100-4.00003-9.

Teng, P. Y., & Kim, W. K., (2018). Review: Roles of prebiotics in intestinal ecosystem of broilers. *Front. Vet. Sci., 5,* 245. doi: 10.3389/fvets.2018.00245.

Thursby, E., & Juge, N., (2017). Introduction to the human gut microbiota. *Biochem. J., 474*, 1823–1836. doi: 10.1042/BCJ20160510.

Tian, Y., Nichols, R. G., Roy, P., Gui, W., Smith, P. B., Zhang, J., Lin, Y., et al., (2018). Prebiotic effects of white button mushroom (*Agaricus bisporus*) feeding on succinate and intestinal gluconeogenesis in C57BL/6 mice. *J. Funct. Foods., 45*, 223–232. doi: 10.1016/j.jff.2018.04.008.

Tupamahu, I. P. C., & Budiarso, T. Y., (2017). The effect of oyster mushroom (*Pleurotus ostreatus*) powder as prebiotic agent on yoghurt quality. *AIP Conf Proc., 1844*, 030006–1–030006–8. doi: 10.1063/1.4983433.

Venturella, G., Ferraro, V., Cirlincione, F., & Gargano, M. L., (2021). Medicinal mushrooms: Bioactive compounds, use and clinical trials. *Int. J. Mol. Sci., 22*, 634. doi: 10.3390/ijms22020634.

Wan, M. W. A., Peng, W., Nam, W. L., Yang, H., Lee, X. Y., Lee, Y. K., Liew, R. K., et al., (2020). A review on valorization of oyster mushroom and waste generated in the mushroom cultivation industry. *J. Hazard. Mater., 400*, 123156. doi: 10.1016/j.jhazmat.2020.123156.

Wang, X., Wang, W., Wang, L., Yu, C., Zhang, G., Zhu, H., Wang, C., et al., (2019). Lentinan modulates intestinal microbiota and enhances barrier integrity in a piglet model challenged with lipopolysaccharide. *Food Funct., 10*(1), 479–489. doi: 10.1039/c8fo02438c.

Whisner, C. M., & Castillo, L. F., (2018). Prebiotics, bone and mineral metabolism. *Calcif. Tissue Int., 102*, 443–479. doi: 10.1007/s00223-017-0339-3.

Xie, J., Liu, Y., Chen, B., Zhang, G., Ou, S., Luo, J., & Peng, X., (2019). *Ganoderma lucidum* polysaccharide improves rat DSS-induced colitis by altering cecal microbiota and gene expression of colonic epithelial cells. *Food Nutr. Res., 63*, 1–12. doi: 10.29219/fnr.v63.1559.

Yang, K., Zhang, Y. J., Cai, M., Guan, R. F., Neng, J., Pi, X. G., & Sun, P., (2020). *In vitro* prebiotic activities of oligosaccharides from the by-products in *Ganoderma lucidum* spore polysaccharide extraction. *RSC Adv., 10*, 14794. doi: 10.1039/C9RA10798C.

Yang, Y., Zhao, C., Diao, M., Zhong, S., Sun, M., Sun, B., Ye, H., & Zhang, T., (2018). The prebiotic activity of simulated gastric and intestinal digesta of polysaccharides from the *Hericium erinaceus*. *Molecules, 23*(12), 3158. doi: 10.3390/molecules23123158.

Ye, J., Wang, X., Wang, K., Deng, Y., Yang, Y., Ali, R., Chen, F., et al., (2020). A novel polysaccharide isolated from *Flammulina velutipes*, characterization, macrophage immunomodulatory activities and its impact on gut microbiota in rats. *J. Anim. Physiol. Anim. Nutr. (Berl)., 104*(2), 735–748. doi: 10.1111/jpn.13290.

Yin, C., Noratto, G. D., Fan, X., Chen, Z., Yao, F., Shi, D., & Gao, H., (2020). The impact of mushroom polysaccharides on gut microbiota and its beneficial effects to host: A review. *Carbohydr. Polym., 250*, 116942. doi: 10.1016/j.carbpol.2020.116942.

Yu, Z. T., Liu, B., Mukherjee, P., & Newburg, D. S., (2013). *Trametes versicolor* extract modifies human fecal microbiota composition *in vitro*. *Plant Foods Hum. Nutr., 68*, 107–112. doi: 10.1007/s11130-013-0342-4.

Zeng, H., Umar, S., Rust, B., Lazarova, D., & Bordonaro, M., (2019). Secondary bile acids and short chain fatty acids in the colon: A focus on colonic microbiome, cell proliferation, inflammation, and cancer. *Int. J. Mol. Sci., 20*(5), 1214. doi: 10.3390/ijms20051214.

Zha, Z., Zhang, Z., Wei, W., Nie, W., Chu, W., Huang, F., Yue, L., et al., (2020). Isolation, structural characterization of polysaccharide from *Cephalosporium sinensis* mycelia and its anti-nephritic effects in adenine-induced CKD rats. *Int. J. Biol. Macromol., 155*, 340–349. doi: 10.1016/j.ijbiomac.2020.03.195.

Zhang, T., Zhao, W., Xie, B., & Liu, H., (2020). Effects of *Auricularia auricula* and its polysaccharide on diet-induced hyperlipidemia rats by modulating gut microbiota. *J. Funct. Foods, 72*, 104038. doi: 10.1016/j.jff.2020.104038.

Zhang, Y. J., Li, S., Gan, R. Y., Zhou, T., Xu, D. P., & Li, H. B., (2015). Impacts of gut bacteria on human health and diseases. *Int. J. Mol. Sci., 16*(4), 7493–7519. doi: 10.3390/ijms16047493.

Zhao, D., Dai, W. J., Tao, H., Zhuang, W., Qu, M., & Chang, Y. N., (2020). Polysaccharide isolated from *Auricularia auricular-judae* (Bull.) prevents dextran sulfate sodium-induced colitis in mice through modulating the composition of the gut microbiota. *J. Food Sci., 85*(9), 2943–2951. doi: 10.1111/1750-3841.15319.

Zhao, R., Hu, Q., Ma, G., Su, A., Xie, M., Li, X., Chen, G., & Zhao, L., (2019). Effects of *Flammulina velutipes* polysaccharide on immune response and intestinal microbiota in mice. *J. Funct. Foods, 56*, 255–264. doi: 10.1016/j.jff.2019.03.031.

Zhu, K. X., Nie, S. P., Tan, L. H., Li, C., Gong, D. M., & Xie, M. Y., (2016). A polysaccharide from *Ganoderma atrum* improves liver function in type 2 diabetic rats via antioxidant action and short-chain fatty acids excretion. *J. Agric. Food Chem., 64*(9), 1938–1944. doi: 10.1021/acs.jafc.5b06103.

CHAPTER 13

Strategies in Artificial Cultivation of Two Entomopathogenic Fungi *Cordyceps militaris* and *Ophiocordyceps sinensis*

PRAKASH PRADHAN[1,2], JAYITA DE[1], and KRISHNENDU ACHARYA[1]

[1]*Molecular and Applied Mycology and Plant Pathology Laboratory, Department of Botany, University of Calcutta, Kolkata, West Bengal, India*

[2]*West Bengal Biodiversity Board, Prani Sampad Bhawan, 5th Floor, LB-2, Sector-III, Salt Lake City, Kolkata, West Bengal, India*

ABSTRACT

Mushrooms are considered a source of various bioactive compounds important to mankind. Entomopathogenic species belonging to *Cordyceps* and *Ophiocordyceps* are not only ecologically important but are also prized indispensable constituents of coveted traditional Chinese medicines. The principal difference between *Ophiocordyceps* and *Cordyceps* being the former produces whole ascospores, while *Cordyceps* produces disarticulated part spores (ascospores) and immersed/superficial perithecia perpendicular to stromal surface at maturity. The notable species among the two genera are *Cordyceps militaris* and *Ophiocordyceps sinensis*, both of which are harvested from nature, however, overharvesting of specimens, especially whose ascospores has not been released into natural habitat, as well as global warming and spatial shift in its distribution, are threating their natural population. Present investigation compiles various methods of artificial cultivation of *C. militaris* and *O. sinensis*, focusing on various cultural-environmental factors, fungal biomass production, bioactive metabolite production as well as their productivity constraints. On account of artificial production feasibility, *C.*

Applied Mycology for Agriculture and Foods: Industrial Applications. Sanjay K. Singh, Deepak Kumar, Md. Shamim, & Rohit Sharma (Eds.)

militaris readily produces fruitbodies on culture medium without dependency on host, whereas, the production of iconic *O. sinensis*-larvae complex has higher technical requirements, long production cycles and increased risk of contamination. During artificial cultivation, variation of productivity with varying strain and variations in media along with degeneration of mycelia due to sub-culturing also needs to be addressed. Further, to minimize pressure on natural sources, mycelia, mycelial extracts and fermented cultural extracts of these species need to be popularized.

13.1 INTRODUCTION

There are over 2,500 edible and medicinal mushrooms reported from wild (Nakalembe et al., 2015) and in 2017 estimated 40 million tons of edible and medicinal mushrooms were produced across the globe (Thakur, 2020). Mushrooms have received much consideration as an abundant source of active biological compounds from pharmacological and medical fraternity (Wasser, 2002). Fungi also play key role in pathogenesis of insects and around 1,200 such entomopathogenic fungi (EPF) have been recorded (Humber, 2000) and some of the well-known species are from Hypocreales (Ascomycota). Its members are mostly pathogenic to insects such as lepidopteran larvae and pupae, and also on Coleopteran, Dipteran, Hemipteran, and Hymenopteran insects and spiders (Baral, 2017). Two members of Subalpine-alpine and Temperate Hypocreales (Ascomycetes) viz. *Ophiocordyceps sinensis* (Berk.) G. H. Sung, J. M. Sung, Hywel-Jones, and Spatafora (Ophiocordycipitaceae) and *Cordyceps militaris* (L.) Link (Cordycipitaceae), respectively, are considered as indispensable constituents of coveted traditional Chinese medicines (Chioza & Ohga, 2014; Qin et al., 2018). The principal difference between *Ophiocordyceps* and *Cordyceps* being the former produces whole ascospores, while *Cordyceps* produces disarticulated into part spores (ascospores) and immersed/superficial perithecia perpendicular to stromal surface at maturity (Sung et al., 2007). Mycobank.org (accessed on 24.05.2021) has 625 listed records of subgeneric taxa under *Cordyceps* Fr., out of which 12 are illegitimate, 32 orthographic variants, 4 are invalid names, while 1 name is uncertain. Regarding *Ophiocordyceps* Petch, there are 309 records of subgeneric taxa in 'mycobank' website, among which 17 are orthographic variants (Kobayasi, 1982; Hywel-Jones, 2002; Liu et al., 2002).

Ophiocordyceps sinensis especially is a biological complex of fruiting body of parasitic fungus and larva and both are considered of medicinal value (Hu et al., 2015). Native occurrence of *O. sinensis* is confined to alpine, sub-alpine meadow and alpine shrubs habitat in high Himalayan mountains of Sichuan, Qinghai, Yunnan, Gansu, and Tibet Autonomous Region in China, Nepal, Bhutan, and India though small population have also been reported from Xinjiang, Heilongjiang, and Inner Mongolia (Sharma, 2004; Yue et al., 2013). The vertical distribution of the host of *O. sinensis* ranges in the cold and arid environment from the lowest elevation of 3,000 m to the highest elevation of 5,100 m, with an optimal elevation range from 4,000 m to 4,800 m (Sharma, 2004; Yue et al., 2013).

In China, *Ophiocordyceps sinensis* is called Dong Chong Xia Cao and the name is likely translated from the Tibetan term Yartsa Gunbu, meaning "summer grass winter worm" (Yue et al., 2013). In winter, the fungal anamorph which is mostly *Hirsutella sinensis*, parasitizes the ghost moth or Himalayan bat moth larvae *Hepialus armoricanus* Oberthür (syn. *Thitarodes armoricanus* Oberthür) [family Hepialidae; order Lepidoptera], and proliferates until the larval cadaver is filled with mycelium; during summer, the stroma arises from the exoskeleton of the dead caterpillar (Sharma, 2004; Baral, 2017; Zhou et al., 2019). Currently, more than 50 species including those of the genus *Hepialiscus*, *Forkalus*, *Bipectilus*, and *Magnificus* as well as *Hepialus oblifurcus* (reported from Kangding and Sichuan), *H. baimaensis* Liang, *H. renzhiensis* Yang, *H. yulongensis* Liang, and *H. deqinensis* Liang (reported from Yunnan) are considered as potential host species of *O. sinensis* (Yue et al., 2013). *Cordyceps militaris* (the type species of *Cordyceps*), is also parasitic and it forms attractive bright orange colored fruitbodies (Seaver, 1911). *Cordyceps militaris* has wide host range, extending to 13 families and 32 spp. of the orders Coleoptera and Lepidoptera and unlike *O. sinensis* it does not have extreme habitat requirements at the natural setting (Shrestha, 2016).

Ophiocordyceps sinensis has many biologically active constituents. Among them, Adenosine ($C_{10}H_{13}N_5O_4$), Cordycepic acid (mannitol; $C_6H_{14}O_6$), Cordycepin and polysaccharides are the major components of *C. sinensis*, besides that *C. sinensis* contains ergosterol, phenylalanine, cyclo-Gly-Pro, glycosides (cyclo-Ala-Leu-rha and phenylalanine-o-glu) and so on (Hu et al., 2015; Zhou et al., 2019). Cordycepin ($C_{10}H_{13}N_5O_3$) – a nucleoside analog (3′-deoxyadenosine) of Adenosine and a nucleoside antibiotic, and Adenosine itself as well are the biomarkers used to distinguish genus *Cordyceps* and *Ophiocordyceps* from other fungi (Chamyuang et al., 2019; Zhou et al., 2019), besides, Cordycepic acid being common to both the species. While fruiting body of *C. militaris* also contains several bioactive metabolites such as Adenosine, Inosine, Pentostatin, Polysaccharides, and Sterols (Yue et al., 2013; Cui et al., 2015). Constituents of *O. sinensis* and *C. militaris* are known to have bioactivity towards anti-tumor, anti-leukemic, anti-metastatic, anti-arrhythmic, hypoglycemic, erythropoietic, immuno-modulating, anti-fatigue, anti-aging, anti-oxidative, anti-inflammatory, anti-bacterial, anti-viral, anti-fungal, anti-protozoal, hepato-protective, nephro-protective, and also has beneficial effects on organ-transplantation, asthma, cough, and cold (Zhou et al., 2009; Yue et al., 2013; Chen et al., 2016; Nguyen et al., 2020) and as a secret medication for maintaining perpetual youthfulness and as a tonic for protecting health, post-illness, and long fatigue (Shrestha et al., 2005).

Due to the strict parasitism and the special geographical environment in which it grows, the output of natural *O. sinensis* cannot increasingly meet market demand and thus, its price is unusually high (Yue et al., 2013). In 2016, the lowest offset price for auction for the lowest grade *O. sinensis* in Sikkim, India was Rs. 5 Lakh/kg (equivalent to USD 7,143/kg @ exchange rate of 1 USD=70 INR) (FEWMD, 2016). For the higher grade in 2015, in the international market, it cost as much as 60,000 USD/kg (Huang & Ohga, 2018). In recent years, wild *O. sinensis* has been excessively excavated, which has led to serious damage to the resource of *O. sinensis*-larval host complex and the environment. *Ophiocordyceps sinensis* resources have significantly declined, rendering them unable to meet the current demand for health care and medicinal use. As a result, investigation of the artificial cultivation of *O. sinensis* appears particularly significant (Yue et al., 2013). Studies have suggested that the chemical constituents of *C. militaris* to be similar as those of *O. sinensis* (Yu et al., 2007). Comparatively the artificial cultivation of *C. militaris*, which also produces Cordycepin and has similar pharmacological activity to *O. sinensis*, is easier than *O. sinensis* (Yue et al., 2013). This review unveils practices and processes involved in artificial cultivation of both *O. sinensis* and *C. militaris*.

13.2 ARTIFICIAL CULTIVATION OF *OPHIOCORDYCEPS SINENSIS*

The process for cultivation of *O. sinensis*-larval complex involves many steps such as isolation of anamorph of fungus, artificial rearing of larvae, infection of larvae by fungus, complete artificial cultivation, semi-natural cultivation, etc. Around 15% of the products of medicinal mushrooms are extracted from the mycelium culture while the rest are derived from the fungal fruiting bodies (Lindequist, 2005). However, unlike *C. militaris*, studies have revealed that so far there is no success of development of fruitbodies of *O. sinensis* from mycelial cultivation. As commercial production of *O. sinensis* through complete artificial cultivation and semi-natural cultivation patterns has not yet been achieved (Yue et al., 2013), studies have been conducted to diminish harvesting pressure on natural *O. sinensis* resources by focusing on cultured mycelia as a substitute for nature collected *O. sinensis* (Baral, 2017). The succeeding subsections discuss general requirements and procedure in artificial culture of *O. sinensis*, rearing of larval host and culturing of *O. sinensis*-larval host complex.

13.2.1 ISOLATION OF ANAMORPH

O. sinensis can be isolated variously from ascospores, fruiting bodies, living larvae, mummified caterpillars, etc. Fresher the material is the better (Zhou et al., 2014). From China, around 30 species belonging to 13 genera, are reported as anamorphs of *O. sinensis*. Many species such as *Cephdosporium sinensis*, *Chrysosporium sinensis*, *Clonostachys rosea*, *Hirsutella hepiali*, *H. sinensis*, *Mortierella hepiali*, *Paecilomyces hepiali*, *P. sinensis*, *Scytalidium hepiali*, *Tolypocladium sinensis*, etc., have been known to be isolated from *O. sinensis*-larval host complex (Zhou et al., 2014). However, several independent molecular studies have suggested that the anamorph of *O. sinensis* to be *Hirsutella sinensis* (Yue et al., 2013; Li et al., 2019).

For fungal isolation, sterilization of fruitbody with alcohol is a must. The sterilized specimen is placed in autoclaved media of water agar and cultured at 20°C. Strains derived from ascospores are regarded as much more advantageous than those derived from fruitbodies (Zhou et al., 2014; Baral, 2017). Preparation of cultures from virulent spores are preferred (Yue et al., 2013). In this method, extremes of stroma of the sterilized fruitbodies are hanged over the surface of the slanted potato dextrose agar (PDA) and the deposited spores are germinated at 20 °C. The emitted ascospores may also be collected through aseptic bags sheathing fruiting bodies or sterilized culture slides below fruitbodies. The ascospores may be isolated by inoculating needle and they may be cultured immediately or preserved at 0–4°C (Yue et al., 2013). For isolation and culture of anamorphic state of *O. sinensis*, Jiang & Yao (2003); and Yue et al. (2013) used 1% peptone PDA and reformative Sabouraud agar (SAB). Yue et al. (2013) used S31 agar, glycerol meal peptone agar and milk agar. While Baral (2017) suggested media as simple as PDA and culture incubation period of 96 h at 28°C. Zhou et al. (2014) may be further consulted for isolation of anamorph from living larvae infected with fungi.

13.2.2 CULTIVATION OF OPHIOCORDYCEPS SINENSIS THROUGH MEDIA CULTURE

Based on the chemical and physical properties, the culturing methods utilized may be of: (i) the liquid-state culture; and (ii) the solid-state culture (Zhou et al., 2014). The primary aspects for successful culture and production of target products are the nutritional components, oxygen, relative humidity (RH) and temperature (Wang et al., 2009). During indoor culturing with light and at the temperature of 8–17°C, one of the H1 strains in S31 media, grew into fruiting bodies with mature perithecium after 7 months (Yue et al., 2013). The ideal growth temperature of *Hirsutella sinensis* is reported to be 18–20°C, and the suitable temperature for mycelial growth being >20°C, but the growth is reported to be at restrained >25°C (Zhou et al., 2014). For various strains of *O. sinensis* from the Tibetan Plateau range the growth temperature are mentioned to range from 4 to 21°C with ideal temperatures of 15–18°C on both liquid and solid culture (Dong & Yao, 2011). For better quality mycelia and faster growth, the addition of magnesium sulfate and potassium dihydrogen phosphate to the media are suggested (Zhou et al., 2014).

13.2.2.1 LIQUID-STATE CULTURE

Liquid state culture may be carried out through rotary shaker/shaker incubator and fermenter (Yan et al., 2014).

13.2.2.2 ROTARY SHAKER/SHAKE INCUBATOR

The higher yield of mycelial biomass (54 g/l) through rotary shaker is reported by Cha et al. (2007) by using media composed of $CaCl_2$ – 0.4% (w/v), K_2HPO_4 – 0.3%, sucrose 2% and yeast extract 0.9% in rotary shaker at 150 rpm for five days at 25°C. Mycelial biomass production (25 g/l) is reported by Yin et al. (2013) who used media with glucose 30 g/l, KH_2PO_4 – 3.0 g/l, $MgSO_4 \cdot 7H_2O$ 1.5 g/l, peptone 15 g/l, potato 200 g/l in 250 ml shake flask. Zhao et al. (2008) used glucose (1.25%), KH_2PO_4 (0.025%), $MgSO_4 \cdot 7H_2O$ (0.0125%), peptone (0.02%), sucrose (1.25%), yeast powder (0.0625%), vitamin B1 (0.0025%) in natural pH and cultural condition of 24°C and culture period of eight days. Shang et al. (2011) suggested media composition of beef extract (0.08%), glucose (2.5%), KH_2PO_4 (0.15%),

$MgSO_4 \cdot 7H_2O$ (0.15%), peptone (0.2%), potato (20%), sucrose (1.5%) in natural pH and cultural condition of 23°C, 130 rpm, and culturing period of four days. Both Zhao et al. (2008); and Shang et al. (2011) obtained mycelial biomass of 19.5 g/l each. Baral (2017) suggested nutrient broth containing peptone (10 g/l), glucose (30 g/l), vitamin B1 – 0.05 g/l, KH_2PO_4 – 1 g/l, $MgSO_4$ – 0.2 g/l, at pH 6.0 with shaking at 120 rpm for 96 h at 28°C. Liu et al. (2007) reported that the polysaccharide content was inhibited by the flask culture, but mycelium weight was increased during 0–16 days of shaking.

13.2.2.3 FERMENTER

The parameters used by Choi et al. (2010) include use of CSL (3%), KH_2PO_4 (0.1%), $MgSO_4$ (0.05%), molasses (0.5%), rice bran (1.5%), at pH 5.5, 25°C, aeration rate of 1 vvm and agitation speed of 150 rpm in 5 l fermenter jar (Choi et al., 2010). This method reported *O. sinensis* exopolysaccharide production of 48.9 g/l in 5–6 days period (Choi et al., 2010). Parameters used by Hsieh et al. (2005) include $(NH_4)_2HPO_4$ (0.5%), corn steep powder (0.5%), KH_2PO_4 – 0.15% (w/v), Sucrose (6.17%), at pH 4.4, 25°C, agitation speed 300 rpm in 5 l Jar fermenter, which led to 3.2 h/l exopolysaccharide production in seven days. Kim & Yun (2005) cultured in media composed of sucrose 20 g/l, corn steep powder 25, g/l $CaCl_2$ – 0.78 g/l, $MgSO_4 \cdot H_2O$ 1.73 (g/l) at 4 pH, 20°C in 5 l Stirred tank fermenter for 16 days which led to production of 20.9 g/l mycelial biomass yield. Yan et al. (2014) may be further consulted for media compositions and fermentation conditions for *O. sinensis*.

13.2.2.4 SOLID-STATE CULTURE

Liu & Cao (1994) studied the solid fermentation of *O. sinensis* mycelium, and their results showed that through solid fermentation, the content of polysaccharide of cultured mycelium was more than that of natural *O. sinensis*, and the amino acid contents were near that of natural ones. In this process, 8 grams of each carbon source (preferably disaccharide) may be added to basal medium consisting of agar 9.8 g in 250 ml distilled water. After which the media is to be autoclaved at 121°C for 30 minutes. After cooling, each petri dish is to be inoculated centrally with an agar plug (5 mm) with active mycelia of *O. sinensis* and incubated at 21°C constant at 60% humidity. The culture to be observed after 21 days of inoculation (Huang & Ohga, 2018).

13.2.3 CULTIVATION OF OPHIOCORDYCEPS SINENSIS – LARVAE COMPLEX

Though extract of the bioactive components may be made from mycelial cultures (Baral, 2017), obtaining *O. sinensis*–larvae complex may be achieved through either the process of artificial cultivation and semi-natural cultivation (Yue et al., 2013). Step wise procedure are detailed as follows with brief biology of the insect host.

13.2.3.1 BIOLOGY OF HOST INSECT

Ophiocordyceps sinensis possesses wide host range and parasitizes more than 60 species of Lepidopteran larvae and 30–40 species of *Thitarodes/Hepialus* larvae (Baral, 2017). The host insects (*Thitarodes/Hepialus*) are holometabolous and has distinct developmental stages such as egg, larva, pupa, and adult, taking 3–4 or even 4–5 years to complete life cycle. The life cycle of *Hepialus* is presented by Zhou et al. (2014); and Li et al. (2019):

1. During June to August, each female lays around 500 eggs. Under the moist and incubated conditions, around 80% of eggs hatch between 30 and 70 days.
2. The larval stage has 6–8 instars (Zhou et al., 2014), to 9 instars (Li et al., 2019), (exiting seventh instar and above (Li et al., 2019)) can be found in the soil all the year around feeding on underground roots (5–25 cm deep in the soil). As larvae grow, the color of the head capsule changes from whitish to pale reddish or yellowish. In natural conditions, the larval survival rate is less than 10% as the larva has to survive many natural enemies. The change from larvae to pupa takes around 2–4 years and it varies from species to species (Zhou et al., 2014).
3. Generally, shift in life stage from larvae to pupa occurs at the end of May, and the pupal stage of around 40 days ranges from June to July, when temperature reaches 10–15°C with 40–45% soil RH at natural habitats (Zhou et al., 2014).
4. During June to August, pupa often emerges at 17:00–20:00 every day. Generally, the number of the adult male insects are lesser than that of the females. While, females mate once, male moths' mate twice or thrice in their lifetime (Zhou et al., 2014). The lifespan of males are known to be less than that of females, but females are also known to die soon after laying eggs (Wang & Yao, 2011; Zhou et al., 2014).

13.2.3.2 FOOD SOURCE OF HOST LARVAE

The soil inhabiting host larvae of *Hepialus* are omnivorous and troglodytic and prefer to feed on the roots of various plants of the family Polygonaceae, such as *Persicaria capitata* (Buch. -Ham. ex D. Don) H. Gross, *P. vivipara* (L.) Ronse Decr., *Polygonum macrophyllum* D. Don and *Rheum pumilum* Maxim.; family Rosaceae, such as *Dasiphora fruticosa* (L.) Rydb. and *Astragalus propinquus* Schischkin.; Cyperaceae, Gentianaceae, Juncaceae (*Juncus leucanthus*), Plantaginaceae (*Cyananthus macrocalyx*), Poaceae, Primulaceae, Ranunculaceae (*Ranunculus brotherusii*), etc. (Huang et al., 1989; Gu et al., 2006; Zhou et al., 2014). Sometimes, inoculation with bacterial strain such as, *Carnobacterium* sp. enhances the larval growth by inducing an improved enzymatic activity in their intestines (Yin et al., 2011).

13.2.3.3 ARTIFICIAL REARING OF HOST LARVAE

The larval hosts of *Ophiocordyceps sinensis* in pasture of Tibetan Plateau complete life cycle in 3–5 years. However, under artificial rearing conditions, this time span could be trimmed to ca. 1–2.2 years (Zhou et al., 2014; Qin et al., 2018). Key concepts of artificial rearing being their procurement from natural fields/their artificially reared source, proper growth medium, requirement of growth nutrition and prevention of larval diseases (Yue et al., 2013). Different insect states require temperature and humidity differently. Favorable temperatures are 15–20°C for incubation and 10–18°C for the larvae period. Temperatures that are too high can cause mutual killing and death, while feeding activities will reduce and growth will retard at lower temperatures. The temperature for the pupal period should be higher than that of the larvae period, which could be controlled at approximately 20°C. Humidity should be maintained at 36%–45% in larval phase, and if grown in soil, the soil should be pre-sterilized and its moisture content is to be kept at 42%–45% (Yue et al., 2013).

The hatching rate of the eggs are reported to be 86.75% at 50% RH and 15°C (Zhou et al., 2014). In soil-less rearing, the nutritional requirements of 5 instar larvae can be met by the artificial feed composed of agar 10 g/1, corn flour 10 g/l, *Rheum pumilum* 5.0 g/l, sorbic acid 0.5 g/l, soybean 15 g/l, wheat bran (WB) 8 g/l and yeast powder 5.0 g/l. Addition of antiseptic agents like sorbic acid and nipagin esters are known to increase the weight of larvae (Zhou et al., 2014).

13.2.3.4 DISEASES OF LARVAL HOST

Diseases in the cultivation process common and involves infection and killing of larvae by many lethal pathogens. Qin et al. (2018) have isolated *Paecilomyces farinosus* (85.92%), *Penicillium polonicum* (5.63%), *Beauveria bassiana* (2.11%) [causes white muscardine disease], *Penicillium commune* (2.82%), *Penicillium expansum* (2.82%), *Tritirachium* sp. (2.11%), *Galactomyces geotrichum* (0.70%), *Fusarium* sp. (0.70%), *Mucor circinelloides* (0.70%), *Mucor hiemalis* (0.70%) from infected cadavers of *Thitarodes xiaojinensis*. Besides, green muscardine, nematodes, mites, and a smaller insect acting as transmitting vectors are also harmful to the rearing. For larval viability, the food and rearing location should be disinfected and sterilized at regular intervals and the soil used for rearing should be exposed in the blazing sun for 1–2 hours (Yue et al., 2013). Fungicidal agents such as azadirachtin and phytoallexin have been used against some muscardine pathogens, however, its use with host larvae of *O. sinensis* under culture conditions needs to be studied (Vyas et al., 1992).

13.2.3.5 INFECTION OF LARVAL HOST

In natural conditions, August or late autumn is regarded as the ideal period for infection of larvae. The underground larvae are mainly infected through ascospores, conidia, and hyphae through channel of skin and intestine. The infection process of larvae is reported to be affected by atmospheric temperature, soil humidity, soil structure and soil temperature (Zhou et al., 2014). The highest infection rates were reported from 3rd to 4th or 4th to 5th instar larvae that are shedding old cuticles and forming new ones (Zhou et al., 2014; Li et al., 2019), while other instars larvae were reported not to be susceptible to infection (Li et al., 2019). In the process of conidial inoculum, at first adhesins facilitates conidial attachment to larval cuticle. After conidial germination, the germ tube extends inside the larval body and enters the hemocoel, ramifies into fusiform hyphae, and fills the hemocoel by yeast-type budding (Li et al., 2019). The infected larva moves 2–5 cm down the soil surface, dies with upward head, becomes stiff and covered with mycelia (Li et al., 2019). After full colonization of the larvae, the fruit-body proliferates out from frontal cortex (between the eyes) of the larval head (Baral, 2017). For molecular dialog involving infection of larvae by *O. sinensis*, Baral et al. (2017) may be referred.

13.2.3.6 COMPLETE ARTIFICIAL CULTIVATION

In this process, the cultured strains are inoculated to reared larvae and after infection they are continued rearing indoors. The harvesting of *Ophiocordyceps sinensis*-larvae complex may be done after 1–2 years. Although this process can increase the larval survival rate and decrease the harvesting period of *O. sinensis*-larvae complex, the cultivation cost is comparatively high (Yue et al., 2013).

13.2.3.7 SEMI-NATURAL CULTIVATION

In this process, the infected larvae are released to natural alpine-subalpine habitats with the aim to harvest after 3–5 years from the released areas. Semi-natural cultivation not only maximizes the use of natural ecological resources but also reduces the cultivation cost drastically, although the cultivation period is too long. The survival rate of released larva are known to depend upon various factors like soil, weather, food availability, parasites, and other natural enemies (Yue et al., 2013; Zhou et al., 2014).

13.3 ARTIFICIAL CULTIVATION OF *CORDYCEPS MILITARIS*

The artificial fruiting of *C. militaris* was first reported by Kobayasi (1941). As fruitbody production in the artificial culture of *O. sinensis* is yet to be developed (Baral, 2017), *C. militaris* provides an easy alternative without dependency to larval host. Shrestha et al. (2005) mentioned that the artificial culture of the Korean isolates produced the stromata with length close to the description of Kobayasi (1941); and Chen (1978) but were comparatively broad. The cultivation of *C. militaris* is reported to undergo through three phases based on growth characteristics viz. mycelial colonization, primordia development, and fruiting body formation, which under favorable conditions take around 5–6 days, 12 days and 45–50 days, respectively after inoculation (Du et al., 2010; Chiang et al., 2017). The fruitbodies of Dowonhongcho and Yedang 3 strains were reported to be developed after 15 days of inoculation (Lee et al., 2017).

Among the strains DT1-DT5 studied by Nguyen et al. (2020), the earliest primordia formation was recorded for strain DT3 with 18 days. The size of the fruiting body of cultured isolates of *C. militaris* from Northwest Himalayas ranged from 4.2 to 7.8 cm (Pathania et al., 2015). While Nguyen et al. (2020) reported strains DT4 (74.23 ± 5.13 mm) and DT3 (72.63 ± 2.62 mm) to exhibit the length of the fruiting body. The succeeding subsections discuss general requirements and procedure in artificial cultivation of *C. militaris*.

13.3.1 ISOLATION OF FUNGAL CULTURE

According to Lin et al. (2017), fresh fruit bodies *C. militaris* are ideal source for the isolation of pure mycelial cultures which could be maintained on potato, glucose, agar (PGA) slants. *Cordyceps militaris*, isolated from the wild, with the associated Lepidopteran larvae in mummified condition, are sealed followed by harvesting. The collected fungus is cleaned with double distilled and deionized water and sterilized with 70% alcohol. Sterilization is preferably done in the presence of a laminar airflow, Bunsen burner and clean petri-plates having the appropriate culture media for growth.

Then with sterilized scalpel stromal head is dissected longitudinally or surface is scraped, and the asci found in the bundle are placed carefully on the media plates. Isolation is done primarily as soon as the germination is confirmed. Germination is tracked and a single ascospore, which has undergone germination, is placed in a clean slant containing media. These inoculated slants are incubated in the dark at 18–20°C for three days at first, followed by uninterrupted light exposure for 18 days and checked for growth, suitable conditions and requirements, after observing for weeks (Kinjo & Zang, 2001). The ascospores from the fruiting bodies can also be obtained by hanging the stromal extremities and waited for the spore to fall, land, and germinate on the medium. The larvae bearing ascospores can be collected by oozing out its body fluids on the agar plate after sterilizing and cutting the pleopod off, followed by cultivation at a temperature of 20°C. After the emergence of the primary clone, the apex of hyphae are cut and placed on the agar. The mycelial characters are observed, studied, and identified. Strains collected from ascospores are beneficial over those isolated from any other part of the fungal tissue, in forming healthy fruiting bodies (Lie et al., 2006). Shrestha et al. (2005) used increased concentration of dextrose, yeast extract and peptone in SDAY media viz. (40 g dextrose, 10 g peptone, 10 g yeast extract and 15 g agar/1,000 ml) to culture isolated ascospores of *C. militaris* for the duration of 3–4 weeks in agar plates. However, they could also be isolated on PDA, incubated at 25°C for seven days (Kang et al., 2014).

Shrestha et al. (2005) based upon studies on Korean isolates has reported that *C. militaris* is a heterothallic and bipolar fungus, and its fruit body production requires two opposite mating type, which should be kept in mind during the cultivation process.

13.3.2 STORAGE CULTURE/STOCK CULTURE

PDA slants (generally 39 g/l) and the storage temperature of at 4–5°C are ideally used for storing stock culture (Das et al., 2010; Kang et al., 2014).

13.3.3 PRE-CULTURE/SEED CULTURE

PDA (39 g/l) is used for preparing slants and plates and kept at 25°C storage temperature (Das et al., 2010). The seed culture is also prepared with basal medium of 20 g/l peptone; 20 g/l sucrose; 1 g/l KH_2PO_4; and 0.5 g/l $MgSO_4 \cdot 7H_2O$, 70 ml of which is poured in 250 ml flask and incubated for 5–7 days at rotary shaker incubator at 150 rpm at 25°C (Chen et al., 2011; Kang et al., 2014). Seed culture may also be prepared in SDAY broth in 100 ml flasks inoculated with 5 pieces of mycelial discs and placed incubated at 25°C for five days at 120 rev/min (Shrestha et al., 2005). For mycelial growth of *C. militaris* the optimum temperature is reported to be 15–20–25°C, and for fruitbody development it is determined as 18–22–25°C (Hung et al., 2009; Lee et al., 2017), while, cultivation at 30°C, stopped both mycelial growth and Cordycepin production (Hung et al., 2009).

13.3.4 SPAWN PREPARATION

Various media compositions have been formulized for isolation of *C. militaris* mycelia and its preservation (Shrestha et al., 2006). Pathania et al. (2015) has reported, for the mycelial growth of *C. militaris*, beef extract (nitrogen source), folic acid (vitamin source), sucrose (carbon source) and zinc chloride (mineral source) to be optimal. Mycelial growth of *C. militaris* strains on solid-state medium could be achieved on SDAY medium (dextrose 20 g/l, peptone 5 g/l, yeast extract 5 g/l, and agar 15 g/l) at 25°C in dark and the mycelial extension diameter (mm) to be measured at 5-day interval from 5 to 25 days (Nguyen et al., 2020). According to Shrestha et al. (2006), the optimal media for the mycelial growth of *C. militaris* are CZYA (30 g/l sucrose, 5 g/l yeast extract, 0.5 g/l KCl, 0.01 g/l KH_2PO_4, 1 g/l K_2HPO_4, 0.5 g/l $MgSO_4 \cdot 7H_2O$, 3 g/l $NaNO_3$, and 20 g/l agar), SDAY and SMAY

(40 g/l maltose, 10 g/l peptone, 10 g/l yeast extract and 15 g/l agar). Das et al. (2010) have suggested to add requisite amount of sawdust (SD) to the medium to enhance the scale of culture.

The morphological characteristics of mycelia such as texture, density, and color could be assessed with visual observations (Nguyen et al., 2020). On the solid-state medium, the texture of mycelial growth is reported to be wooly/floccose, which is initially creamy in color which turns into light brown at maturity and becomes compact on 25th day (Pathania et al., 2015; Nguyen et al., 2020). However, Nguyen et al. (2020) while studying five strains (DT1–DT5) and their colony diameter, has reported significant differences ($P < 0.05$) in the growth rate among the studied strains.

13.3.5 SUBSTRATE PREPARATION

Cereals are widely used as substrate for obtaining fruit bodies of *Cordyceps militaris* (Wu et al., 2014). Shells of cottonseed, poplar SD, corncob residues, rice as well as spent substrates of *Flammulina velutipes* can be used as substrates and evaluated for the growth of fruit bodies. These substrates can be individually ground, mixed with rice bran and WB in 8:1:1 ratio (w/w/w). This mixture is moistened using peptone (5 g/l) and glucose (20 g/l) containing nutrient medium. The moisture content must be maintained at 650 g kg^{-1}. Substrates weighing 500 g is put inside glass bottles or about 2.5 kg in plastic bags, uniformly pressed, sealed using polypropylene plastic and autoclaved at 121°C at 15 pounds $inch^{-2}$ for half an hour (Li et al., 2016).

Ceramic beads measuring approximately 1 cm can be used in place of organic substrates for the production of *Cordyceps militaris* fruiting bodies. This modern technique requires about 350 g ceramic beads put inside a plastic bottle with 150 ml nutrient solution. On a large scale, about 1.8 kg of ceramic beads can be filled in plastic bags with 750 ml of nutrient solution. This solution generally contains 50 g sodium carboxymethyl cellulose, 20 g potato powder, 20 g glucose, 2 g peptone, 2 g yeast extract, 0.5 g $MgSO_4.7H_2O$ and 0.5 g KH_2PO_4 in 1 liter distilled water. Moisture content is to be maintained at 30%.

Wood based substrates means large felling of trees and agricultural waste-based substrate may involve heavy metal accumulation, hence use of ceramic/glass beads are suggested by many (Brunnert & Zadražil., 1983; Fischer et al., 1995; Chiu et al., 2000). These beads provide a controllable, clean, and solid substrate environment for mycelial growth and colonization (Redecker et al., 1995). Sometimes, these beads are provided with holes on their smooth surface in order to supply optimum moisture as well as aeration to the growing mycelia in the vessels used for cultivation (Badham, 1989; Ohga et al., 1990).

13.3.6 CEREAL GRAIN-BASED CULTURE

Cereal grains such as husked rice, brown rice, wheat grains, German millet, and common millet have been used as substrate for cereal grain-based culture. Compared to other substrates, brown rice, German millet, and common millet are known to exhibit and induce higher stromata lengths of around 50 mm (Kim et al., 2010). The results of Tao et al. (2020) with the studies on CM3 strain *C. militaris* showed that the rice produced yields of 23.19 g/bottle with biological efficiency of 62.26% and wheat produced yields of 19.07 g/bottle with biological efficiency of 54.48%. For the culture of *C. militaris*, most of the studies have suggested rice to be ideal base for seed grain based medium (Kim et al., 2010; Nguyen et al., 2020). Rice, especially brown rice showed the higher yield of fruiting bodies with 6–7 g/bottle (Kim et al., 2010).

Rice medium may be prepared in 300 ml jars by mixing nitrogen source, vitamins, auxiliary trace elements in distilled water (1:1). The medium may be autoclaved for 30 min at 121°C. Cooled, sterile medium may be inoculated with 5 ml seed culture and incubated in the dark at 25°C, over 70% RH for 14–20 days (Das et al., 2010; Chen et al., 2011).

On the other hand, after the inoculation in dark for 2 weeks, the medium may be exposed to 12 h exposure/day at 600±20 lux light intensities emitted through incandescent lighting system till 45th day, and the nutrients in jar may be replenished by adding PDB every 8th day. This method is suggested by Chen et al. (2011) to produce fruit body of *C. militaris* with the higher amount of Cordycepin and mannitol.

Cultivation of *Cordyceps militaris* in brown rice medium may be carried out in 1,000 ml polypropylene bottle by adding brown rice 50–60 g, silkworm pupae powder 10 g, in distilled water (90 ml), which may be autoclaved at 121°C for 20 min (Shrestha et al., 2005, 2012a; Nguyen et al., 2020).

The inoculated liquid spawn may be incubated in the dark at 22°C with 60–70% RH in dark for 15–20 days (Chen et al., 2011; Nguyen et al., 2020). After complete colonization of the medium, the strains are to be cultured at 20°C, 85–90% RH, and light/dark cycle of 12:12 h to initiate fruitbody formation and growth (Nguyen et al., 2020). Maintenance of high humidity of 70–85–90% has been recommended for the fructification of *C. militaris* (Shrestha et al., 2012b; Nguyen et al., 2020).

13.3.7 SHAKING CULTURE

Shaking culture is a type of submerged culture where, 50–100 ml of seed culture liquid medium poured in shake flask (250–500 ml) is inoculated with 5 mm^2 agar plug from *C. militaris* pre-culture plate and incubated for 5 – 7 days at 25°C on a rotary shaker at 50–150 rpm, or till the carbon source of media is depleted (Das et al., 2010). Strains (DT1–DT5) of *C. militaris* has been reported to grow mycelium on (peptone – 1 g/l, yeast extract – 1 g/l, $MgSO_4 \cdot 7H_2O$ – 1 g/l, KH_2PO_4 – 2 g/l) liquid media (Nguyen et al., 2020).

13.3.8 SUBMERGED CULTURE

Submerged culture is advantageous over solid culture and is becoming increasingly popular as it saves time, enhances the content of bioactive compounds and is efficient in the production of biomass (Jin et al., 2017). Here, open/closed fermenters are used where *C. militaris* is cultured in liquid medium with ample agitation and aeration (Das et al., 2010).

The mycelial growth in submerged culture is affected by many factors like rate of aeration, agitation speed, culture medium, initial pH, inoculation size, etc. Generally, under fermentation conditions, filamentous, and/or pelleted forms could be observed as growth forms and *C. militaris* is known to produce both filamentous and pelleted forms in submerged culture conditions, however, size of the pellet during incubation and culture needs to be smaller to facilitate transportation of oxygen and nutrients to the pellet core (Park et al., 2002; Turlo, 2014; Jin et al., 2017).

For submerged culturing of *C. militaris*, 10 pieces of 0.5 cm^2 of agar plugs are excised and homogenized in sterile water (100 ml). They are inoculated to 500 ml flask containing 250 ml in the medium (20 g/l glucose, 1 g/l peptone, 1 g/l yeast extract, 1 g/l $MgSO_4 \cdot 7H_2O$, and 2 g/l KH_2PO_4) and kept on a rotary shaker incubator (120 rpm) at 22C. The density, size, and morphology of pellets are recorded for 1–25 days.

13.3.9 STATIC CULTURE/SURFACE LIQUID CULTURE

Surface liquid culture is carried out by inoculating 500 ml (8.5 cm × 14.0 cm) or 1,000 ml (11 cm × 15 cm) culture bottles (Das et al., 2010; Kang et al., 2014). When seed cultures are used from slants and plates with actively growing *C. sinensis*, inoculation may be performed @ 1 cm disk of seed culture per 500 ml bottle, while 10% (v/v) of liquid seed culture may be inoculated to 1,000 ml culture bottles.

Kang et al. (2014) has suggested the use of 20 g/l basal media; 20 g/l peptone; 1 g/l KH_2PO_4; and 0.5 g/l $MgSO_4 \cdot 7H_2O$ at natural pH, which is to be autoclaved at 121°C for 30 min. Das et al. (2010) have used working volume of 100 ml liquid culture medium for 500 ml bottles, however, Kang et al. (2014) reported higher Cordycepin production with a working volume of 700 ml per 1,000 ml (11 cm × 15 cm) bottle. Sterilized cotton plugs are suggested to be placed at the lid holes for aeration and the bottles are incubated at 25°C, undisturbed for 35 days. After that, filtrate may undergo analysis for targeted bioactive materials (Masuda, 2006; Das et al., 2010; Kang et al., 2014).

13.3.10 CONTINUOUS CULTURE/REPEATED BATCH CULTURE

For bio-industrial application, one cycle cell culture as mentioned above may be hinderance for gain in productivity. A repeated batch culture or continuous culture promises higher prospects and productivity for pharmaceutical uses. Repeated batch culture involves constant removal of spent medium and constant replenishment with fresh medium to provide actively growing mycelia with their preferred environment and nutrition (Das et al., 2010). For exopolysaccharides production in stirred-tank fermenter, 150 rpm agitation rate and 1.5 vvm aeration rate were found to be most effective and enhancement of production of *C. militaris* to the maximum of 5.713 g/l, with the productivity of 476 mg/l/day were reported (Wang et al., 2019).

13.3.11 CULTURE MEDIA

Mushroom growth is known to be noticeably affected by nutrients (Nguyen et al., 2020). Both solid and liquid media are devised for culturing *C. militaris*. For solid media, beech wood meal, brown rice, cob particles, common millet cottonseed shells, German millet, husked rice, rice, and WB, wheat grains, etc., could be used as the basal substrate and nutritional sources (Das et al., 2010; Kim et al., 2010). Depending upon culture needs and the use of variable concentration of media constituents, liquid media may be classified into basal media, enriched media, and enriched media with additive(s). Enriched media is designed with optimum media concentration aiming towards attaining higher production, while basal media have the bare minimum nutrient constituents for sustenance of fungal growth. Basal media is reported to have lower Cordycepin production while optimized media for *C. militaris* is reported to produce higher level (6.84 g/l) of Cordycepin (Das et al., 2008, 2010).

13.3.11.1 BASAL MEDIA

Yeast extract 7.5 g/l and Peptone 2.5 g/l as Nitrogen source and Glucose 20 g/l as carbon source has been suggested by Das et al. (2010).

13.3.11.2 ENRICHED MEDIA

Das et al. (2010) suggested following media based on carbon, nitrogen sources and other micronutrients following Vogel's medium: 93.8 g/l Yeast extract, 86.2 g/l Glucose and micronutrients diluted to 1/10 concentration viz. 0.01 g/l $CaCl_2 \cdot 2H_2O$; 0.2 g/l NH_4NO_3; 0.02 g/l $MgSO_4 \cdot 7H_2O$; 0.28 g/l $NaOC(COOH)(CH_2COONa)_2 \cdot 2H_2O$; 0.28 g/l KH_2PO_4; 0.025×10^{-3} g/l $CuSO_4 \cdot 5H_2O$; 0.10×10^{-3} g/l $Fe(NH_4)_2(SO_4)2 \cdot 6H_2O$; 0.46×10^{-3} g/l Citric acid; 0.50×10^{-3} g/l $ZnSO_4 \cdot 7H_2O$; 5.0×10^{-6} g/l H_3BO_3; 5.0×10^{-6} g/l $MnSO_4 \cdot (4–5) H_2O$; 5.0×10^{-6} g/l $Na_2MoO_4 \cdot 2H_2O$.

13.3.11.3 ENRICHED MEDIA WITH ADDITIVE(S)

In addition to above enriched media, addition of Adenosine @ 6 g/l is reported to have Cordycepin production of 8.57 g/l from *C. militaris* (Das et al., 2009). Kang et al. (2014) has observed increased production with addition of 10 mg/L of Vitamin B1, NAA, and Vitamin B11. When Vitamin B1 was used, Cordycepin production reached 1,159±109 mg/l. Other suggested additives for increased Cordycepin production are adenine, glycine, L-alanine, and Hypoxanthine (Kang et al., 2014). Cordycepin production response of 1451.43 mg/l was predicted through canonical analysis with the optimized medium constituents of 24.7 g/l sucrose, 1.11 g/l $K_2HPO_4 \cdot 3H_2O$, and 0.90g/l $MgSO_4 \cdot 7H_2O$ (Kang et al., 2014). The optimized and enriched media with additive suggested by Kang et al. (2014) for liquid static culture include 5.45 g/l hypoxanthine; 1.11 g/l $K_2HPO_4 \cdot 3H_2O$; 12.23 g/l L-alanine; 0.90 g/l $MgSO_4 \cdot 7H_2O$; 20 g/l peptone; 24.7 g/l sucrose and 10 mg/l Vitamin B1, which produced average of 2,006 ± 34 mg/l Cordycepin for working volume of 700 ml per 1,000 ml bottle.

13.4 COMPARISON OF BIOACTIVE METABOLITES BETWEEN WILD AND CULTURED SPECIMENS

The improvement of biological efficiency needs optimization of several factors like humidity, light conditions, minerals, nutrition source, pH, temperature, etc. (Chamyuang et al., 2019). However, the bioactive metabolite has been found to vary widely among the strain used and choice of culture media (Hung et al., 2009). The Cordycepin content in *C. sinensis* ranges from 3 μg/ml in wild to 22.0 μg/ml in cultivated specimens which is known to differ significantly ($P<0.01$) among sampling years and different collections in a single year (Zhou et al., 2019). Further, levels of Adenosine have also been reported to be higher in cultured specimens than that of wild (Ning et al., 2006; Wang et al., 2015). Tao et al. (2020) reported that Adenosine content of CM9 strain of *C. militaris* cultivated on tussah pupae was 2.62 mg/g, while the Adenosine content of CM3 cultured on rice was 2.51 mg/g.

Hung et al. (2009) showed that Cordycepin production was strain dependent with the yields of 544.82 ± 99.80 mg/l (36.32 ± 6.65 mg/l/d) and 587.68 ± 16.82 mg/l (39.18 ± 1.12 mg/l/d) from *C. militaris* BCC2816 and BCC2819, respectively, at 25°C, while other *C. militaris* strains studied produced significantly lower Cordycepin yields. Inorganic salts like KCL, $K_2HPO_4 \cdot 3H_2O$, KH_2PO_4, $MgSO_4 \cdot 7H_2O$, and $MgCl_2 \cdot 6H_2O$ have been recognized as useful media components for Cordycepin production by *C. militaris*, while 1120 ± 105 mg/l Cordycepin yield has been observed using $K_2HPO_4 \cdot 3H_2O$ (Mizuno, 1999; Masuda, 2006, 2007; Das et al., 2010; Kang et al., 2014).

Although, the biomass was reported to be lower in comparison to sucrose. Higher Cordycepin titer of 249.74 ± 12.71 mg/l and Cordycepin production yield of 0.026 ± 0.001 g/g were reported by (Wongsa et al., 2020), when cultured in glucose. On the other hand, lactose (823.72 ± 85.64 mg/l) and sucrose (843.63 ± 66.70 mg/l) were reported to produce higher levels of Cordycepin than glucose, however, not much difference of Cordycepin production yield of *C. militaris* was reported between culture of xylose (0.024 g/g) and glucose (0.026 g/g) (Kang et al., 2014).

Though a poor source of carbon, xylose is a structural precursor of Cordycepin, and xylose-based medium has yielded of 0.094 ± 0.002 g/g of extracellular Cordycepin on dry biomass (Vongsangnak et al., 2017; Wongsa et al., 2020). Through optimal culture conditions (glucose 10 g/l, K_2HPO_4 – 1.0 g/l, $MgSO_4 \cdot 7H_2O$ 1.0 g/l, peptone 10 g/l, α-naphthylacetic acid 1.0 mg/l), the average content of Cordycepin isolated from fruiting bodies of *C. militaris* reached 9.17 ± 0.09 mg/g (Wen et al., 2014).

C. militaris cultured in PDB has shown Cordycepin content of 1.08± 0.73 mg/g, brown rice medium has shown 2.98 ± 1.41 mg/g, silkworm pupae medium has shown 4.17 ± 1.66 mg/g, CM9 strain on wheat has shown 5.41 mg/g and CM4 strain cultured on wheat has shown 5.68 mg/g (Guo et al., 2016; Kang et al., 2017; Lee et al., 2017; Tao et al., 2020).

The developed strains KSP8 (Kang et al., 2017) and Haizhou 1 (Du et al., 2010) has shown increased content of Cordycepin of 6.63 mg/g and 24.98 mg/g, respectively. The amount of Cordycepin in the strains DT1–DT5 available from Mushroom Research and Development Center, Vietnam and cultured in rice based medium varied from 0.24 mg/g to 1.68 mg/g in DT5 and DT3 strains, respectively (Nguyen et al., 2020). The Cordycepin content of the *C. sinensis* strain TBRC7358 reported Cordycepin production of 235.39 ± 5.17 mg/l which was higher than 149.50 ± 15.71 mg/l in TBRC6039 (Raethong et al., 2018; Wongsa et al., 2020). Hung et al. (2009) has reported good results of Cordycepin production at 25°C, while it stopped at 30°C.

Zhou et al. (2019) reported that the amount of mannitol in cultured *C. sinensis* in the same year was significantly different ($P<0.01$) which ranged between 128 µg/ml in C2 strain to 514 µg/ml in C4 strain. Further, Zhou et al. (2019) also showed the mean content of mannitol in cultured samples (437.58 µg/ml) to be lower than the wild samples (502.10 µg/ml). Guan et al. (2010); and Zhou et al. (2019) found that wild samples to contain more freely available mannitol than cultured samples, similarly, available glucose was more in wild samples than the cultured ones.

Wu et al. (2014) reported that polysaccharides derived from cultured *C. militaris* and natural *C. sinensis* as well as various batches of cultured *C. militaris* were highly similar. Polysaccharide was reported to be lower in wild specimens than those from culture (Wang et al., 2015). Conversely, Zhou et al. (2019), detected higher contents in the wild samples. Hu et al. (2015) reported high activity of glycosidases in wild *C. sinensis*, than in cultured samples.

13.5 CARBON AND NITROGEN SOURCES AND C:N RATIO

Study on transcriptomic response of *C. militaris* to various sources of carbon has shown that culturing of *C. militaris* in favorable carbon source leads to up-regulation genes associated for Cordycepin biosynthesis (Raethong et al., 2018) and the similar finding of dependence of *C. militaris* on favorable carbon source for growth and Cordycepin production was confirmed by

Wongsa et al. (2020). Cereal grains have been utilized as the primary source of carbon for large scale cultivation of *C. militaris* (Lin et al., 2017). Corn cob fragments, shells of cottonseed, and grains of corn, millet, rice, sorghum, and wheat are among the most used substrates for production of *C. militaris* fruitbodies and its metabolites (Lin et al., 2017; Shrestha et al., 2012b).

Observably, *C. militaris* yielded the maximum specific growth rate (μ_{max}) on sucrose with μ_{max} of 0.0114/h, followed by glucose (μ_{max} = 0.0100/h) and fructose (μ_{max} = 0.0098/h), however, significant reductions in growth were observed on xylose (μ_{max} = 0.0022/h) and arabinose (μ_{max} = 0.0051/h) which demonstrates that xylose and arabinose were less favored to the growth of *C. militaris* (Shang et al., 2016; Raethong et al., 2018). However, if the anticipated extra-cellular metabolic product is Cordycepin, then, xylose hold promises as an alternative carbon source (Wongsa et al., 2020). Kang et al. (2014) reported that peptone (most favored nitrogen source from the study), beef extract, casein, NH_4NO_3 and yeast extract to be favorable for Cordycepin production.

For the production of exopolysaccharides at the rate of 1.686 g/l in a submerged culture, Wang et al. (2019) determined yeast extract at the rate of 10 g/l and glucose at the rate of 40 g/l to be the optimal. Sucrose and Glucose to be favorable carbon sources has also been confirmed by Wongsa et al. (2020) who reported specific growth rates (μ_{max}) of *C. militaris* strain TBRC7358 at similar values of 0.0075–0.0079/h for both Sucrose and Glucose. However, the rate of growth was reported to be slow in xylose (0.00375/h). Among the preferred carbon source, glucose led to the extracellular Cordycepin production by *C. militaris* at 0.1090 ± 0.0124 g/l (Raethong et al., 2020). The contents of Cordycepin, Niacin, and Potassium isolated from the fruiting bodies of *C. militaris* cultivated on soy powder media were reported by Cho et al. (2010) to be higher than those obtained from the strain cultivated on other media.

Raethong et al. (2020) observed that at CN limited condition, metabolic model iNR1329 of *C. militaris*, is able to utilize various complex CN substrates, like chitin, glucosamine, gamma-aminobutyrate (GABA), and some amino acids for growth. GABA as a potential source of CN is also supplemented by the fact that rice grains have good GABA content, which is already been widely used in the cultivation of *C. militaris* (Raethong et al., 2020). *Cordyceps militaris* may also utilize aspartate, glutamate, serine, and isoleucine as alternative CN source, however iNR1329 is also reported to use ammonia, asparagine, glutamine, glycine, and urea as single sources of nitrogen, while utilizable single carbon sources are reported as glucose, sucrose or xylose (Raethong et al., 2020).

For *Ophiocordyceps sinensis*, the content examination of polysaccharides and mycelium weight by different nitrogen sources showed that NH_4NO_3 was the appropriate source for culturing *O. sinensis* (Liu et al., 2007). Huang & Ohga (2018) showed that among the monosaccharides (Fructose, Glucose), disaccharides (Maltose, Sucrose), polysaccharides (Maltose, Malt extract) used as the carbon source for mycelial growth of *O. sinensis*, the higher rate of mycelial growth in terms of colony diameter (cd) and whiteness index (wi) were obtained from maltose (cd = 83.56 mm/wi = 57.06) followed by sucrose (cd = 81.73 mm/wi = 55.97) and dextrin (cd = 77 mm/wi = 53.18). Huang & Ohga (2018) showed that out of beef extract, soy extract, whey protein, yeast extract to be the complex organic nitrogen sources, whey protein showed the higher mycelial growth of *O. sinensis* with the whiteness index (wi) of 48.77 and colony diameter (cd) of 81.16 mm. Moreover, Soy protein (wi = 46.89, cd = 62.87 mm) and yeast extract (wi=42.65, cd=53.78 mm) is reported to give less diameter of colony (cd = 66.48 mm) than control, but more white colonies (wi=39.08).

The carbon and nitrogen (C:N) ratio were reported to be correlated with Cordycepin production and growth (Park et al., 2001; Mao et al., 2005). Using sucrose and peptone, Huang & Ohga (2018) showed that the optimal carbon and nitrogen ratio for culturing *O. sinensis* to be of 18:1. For the culture medium designed for mycelial growth of *O. sinensis*, optimum C:N ratio of 12:1 was reported by Dong & Yao (2005). For *C. militaris*, Raethong et al. (2020) has shown C:N ratio of 12.7:1 to be optimal for *C. militaris*. While C:N ratio of 8:1 to 8.3:1 has been shown to increase extracellular Cordycepin production by *C. militaris* (0.3776 g/l), and increased growth rate (0.0174 1/h) (Raethong et al., 2020).

13.6 LIGHT CONDITIONS

In the cultivation process of *C. militaris*, studies have shown that light plays important role in determining formation of pigment, production of Cordycepin, Carotenoid, and stroma (Chiang et al., 2017; Shrestha et al., 2012b). Light intensity of 500 to 1,000 lux used in 12:12 h dark/light cycle at 25°C were found to be optimal for induction of fruitbody formation from the mycelial stage (Hong et al., 2010; Chen et al., 2011). Dong et al. (2013) showed that pink light (1/3, 450–460 nm + 2/3, 620–630 nm) increased dried matter contents and bio-efficiency of *C. militaris* fruit bodies to (40.06/6.77%) in comparison to daylight (22.69/4.55%), blue light (24.44/4.86%) and red light (22.06/4.17%), respectively. Pink light is reported to enhance accumulation of Cordycepin and Carotenoids, while higher adenosine content is reported to be accumulated under red light.

13.7 PRODUCTION CONSTRAINTS

The distributional range of *Hepialus* is mainly affected by food, vegetation, soil structure, temperature, and humidity (Yang et al., 2010). Global climatic fluctuations and their effect on the natural habitat of *O. sinensis* have had a detrimental impact on the life cycle and thus population of *O. sinensis* (Baral, 2017). Climate change has also reported to polarization of distribution of *O. sinensis* in the region above 4,600 m altitude with smaller poor quality fruitbodies (Yang et al., 2010). Harvesting of immature *O. sinensis* (prior to spores' maturity and dispersal) and during harvesting season, mass migration of harvesters at once to the collection sites are ascribed to decrease in interactions between host and insects and rapid decline and deterioration of *O. sinensis* population, primarily due to gradual reduction in overall population of spores (Baral, 2017).

Although, average survival rate of host larvae in experimental conditions are reported to be 01–7.7%, large scale rearing of larvae has also been accomplished (Zhou et al., 2014; Li et al., 2019), however, two principal limiting factors for cultivation of *O. sinensis* at large scale are reported to be: (i) low infection rate of host by anamorph of *O. sinensis*; and (ii) low rate of induction of primordium after infection (Li et al., 2019).

For *C. militaris* fruitbody production and levels of bioactive compounds in fruitbodies, both strain and substrate are found to be crucial (Tao et al., 2020). Even when the cultured hypha of *O. sinensis* strains from three different origins were studied in different culture media, there were discrepancies among different culture media, hyphal germ initiation time of different strains, diameters of hyphae at 40 days, and hypha morphology Fang et al. (2010). Identification of anamorphs and species of fungal counterparts *O. sinensis* is challenging as in the wild some readily produce anamorphs, but they don't in culture conditions. However, the use of robust molecular biology approaches has been helpful in solving this type of problem (Baral, 2017).

Ophiocordyceps sinensis is a slow growing psychrophilic fungus (Li et al., 2019) and its host larvae of ghost moth or Himalayan bat moth have three to five years life cycle, that exclusively live under soil in the alpine-subalpine pastures and shrublands (Zhou et al., 2019). This long lifespan is artificially shortened to a minimum of ca. one year under controlled environments, which is considered yet long enough for large scale rearing considering the susceptibility of the insect to numerous pathogens. Though shortening of lifespan has been productive in terms of rapid growth, it is detrimental for accumulation of nutrients and dry matter content in larvae (Qin et al., 2018). Under long-term artificial feeding, reproductive degeneration including the sex ratio imbalance and decline in fertility has been observed in the host *Hepialus*. To counter reproductive degeneration, hybrid breeding of the host insect has been suggested (Li et al., 2016).

Although *C. militaris* is the most ideal substitute of *O. sinensis* for bioactive metabolites, *C. militaris* cultivation industry faces several challenges, the most prominent being degeneration of strain (Shrestha et al., 2012b; Yin et al., 2017). The degeneration of *C. militaris* is reported after third sub-cultured generation and is reported to be due to genes in charge of biosynthesis of toxins, chromosome remodeling, DNA methylation and metabolism (Yin et al., 2017), and degenerated *C. militaris* strains are known to exhibit characteristics such as slow mycelial growth with reduced mycelial density, inability to form primordia, excessive aerial hyphae, poor development of stroma and fruiting bodies with irregular shapes and sizes, etc. (Shrestha et al., 2012b; Sun et al., 2018; Wang et al., 2017).

13.8 CONCLUSION

At the current state of affairs, cultured strains and reared larval breed may not be viable for a very long period; and time and again, cultivators would have to resort to natural sources for fresh biological inputs for artificial culture. However, overharvesting of specimens, especially whose ascospores have not been released into natural habitat, as well as global warming and spatial shift in its distribution, are threatening population of both *O. sinensis* and its larval host. Adequate legislative measures are hence needed for alpine-subalpine pastoral management and harvesting of *O. sinensis*-larval complex. On the commercial production front, *O. sinensis* may not have achieved the status what *C. militaris* have, primarily due to the former being considered and demanded as the picturesque *O. sinensis*-larvae complex, and the cultured mycelial mass of *O. sinensis* devoid of larval host, and fruitbody less mycelium, has not been paid the attention it deserves. The commercial cultivation of *O. sinensis*, requires higher technical requirements, long production cycles and increased risk of contamination. Consequently, it is vital to better understand the natural habitat conditions for emulating it to artificial culture conditions, enhanced culture and breeding techniques for larvae, mechanisms, and approaches of fungal parasitism on larvae. To minimize pressure on natural sources, mycelia, mycelial extracts and fermented cultural extracts of *O. sinensis* may be popularized.

Unlike *O. sinensis*, at artificial culture conditions *C. militaris* readily produces fruitbodies with minimum effort and further, its non-dependency upon larvae, and bioactive constituents nearly equivalent to that of *O. sinensis* has made it a suitable substitute for *O. sinensis*. For *C. militaris* cultivation, the variation of productivity with varying strain and variations in media along with degeneration of mycelia due to sub-culturing needs to be addressed. Further, it is evident that not any strain would produce the highest yield in any culture media, hence commercial ventures would require screening of both strain and compatible media beforehand. Ion beam irradiated mutants of *Cordyceps militaris* are reported to give higher productivity, hence studies should also be focused on breeding of strains. For both *O. sinensis* and *C. militaris*, the screening of a few industry-specific highly virulent, fast growing, disease resistant and highly productive strains, and accordingly development of enhanced media for the screened strains for industrial use needs to be prioritized.

Heavy metals such as Pb, Hg, and Cd are cause of concern as leachates from organic substrates and suitable protocols need to be developed. Besides fruitbodies and mycelial mass, the ultimate bioactive metabolites of *O. sinensis* and *C. militaris* are Cordycepin, Mannitol, Polysaccharides, Adenosine, and other relevant compounds, therefore their direct extraction from culture filtrate will be crucial towards development of *Cordyceps* based green pharmacognostic and pharmacological processes.

KEYWORDS

- **artificial culture**
- **Ascomycota**
- **bioactive compounds**
- **cordycipitaceae**
- **ophiocordycipitaceae**
- **traditional Chinese medicine**

REFERENCES

Badham, E. R., (1989). Influence of water potential on growth of shiitake mycelium. *Mycologia, 81*, 464–468.

Baral, B., (2017). Entomopathogenicity and biological attributes of Himalayan treasured fungus *Ophiocordyceps sinensis* (Yarsagumba). *J. Fungi., 3*, 4. doi: 10.3390/jof3010004.

Brunnert, H., & Zadražil, F., (1983). The translocation of mercury and cadmium into the fruiting bodies of six higher fungi: A comparative study on species specificity in five lignocellulolytic fungi and the cultivated mushroom *Agaricus bisporus. Eur. J. Appl. Microbiol., 17*, 358–364. doi: 10.1007/BF00499504.

Cha, S. H., Lim, J. S., Yoon, C. S., Koh, J. H., Chang, H. I., & Kim, S. W., (2007). Production of mycelia and exo-biopolymer from molasses by *Cordyceps sinensis* 16 in submerged culture. *Bioresour. Technol., 98*, 165–168.

Chamyuang, S., Owatworakit, A., & Honda, Y., (2019). New insights into cordycepin production in *Cordyceps militaris* and applications. *Ann. Transl. Med., 7*(3), 9–11.

Chen, X. X., Chen, P., Huang, Y., Mo, M. H., Xu, J., & Zheng, S. C., (2016). The mechanisms of pharmacological activities of *Ophiocordyceps sinensis* Fungi. *Phytother. Res., 30*, 1572–1583. doi: 10.1002/ptr.5673.

Chen, Y. S., Liu, B. L., & Chang, Y. N., (2011). Effects of light and heavy metals on *Cordyceps militaris* fruit body growth in rice grain-based cultivation. *Korean. J. Chem. Eng., 28*(3), 875–879. doi: 10.1007/s11814-010-0438-6.

Chen, Z. C., (1978). Notes on new Formosan forest fungi VI. Genus *Cordyceps* and their distribution in Taiwan. *Taiwania., 23*, 153–162.

Chiang, S., Liang, Z., Wang, Y., & Liang, C., (2017). Effect of light emitting diodes on the production of cordycepin, mannitol and adenosine in solid-state fermented rice by *Cordyceps militaris. J. Food Compos. Anal., 60*, 51–56.

Chioza, A., & Ohga, S., (2014). A review on fungal isolates reported as anamorphs of *Ophiocordyceps sinensis. J. Mycol., 2014*, 913917.

Cho, S. M., Park, H. J., Seo, G. S., & Hong, J. D., (2010). Effect of media composition on the cordycepin and content nutritional components of *Cordyceps militaris. Kor. J. Mycol., 37*(2), 161–166.

Choi, J. W., Ra, K. S., Kim, S. Y., Yoon, T. J., Yu, K. W., Shin, K. S., Lee, S. P., & Suh, H. J., (2010). Enhancement of anti-complementary and radical scavenging activities in the submerged culture of *Cordyceps sinensis* by addition of *Citrus* peel. *Bioresour. Technol., 101*, 6028–6034.

Cui, J. D., (2015). Biotechnological production and applications of *Cordyceps militaris*, a valued traditional Chinese medicine. *Crit. Rev. Biotechnol., 35*(4), 475–484.

Das, S. K., Masuda, M., Hatashita, M., Sakurai, A., & Sakakibara, M., (2008). A new approach for improving cordycepin productivity in surface liquid culture of *Cordyceps militaris* using high-energy ion beam irradiation. *Lett. Appl. Microbiol., 47*, 534–538.

Das, S. K., Masuda, M., Hatashita, M., Sakurai, A., & Sakakibara, M., (2010). Optimization of culture medium for cordycepin production using *Cordyceps militaris* mutant obtained by ion beam irradiation. *Process Biochem., 45*, 129–132.

Das, S. K., Masuda, M., Sakurai, A., & Sakakibara, M., (2009). Effects of additives on cordycepin production using a *Cordyceps militaris* mutant induced by ion beam irradiation. *Afr. J. Biotechnol., 8*(13), 3041–3047.

Das, S. K., Masuda, M., Sakurai, A., & Sakakibara, M., (2010). Medicinal uses of the mushroom *Cordyceps militaris*: Current state and prospects. *Fitoterapia, 81*, 961–968. doi: 10.1016/j.fitote.2010.07.010.
Dong, C. H., & Yao, Y. J., (2005). Nutritional requirements of mycelial growth of *Cordyceps sinensis* in submerged culture. *J. Appl. Microbiol., 99*, 483–492.
Dong, C. H., & Yao, Y. J., (2011). On the reliability of fungal materials used in studies on *Ophiocordyceps sinensis*. *J. Ind. Microbiol. Biotechnol., 38*, 1027–1035.
Dong, J. Z., Lei, C., Zheng, X. J., Ai, X. R., Wang, Y., & Wang, Q., (2013). Light wavelengths regulate growth and active components of *Cordyceps militaris* fruit bodies. *J. Food Biochem., 37*, 578–584. doi: 10.1111/jfbc.12009.
Du, A. L., Zhang, X., & Zhang, H. Z., (2010). A new high cordycepin *Cordyceps militaris* cultivar 'Haizhou 1'. *Acta Hortic., 37*, 1373–1374.
Fang, X., Zhu, Z. X., Wei, K. L., Zhang, S. X., Nan, Z., & Wei, T., (2010). Effects of different culture media on the hyphal growth of *Cordyceps sinensis* strains. *Chin. J. Microecol., 22*, 534–536.
FEWMD. Public auction notice of Department of Forest, Environment & Wildlife Management (L.No.284/CF(T)/FEWMD, dated 09.07.2016). http://sbbsikkim.nic.in/pdf/news/auction.pdf (accessed on 13 January 2023).
Gu, D. X., Zhang, G. R., Wang, J. H., & Liu, X., (2006). A review and prospect on the studies of *Cordyceps sinensis* (Berk) Sacc. *J. Chin. Inst. Food Sci. Technol., 6*, 137–141.
Guan, J., Yang, F. Q., & Li, S. P., (2010). Evaluation of carbohydrates in natural and cultured *Cordyceps* by pressurized liquid extraction and gas chromatography coupled with mass spectrometry. *Molecules, 15*(6), 4227–4241. doi: 10.3390/molecules15064227.
Guo, M., Guo, S., Huaijun, Y., Bu, N., & Dong, C. H., (2016). Comparison of major bioactive compounds of the caterpillar medicinal mushroom, *Cordyceps militaris* (*Ascomycetes*), fruiting bodies cultured on wheat substrate and pupae. *Int. J. Med. Mushrooms, 18*(4), 327–336.
Hong, I. P., Kang, P. D., Kim, K. Y., Nam, S. H., Lee, M. Y., Choi, Y. S., Kim, N. S., et al., (2010). Fruit body formation on silkworm by *Cordyceps militaris*. *Mycobiology, 38*(2), 128–132. doi: 10.4489/MYCO.2010.38.2.128.
Hsieh, C. Y., Tsai, M. J., Hsu, T. H., Chang, D. M., & Lo, C. T., (2005). Medium optimization for polysaccharide production of *Cordyceps sinensis*. *Appl. Biochem. Biotechnol., 120*, 145–157.
Hu, H., Xiao, L., Zheng, B., Wei, X., Ellis, A., & Liu, Y. M., (2015). Identification of chemical markers in *Cordyceps sinensis* by HPLC-MS/MS. *Anal. Bioanal. Chem., 407*(26), 8059–8066. doi: 10.1007/s00216-015-8978-6.
Huang, P. Y., & Ohga, S., (2018). Nutritional requirements for mycelial growth of *Ophiocordyceps sinensis* on agar media. *Adv. Microbiol., 8*, 821–829. doi: 10.4236/aim.2018.811054.
Huang, T. F., Fu, S. Q., & Luo, Q. M., (1989). Experiments on the feeding habits of *Hepialus armoricanus* larva in Kangding. *Sichuan J. Zool., 8*, 8–10.
Humber, R. A., (2000). Fungal pathogens and parasites of insects. In: Priest, F. G., & Goodfellow, M., (eds.), *Applied Microbial Systematics*. Kluwer Academic Publishers, Dordrecht.
Hung, L. T., Keawsompong, S., Hanh, V. T., Sivichai, S., & Hywel-Jones, N. L., (2009). Effect of temperature on cordycepin production in *Cordyceps militaris*. *Thai J. Agric. Sci., 42*(4), 219–225.
Hywel-Jones, N. L., (2002). Multiples of eight in *Cordyceps* ascospores. *Mycol. Res., 106*, 2–3.
Jiang, Y., & Yao, Y. J., (2003). Anamorphic fungi related to *Cordyceps sinensis*. *Mycosystema., 22*, 161–176.

Jin, J., Zhong, C., Qin, Y., Cai, Y., Zhen, L., Shen, B., Chen, L., et al., (2017). A new cordycepin-producing caterpillar fungus *Ophiocordyceps xuefengensis* with artificial infection to the host, cultivation of mycelia and stromata. *FEMS Microbiol. Lett., 364*(20), 1–8. doi: 10.1093/femsle/fnx181.

Kang, C., Wen, T. C., Kang, J. C., Meng, Z. B., Li, G. R., & Hyde, K. D., (2014). Optimization of large-scale culture conditions for the production of cordycepin with *Cordyceps militaris* by liquid static culture. *Sci. World J., 2014*, 510627. doi: 10.1155/2014/510627.

Kang, N., Lee, H. H., Park, I., & Seo, Y. S., (2017). Development of high cordycepin-producing *Cordyceps militaris* strains. *Mycobiology., 45*(1), 31–38.

Kim, H. O., & Yun, J. W., (2005). A comparative study on the production of exopolysaccharides between two entomopathogenic fungi *Cordyceps militaris* and *Cordyceps sinensis* in submerged mycelial cultures. *J. Appl. Microbiol., 99*(4), 728–738. doi: 10.1111/j.1365-2672.2005.02682.x.

Kim, S., Shrestha, B., Sung, G., Han, S., & Sung, J., (2010). Optimum conditions for artificial fruiting body formation of *Cordyceps cardinalis*. *Mycobiology, 38*(2), 133–136.

Kinjo, N., & Zang, M., (2001). Morphological and phylogenetic studies on *Cordyceps sinensis* distributed in southwestern China. *Mycoscience, 42*, 567–574. doi: 10.1007/BF02460956.

Kobayasi, Y., (1941). The genus *Cordyceps* and its allies. *Sci. Rep. Tokyo Bunrika Daigaku. Sect. B., 5*, 53–260.

Kobayasi, Y., (1982). Keys to the taxa of the genera *Cordyceps* and *Torrubiella*. Transactions of the *Mycol. Soc. Jpn., 23*, 329–364.

Lee, H. H., Kang, N., Park, I., Park, J., Kim, I., Kim, J., Kim, N., et al., (2017). Characterization of newly bred *Cordyceps militaris* strains for higher production of cordycepin through HPLC and URP-PCR analysis. *J. Microbiol. Biotechnol., 27*(7), 1223–1232. doi: 10.4014/jmb.1701.01043.

Li, Q. P., He, Y., Liu, J. M., Xia, J. M., Li, W. J., & Liu, X. Z., (2016). Hybrid breeding of high quality of *Hepialus* sp., the host of *Ophiocordyceps sinensis*, and prevention of the host insect reproductive degradation. *Mycosystema, 35*, 456–466.

Li, X., Liu, Q., Li, W., Li, Q., Qian, Z., Liu, X., & Dong, C., (2019). A breakthrough in the artificial cultivation of Chinese *Cordyceps* on a large scale and its impact on science, the economy, and industry. *Crit. Rev. Biotechnol., 39*(2), 181–191. doi: 10.1080/07388551.2018.1531820.

Lin, Q., Long, L., Wu, L., Zhang, F., & Wu, S., (2017). Evaluation of different agricultural wastes for the production of fruiting bodies and bioactive compounds by medicinal mushroom. *J. Sci. Food Agric., 97*(10), 3476–3480.

Lindequist, U., Niedermeyer, T. H. J., & Jülich, W. D., (2005). The pharmacological potential of mushrooms. *Evid. Complement. Altern. Med., 2*(3), 285–299. doi: 10.1093/ecam/neh107.

Liu, Y. W., Liu, N., Su, J. N., Han, B., & Tian, X. R., (2007). Effect of culture condition on mycelium weight and polysaccharides content in the mycelium of *Cordyceps sinensis*. *J. Hebei Univ. (Nat. Sci. Ed.)., 24*, 70–73.

Liu, Z. M., & Cao, Y. S., (1994). Study on solid fermentation cultured *Cordyceps sinensis* mycelium. *Henan Sci., 12*, 335–338.

Liu, Z. Y., Liang, Z. Q., Liu, A. Y., Yao, Y. J., Hyde, K. D., & Yu, Z. N., (2002). Molecular evidence for teleomorph-anamorph connections in *Cordyceps* based on ITS-5.8S rDNA sequences. *Mycol. Res., 106*, 1100–1108.

Mao, X. B., Eksriwong, T., Chauvatcharin, S., & Zhong, J. J., (2005). Optimization of carbon source and carbon/nitrogen ratio for cordycepin production by submerged cultivation of medicinal mushroom *Cordyceps militaris*. *Process Biochem., 40*, 1667–1672.

Masuda, M., Urabe, E., Honda, H., Sakurai, A., & Sakakibara, M., (2007). Enhanced production of cordycepin by surface culture using the medicinal mushroom *Cordyceps militaris*. *Enzym. Microb. Technol., 40*, 1199–1205.
Masuda, M., Urabe, E., Sakurai, A., & Sakakibara, M., (2006). Production of cordycepin by surface culture using the medicinal mushroom *Cordyceps militaris*. *Enzym. Microb. Technol., 39*, 641–646.
Mizuno, T., (1999). Medicinal effects and utilization of *Cordyceps* (Fr.) link (*Ascomycetes*) and *Isaria* Fr. (mitosporic fungi) Chinese caterpillar fungi, "tochukaso" (review). *Int. J. Med. Mushrooms., 1*, 251–261.
Nakalembe, I., Kabasa, J. D., & Olila, D., (2015). Comparative nutrient composition of selected wild edible mushrooms from two agro-ecological zones, Uganda. *SpringerPlus, 4*, 433. doi: 10.1186/s40064-015-1188-z.
Nguyen, L. T., Le, V. V., Nguyen, B. T. T., Ngo, N. X., Nguyen, H. T. T., Nguyen, Q. D., & Mulla, S., (2020). Cultural characteristics and cordycepin production of some *Cordyceps militaris* strains under artificial cultivation conditions. *J. Biotechnol. Comput. Biol. Bionanotechnol., 101*(2), 135–145. doi: 10.5114/bta.2020.94772.
Ning, L., Shu-shana, D., Xue-meia, N., Wen-shenga, Z., En-Ruob, G., & Qin, L., (2006). Determination of nucleosides in natural *Cordyceps sinensis* and cultured *Cordyceps* by RP-HPLC. *Chin. Pharm. J., 41*(12), 948–951.
Park, J. P., Kim, S. W., Hwang, H. J., & Yun, J. W., (2001). Optimization of submerged culture conditions for the mycelial growth and exo-biopolymer production by *Cordyceps militaris*. *Lett. Appl. Microbiol., 33*(1), 76–81.
Park, J. P., Kim, Y. M., Kim, S. W., Hwang, H. J., Cho, Y. J., Lee, Y. S., Song, C. H., & Yun, J. W., (2002). Effect of aeration rate on the mycelial morphology and exo-biopolymer production in *Cordyceps militaris*. *Process Biochem., 37*(11), 1257–1262.
Pathania, P., Joshi, M., & Sagar, A., (2015). Morphological, physiological and molecular studies on wildly collected *Cordyceps militaris* from Northwest Himalayas, India. *Eur. J. Biotechnol. Biosci., 3*(1), 53–62.
Qin, Q. L., Zhou, G. L., Zhang, H., Meng, Q., Zhang, J. H., Wang, H. T., Miao, L., & Li, X., (2018). Obstacles and approaches in artificial cultivation of Chinese *Cordyceps*. *Mycology, 9*(1), 7–9. doi: 10.1080/21501203.2018.1442132.
Raethong, N., Laoteng, K., & Vongsangnak, W., (2018). Uncovering global metabolic response to cordycepin production in *Cordyceps militaris* through transcriptome and genome-scale network-driven analysis. *Sci. Rep., 8*, 9250.
Raethong, N., Wang, H., Nielsen, J., & Vongsangnak, W., (2020). Optimizing cultivation of *Cordyceps militaris* for fast growth and cordycepin overproduction using rational design of synthetic media. *Comput. Struct. Biotechnol. J., 18*, 1–8. doi: 10.1016/j.csbj.2019.11.003.
Redecker, D., Thierfelder, H., & Werner, D., (1995). A new cultivation system for arbuscular mycorrhizal fungi on glass beads. *Angew. Bot., 69*, 189–191.
Seaver, F. J., (1911). The *hypocreales* of North America–IV. *Mycologia, 3*, 207–230.
Shang, C. R., Bai, L. F., Wang, J. Z., & Zheng, J. K., (2011). Optimizing on liquid culture media of *Cordyceps sinensis* mycelia. *Shandong Chem. Ind., 40*, 72–74.
Shang, Y., Xiao, G., Zheng, P., Cen, K., Zhan, S., & Wang, C., (2016). Divergent and convergent evolution of fungal pathogenicity. *Genome Biol. Evol., 8*(5), 1374–1387. doi: 10.1093/gbe/evw082.
Sharma, S., (2004). Trade of *Cordyceps sinensis* from high altitudes of the Indian Himalaya: Conservation and biotechnological priorities. *Curr. Sci., 86*(12), 1614–1619.

Shrestha, B., Han, S. K., Lee, W. H., Choi, S. K., Lee, J. O., & Sung, J. M., (2005). Distribution and *in vitro* fruiting of *Cordyceps militaris* in Korea. *Mycobiology, 33*(4), 178–181.

Shrestha, B., Han, S. K., Sung, J. M., & Sung, G. H., (2012a). Fruiting body formation of *Cordyceps militaris* from multi-ascospore isolates and their single ascospore progeny strains. *Mycobiology, 40*(2), 100–106.

Shrestha, B., Lee, W. H., Han, S. K., & Sung, J. M., (2006). Observations on some of the mycelial growth and pigmentation characteristics of *Cordyceps militaris* isolates. *Mycobiology, 34*(2), 83–91.

Shrestha, B., Tanaka, E., Hyun, M. W., Han, J. G., Kim, C. S., Jo, J. W., Han, S. K., et al., (2016). Coleopteran and lepidopteran hosts of the entomopathogenic genus *Cordyceps sensu lato. J. Mycol., 2016*, 7648219. doi: 10.1155/2016/7648219.

Shrestha, B., Zhang, W., & Zhang, Y., (2012b). The medicinal fungus *Cordyceps militaris*: Research and development. The medicinal fungus *Cordyceps militaris*: Research and development. *Mycol. Prog., 11*(3), 599–614.

Sun, H., Hu, T., Guo, Y., & Liang, Y., (2018). Preservation affects the vegetative growth and fruiting body production of *Cordyceps militaris. World J. Microbiol. Biotechnol., 34*(11), 166.

Sung, G. H., Hywel-Jones, N. L., Sung, J. M., Luangsa-Ard, J. J., Shrestha, B., & Spatafora, J. W., (2007). Phylogenetic classification of *Cordyceps* and the clavicipitaceous fungi. *Stud. Mycol., 57*, 5–59. doi: 10.3114/sim.2007.57.01.

Tao, S. X., Xue, D., Lu, Z. H., & Huang, H. L., (2020). Effects of substrates on the production of fruiting bodies and the bioactive components by different *Cordyceps militaris* strains (*Ascomycetes*). *Int. J. Med. Mushrooms., 22*(1), 55–63.

Thakur, M. P., (2020). Advances in mushroom production: Key to food, nutritional and employment security: A review. *Indian Phytopathol., 73*, 377–395. doi: 10.1007/s42360-020-00244-9.

Turlo, J., (2014). The biotechnology of higher fungi – current state and perspectives. *Folia Biol. Oecol., 10*, 49–65.

Vongsangnak, W., Raethong, N., Mujchariyakul, W., Nguyen, N. N., Leong, H. W., & Laoteng, K., (2017). Genome-scale metabolic network of *Cordyceps militaris* useful for comparative analysis of entomopathogenic fungi. *Gene., 626*, 132–139.

Vyas, R. V., Jani, J. J., & Yadav, D. N., (1992). Effect of some natural pesticides on entomogenous muscardine fungi. *Indian J. Exp. Biol., 30*(5), 235–436.

Wang, C. C., Wu, J. Y., Chang, C. Y., Yu, S. T., & Liu, Y. C., (2019). Enhanced exopolysaccharide production by *Cordyceps militaris* using repeated batch cultivation. *J. Biosci. Bioeng., 127*(4), 499–505. doi: 10.1016/j.jbiosc.2018.09.006.

Wang, F., Song, X., Dong, X., Zhang, J., & Dong, C., (2017). DASH type cryptochromes regulate fruiting body development and secondary metabolism differently than CmWC-1 in the fungus *Cordyceps militaris. Appl. Microbiol. Biotechnol., 101*, 4645–4657.

Wang, J., Kan, L., Nie, S., Chen, H., Cui, S. W., Phillips, A. O., Phillips, G. H., et al., (2015). A comparison of chemical composition, bioactive components and antioxidant activity of natural and cultured *Cordyceps sinensis. LWT—Food Sci. Technol., 63*(1), 2–7. doi: 10.1016/j.lwt.2015.03.109.

Wang, X. L., & Yao, Y. J., (2011). Host insect species of *Ophiocordyceps sinensis*: A review. *Zookeys, 127*, 43–59.

Wang, Y. C., Ko, W. S., & Chang, C. C., (2009). *Methods for Making and Compositions Comprising Fermentation Products of Cordyceps Sinensis.* Google Patents. US Patent App. 12/461, 304.

Wasser, S. P., (2002). Medicinal mushrooms as a source of antitumor and immunomodulating polysaccharides. *Appl. Microbiol. Biotechnol., 60*, 258–274.
Wen, T. C., Li, G. R., Kang, J. C., Kang, C., & Hyde, K. D., (2014). Optimization of solid-state fermentation for fruiting body growth and cordycepin production by *Cordyceps militaris*. *Chiang Mai J. Sci., 41*(4), 858–872.
Wongsa, B., Raethong, N., Chumnanpuen, P., Wong-ekkabut, J., Laoteng, K., & Vonsangnak, W., (2020). Alternative metabolic routes in channeling xylose to cordycepin production of *Cordyceps militaris* identified by comparative transcriptome analysis. *Genomics, 112*(1), 629–636. doi: 10.1016/j.ygeno.2019.04.015.
Wu, D. T., Xie, J., Wang, L. Y., Ju, Y. J., Lv, G. P., Leong, F., Zhao, J., & Li, S. P., (2014). Characterization of bioactive polysaccharides from *Cordyceps militaris* produced in China using saccharide mapping. *J. Funct. Foods, 9*, 315–323. doi: 10.1016/j.jff.2014.05.005.
Yan, J. K., Wang, W. Q., & Wu, J. Y., (2014). Recent advances in *Cordyceps sinensis* polysaccharides: Mycelial fermentation, isolation, structure and bioactivities: A review. *J. Funct. Foods, 6*, 33–47. doi: 10.1016/j.jff.2013.11.024.
Yang, D. R., Peng, Y. Q., Chen, J. Y., Cao, Y. Q., & Yang, P., (2010). The distribution pattern of *Cordyceps sinensis* in China and the response to environmental changes. *Chin J. Grassl., 32*(suppl), 22–27.
Yin, H., Qiao, C., Qin, S., Tang, W., & Jia, S., (2013). Optimization of fermentation medium for *Cordyceps sinensis*. *Food Res. Dev., 34*, 5–8 [in Chinese].
Yin, J., Xin, X., Weng, Y., & Gui, Z., (2017). Transcriptome-wide analysis reveals the progress of *Cordyceps militaris* subculture degeneration. *PloS One, 12*(10), 1–14.
Yu, R. M., Yang, W., Song, L. Y., Yan, C. Y., Zhang, Z., & Zhao, Y., (2007). Structural characterization and antioxidant activity of a polysaccharide from the fruiting bodies of cultural *Cordyceps militaris*. *Carbohydr. Polym., 70*, 430–436.
Yue, K., Ye, M., Lin, X., & Zhou, Z., (2013). The artificial cultivation of medicinal caterpillar fungus, *Ophiocordyceps sinensis* (*Ascomycetes*): A review. *Int. J. Med. Mushrooms, 15*(5), 425–434.
Zhao, R., & Guo, C. J., (2008). Optimizing on liquid culture media of *Cordyceps sinensis* mycelia. *J. Tianjin Normal University (Nat. Sci. Ed.), 28*, 8–11.
Zhou, X. W., Gong, Z. G., Lin, L. J., Su, Y., & Tang, K., (2009). Cordyceps fungi: Natural products, pharmacological functions and developmental products. *J. Pharm. Pharmacol., 61*, 279–291. doi: 10.1211/jpp.61.03.0002.
Zhou, X. W., Li, L. J., & Tian, E. W., (2014). Advances in research of the artificial cultivation of *Ophiocordyceps sinensis* in China. *Crit. Rev. Biotechnol., 34*(3), 233–243. doi: 10.3109/07388551.2013.791245.
Zhou, Y., Wang, M., Zhang, H., Huang, Z., & Ma, J., (2019). Comparative study of the composition of cultivated, naturally grown *Cordyceps sinensis*, and stiff worms across different sampling years. *PLoS One, 14*(12), e0225750. doi: 10.1371/journal.pone.0225750.

CHAPTER 14

Role of Filamentous Fungi in the Production of Antibiotics or Antimicrobial Agents

HIMANKI DABRAL[1,2], HEMANT R. KUSHWAHA[3], and ANU SINGH[2,3]

[1]*School of Agriculture Sciences, Shri Guru Ram Rai University, Dehradun, Uttarakhand, India*

[2]*Samvet Bharat, Dehradun, Uttarakhand, India*

[3]*School of Biotechnology, Jawaharlal Nehru University, New Delhi, India*

ABSTRACT

Filamentous fungi are rich source of production of medicinal compounds at the industrial level as high levels of proteins and metabolites secreted by fungi in their culture medium possess efficient biomedical properties. Fungal species generally possess specific biomedical activities, however, out of a total 1.5 million only ~1,800 fungal species are described to have biomedical activities and prospects to explore new compounds for new biomedical and industrial applications. Initially, filamentous fungi (e.g., *Trichoderma, Aspergillus, Rhizopus*, and *Penicillium* species) were explored for their extraordinary extracellular enzyme synthesis mechanisms and protein secretion machinery for producing single cell proteins for large scale industrial production. Recent developments in the fields of genomics and proteomics offers new opportunities to biotechnologists to combine all the information to extend the bio-manufacturing of recombinant proteins of fungal origins in filamentous fungi. This can be achieved by facilitating filamentous fungi to express multiple proteins encoding genes and identifying new biosynthetic pathways to produce new proteins as well as secondary metabolites.

Applied Mycology for Agriculture and Foods: Industrial Applications. Sanjay K. Singh, Deepak Kumar, Md. Shamim, & Rohit Sharma (Eds.)

Similarly, bacterial cells are also used as bio-factory to express proteins of fungal origin, this method reduces the time of production as compared to fungal method.

14.1 INTRODUCTION

Globally healthcare system is facing challenges due to the antimicrobial resistance exhibited by several classes of bacteria against existing antibiotics for example development of resistant of methicillin against *Staphylococcus aureus,* vancomycin against *enterococc*i. Sometimes the ineffectiveness of the newly developed antibiotic against the causal organism leads to a situation like a pandemic or epidemics (Svahn et al., 2012). This can be overcome by identification, isolation, and exploration of the other microorganism present in the environment as well as their biological active substances. One such microorganism is fungi which represent a valuable source of medicinally active compounds (Hyde et al., 2019). Both Ascomycetes and Basidiomycetes division have been found to be potent source of active compounds but reported to be less utilized for the purpose (Lindequist et al., 2005; Blackwell, 2011), because some of the species are still not studied or not feasible for culture. But now this becomes more approachable with the availability of whole genome sequencing technique which can locate and identify the gene clusters for bioactive compounds (Hyde et al., 2019). This chapter encloses the medicinal importance of the fungi and how they are utilized for the antibiotic production by exploring various scientific advancements to cope with future challenges in our health care system.

14.1.1 THERAPEUTIC SIGNIFICANCE OF FUNGI

14.1.1.1 HISTORY

In many of the historical accounts' fungi have been utilized as a curative agent (Mizuno et al., 1995a). It has also been observed that with exploring fungal diversity many important properties have identified like consumption of *Psilocybe* species (Schultes et al., 2001). Mushroom identified originally by Asian civilization as a health promoting substance that boost the immune system, and later it has used against several diseases (Mizuno et al., 1995b; Hatvani, 2001; Gregori et al., 2007). Both Ascomycota and Basidiomycota represent vast source of fungal diversity (Blackwell, 2011). Fungal species of Basidiomycetes have displayed broad spectrum of medicinal activity like antioxidant, antifungal, antiviral, antibacterial, anti-inflammatory, antitumor, and immuno-modulating effects (Gunde-Cimerman, 1999; Wasser et al., 1999c; Zaidman et al., 2005; Gregori et al., 2007; Shamaki et al., 2013; Badalyan et al., 2019; Sandargo et al., 2019a), yet still being less explored. Similarly, fungi from Ascomycetes division also reported to exhibit antitumor and hypoglycemic activity (Sone et al., 1985; Kiho et al., 1993). Mushroom have been known for their medicinal value about 100 species of mushrooms have been utilized by China in their traditional system of medicine (Smith et al., 2002). Some of the commonly used species of mushroom are presented in (Table 14.1) (adapted from Stamets, 2000).

TABLE 14.1 Some Important Medicinal Mushrooms Used for Therapeutic Purpose

Mushroom	Common Name	Species
Polypore	Turkey Tail	*Trametes versicolor*
Gilled	Enoki	*Flammulinave lutipes*
Polypore	Reishi	*Ganoderma lucidum*
Polypore	Maitake	*Grifola frondosa*
Beech	Shiitake	*Lentinu sedodes*
Gilled	Split Gill	*Schizophyllum commune*
Jelly	White Jelly	*Tremella fuciformis*
Oyster	Oyster	*Pleurotus ostreatus*

14.1.1.2 FUNGAL DIVERSITY

Fungi represent the second most abundant group of organisms after insect (Purvis & Hector, 2000). They generally grow on various substrates like soil, wood, and other communities and form a network of thin filament. According to Hawksworth (1991) the available fungal species accounts of about 1.5 million out of which detail description of only 70–1,00,000 species is available (Kirk et al., 2011). The average number of fully explained species increases every year for example 1229 species were reported during 1980 to 1989, then 1097 more species were described during 1990 to 1999 (Hawksworth, 2001) and further 1196 were described during 1999 to 2009 (Hibbett et al., 2011). In continuation during 2009–2012 the average increase of newly described species was 1430 (Dai et al., 2015). Hawksworth & Luecking in

2017 reported worldwide fungal diversity ranging from 2.2 to 3.8 million. Later on, more than 5.1 million fungal species were recorded worldwide using new tools of molecular screening (O'Brien et al., 2005; Blackwell, 2011; Bass et al., 2011) which further increases up to 12 million (11.7–13.2) species (Bing et al., 2019).

The use of fungi active compounds for therapeutic purpose or for antibiotic production depend upon the fact that fungi and animal both groups share common types of microbial pathogens example *Escherichia coli* and *Staphylococcus aureus* so it would be possible to utilize fungi defense mechanism for the development of potent antibiotics (Zjawiony, 2004). However, fungi are divergent from plants in several characteristics like body structure, composition of cell wall and reproduction modes (Whittaker, 1969); therefore put into separate kingdoms. Beside that they are heterotrophic in nature, lack chloroplast and uptake nutrients from the environment (Sharma, 2004). There are several divisions under kingdom fungi like Basidiomycota, Ascomycota, Blastocladiomycota, Chytridiomycota, and Zygomycota, respectively out of which Basidiomycetes group found to be highly evolved contributing many strains Laetiporus, Panus, Auricularia, Ganoderma, Grifola, Lentinus (Lentinula), Pleurotus, Trametes, Hericium, Coprinus, Flammulina, Schizophyllum, and Tramella commonly utilized for pharmaceutical preparation (Polishchuk et al., 2009). On the other hand, division Ascomycota has been less explored for another therapeutic purpose except anti-hypercholesterolemic activity (Ainsworth, 2008; Blackwell, 2011). As described by Hyde et al., 2019, fungi have great potential to be used for various biotechnological and industrial applications, as shown in Table 14.2.

TABLE 14.2 Different Categories of Biotechnological and Industrial Applications of Fungi

Category		Industrial and Biotechnological Applications
1.	Biomedical usage for humans	Antibacterial antibiotics
		Antimycotics
		Biofilm inhibitors
		Anti-cancer agents
		Anti-diabetes
		Improving nerve functioning
		Traditional Chinese Medicine
		Cardiovascular disease control
		Antiviral agents
		Immunosuppressive and immunomodulatory agents

TABLE 14.2 *(Continued)*

Category		Industrial and Biotechnological Applications
2.	Medicinal usage against plant disease	Biocontrol of plant disease using endophytes
		Biocontrol of insects
		Biocontrol of nematodes and fungal nematicides
		Biocontrol of weeds and herbicides
		Fungal antagonists
		Bio control of rusts and smuts
3.	Enhancing crops and forestry	Biofertilizers
		Arbuscular mycorrhizae
		Ectomycorrhizal fungi in forestry
		Orchid mycorrhizae and endophytes in biotechnology
		Growth promoting hormones
		Mitigating abiotic stress
4.	Food and beverages from fungi	Edible mushrooms
		Food coloring
		Food flavoring
		Wine, beer and spirits
		Functional foods and nutraceuticals
		Probiotic
5.	Other applications	Agricultural waste disposal
		Mycoremediation
		Mycofumigation
		Biomass to biofuel
		Fungal degradation of plastics
		Degradation of polycyclic aromatic hydrocarbon
6.	Commodities: Raw material for other products	Cosmetics
		Agarwood
		Fungal enzymes
		Preservatives
		Organic acids
		Textile dyes

14.1.1.3 SECONDARY METABOLITES PRODUCED BY FUNGI

Secondary metabolites are basically compounds of low molecular weight produced by fungi not utilized for their own growth and reproduction like primary metabolites (De Silva et al., 2012) but they are found to play important role in regulating survival of the host organism from which it is produced through interacting with the environment (Mosunova et al., 2020). There are few classes of fungal derived secondary metabolites based on their precursor type and enzyme used in their biosynthetic pathway (Keller, 2019) terpenes, polyketides, indole alkaloids and non-ribosomal peptides as given in Figure 14.1—are synthesized by following enzymes, respectively terpene cyclases, polyketides synthases, dimethyl tryptophan synthases and non-ribosomal peptide synthetases by utilizing several biosynthetic pathways (Keller et al., 2005). With the advancement of genome sequencing technique more biosynthetic gene cluster have been identified that encode gene for more than one enzyme leading to the manufacturing of hybrid secondary metabolites, e.g., meroterpenoids productions through polyketide synthase (PKS) and terpene cyclase (Itoh et al., 2010). On the other hand, these clusters also encode genes for proteins like ABC transporters which help in the export and protection of secondary metabolites (Gardiner et al., 2005; Wiemann et al., 2009; Brown et al., 2015; Dolan et al., 2015) and provide opportunity for the large-scale production of therapeutic compounds.

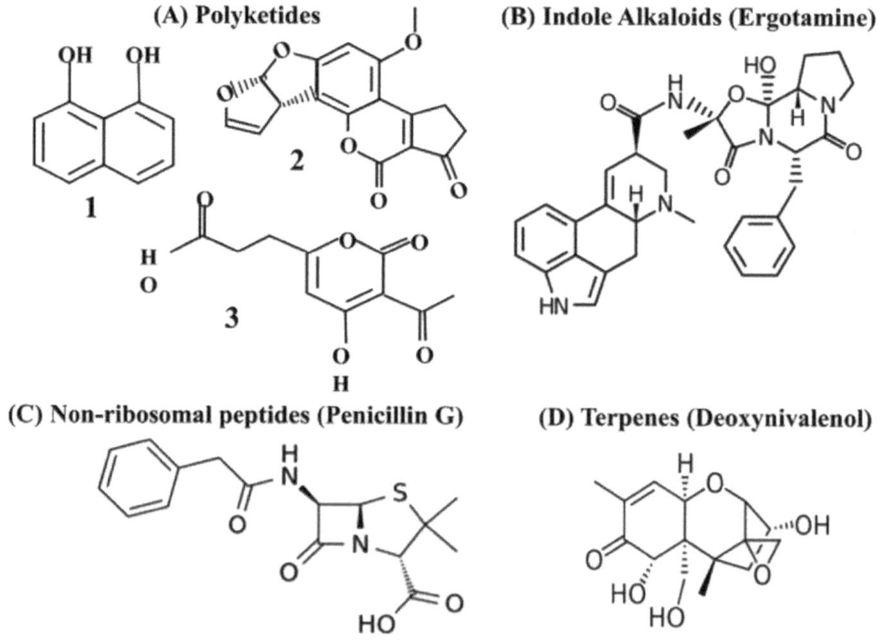

FIGURE 14.1 Some of fungus derived secondary metabolites. (A) Polyketides: (1) 1,8-dihydroxynaphthalene; (2) alatoxin B1; and (3) cyperone B1; (B) indole alkaloids (ergotamine); (C) non-ribosomal peptides (penicillin G); and (D) terpenes (deoxynivalenol).

14.1.1.3.1 Antibiotic Production

The therapeutic significance of fungus was first recognized by Sir Alexander Fleming from the discovery of broad-spectrum antibiotic 'penicillin' obtained from fungus *Penicillium rubens* (Houbraken et al., 2011). This further led to the discovery of another antibiotic cyclosporine A produced by fungus *Tolypocladium inflatum,* successfully used as immunosuppressant during organ transplantation (Colombo & Ammirati, 2011). Monocholin k is another fungal derived drug that contributes annual turn-over of US $50 billion (Hyde et al., 2019). Utilization of fungal protective strategies against another microorganism can help in developing therapeutic strategies for pharmaceutical products.

14.1.1.3.2 Immuno-Suppressants

They represent a class of drugs that are used during organ transplantation to suppress host immune response against the transplanted organ. Some of the most commonly used immunosuppressant drugs include Cyclosporin A isolated from *Tolypocladium inflatum* inhibits interleukin-2 production (Carlile et al., 2000; Kavanagh, 2011). It is an active component of several drugs like Neoral® and Restasis. Similarly, another metabolite Mycophenolic acid produced by *P. roquefortii* (Del-Cid et al., 2016) is the active component of several drugs namely Myfortic® and CellCept® which have cytostatic effect on lymphocytes (Allison & Eugui, 2000).

14.1.1.3.3 Statins

Several statins have been derived from fungal species like lovastatin from *Aspergillus terreus*, mevastatin from *Penicillium citrinum* which can convert to pravastatin via biotransformation (Endo et al., 1976; Manzoni et al., 1999). These statins molecules prevent the activity of 3-hydroxy-3-methylglutaryl-coenzyme A (HMG-CoA) reductase, a vital enzyme used in the conversion of HMG-CoA to mevalonate during the process of cholesterol biosynthesis in liver, therefore act as hypocholesterolemic agents (Shu, 1998; Quinn, 1999).

14.1.1.3.4 Alkaloids

Alkaloids are naturally occurring nitrogen containing heterocyclic compounds obtained from several fungal species like Psilocybe and Claviceps species

(Hoffmann, 2003). These secondary metabolites such as propylamine and ergotamine isolated from *Claviceps purpure* has been utilized for the development of medicine. Indole alkaloid have been reported to exhibit antiviral property (Peng et al., 2013; Li et al., 2014; Zhao et al., 2017).

14.1.1.3.5 Terpenes

Ascomycota and Basidiomycota represent diverse source of secondary metabolites specially terpenoids. These compounds are based on isoprene structures. They occur in several forms such as diterpenes, triterpenes, tetraterpenes, sesquiterpenes, and carotenoids. They have been demonstrated for their anticancer property (Lin et al., 2003). It has been reported that fungi gained genes for terpene biosynthesis from bacteria through horizontal gene transfer (Jia et al., 2019).

14.1.1.3.6 Polyketides

These are the most abundant secondary metabolites of fungal origin which are derived by different classes of PKS enzyme using various pathways. Most of the polyketides are synthesized by using type I PKSs, which are most abundant in fungal genome (Herbst et al., 2018). These polyketides have been used effectively as anticancer, anti-cholesterol, and antibiotic agents. Similarly, polyketide like Bikaverin isolated from Fusarium species found to demonstrated antitumor activity (Limon, 2010) and antibacterial activity against *Ralstonia solanacearum* (Spraker et al., 2018). Monacolin K, citrilin, and azaphilone polyketides derived from *Monascus* spp. are used to control bad cholesterol (Endo et al., 1985; Endo et al., 1986; Akihisa et al., 2005).

14.1.1.3.7 Pigments

Milanin have been found as a natural pigment obtained from fungal spores. It uses as a protective agent has been first reported in the case of albino mutants in maize. Later it is being utilized as an antioxidant and UV protector for human skin (Zao et al., 2016). It is a product of l-3,4-dihydroxyphenylalanine (l-DOPA) pathways (Keller et al., 2019). There are several pigments which have mostly derived from *Monascus* spp. to have medicinal purpose namely monascin, ankaflavin, monascorubin, rubropunctatin, monascropunctamine (Jung et al., 2003) commonly utilized as antimicrobial (Kim et al., 2006), antivital (Sun et al., 2012), anticancer (Yasukawa et al., 1994), and Hperlipidemic agent (Lee et al., 2006).

14.1.1.3.8 Lipids

Lipids are an important component of fungal cell membrane and a source of energy for various metabolic pathways. The typical fatty acid present in the phospholipid membrane includes stearic and palmitic acids along with their derivatives oleic, linoleic, linolenic, and palmitoleic acids (Suutari, 1995). These unsaturated fatty acids retain antibacterial activity against bacteria (both gram-positive and gram-negative) (Kabara et al., 1972; Bergsson et al., 2011).

14.1.1.3.9 Other Bioactive Metabolites

Other vital bioactive metabolites include phenols, lectins, sterols, and polysaccharides with a significant effect on immune response. Like inhibition of HIV-1 reverse transcriptase enzyme using lectin isolated from *Pleurotus citrinopileatus* (Li et al., 2008).

14.2 APPLICATION IN HUMAN HEALTH

It has been clear from the above-mentioned section that fungi are valuable source of bioactive compounds that is being utilized in several pharmaceutical preparation and these metabolites are also helpful in overcoming multidrug resistance against several bacterial and fungal pathogens. Many fungal species like *L. edodes*, *G. frondosa*, *A. bisporus* and *P. ostreatus* are found to be good source of certain vitamins such as niacin, flavin, and pyridoxine (Solomko et al., 1988). They are also found to valuable source of trace minerals like selenium, proteins like hydrophobins, organic acid like malate, fumarate, and ascorbate (Dikeman et al., 2005; Valentao et al., 2005; Iwalokun et al., 2007). Various bioactive compounds of fungal species with biomedical applications are listed in Table 14.3 (adapted from Helen A. Smith; Thesis, 2014).

TABLE 14.3 Various Bioactive Compounds of Fungal Species with Biomedical Applications

Bioactive Compounds	Species	Source	Biomedical Effect
β-glucan	*Ganoderma lucidum*	Fruits bodies	Antitumor, lipocholesteric
Ganomycin	*Ganoderma pfeifferi*	Mycelium	Antibacterial
β-(1,3)-D-glucan	*Grifola frondosa*	Fruit bodies, mycelium	Antitumor, antiviral, antimicrobial
Aqueous extract	*Tramella fuciformis*	Fruit bodies	Reduction in thrombus size, decrease in blood pressure
Ethanolic extract	*Tramellam esenterica*	Culture liquid	Increase immune response
Lovastatin	*Pleurotus ostreatus*	Fruit bodies	Reduction in cholesterol
Lentinan	*Lentinus edodes*	Fruit bodies, mycelium	Antiviral, Antitumor, immune modulator
Krestin, β-D-glycoprotein	*Schizophyllum commune*	Culture liquid	Antitumor
Aqueous ethanolic extract	*Agaricus blazei*	Fruit bodies	Antiviral

14.2.1 ANTIBACTERIAL ACTIVITY

Several metabolites from Basidiomycetes and Ascomycetes have displayed antibacterial activity against large no of bacterial strain (Polishchuk et al., 2009). Several researchers have demonstrated antibacterial action of lentinan and cortinelin isolated from Lentinus edodes against bacteria (both gram-positive and gram-negative) (Jong et al., 1991; Hirasawa et al., 1999; Polishchuk et al., 2009). Antibacterial action of terpenoids, hydroquinone from *Ganoderma pfeifferi* against *S. aureus* (Mothana et al., 2000) have also been reported. Antitumor and hypoglycemic activity of *Cordeyceps militari* have also been observed (Park et al., 2002). The anti-inflammatory effect of fungal polysaccharides chitin has also been reported (Rodrigues et al., 2011).

14.2.2 ANTIVIRAL ACTIVITY

Due to high rate of mutation exhibited by viral genome, the development of specific antiviral drugs is quite challenging (De Palma et al., 2008). Antiviral drugs work by targeting the viral coat protein or the host component that the virus uses for self-replication (Clercq, 2002). Many bioactive compounds isolated from Basidiomycetes possess antiviral properties and has been tested for human immunodeficiency virus (HIV) and herpes simplex virus (HSV) (Wasser et al., 1999). Polysaccharides, proteins, amino acids from Basidiomycota exhibited antiviral activity (Faccin et al., 2007; Razumov et al., 2010; Yamamoto et al., 2013), similarly polyketides isolated from order *Hypocreales*, *Pezizale* exhibited antiviral activity (Pérez et al., 2014; Jia et al., 2015; Pang et al., 2018).

14.2.3 ANTITUMOR ACTIVITY

Fungi represent the diverse source of anti-carcinogenic compounds. It was first demonstrated against sarcoma S-180 by (Lucas et al., 1957). It has been reported that about 650 species belong to Basidiomycota were found to possess to have anti-tumor compounds (Polishchuk et al., 2009). Cordycepin isolated from *Cordyceps militaris* also exhibited anticancer activity. Similarly, several other bioactive compounds namely Perybisin and macrosphelide from Periconiabissoides have been reported to inhibit the adhesion of human leukemia HL-60 cells (Evidente et al., 2014).

14.2.4 ANTIHYPERTENSIVE ACTIVITY

Hypertension and high blood pressure are now the most common health problem in our society due to unhealthy lifestyle. This problem often leads to the development of other diseases like disease, renal failure and diabetes Hypertension is one of the extremely important public health issue. Several fungi like *P. ostreatus* is produces antihypertensive compounds like lovastatin, which act as cholesterol lowering agent (Wasser et al., 1999). Bobek et al. (1991a). Due to low fat and high fiber content *Pleurotussajor-caju* has been used in treating diabetes (Kanagasabapathy et al., 2012).

14.3 BIOMASS PRODUCTION FROM FILAMENTOUS FUNGI

The importance of the fungal secondary metabolites has been well known from the above-discussed sections. It has also been discussed that there are certain fungal species like *Aspergillus*, *Penicillium*, *Trichoderma* which are mostly utilized for the extraction of the bioactive medicinal components. There biomass is produced in a suitable bioreactors following different types of fermentation methods. There are three fermentation methods which are commonly used on the basis of the used substrate, i.e., solid state fermentation (SSF), submerged fermentation (SmF), and biofilm fermentation.

14.3.1 SOLID STATE FERMENTATION (SSF)

In this type of fermentation fungal biomass is produced on the surface of a solid substrate with no water using any one of these bioreactors rotating drum, packed bed bioreactor or tray bioreactors. Selection of the bioreactors for the fungal biomass production depends upon the type of regulation system required for their growth and higher production. The use of solid-state fermentation has the advantage of higher production of secondary metabolite and condition within the bioreactors can be regulated with flexibility of different size matrix, composition, porosity, and water holding capacity (Kumar et al., 1987; Hölker et al., 2004).

14.3.1.1 TYPES OF BIOREACTORS

1. **Tray Bioreactor:** These are the traditional type of bioreactors where solid substrate is placed or fermented on perforated trays with no

mixing. These are kept in a room having controlled temperature and humidity. It has been reported that forced air circulation was beneficial for controlling environmental condition as indicated in Figure 14.2.

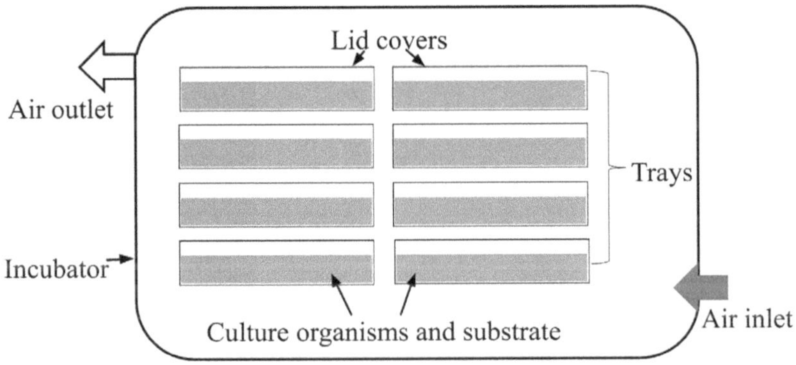

FIGURE 14.2 Schematic representation of tray bioreactor.

2. **Rotating Drum Bioreactor:** These are horizontal shaped cylinder filled with substrate where mixing is done either semi-continuous or continuous mode and heat generated is released through headspace (Figure 14.3). It has been demonstrated that by using intermittent mixing more uniform growth of the mycelium was observed (Musoni et al., 2015).

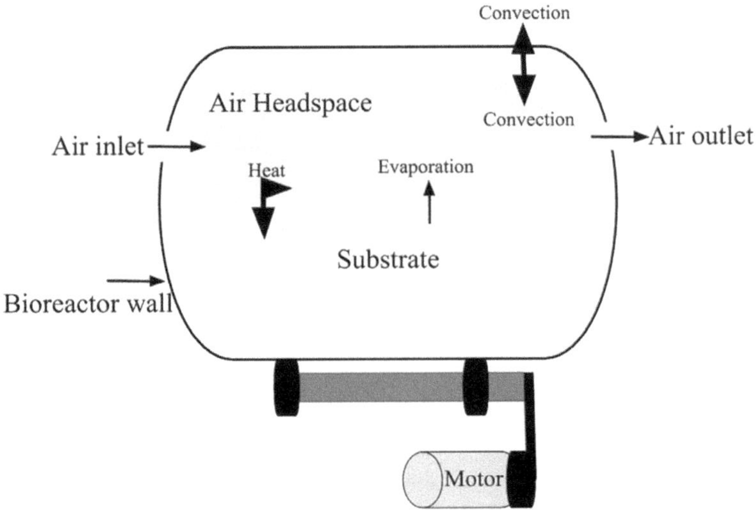

FIGURE 14.3 Schematic representation of rotating drum bioreactor.

3. **Packed Bed Bioreactors:** These are made up of glass or plastic column with perforated base indicated in Figure 14.4. Air is applied to forcefully to the base of the column to achieve effective heat control (Mitchell et al., 2010). Further, it has been demonstrated that forced air with moisture content was effective in maintaining the temperature gradient of the column (Assamoi et al., 2008).

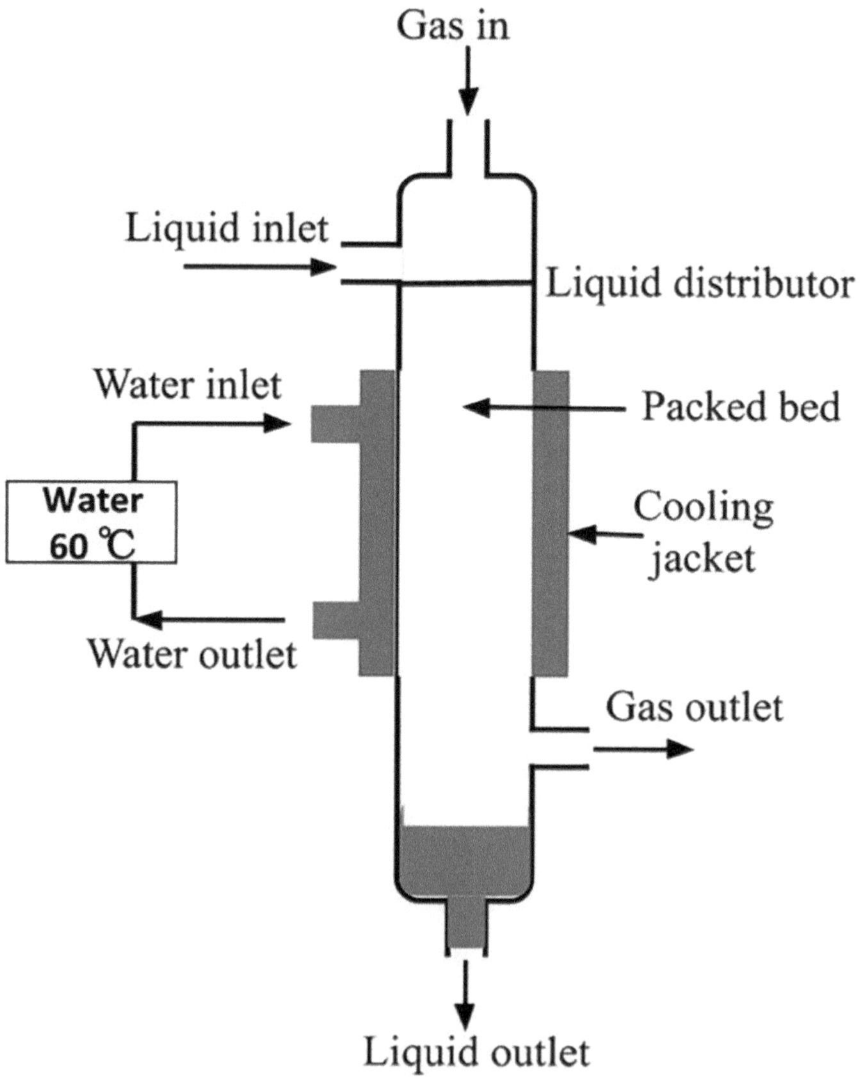

FIGURE 14.4 Schematic representation of packed bed bioreactors.

14.3.2 SUBMERGED FERMENTATION (SMF)

In this method, the fungus is grown in liquid media like broth or molasses is used. After depletion of the nutrients, they are replaced with a constant amount of the fresh media, and extraction is done in the same liquid medium only. The most commonly used bioreactors for SmF includes continuous stirred-bioreactors (Figure 14.5), fluidized-bed reactors and plug-flow reactor. No of antibiotics have been developed by using both SSF and SmF technique given in Table 14.4 (adapted from Subramaniyam et al., 2012).

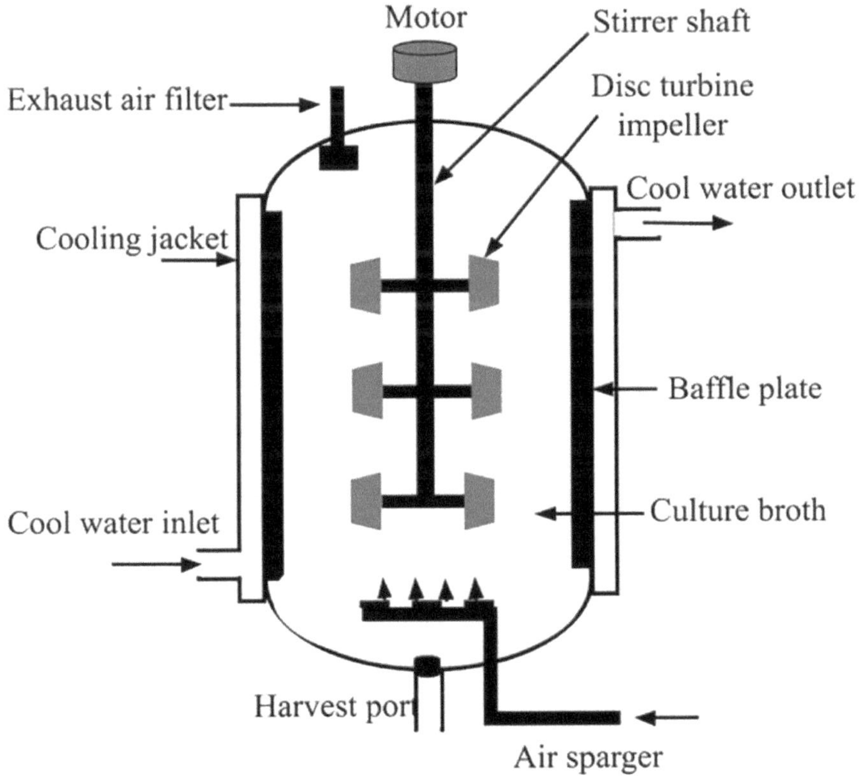

FIGURE 14.5 Schematic representation of stirred tank bioreactor.

14.3.3 BIOFILM FERMENTATION

In this fermentation mycelium is grown on an inert solid support made up of Teflon glass or plastic which is then placed or integrated in a bioreactor

TABLE 14.4 Production of Different Antibiotics Using Various Fermentation Technique

Antibiotic	Organism	Substrate		Productivity	
		SSF	SmF	SSF	SmF
Cyclosporin A	*Trichoderma cylindrosporium*	Wheat bran, agro-industrial residue	Medium containing glucose, casein acid hydrolysate, peptone, and malt extract	1.4 mg/g	0.08 mg/mL
Cephamycin C	*Nocardia lactamdurans*	Soybean flour	Production media	15.75 mg/gds	13.65 mg/mL
Tetracycline S	*Streptococcus viridifaciens*	Sweet potato residue	Cellulosic substrates	2129 μg/g 300 μg/g	N/A
Griseofulvin	*Penicillium griseofulvum*	Rice bran	Production media	9.732 mg/g	0.1 mg/mL
Surfactin	*Bacillus subtilis* RB14	Soybean curd residue	Semisynthetic and synthetic media	200–250 mg/kg	250 mg/L
Mycophenolic acid	*Penicillium brevicompactum*	Pearl barley, wheat bran and rice	Mannitol	6.1 mg/g	1.2 mg/g

to make the cell immobilized (Muffler et al., 2014). Biofilm reactors are available in various designs selected as per the requirement of the biofilm (Figure 14.6).

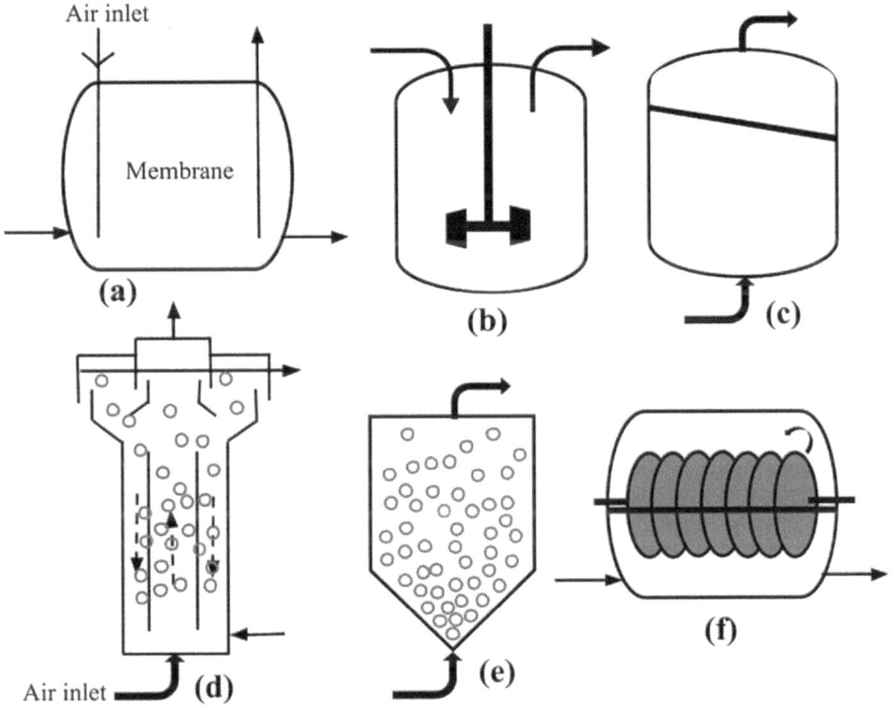

FIGURE 14.6 Common designs of biofilm reactors: (a) membrane biofilm reactor; (b) stirred tank reactor; (c) fixed bed reactor; (d) airlift reactor; (e) fluidized bed reactor; and (f) rotating disk reactor.

14.4 PRODUCTION OF RECOMBINANT PROTEINS USING FUNGAL HOST

14.4.1 FUNGI FOR THE PRODUCTION OF RECOMBINANT PROTEINS

Fungi are found to be valuable source of natural products or secondary metabolites utilized in the development of valuable medicinal product. There are several fungal species which are capable of producing proteins in larger quantities (Iwashita, 2002; Wang et al., 2005) and have been utilized as s secondary host for the expression of heterologous proteins. In the last few decades, the use of fungal cell factories for the development of valuable drugs have been promoted due to the availability of genome sequencing technique which help in the localization and identification of the concerned gene clusters (Weld et al., 2006; Brakhage & Schroeckh, 2011; Wiemann & Keller, 2013; Yaegashi et al., 2014). These gene clusters are basically group of genes that encode several enzymes that is responsible for the development of the concerned secondary metabolite. Then these gene clusters are identified using suitable bioinformatics tool and expressed in secondary host (Lazarus et al., 2014) by gene cloning. Later it was observed recombinant proteins expressed in fungal genome degraded by their homologous proteases. Whereas improper folding and processing of these proteins like improper glycosylation often leads to their degradation (Archer & Peberdy, 1997; Gouka et al., 1997). Therefore, by using genomics and proteomics studies expression of genes encoding multiple proteins have been achieved in the fungal host like overexpression of fungal chaeperons in *Aspergillus niger* along with the gene of interest (Nevalainen et al., 2004. On the other hand, mutagenesis has also contributed to the development of more potent fungal strains utilized for heterologous protein expression (Peterson & Nevalainen, 2012).

14.4.2 BIOSYNTHETIC GENE CLUSTERS

The synthesis of fungal secondary metabolite is governed by genes present in the gene clusters. They encode enzymes like peptide synthetase and PKS involved in the production of the secondary metabolite. The smaller gene clusters contain 2 or 4 genes like for the production of valactamide (Clevenger et al., 2017) or it ranges up to 20 genes example for the production of aflatoxin (Yu et al., 2004). Beside that these small gene cluster in addition to synthase and synthetase enzyme they also contain tailoring enzymes whereas larger gene clusters also contain genes for transcription factors involved in regulation (Neubauer et al., 2016 and Lebar et al., 2018).

14.4.2.1 REGULATION OF BIOSYNTHETIC GENE CLUSTERS

The genes cluster present in the fungal genome encoding various enzyme for secondary metabolite production are regulated at various levels to get the functional product. The levels are as in succeeding sections.

14.4.2.2 REGULATION AT THE LEVEL OF TRANSCRIPTION

Several transcription factors are found to be present within the fungal gene clusters that regulate their transcription. Most common transcriptional factor is C6-zinc cluster protein that recognize sequence specific palindrome present in the promoter region of the gene cluster (Fernandes et al., 1998) and also allow the regulation of genes within cluster and transcription of genes of other metabolic pathways (Wiemann et al., 2014). Another transcriptional factor, i.e., velvet complex has been well studied in *A. nidulans* it was found to affect the production of secondary metabolites in every fungal genus (Bayram et al., 2008).

14.4.2.3 EPIGENETIC REGULATION OF GENE EXPRESSION

It changes the rate of gene expression without changing the gene sequence through chemical modification of histones. It has been reported from that by deletion of gene hdaA for histone deacetylase enzyme resulted in the increase expression of expression of secondary metabolites in *A. nidulans* (Shwab et al., 2007).

14.4.2.4 ENVIRONMENTAL REGULATION OF GENE CLUSTERS

It has been reported in several studies that both the epigenetic control and transcription factor are expressed in response to certain environmental stimulus like which in turn regulate the expression of the concerned gene cluster. Like repression of aflatoxin and sterigmatocystin mycotoxin producing gene cluster by exposure to white light (Bayram et al., 2008) whereas the expression of mycotoxins from Alternaria alternate are affected by blue light (Pruss et al., 2014). Similarly, the expression of polyketide dihydrotrichotetronine is regulated by the activation of Cre A transcription factortowards photoresponse (Monroy et al., 2017).

14.4.2.5 HETEROLOGOUS EXPRESSION OF RECOMBINANT PROTEINS

Aspergillus species are most commonly used as an expression system for the expression of recombinant proteins. Among these species, *Aspergillus nidulance* has been used as a model organism for heterologous expression from other species like, transfer of geodin cluster genes of *A. terreus* into *A. nidulanse* to achieve an increased level of expression of secondary metabolite (Nielsen et al., 2013). The expression of pravastatin which has anti-cholesterol activity has been achieved by co-expression of compactin gene cluster from *Penicillium citrinum*, and cytochrome P450 encoding gene from *Amycolatopsis orientalis* in the expression system (Mclean et al., 2015). Another *Aspergillus* species commonly used is *A. oryzae* where a high level of expression of polyketide tenellin from *Beauveria bassiana* has been achieved (Heneghan et al., 2010). Similarly, increased expression of siderophore ferrirdodin gene from *Fusarium sacchari* into *A. oryzae* has been achieved (Munawar et al., 2013). Recently Bailey et al., 2016, successfully demonstrated the heterologous expression of antibiotic pleuromutilin in *A. oryzae*.

14.4.3 BACTERIAL CELL FOR PRODUCTION OF RECOMBINANT PROTEINS VIA TRANSFORMING FUNGAL GENES

Bacterial cells particularly Escherichia coli (*E. coli*) is widely utilized as host to express recombinant proteins for large scale production. Earlier, there were some limitations associated with the usage of *E. coli* for over-expression of recombinant proteins such as aggregation of recombinant proteins as inclusion bodies in the periplasm/cytoplasm, error in protein folding, insolubility of the recombinant protein, or shortage/unavailability of required chaperones, etc. However, researchers have developed some strategies to overcome these limitations for example: (i) controlled rate of protein synthesis via changing the promoter sequence which can regulate the expression level and adjusting the physical parameters of growth conditions (pH and temperature of the culture medium); (ii) facilitating maximum secretion of recombinant proteins into periplasm resulting in higher solubility, correct folding and downstream processing of recombinant protein; and (iii) co-expression of required chaperone genes (majorly heat shock proteins) to promote proper folding of recombinant proteins (Le et al., 2011). The basic methodology for producing recombinant proteins in *E. coli* is shown in Figure 14.7.

E. coli is used extensively to produce some of the fungal enzymes which are potentially used in industrial application. One of the major example is xylanase, a class of hydrolytic enzymes used to degrade linear polysaccharide xylan into xylose by cleaving heterogeneous β-1,4-glycoside linkage (Bhardwaj et al., 2019). Various microorganisms such as bacteria, yeast, and majorly fungi possess these xylanases and helps nature to biodegrade biomass. Fungal genera mainly include Pichia, Chrysosporium, Fusarium, Versicolor, Trichoderma, Coriolus, Phanerochaete, and Aspergillus. Xylanase has major applications in industries like paper industry, food, and feed, textiles, bio-bleaching of wood pulp, extraction of plant oils, coffee, starch, etc. Primarily xylanase synthesis is performed by fungal culture solid-state fermentation, a high productivity at low cost process. But, advancement in recombinant DNA technology offers to use other organisms for xylanase production at industrial level, e.g., bacteria and yeast along with fungi (Singh et al., 2019).

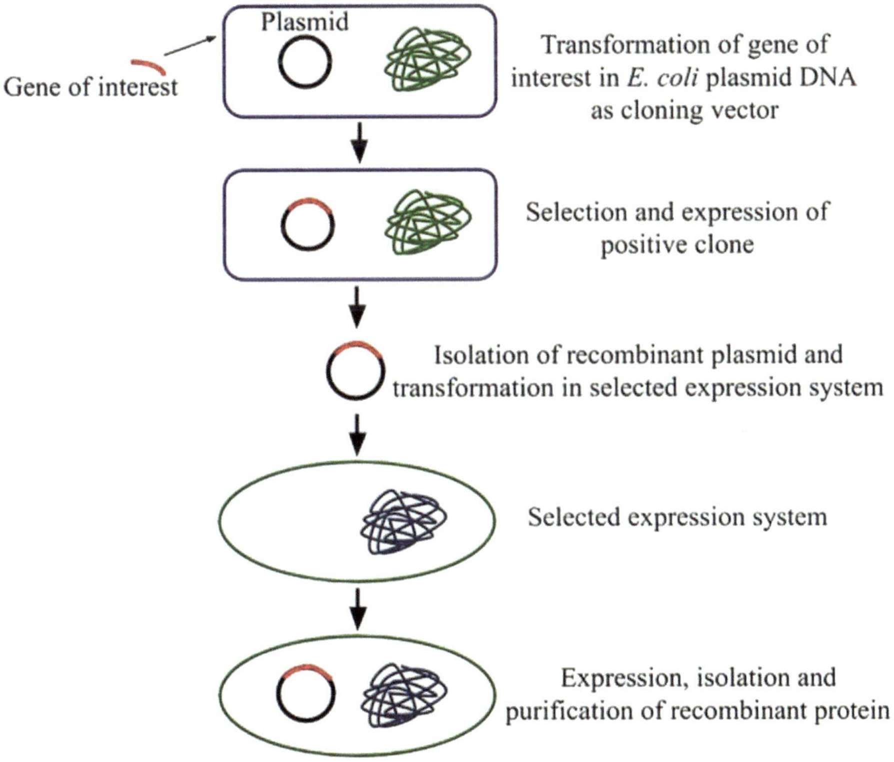

FIGURE 14.7 Basic methodology to produce recombinant proteins in *E. coli*.

E. coli based cloning and expression of xylanases is performed by many researchers with gradual modifications to obtain recombinant protein with optimal activities. Le et al. (2011) demonstrated a strategy to overcome formation of inclusion-body while expressing fungal xylanase in *E. coli*, where an expression plasmid pHsh possessing a synthetic heat-shock promoter was used and the expression of target gene was regulated by an alternative sigma factor (σ32). The recombinant xylanase was expressed in soluble form indicating three essential factors for such expression: the pHsh promoter was controlling the target gene, recombinant protein was exported to the periplasm and upshift in temperature induced gene expression. Similarly, other strategies were developed, e.g., xylanase require N-glycosylation while *E. coli* have ability to do O-glycosylation only. Therefore, a glycosylated β-Xylosidase gene was explored from Phytophthora thermophile (*P. thermophila*) a thermophilic fungus and expressed in *E. coli* system, the extracellular expression of optimally active recombinant protein indicates non-essentiality of glycosylation (Mustafa et al., 2016). Another example of expressing recombinant fungal protein in *E. coli* is manganese peroxidases (MnPs), an oxidative enzyme used for lignin biodegradation and secreted naturally only by *C. subvermispora.* Lin MI et al. (2018) demonstrated a novel method to express soluble MnPs in *E. coli*, where the genes of MnPs were codon-optimized and a cold shock promoter was used as control followed by the continuous addition of hemin during growth to maintain a suitable level of heme, which results in 20-fold higher yield. Later, some chaperones were co-expressed which improves the solubility of the expressed recombinant protein. The purified MnPs characterized to compare the heme-to-protein ratio as generally found in native MnPs, the results were highly satisfying. It is evident that various factors such as fast growth on low-cost substrate, simple transformation strategy, simple isolation and purification methods led to choose *E. coli* as a potential recombinant expression platform to produce fungal enzymes for large scale industrial applications.

14.5 MUSHROOMS: PRODUCTION STATUS AND MARKET VALUE

The global market for mushrooms as per geographical region is categorized into Europe, Latin America, North America, East Asia, South Asia, Oceania, the Middle East & Africa (MEA), and Emerging Countries. The worldwide increasing demand of edible and medicinal fungi has encouraged researchers to isolate and identify new species having therapeutic properties and immunity enhancer. As an example, Chang et al. (1992), explored a new species of mushroom which is a macro-fungus with a characteristic fruiting body (may behypogeous or epigeous), and its size is large enough to see with naked eyes and can be handpicked easily. Till 2005, overall estimated species of mushrooms were 1,40,000 while only 22000 were reported and only 10% were completely explained, this showed a huge scope to investigate new species (Lindequist et al., 2005). During the past two decades it is noticed that China was the leading producer of edible mushrooms worldwide with a production of ~64% and the oyster mushroom production reached over 85%. As per reports during 2001 to 2002, 7% of global supply of mushroom was produced by U.S. with *A. bisporus* majorly (90%) and *Morchella, Hypsizygus, Pleurotus, Lentinula, Grifola, Flammulina, and Herium* the other contributors. The demand of production of button and oyster mushroom is increasing, due to increasing choice of vegan food worldwide and mushrooms contains antioxidants, low fat and lipid content, negligible sodium and gluten, more potassium, selenium, magnesium, calcium, phosphorous, and vitamins which helps in maintaining the good health (Adejoye et al., 2006).

Earlier global trends show continuous increase in total mushroom production for the past 50 years, i.e., from ~3,40,000 kilos in 1965 to ~406 million kilos during 2012–2013, where top sales were $1.1 billion (Chang, 1999; U.S.D.A., 2013). The market value of mushroom production further increased drastically to $35 billion till 2015 with an expected growth of 9.2% between 2016 and 2021 which indicates the market value going to be raised by $60 billion with the end of year 2021. Raman et al., 2018 presented the list of leading mushroom producing countries during 2010–2016, which shows that India was not among the top 10 countries, it's on 11th rank (Table 14.5).

India is mostly transporting mushrooms to nearby countries only, because of the short shelf life of mushrooms. However, mushroom cultivation is increasing in Indian states still there is a huge need for new technologies which can help in increasing the shelf life of mushrooms to transport it easily within the nation and export to other countries as well. Various types of mushroom species are found in India still people are not much interested to consume mushroom on routine basis, though, after knowing about the health benefits of mushroom, it's per capita consumption is increasing, particularly fresh mushrooms. Data published by Directorate of Mushroom Research ICAR, Solan (India) in 2011, shows that mushroom consumption in India is about 30–40 g compared to the consumption of 2–3 kg of mushroom in Europe and US (Figure 14.8) (Raman et al., 2018). Consumption of mushrooms can be increased in India via creating more awareness of health benefits of edible mushrooms as fresh as well as in other forms.

TABLE 14.5 Global Mushroom Production 2010–2016 (tons)

Country	2010	2011	2012	2013	2014	2015	2016
China	4,826,000	5,658,972	6,527,965	7,068,102	7,591,140	8,026,152	7,786,368
Italy	6,84,401	7,61,858	1,016,886	6,04,525	6,00,114	5,94,835	6,83,620
USA	3,59,469	3,90,902	4,02,904	4,08,157	4,32,100	4,29,562	4,19,630
Netherlands	2,66,000	3,04,000	3,07,000	3,23,000	3,10,000	3,10,000	3,00,000
Poland	2,30,000	2,20,000	2,28,300	2,37,069	2,45,747	2,52,944	2,60,140
Spain	1,33,000	1,46,100	1,47,440	1,49,700	1,49,854	2,18,795	1,97,010
Canada	78,452	84,202	1,37,597	1,32,448	1,34,545	1,18,642	1,33,935
France	1,19,373	1,15,696	1,16,602	1,04,461	1,08,671	1,01,135	1,01,949
United Kingdom	69,300	70,740	78,580	85,484	94,857	1,03,197	99,813
Ireland	54,500	57,700	69,600	67,700	69,600	72,200	70,000
India	40,600	40,600	30,000	17,000	28,000	33,699	29,992
Republic of Korea	26,250	24,787	26,000	26,771	26,860	26,292	26,158
All other	5,369,166	6,227,659	7,094,714	7,665,730	8,216,306	8,659,997	8,480,171
Total Production	**12,256,511**	**14,103,216**	**16,183,588**	**168,90,147**	**18,007,794**	**18,947,450**	**18,588,786**

Note: Bata also include truffles which are included for a small share of total production.

Source: FAO, FAOSTAT production database.

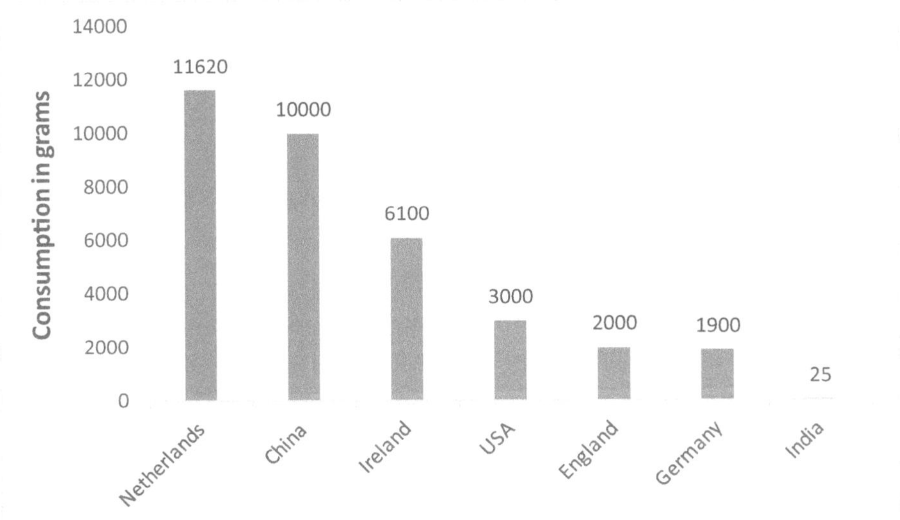

FIGURE 14.8 Mushroom consumption on a global level.
Source: Directorate of Mushroom Research ICAR, Solan (India) (2011).

Among all edible mushrooms, the farming and consumption of button mushroom is rising worldwide along with its market share accounted as largest in 2019. Various research institutes are working on button mushroom to explore its potential anti-cancer activities, estimating huge demand in the future. China Business Research Institute reported that the annual yield of mushroom in China reached to 38.42 million tons in 2017, which represented about 75% of global production. China is the leading producer and exporter of mushroom as exported value was recorded to be $3.8 billion in 2018. Mushroom consumption and cultivation are increasing in Asian countries significantly as the vegan population is increasing, leading to the growing market for mushrooms (https://www.marketsandmarkets.com/Market-Reports/mushroom-cultivation-market-69443268.html).

Surveys show that pharmaceutical industries are interested drastically to produce medicines from natural ingredients of these medicinal fungi. Although the shelf life of mushrooms is less, some of the options rising in the medicinal fungi market these days are processed mushrooms as whole, dry powder, extract, etc. Based on the usage, this market is divided as household usage (fresh, pickles, sausages), commercial (in restaurants, processed foods) and industries. Industries can be further divided into food processing (dietary supplements, beverages, cosmetics, personal care, etc.), and pharmaceutical (antibiotics, antimicrobials, anticancer, immune enhancers, antioxidants, etc.). Mushroom cultivation is simple and seasonal yet profitable business as these can be grown in small areas like lawns, woodlands, polytunnels, soil bags on shelves in darkroom (with chilling chambers), this minimal use of land area is positively affecting the market and encouraging mushroom farming.

14.6 FUTURE PROSPECTS

It is evident that various types of pharmacological activities have shown by many fungi which is enormously broad, some of the examples include anti-bacterial, antiviral, anti-parasitic, antioxidant, antitumor, immune-modulator, blood pressure, hypercholesterolemia, and diabetes. Researchers are much more interested in exploring new species as good nutritious source for dietary supplements, cosmetics, and novel bioactive molecules for biomedical usage. However, drug discovery from natural sources has some bottlenecks as it is quite difficult to discover natural product drug candidates and the paradigm in the pharma industry. A series of fractionation steps is performed for the extraction of a bioactive compound, but the yield is very low, even sometimes known compound is extracted due to duplication (Li et al., 2011). On the other hand, application of fungal polysaccharides (e.g., Krestin, Schizophyllan, and Lentinan) as immunomodulatory and anticancer medicine is restricted due to the fact that fungal polysaccharides with high molecular weight (0.1 to 0.5 million Da) are not digested properly in human body. Therefore, research is going on to develop better drug discovery mechanisms and exploring low molecular weight compounds as potential drug candidate (Goutam, 2011; Zaidman et al., 2005). Basically major research areas for utilizing maximum fungal species with high yield of bioactive product comprises: (i) alternative hosts possibly approved by the Food and Drug Administration board; (ii) improved molecular techniques for expression and secretion of non-fungal proteins; (iii) upgraded metabolic engineering to manipulate targeted biosynthetic pathways and (iv) significant analysis of gene expression via genomics, proteomics, and metabolomics. Multidisciplinary approach of functional genomics, proteomics, metabolomics along with bioinformatics and systems biology can lead to screen potential fungal compounds to be used as potential drug candidate (Hyde et al., 2019).

Functions and interactions of any gene or protein for any specific use is studied under functional genomics via genome or RNA sequencing. Newly developed techniques like PacBio with Illumina guided to explore fully closed genomes and increase the number of fungal genome databases FungiDB and

Ensembl Fungi. Biosynthesis of secondary metabolites is now feasible by the application of OMICS and bioinformatics along with synthetic biology. Annotation tools like FunGeneClusters and antiSMASH have empowered researchers to express heterologous gene clusters in well-established host systems like *A. niger* (Blin et al., 2017; Vesth et al., 2016; Boecker et al., 2018) for biosynthesis of industrially important secondary metabolites to be used as antibiotics (Lin et al., 2019). With the application of genome mining, it is now expected that biosynthesis of secondary metabolites of novel classes will be performed in coming future with very high yields and in a short time span (Dilokpimol et al., 2018; Hyde et al., 2019). Also, recombinant technologies and metabolic engineering of biosynthetic pathways can lead towards the production of new proteins and metabolites as target of interest. The field is open for new achievements as a vast variety of fungal species is still unexplored.

KEYWORDS

- **biosynthetic pathways**
- **fermentation**
- **herpes simplex virus**
- **human immunodeficiency virus**
- **metabolic engineering**
- **polyketide synthase**
- **solid state fermentation**

REFERENCES

Adejoye, O. D., Adebayo-Tayo, B. C., Ogunjobi, A. A., Olaoye, O. A., & Fadahunsi, F. I., (2006). Effect of carbon, nitrogen and mineral sources on growth of *Pleurotus florida*, a Nigeria edible mushroom. *African Journal of Biotechnology, 5*(14), 1355–1359. https://doi.org/ 10.5897/AJB2006.000-5061.

Ainsworth, G. C., (2008). *Ainsworth & Bisby's Dictionary of the Fungi*. CABI Europe.

Akihisa, T., Tokuda, H., Yasukawa, K., Ukiya, M., Kiyota, A., Sakamoto, N., Suzuki, T., Tanabe, N., & Nishino, T., (2005). Azaphilones, furanoisophthalides, and amino acids from the extracts of monascuspilosus-fermented rice (red-mold rice) and their chemopreventive effects. *Journal of Agricultural and Food Chemistry, 53*(3), 562–565. https://doi.org/10.1021/jf040199p.

Allison, A. C., & Eugui, E. M., (2000). Mycophenolate mofetil and its mechanisms of action. *Immunopharmacology, 47*(2, 3), 85–118. https://doi.org/10.1016/s0162 3109(00)00188-0.

Archer, D. B., & Peberdy, J. F., (1997). The molecular biology of secreted enzyme production by fungi. *Critical Reviews in Biotechnology, 17*(4), 273–306. https://doi.org/10.3109/07388559709146616.

Assamoi, A. A., Destain, J., Delvigne, F., Lognay, G., & Thonart, P., (2008). Solid-state fermentation of xylanase from *Penicillium canescens* 10-10c in a multi-layer-packed bed reactor. *Applied Biochemistry and Biotechnology, 145*(1–3), 87–98. https://doi.org/10.1007/s12010-007-8077-z.

Bass, D., & Richards, T. A., (2011). Three reasons to re-evaluate fungal diversity 'on earth and in the ocean.' *Fungal Biology Reviews, 25*(4), 159–164. https://doi.org/10.1016j.fbr.2011.10.003.

Bayram, O., Krappmann, S., Bok, W. J., Helmstaedt, K., Valerius, O., Stromeyer, S. B., Kwon, N. J., et al., (2008). VelB/VeA/LaeA complex coordinates light signal with fungal development and secondary metabolism. *Science, 320*, 1504–1506. https://doi.org /10.1126/science.1155888.

Bergsson, G., Hilmarsson, H., & Thormar, H., (2011). Antibacterial, antiviral and antifungal activities of lipids. In: *Lipids and Essential Oils as Antimicrobial Agents* (pp. 47–80).

Bhardwaj, N., Kumar, B., & Verma, P., (2019). A detailed overview of xylanases: An emerging biomolecule for current and future prospective. *Bioresources and Bioprocessing, 6*, 40. https://doi.org/10.1186/s40643-019-0276-2.

Bing, W., Hussain, M., Zhang, W., Stadler, M., Liu, X., & Xiang, M., (2019). Current insights into fungal species diversity and perspective on naming the environmental DNA sequences of fungi. *An International Journal on Fungal Biology, 10*(3), 127–140. https://doi.org /10.1080/21501203.2019.1614106.

Blackwell, M., (2011). The fungi: 1, 2, 3. 5.1 million species? *American Journal of Botany, 98*(3), 426–438. https://doi.org /10.3732/ajb.1000298.

Bobek, P., Ginter, E., Jurcovicova, M., & Kuniak, L., (1999a). Cholesterol lowering effect of the mushroom *Pleurotus ostreatus* in hereditary hypercholesterolemic rats. *Annals of Nutrition and Metabolism, 35*(4), 191–195. https://doi.org /10.1159/000177644.

Brakhage, A. A., & Schroeckh, V., (2011). Fungal secondary metabolites—Strategies to activate silent gene clusters. *Fungal Genet Bio., 48*, 15–22. https://doi.org10.1016/j.fgb.2010.04.004.

Brown, D. W., Lee, S. H., Kim, L. H., Ryu, J. G., Lee, S., Seo, Y., Kim, Y. H., et al., (2015). Identification of a 12-gene fusaric acid biosynthetic gene cluster in *Fusarium* species through comparative and functional genomics. *Molecular Plant Microbe Interactions, 28*(3), 319–332. https://doi.org10.1094/MPMI-09-14-0264-R.

Carlile, M. J., Watkinson, S. C., & Gooday, G. W., (2000). *The Fungi* (pp. 1–565). Academic Press, London.

Chang, S. T., & Buswell, J. A., (1996). Mushroom nutriceuticals. *World Journal of Microbiology and Biotechnology, 12*(5), 473–476. https://doi.org10.1007/BF00419460.

Chang, S. T., & Miles, P. G., (1992). Mushroom biology — A new discipline. *Mycologist, 6*(2), 64–65. https://doi.org10.1016/S0269-915X(09)80449-7.

Cid, A. D., Dura´n, C. G., Vaca, I., Rojas-Aedo, J. F., Garcı´a-Rico, R. O., Levica´n, G., & Cha´vez, R., (2016). Identification and functional analysis of the mycophenolic acid gene cluster of *Penicilliumroqueforti. PloS One, 11*(1), e0147047. https://doi.org/ 10.1371/journal.pone.0147047.

Clercq, E. D., (2002). Strategies in the design of antiviral drugs. *Nature Reviews Drug Discovery, 1*(1), 13–25. https://doi.org/10.1038/nrd703.

Clevenger, K. D., Bok, J. W., Ye, R., Miley, G. P., Verdan, M. H., Velk, T., Chen, C., et al., (2017). A scalable platform to identify fungal secondary metabolites and their gene clusters. *Nature Chemical Biology, 13*(8), 895–901.https://doi.org/10.1038/nchembio.2408.

Colombo, D., & Ammirati, E., (2011). Cyclosporine in transplantation – a history of converging timelines. *Journal of Biological Regulators and Homeostatic Agents, 25*(4), 493–504.

Dai, Y. C., Cui, B. K., Si, J., He, S. H., Hyde, K. D., Yuan, H. S., Liu, X. Y., & Zhou, L. W., (2015). Dynamics of the worldwide number of fungi with emphasis on fungal diversity in China. *Mycological Progress, 14*(62), 1–9. https://doi.org/ 10.1007/s11557-015-1084-5.

De Silva, D. D., Rapior, S., Fons, F., et al., (2012). Medicinal mushrooms in supportive cancer therapies: An approach to anti-cancer effects and putative mechanisms of action. *Fungal Diversity, 55*, 1–35. https://doi.org/10.1007/s 13225-012-0151-3.

Dikeman, C. L., Bauer, L. L., Flickinger, E. A., & Fahey, G. C. Jr., (2005). Effects of stage of maturity and cooking on the chemical composition of select mushroom varieties. *Journal of Agricultural and Food Chemistry, 53*(4), 1130–1138. https://doi.org/ 10.1021/jf048541l.

Dolan, S. K., O' Keffee, G., Jones, G. W., & Doyle, S., (2015). Resistance is not futile: Gliotoxin biosynthesis, functionality and utility. *Trendsin Microbiology, 23*(7), 419–428. https://doi.org/ 10.1016/j.tim.2015.02.005.

Endo, A., Hasumi, K., & Negishi, S., (1985). Monacolins J and L, new inhibitors of cholesterol biosynthesis produced by *Monascus ruber. Journal of Antibiotics, 38*(3), 420–422. https:// doi.org/ 10.7164/antibiotics.38.420.

Endo, A., Hasumi, K., Yamada, A., Shimoda, R., & Takeshima, H., (1986). The synthesis of compactin (ML-236B) and monacolin K in fungi. *Journal of Antibiotics (Tokyo), 39*(11), 1609, 1610. https://doi.org/doi.org/10.7164/antibiotics.39.1609.

Endo, A., Kuroda, M., & Tanzawa, K., (1976). Competitive inhibition of 3-hydroxy-3-methylglutaryl coenzyme A reductase by ML-236A and ML-236B fungal metabolites, having hypocholesterolemic activity. *FEBS Letters, 72*, 323–326. https://doi.org/ 10.1016/ 0014-5793(76)80996-9.

Evidente, A., Kornienko, A., Cimmino, A., Andolfi, A., Lefranc, F., Mathieu, V., & Kiss, R., (2014). Fungal metabolites with anticancer activity. *Nature Product Reports, 31*(5), 617–627. https://doi.org/ 10.1039/c3np70078j.

Faccin, L. C., Benati, F., Rincão, V. P., Mantovani, M. S., Soares, S. A., Gonzaga, M. L., Nozawa, C., & CarvalhoLinhares, R. E., (2007). Antiviral activity of aqueous and ethanol extracts and of an isolated polysaccharide from *Agaricu sbrasiliensis* against poliovirus type 1. *Letters in Applied Microbiology, 45*, 24–28. https://doi.org/10.1111/j.1472-765X.2007.02153.x.

Fernandes, M., Keller, N. P., & Adams, T. H., (1998). Sequence specific binding by *Aspergillus nidulans* AflR, a C6 zinc cluster protein regulating mycotoxin biosynthesis. *Molecular Microbiology, 28*, 1355–1365. https://doi.org/ 10.1046/j.1365-2958.1998.00907.x.

Gardiner, D. M., Jarvis, R. S., & Howlett, B. J., (2005). The ABC transporter gene in the sirodesmin biosynthetic gene cluster of *Leptosphaeria maculans* is not essential for sirodesmin production but facilitates self-protection. *Fungal Genetics and Biology, 42*(3), 257–263. https://doi.org/ 10.1016/j.fgb.2004.12.001.

Gouka, R. J., Punt, P. J., & Van, D. H. C. A., (1997). Efficient production of secreted proteins by *Aspergillus:* Progress, limitations and prospects. *Applied Microbiology and Biotechnology, 47*(1), 1–11. https://doi.org/10.1007/s002530050880.

Goutam, B., (2011). Natural products in drug discovery: Impacts and opportunities-an assessment. In: Goutam, B., (ed.), *Bioactive Natural Products: Opportunities and Challenges in Medicinal Chemistry* (pp. 1–199). World Scientific. https://doi.org/ 10.1142./9789814335386_001.

Gregori, A., Svagelj, M., & Pohleven, J., (2007). Cultivation techniques and medicinal properties of *Pleurotus spp. Food Technology and Biotechnology, 45*, 238–249.

Gunde-Cimerman, N., (1999). Medicinal value of the genus *Pleurotus* (Fr.) P. Karst. (*Agaricaless.* L., *Basidiomycetes*). *International Journal of Medicinal Mushrooms, 1*(1), 69–80. https://doi.org/10.1615/IntJMedMushrooms.v1.i1.50.

Hatvani, N., (2001). Antibacterial effect of the culture fluid of *Lentinusedodes* mycelium grown in submerged liquid culture. *International Journal of Antimicrobial Agents, 17*(1), 71–74. https://doi.org/10.1016/s0924-8579(00)00311-3.

Hawksworth, D. L., & Luecking, R., (2017). Fungal diversity revisited: 2.2 to 3.8 million species. *Microbiology Spectrum, 5*(4). https://doi.org/10.1128/microbiolspec.FUNK-0052-2016.

Hawksworth, D. L., (1991). The fungal dimension of biodiversity: Magnitude, significance, and conservation. *Mycological Research, 95*(6), 641–655. https://doi.org/10.1016/S0953-7562 (09)80810-1.

Hawksworth, D. L., (2001). The magnitude of fungal diversity: The 1.5 million species estimate revisited. *Mycological Research, 105*(12), 14221432. https://doi.org/10.1017/S0953756201004725.

Helen, A. S., (2014). *Production of Antimicrobials and Antioxidants from Filamentous Fungi.* Thesis. National University of Ireland Maynooth.

Heneghan, M. N., Yakasai, A. A., Halo, L. M., Song, Z., Bailey, A. M., Simpson, T. J., Cox, R. J., & Lazarus, C. M., (2010). First heterologous reconstruction of a complete functional fungal biosynthetic multigene cluster. *Chembiochem,* 111508–111512. https://doi.org/10.1002/cbic.201000259.

Herbst, D. A., Townsend, C. A., & Maier, T., (2018). The architectures of iterative type I PKS and FAS. *Nature Product Reports, 35*(10), 1046–1069. https://doi.org/10.1039/c8np00039e.

Hibbett, D. S., Ohman, A., Glotze, D., Nuhn, M., Kirk, P., & Nilsson, R. H., (2011). Progress in molecular and morphological taxon discovery in fungi and options for formal classification of environmental sequences. *Fungal Biology Reviews, 25*(1), 38–47. https://doi.org/10.1016/j.fbr.2011.01.001.

Hirasawa, M., Shouji, N., Neta, T., Fukushima, K., & Takada, K., (1999). Three kinds of antibacterial substances from *Lentinus edodes* (Berk.) sing. (Shiitake, an edible mushroom). *Internation Journal of Antimicrobial Agents, 11*(2), 151–157. https://doi.org/10.1007/BF03175028.

Hölker, U., Höfer, M., & Lenz, J., (2004). Biotechnological advantages of laboratory-scale solid-state fermentation with fungi. *Applied Microbiology and Biotechnology, 64*(2), 175–186. https://doi.org/ 10.1007/s00253-003-1504-3.

Houbraken, J., Frisvad, J. C., & Samson, R. A., (2011). Fleming's penicillin producing strain is not *Penicillium chrysogenum* but *P. rubens. IMA Fungus, 2*(1), 87–95. https://doi.org/0.5598/imafungus.2011.02.01.12.

https://www.marketsandmarkets.com/Market-Reports/mushroom-cultivation-market-69443268.html (accessed on 13 January 2023).

Hyde, K., Xu, J., Rapior, S., Jeewon, R., Lumyong, S., Niego, A. G., Abeywickrama, P. D., & Stadler, M., (2019). The amazing potential of fungi: 50 ways we can exploit fungi industrially. *Fungal Diversity, 97*(1), 1–136. https://doi.org/ 10.1007/s13225-019-00430-9.

Itoh, T., Tokunaga, K., Matsuda, Y., Fujii, I., Abe, I., Ebizuka, Y., & Kushiro, T., (2010). Reconstitution of a fungal meroterpenoid biosynthesis reveals the involvement of a novel family of terpene cyclases. *Nature Chemistry, 2*(10), 858–864. https://doi.org/10.1038/nchem.764.

Iwalokun, B. A., Usen, U. A., Otunba, A. A., & Olukoya, D. K., (2007). Comparative phytochemical evaluation, antimicrobial and antioxidant properties of *Pleurotus ostreatus*. *African Journal of Biotechnology, 6*(15), 1732–1739. https://doi.org/ 10.5897/AJB2007.000-2254.

Jia, Q., Chen, X., Köllner, T. G., Rinkel, J., Fu, J., Labbé, J., & Chen, F., (2019). Terpene synthase genes originated from bacteria through horizontal gene transfer contribute to terpenoid diversity in fungi. *Scientific Reports, 9,* 9223. https://doi.org/10.1038/s41598-019-45532-1.

Jia, Y. L., Guan, F. F., Ma, J., Wang, C. Y., & Shao, C. L., (2015). Pestalotiolide A, a new antiviral phthalide derivative from a soft coral-derived fungus *Pestalotiopsis* sp. *Natural. Product. Sciences, 21*(4), 227–230. https://doi.org/ 10.20307/nps.2015.21.4.227.

Jung, H., Kim, C., Kim, K., & Shin, C. S., (2003). Color characteristics of *Monascus* pigments derived by fermentation with various amino acids. *Journal of Agricultural and Food Chemistry, 51*(5), 1302–1306. https://doi.org/10.1021/jf0209387.

Kabara, J. J., Conley, A. J., & Truant, J. P., (1972). Relationship of chemical structure and antimicrobial activity of alkyl amides and amines. *Antimicrobial Agents and Chemotherapy, 2*(6), 492–498. https://doi.org/10.1128/aac.2.6.492.

Kanagasabapathy, G., Kuppusamy, U. R., AbdMalek, S. N., Abdulla, M. A., Chua, K. H., & Sabaratnam, V., (2012). Glucan-rich polysaccharides from *Pleurotus sajor-caju* (Fr.) Singer prevents glucose intolerance, insulin resistance and inflammation in C57BL/6J mice fed a high-fat diet. *BMC Complementary and Alternative Medicine, 12*(261). https://doi.org/10.1186/1472-6882-12-261.

Kavanagh, K., (2011). *Fungi: Biology and Applications* (pp. 1–384). Wiley.

Keller, N. P., (2019). Fungal secondary metabolism: Regulation, function and drug discovery. *Nature Reviews Microbiology, 17*(3), 167–180. https://doi.org/10.1038/s41579-018-0121-1.

Keller, N. P., Turner, G., & Bennett, J. W., (2005). Fungal secondary metabolism: From biochemistry to genomics. *Nature Reviews Microbiology, 3*(12), 937–947. https://doi.org/10.1038/nrmicro1286.

Kiho, T., Hui, J., Yamane, A., & Ukai, S., (1993). Polysaccharides in fungi. XXXII. Hypoglycemic activity and chemical properties of a polysaccharide from the cultural mycelium of *Cordyceps sinensis*. *Biological and Pharmaceutical Bulletin, 16*(12), 1291–1293.https://doi.org/ 10.1248/bpb.16.1291.

Kirk, P. M., Stalpers, J. A., Minter, D. W., & Cannon, P. F., (2011). *Dictionary of the Fungi.* CABI Publishing, CAB International, Wallingford, UK.

Kumar, P. K. R., & Lonsane, B. K., (1987). Gibberellic acid by solid state fermentation: Consistent and improved yields. *Biotechnology and Bioengineering, 30*(2), 267–271. https://doi.org/ 10.1002/bit.260300217.

Lazarus, C. M., Williams, K., & Bailey, A. M., (2014). Reconstructing fungal natural product biosynthetic pathways. *Nature Product Report, 31*(10), 1339–1347. https://doi.org/https://doi.org/10.1039/C4NP00084F.

Le, Y., Peng, J., Wu, H., Sun, J., & Shao, W., (2011). An approach to the production of soluble protein from a fungal gene encoding an aggregation-prone xylanase in *Escherichia coli*. *PloS One, 6*(4), e18489. https://doi.org/10.1371/journal.pone.0018489.

Lebar, M. D., et al., (2018). Identification and functional analysis of the aspergillic acid gene cluster in *Aspergillus flavus*. *Fungal Genetics and Biology, 116*, 14–23. https://doi.org/10.1016/j.fgb.2018.04.009.

Li, J. W. H., & Vederas, J. C., (2011). Drug discovery and natural products: End of era or an endless frontier? *Biomeditsinskaya Khimiya, 57*(2), 148–160. https://doi.org/10.18097/pbmc20115702148.

Limón, M. C., Rodríguez-Ortiz, R., & Avalos, J., (2010). Bikaverin production and applications. *Applied Microbiology and Biotechnology, 87*(1), 21–29. https://doi.org/10.1007/s00253-010-2551-1.

Lin, M. I., Nagata, T., & Katahira, M., (2018). High yield production of fungal manganese peroxidases by *E. coli* through soluble expression, and examination of the activities. *Protein Expression and Purification, 145*, 45–52. doi: 10.1016/j.pep.2017.12.012.

Lin, S. B., Li, C. H., Lee, S. S., & Kan, L. S., (2003). Triterpene-enriched extracts from *Ganodermalucidum* inhibit the growth of hepatoma cells via suppressing protein kinase C, activating mitogen-activated protein kinases and G2-phase cell cycle arrest. *Life Sciences, 72*(21), 2381–2390. https://doi.org/10.1016/s0024-3205(03)00124-3.

Lin, Y. L., Wang, T. H., Lee, M. H., & Su, N. W., (2008). Biologically active components and nutraceuticals in the *Monascus*-fermented rice: A review. *Applied Microbiology and Biotechnology, 77*(5), 965–973. https://doi.org/10.1007/s00253-007-1256-6.

Lindequist, U., Niedermeyer, T. H. J., & Julich, W. D., (2005). The pharmacological potential of mushrooms. *Evidence-Based Complementary and Alternative Medicine, 2*(3), 285–299. https://doi.org/10.1093/ecam/neh107.

Lucas, E. H., Ringler, R. L., Byerrum, R. U., Stevens, J. A., Clarke, D. A., & Stock, C. C., (1957). Tumor inhibitors in *Boletus edulis* and other holobasidiomycetes. *Antibiotics and Chemotherapy, 7*(1), 1–4.

Manan, M. A., & Webb, C., (2017). Design aspects of solid-state fermentation as applied to microbial bioprocessing. *Journal of Applied Biotechnology and Bioengineering, 4*(1), 511–532. https://doi.org/10.15406/jabb.2017.04.00094.

McLean, K. J., Hans, M., Meijrink, B., Van, S. W. B., Vollebregt, A., Tee, K. L., Van Der, L. J. M., et al., (2015). Single-step fermentative production of the cholesterol-lowering drug pravastatin via reprogramming of *Penicillium chrysogenum*. *Proceeding of National Academy of Sciences USA, 112*(9), 2847–2852. https://doi.org/10.1073/pnas.1419028112.

Mitchell, D. A., Cunha, L. E. N., Machado, A. V. L., Luz, Jr. L. F. De. L., & Krieger, N., (2010). A model-based investigation of the potential advantages of multi-layer packed beds in solid-state fermentation. *Biochemical Engineering Journal, 48*(2), 195–203. https://doi.org/10.1016/j.bej.2009.10.008.

Mizuno, T., Sakai, T., & Chihara, G., (1995b). Health foods and medicinal usages of mushrooms. *Food Reviews International, 11*, 69–81. https://doi.org/10.1080/87559129509541020.

Monroy, A. A., Stappler, E., Schuster, A., Sulyok, M., & Schmoll, M. A., (2017). CRE1-regulated cluster is responsible for light dependent production of dihydrotrichotetronin in *Trichoderma reesei*. *PLOS One, 12*, e0182530.https://doi.org/10.1371/journal.pone.0182530.

Mosunova, O., Navarro-Muñoz, J. C., & Collemare, J., (2021). The biosynthesis of fungal secondary metabolites: From fundamentals to biotechnological applications. *Encyclopedia of Mycology, 2,* 458–476. https://doi.org/10.1016/B978-0-12-809633-8.21072-8 (accessed on 22 February 2023).

Mothana, R. A. A., Jansen, R., Juelich, W. D., & Lindequist, U., (2006). Ganomycins A and B, new antimicrobial farnesyl hydroquinones from the basidiomycete *Ganoderma pfeifferi*. *Journal of Natural Products, 63*(3), 416–418.

Muffler, K., Lakatos, M., Schlegel, C., Strieth, D., Kuhne, S., & Ulber, R., (2014). Application of biofilm reactors in white biotechnology. *Advances in Biochemical Engineering Biotechnol., 146*, 123–161. https://doi.org/ 10.1007/10_2013_267.

Munawar, A., Marshall, J. W., Cox, R. J., Bailey, A. M., & Lazarus, C. M., (2013). Isolation and characterization of a ferrirhodin synthetase gene from the sugarcane pathogen *Fusarium sacchari*. *Combining Chemistry and Biology, 14*(3), 388–394.https://doi.org/10.1002/cbic.201200587.

Musoni, M., Destain, J., Thonart, P., & Bahama, J. P., (2015). Bioreactor design and implementation strategies for the cultivation of filamentous fungi and the production of fungal metabolites: From traditional methods to engineered systems. *Biotechnology, Agronomy, Society and Environment, 19*(4), 430–442.

Mustafa, G., Kousar, S., Rajoka, M. I., & Jamil, A., (2016). Molecular cloning and comparative sequence analysis of fungal β-xylosidases. *AMB Express, 6*(1), 30. https://doi.org/10.1186/s13568-016-0202-3.

Neubauer, L., Dopstadt, J., Humpf, H. U., & Tudzynski, P., (2016). Identification and characterization of the ergochrome gene cluster in the plant pathogenic fungus *Claviceps purpurea*. *Fungal Biology and Biotechnology, 3*(2). https://doi.org/10.1186/s40694-016-0020-z.

Nevalainen, H., Penttilä, M., & Te'o, V. J. S., (2004). Application of genetic engineering for strain development in filamentous fungi. In: Arora, D. K., Bridge, P. D., & Bhatnagar, D., (eds.), *Handbook of Fungal Biotechnology (Mycology Series 20)* (pp. 193–208). Marcel Dekker, New York.

Nielsen, M. T., Nielsen, J. B., Anyaogu, D. C., Anyaogu, D. C., Holm, D. K., Nielsen, K. F., Larsen, T. F., & Mortensen, U. H., (2013). Heterologous reconstitution of the intact geodin gene cluster in *Aspergillus nidulans* through a simple and versatile PCR based approach. *PLoS One, 8*(8), e72871. doi: 10.1371/journal.pone.0072871.

O'Brien, H. E., Parrent, J. L., Jackson, J. A., Moncalvo, J. M., &Vilgalys, R., (2005). Fungal community analysis by large-scale sequencing of environmental samples. *Applied Environmental Microbiology, 71*(9), 5544–5550. https://doi.org/ 10.1128/AEM.71.9.5544-5550.2005.

Park, J. P., Kim, Y. M., Kim, S. W., Hwang, H. J., Cho, Y. J., Lee, Y. S., Song, C. H., & Yun, J. W., (2002). Effect of agitation intensity on the exo-biopolymer production and mycelial morphology in *Cordyceps militaris*. *Letters in Applied Microbiology, 34*(6), 433–438. https://doi.org/10.1046/j.1472-765X.2002.01126.

Peng, J., Lin, T., Wang, W., Xin, Z., Zhu, T., Gu, Q., & Li, D., (2013). Antiviral alkaloids produced by the mangrove-derived fungus *Cladosporium* sp. PJX-41. *J. Nat. Prod., 76*(6), 1133–1140. https://doi.org/10.1021/np400200k.

Pérez, M., Soler-Torronteras, R., Collado, J. A., Limones, C. G., Hellsten, R., Johansson, M., et al., (2014). The fungal metabolite galiellalactone interferes with the nuclear import of NF-κB and inhibits HIV replication. *Chemical Biology Interaction, 5*(214), 69–76. 10.https://doi.org/10.1016/j.cbi.2014.02.012.

Polishchuk, E. N., & Kovalenko, A. G., (2009). Biological activities of glycopolymers from basidiomycetes mushrooms. *Biopolymers and Cell, 25*(3), 181–193. https://doi.org10.7124/bc.0007DB.

Quinn, R. J., (1999). High-throughput screening in natural product drug discovery in Australia utilizing Australia's biodiversity. *Drug Development Research, 46*, 250–254. https://doi.org/10.1002/(SICI)1098-2299(199903/04)46:3/4<250::AID-DDR9>3.0.CO;2-8.

Raman, J., Lee, S. K., Im, J. H., Oh, M. J., Oh, Y. L., & Jang, K. Y., (2018). Current prospects of mushroom production and industrial growth in India. *Journal of Mushrooms, 16*(4), 239–249.

Razumov, I. A., Kosogova, T. A., Kazachinskaia, E. I., Puchkova, L. I., Shcherbakova, N. S., Gorbunova, I. A., Mikhaĭlovskaia, I. N., et al., (2010). Antiviral activity of aqueous extracts and polysaccharide fractions from mycelium and fruit bodies of higher fungi. *Antibiot Khimioter, 55*(9, 10), 14–18. Russian. PMID: 21400748.

Rodrigues, M. L., Nimrichter, L., Cordero, R. J., & Casadevall, A., (2011). Fungal polysaccharides: Biological activity beyond the usual structural properties. *Frontiers in Microbiology, 2*, 171. https://doi.org/10.3389/fmicb.2011.00171.

Ruiza, H. A., Rodríguez-Jasso, R. M., Rodríguez, R., Juan, C., Contreras-Esquivel, & Aguilar, C. N., (2012). Pectinase production from lemon peel pomace as support and carbon source in solid-state fermentation column-tray bioreactor. *Biochemical Engineering Journal, 65*, 90–95. https://doi.org 10.1016/j.bej.2012.03.007.

Schultes, R. E., Hofmann, A., & Rätsch, C., (2001). *Plants of the Gods: Their Sacred, Healing, and Hallucinogenic Powers*. Inner Traditions/Bear.

Shamaki, B. U., Abdul-Dahiru, E., Abdulrahman, F.I., Ogbe, A. O., & Sandabe, U. K. (2013). *In vitro* Inhibition of neuraminidase activity of influenza virus (H5N2) by methanolic soluble fraction of *Ganoderma lucidum* extract. *International Journal of Modern Biology and Medicine*, *4*(3), 147–154.

Sharma, P. D., (2004). *The Fungi*. Rastogi Publications.

Shwab, E. K., Bok, J. W., Tribus, M., Galehr, J., Graessle, S., & Keller, N. P., (2007). Histone deacetylase activity regulates chemical diversity in *Aspergillus*. *Eukaryotic Cell, 6*(9), 1656–1664. https://doi.org/10.1128/EC.00186-07.

Singh, S., Sidhu, G. K., Kumar, V., Dhanjal, D. S., Datta, S., & Singh, J., (2019). Fungal xylanases: Sources, types, and biotechnological applications. In: *Recent Advancement in White Biotechnology Through Fungi* (pp. 405–428).

Smith, J. E., Rowan, N. J., & Sullivan, R., (2002). Medicinal mushrooms: A rapidly developing area of biotechnology for cancer therapy and other bioactivities. *Biotechnology Letters, 24*, 1839–1845. https://doi.org/10.1023/A:1020994628109.

Solomko, E. F., & Eliseeva, G. S., (1988). Biosynthesis of vitamins B by the fungus *Pleurotus ostreatus* in a submerged culture. *PrikladnaiaBiokhimiia I Mikrobiologiia, 24*(2), 164–169. PMID: 3290883.

Sone, Y., Okuda, R., Wada, N., Kishida, E., & Misaki, A., (1985). Structures and antitumor activities of the polysaccharides isolated from fruiting body and the growing culture of mycelium of *Ganoderma lucidum*. *Agricultural and Biological Chemistry, 49*, 2641–2653. doi: 10.1271/BBB1961.49.2641.

Stamets, P., (2000). *Growing Gourmet and Medicinal Mushrooms* (3rd edn., pp. 1–574). Ten Speed Press, Emeryville.

Subramaniyam, R., & Vimala, R., (2012). Solid state and submerged fermentation for the production of bioactive substances: A comparative study. *International Journal of Science & Nature, 3*(3), 480–486.

Suutari, M., (1995). Effect of growth temperature on lipid fatty acids of four fungi (*Aspergillus Niger, Neurospora crassa, Penicillium chrysogenum,* and *Trichoderma reesei*). *Archives of Microbiology, 164*(3), 212–216. doi: 10.1007/BF02529973.

Ting, C. H., (1999). World production of cultivated edible and medicinal mushrooms in 1997 with emphasis on *Lentinus edodes* (Berk.) sing, in China. *International Journal of Medicinal Mushrooms, 1*, 291–300. https://doi.org/10.1615/intjmedmushr.v1.i4.10.

Valentão, P., Andrade, P. B., Rangel, J., Ribeiro, B., Silva, B. M., Baptista, P., & Seabra, R. M., (2005). Effect of the conservation procedure on the contents of phenolic compounds and organic acids in chanterelle (*Cantharellus cibarius*) mushroom. *J Agric Food Chem., 53*(12), 4925–4931. doi: 10.1021/jf0580263.

Wang, L., Ridgway, D., Gu, T., & Moo-Young, M., (2005). Bioprocessing strategies to improve heterologous protein production in filamentous fungal fermentations. *Biotechnology Advances, 23*(2), 115–129. doi: 10.1016/j.biotechadv.2004.11.001.

Wasser, S. P., & Weis, A. L., (1999a). General description of the most important medicinal higher basidiomycetes mushrooms. *International Journal of Medicinal Mushrooms, 1*, 351–370. doi: 10.1615/INTJMEDMUSHR.V1.I4.80.

Weld, R., Plummer, K., Carpenter, M., et al., (2006). Approaches to functional genomics in filamentous fungi. *Cell Research, 16*, 31–44. https://doi.org/10.1038 /sj.cr.7310006.

Wiemann, P., & Keller, N. P., (2014). Strategies for mining fungal natural products. *Journal of Industrial Microbiology & Biotechnology, 41*(2), 301–313. doi: 10.1007/s10295-013-1366-3.

Wiemann, P., Willmann, A., Straeten, M., Kleigrewe, K., Beyer, M., Humpf, H. U., & Tudzynski, B., (2009). Biosynthesis of the red pigment bikaverin in *Fusarium* fujikuroi: Genes, their function and regulation. *Molecular Microbiology, 72*(4), 931–946. doi: 10.1111/j.1365-2958.2009.06695.x.

Yaegashi, J., Praseuth, M. B., Tyan, S. W., Sanchez, J. F., Entwistle, R., Chiang, Y. M., Oakley, B. R., & Wang, C. C., (2013). Molecular genetic characterization of the biosynthesis cluster of a prenylated isoindolinone alkaloid aspernidine A in *Aspergillus nidulans*. *Organic Letters, 15*(11), 2862–2865. doi: 10.1021/ol401187b.

Yamamoto, K. A., Galhardi, L. C., Rincão, V. P., SoaresSde, A., Vieira, I. G., Ricardo, N. M., Nozawa, C., & Linhares, R. E., (2013). Antiherpetic activity of an *Agaricus brasiliensis* polysaccharide, its sulfated derivative and fractions. *International Journal of Biological Macromolecules, 52*, 9–13. doi: 10.1016/j.ijbiomac.2012.09.029.

Zaidman, B. Z., Yassin, M., Mahajna, J., & Wasser, S. P., (2005). Medicinal mushroom modulators of molecular targets as cancer therapeutics. *Applied Microbiology and Biotechnology, 67*(4), 453–468. doi: 10.1007/s00253-004-1787-z.

Zhao, J., Lu, J., Shen, Y., Tan, Z., Zhang, M., Chen, R., et al., (2017). Stachybotrysams A-E, prenylated isoindolinone derivatives with anti-HIV activity from the fungus *Stachybotrys chartarum*. *Phytochemistry Letters, 20*, 289–294. https://doi.org/10.1016/j.phytol.2017.04.031.

Zhao, L., Kim, J. C., Paik, M. J., Lee, W., & Hur, J. S., (2016). A multifunctional and possible skin UV protectant, (3R)-5-hydroxymellein, produced by an endolichenic fungus isolated from *Parmotrema austrosinense*. *Molecules, 22*(1), 26. doi: 10.3390/molecules22010026.

CHAPTER 15

Advances in Fungal Enzymes and Their Applications

KAKOLI DUTT[1] and GAUTAM KUMAR MEGHWANSHI[2]

[1]*Department of Biosciences, Banasthali Vidyapith, Rajasthan, India*

[2]*Department of Microbiology, M.G.S. University, Bikaner, Rajasthan, India*

ABSTRACT

Microorganisms are associated with humans since the very beginning where they have been used for production of different fermented beverages and foods like beer, wine, whisky, sake, koje, kefir, cheese, yogurt, idli, bread, mushrooms, ogi, rabadi, and many more. Among the microbial diversity of commercial significance, fungi have carved their own singular niche. Not only are they relevant in agriculture, animal feed, health supplements for humans, biodegradation, and nutrient recycling, fungi are needed for production of antibiotics, drugs, different industrial enzymes like amylase, protease, lipase, milk clotting proteases, etc. Developments in fungal biotechnology during the last century have established them as an indispensable part of the white biotechnology. A fungus produces a vast diversity of enzymes with different properties and catalytic potentials. Among them the extracellularly secreted hydrolytic enzymes of fungi are harnessed at commercial scale for use in different commercial sectors based upon hydrolysis of various types of natural complex substrates like starch, cellulose, lignocelluloses, proteins, and lipids. These find further applications in industries like detergent formulations, textile processing, leather processing, polymer synthesis, pharmaceutical synthesis, synthesis of pesticides, waste remediation, biofuels production, and many more.

Applied Mycology for Agriculture and Foods: Industrial Applications. Sanjay K. Singh, Deepak Kumar, Md. Shamim, & Rohit Sharma (Eds.)

15.1 INTRODUCTION

Enzymes are mostly proteins (except ribozymes) with catalytic activity, encoded by the genomes of all living beings and even viruses. They are inevitable for the survival, growth, and reproduction of all organisms (including plants, animals and microorganisms), as the life supporting biochemical processes cannot occur without them. Enzymes possess numerous yet exquisite characteristics such as viability and functionality outside of the producing cell or organism, ability to catalyze a plethora of biochemical reactions and capability of transforming unlimited range of both natural and artificial substrates into useful products which are backbone the modern human lifestyle. They can catalyze all sorts of biochemical/ chemical reactions such as oxidation-reduction, group transfer, hydrolysis, bond cleavage (i.e., cleavage of C–C, C–O, C–N, and other bonds leading to double bonds), isomerization, bond formation (coupled to ATP hydrolysis) and translocation (movement of a molecule usually across a cell membrane). Thus, they find applications in many industrial sectors such as pharmaceuticals, therapeutics, food processing, detergent, biofuels, paper, and pulp, polymer, pesticide, agrochemicals, and many others (Saxena et al., 2005; Meghwanshi, 2018; Meghwanshi & Vashishtha, 2018; Meghwanshi et al., 2020).

Although, enzymes have been sourced from plants (papain, bromelain, etc.), and animals (rennet, trypsin, pepsin, etc.); presently majority of the technical enzymes are sourced from microorganisms, especially fungi and bacteria (Meghwanshi et al., 2020). Both fungi and bacteria are highly exploited for enzymes like amylases, protease, lipases, etc. However, there are many enzymes which are preferentially produced from fungi such as phytases, glucose oxidases, lignin peroxidases, xylanases, cellulases, hemicellulases, mannanases, pectinases, and β-fructofuranosidase, L-Asparaginase, fructosyltransferase, inulinase, and glucose isomerase. The predilection for fungal enzymes is due to their easy identification, convenient culturing, and simple downstream processing along with amazing enzymatic spectra and amenability to genetic manipulation (Meghwanshi & Vashishtha, 2018; Kango et al., 2019). Progress in fungal omics has the potential to further develop the role of enzymes in industrial biotechnology (Peciulyte et al., 2017; Kango et al., 2019). In this chapter, an attempt has been made to present the significance of some commercially relevant groups of fungal enzymes.

15.2 THE CURRENT BIOTECHNOLOGICAL STATUS

There have been several key developments which have shaped the way of microbial use by humanity. In fact, one may call them to be the earliest domesticated organisms. If we observe the uses of fungi alone, the application of yeast in bread making and wine making preceded all written records. It is interesting that Babylonians (circa 6000 BC) and Egyptians (circa 5000 BC) documented the production of various fermented food like wine, beer, and bread, all dependent of yeast fermentation (Kavanaugh, 2005). In the modern times, identifying those responsible for such processes has been quite an eyeopener. We find today that rather than using the whole organism, i.e., fungi or yeast we can work out a plethora of industrial, technical, and food processing based commercial activities very efficiently by using harvested enzymes from them. The first enzyme produced industrially was amylase found in several microorganisms and sourced first from fungi in 1894 (Crueger & Crueger, 1984; Saranraj & Stella, 2013), since then fungi based global enzyme market has flourished with old and new applications constantly boosting their commercial significance. Today, the industrial fungal enzymes dominate the biotechnology market (Lange et al., 2020). Fungal enzymes are widely used in making of various fermented products like bread, beer, and wine (*Saccharomyces cerevisiae*), tempeh (*Rhizopus oryzae* or *R. oligosporus* proteases), and miso/soy sauce/tamari (amylases from *Aspergillus oryzae* or *Aspergillus sojae*). These are responsible for the unique characteristics and taste of many cheeses such as brie, camembert, and blue vein cheese. Currently, the fungal enzymes find massive application in lignocellulose biotechnology for and in the conversion of plant and other types of biomasses to ethanol and biodiesel (Rodionova et al., 2017; Seon et al., 2020).

The exploration of fungal diversity has just started, and a huge range of metabolites including enzymes have been identified from a small group having prominent significance in biotechnology. These few fungal genera make up for the vast majority of metabolites harvested from microbes. It is interesting to observe the enhancement of fungal product diversity as new fungal genera and species continue to be isolated (Lenge et al., 2020). These fungi are resourced producers of a variety of native as well as recombinant enzymes. Currently, abundant information is available as research articles and genomic and protein databases which gives information about the range of metabolites or heterologously expressed biomolecules which can be produced by fungi (Li et al., 2016; Nielsen et al., 2017). This point is exemplified by the expression of human insulin in *Saccharomyces cerevisiae*, and production of a range of enzymes (both single and multiple enzymes) and other biomolecules in recombinant hosts like *Aspergillus* sp. and *Trichoderma* sp. (Arnau et al., 2020).

Fungal enzymes have been commercialized for different industrial applications. Amylases have been exploited for food, detergent, textiles, and bio-ethanol industries while the lipolytic enzymes find use in food and dairy, leather, pharmaceutical, detergent, pulp, and paper, polymer, biodiesel, and waste management (Meghwanshi & Vashishtha, 2018; Verma et al., 2021). Fungal proteolytic enzymes are used in many industries such as detergent, dairy, leather, food, animal feed and pharmaceutical; besides they are also used in waste recycling (da Silva, 2017; Banerjee & Ray, 2017). Among the different carbohydrases obtained from fungi, cellulases are used in manufacturing, detergent aid, coffee processing and animal feeds additive (Panchapakesan & Shankar, 2016; Ahmed & Bibi, 2018) while pectinases from fungi are used in extraction of clear fruit juices and cloudy juices, coffee, cocoa, and tea fermentation, preparation of jams and jellies and also in pickling of vegetables (Meghwanshi, 2018; Kubra et al., 2018; Sudeep et al., 2020). In industries where fruit juice maceration, biobleaching, wastepaper deinking, feed improvement, processing of fibers and biomass saccharification are carried out, hemicelluloses are utilized (Kango et al., 2019). β-glucosidases (BGLs) find use in the production of bioethanol from plant biomass. Other uses like flavor improvement in fruits such as grapes, yellow plum, mango, and strawberry and deinking of wastepaper are dependent on transglycosylation and reverse hydrolysis reactions which form a variety of oligosaccharides, aryl-, and alkyl-β-D-glycosides with wide applications (Ahmed et al., 2017). β-Mannanases have been used in different industries such as animal feed, food, biorefinery, textile, detergent, and paper and pulp (Dawood & Ma, 2020). Fungal xylanases are used in animal feed improvement, pharmaceutical analysis, food industry, pulp, and paper industry, baking industry and vegetables extracts (Polizeli et al., 2005). In pharmaceuticals, fungal L-Asparaginase is of great potential as it can find use in the treatment of various cancers such as acute lymphoblastic and myelocytic leukemia, lymphocytic leukemia, Hodgkin's lymphoma, etc. Another interesting application of this enzyme is in mitigation of acrylamide in food (Saxena et al., 2015; Agrawal & Kango, 2019; Kango et al., 2019). L-Glutaminase known to exhibit antitumor properties finds use in clinical analysis and since it can also work as flavor enhancer in food industry; this enzyme is utilized in the manufacture of various metabolites (Binod et al., 2017). Fructo-oligosaccharides (FOS) a group of prebiotics with several health benefits like being bifidogenic, non-cariogenic, hypolipidemic, and helps in ion absorption through gut is synthesized using fructosyltransferase (Kango et al., 2019). Fungal inulinases are used for the production of ultra-high fructose syrup, bioethanol, and single cell oil (SCO) (Rawat et al., 2017) with similar uses shown by the fungal Glucose isomerases along with xylitol purification (Desai et al., 2016). Heterologous overexpression in suitable hosts, immobilization on novel matrices, and tailoring of fungal enzymes are currently becoming the leading methodology in enhancing various biotechnological exploitations of enzymes and other metabolites (Meghwanshi & Vashishtha, 2018).

15.3 FUNGI USED IN THE FIELD/AREA/BIOTECHNOLOGY

Out of the great diversity of fungal enzymes known around 75% are hydrolytic in nature. In sectors like food and beverages, animal feed, and pharmaceuticals, various carbohydrases like cellulase, lactase, amylase, pectinase, and mannanases find prominent use (Table 15.1). Arnau et al. (2020), have reported a number of fungal spp. for the production of various enzymes, for instance *Aspergillus niger* for the production of aminopeptidase, cellulase, glucanase (beta, endo-, and exo-beta), glucoamylase, glucose oxidase, lactase or β-D-galactosidase, pectinase, and xylanase; *Aspergillus oryzae* for amino-peptidase (leucyl), lipase, protease, tannase; *Aspergillus melleus* for AMP deaminase; *Penicillium funiculosum* for cellulase and xylanase (Walsh et al., 2012); *Trichoderma reesei* for cellulase, glucanase (exo-beta) and xylanase; *Rhizopus oryzae* and *Rhizopus niveus* for lipase; *Penicillium citrinum* for ribonuclease; *Humicola insolens* for glucanase (beta) and xylanase; *Humicola lanuginose* for lipase (Meghwanshi & Vashishtha, 2018); *Rhizomucor miehei* for protease (mucorpepsin) and lipase (Mensink et al., 2017; Meghwanshi & Vashishtha, 2018); *Talaromyces emersonii* for glucanase (beta) and *Chaetomium erraticum* for dextranase. Carbohydrases are also relevant in textile, paper, and biofuels. Increasing pharmaceutical and nutraceutical presence in Asia Pacific also shows an increase in carbohydrase application. Sources of fungal α-Amylase are found mostly in *Aspergillus* species and few species of *Penicillium, P. brunneum, P. expansum* MT-1 *and P. fellutanum* (Erdal et al., 2010; Sundarram et al., 2014). There are some interesting reports of several endophytic fungi like *Cylindrocephalium* sp., *Mycelia sterili*a, *Colletotrichum* sp. as potential amylase producers (Sunitha et al., 2012).

Among yeasts, *Arxula adeninivorans*, *Candida japonica*, *Filobasidium capsuligenum*, *Lipomyces* sp., *Saccharomycopsis* sp., *Schwanniomyces* sp. (Knox et al., 2004) have been cited as amylase producers. An advantage of fungal amylases over the bacterial counterpart is that they are more stable, and this is why α-amylase from *Aspergillus oryzae* was the first commercialized microbial enzyme produced on large scale by solid-state cultivation. It is interesting that *Aspergillus* spp. have taken the limelight among all commercially utilized fungi probably because not only it is ubiquitous and cosmopolitan, but it also has non-fastidious nutritional requirement (Abu et al., 2005). *Aspergillus niger* has been reported to produce many other enzymes like aminopeptidase, arabinofuranidase, catalase, cellulose, beta glucanase, beta galactosidase, pectinase, lysophospholipase, and xylanase. Several species of *Aspergillus* like *Aspergillus melleus, A. niger, A. sojae, A. sydowi, A. terreus, A. nidulans* HA10, *A. funiculosus, A. tamarii, Cephalosporium, Conidiobolus coronatus, Penicillium griseofulvin, P. expansum, Rhizopus oligosporus, R. oryzae, Humicola* sp., *Trichoderma koningii* and *Neurospora crassa* have been reported as key enzyme producers (Sharma, 2019).

Lipases also show variable uses in industries like dairy, biofuels, pharmaceuticals, and detergents. Mostly lipase producing filamentous fungi include *Rhizopus, Rhizomucor, Aspergillus, Mucor, Penicillium, Geotrichum, Ashbya, Fusarium, Humicola, Beauveria, Candida, Alternaria, Eurotrium, Acremonium,* and *Ophiostoma* (Godfredson et al., 1990; Neugnot et al., 2002; Basheer et al., 2011; Costa et al., 2012; de Almeida et al., 2013; Chandra & Enespa, 2019a, b).

Two thirds of the commercial hydrolases are proteases which find usage in industries ranging from food, detergent, pharmaceutical, silk, leather processing, waste management to recovering silver from used X-ray films as well as in protein synthesis. Among all the different roles of proteases, the fungal proteases are main players as not only they are higher producers but there is a distinct advantage of downstream processing (Gupta et al., 2002; Johnvesly & Naik, 2001; Qing et al., 2013; Rao et al., 1998; Hajji et al., 2010; Savitha et al., 2011; Abidi et al., 2011). The high applicability of proteases in food processing stems from improving the digestibility, sensory, and nutritional quality of food along with reducing allergenic compounds (Tavano, 2013). A very prominent role is played by *Mucor miehei* and *Endothia parasitica* in cheese making replacing the traditionally used calf chymosin. Other significant applications are there in beer haze clarification, modification of dough viscoelastic properties and production of protein rich foods like hydrolysis of soy protein, gelatin, casein, and whey protein,

protein recovery from meat and its tenderization (Ward, 2011; Kumari et al., 2012; Tomar et al., 2008; Demain & Adrio, 2008).

Many deep-sea fungi such as *Aspergillus terreus, Acremonium butyric* and *Beauveria brigniartii* have been reported for producing psychrophilic proteases (Srilakshmi et al., 2015). High temperature active proteases from thermophilic fungi such as *Thermomyces lanuginosus, Thermomyces ibadanensis, Thermoascus aurantiacus, Canariomyces thermophila, Talaromyces, Dactylomyces thermophiles* and *Myriococcum thermophilum* with commercial applications have been reported by Keil (1992); and Srilakshmi et al. (2015). Strains of yeast, including *Candida lipolytica, Yarrowia lipolytica* and *Aureobasidium pullulans* are used for large scale protease production (Erjavec, 2012; Srilakshmi et al., 2015). The genus *Aspergillus* alone contributes more than 25% of proteases from fungal sources. The important species are *Aspergillus niger, A. candidus, A. fumigatus, A. flavus, A. melleus, A. sojae, A. oryzae, A. sydowi*, and *A. sulfurous* (Srilakshmi et al., 2015). Strains of *Aspergillus, Penicillium, Trichoderma, Phanerochaete*, and *Fusarium* among fungi and yeasts (*Yarawia lipolytica* and *Saccharomyces cerevisiae*) have also been reported as pectinase producers by Texeira et al. (2011); Heerd et al. (2012); Ruiz et al. (2012); Dange & Harke (2018), El-Enshasy et al. (2018).

Among other hydrolytic enzymes, tannases which are inducible enzymes are produced from *Aspergillus* sp. where *A. fumigatus, A. versicolor, A. flavus*, and *A. caespitosum* are most notable. Other significant tannase producing genera are *Penicillium, Fusarium*, and *Trichoderma*. These enzymes are relevant in winery where they prevent phenol induced mediarization of wine. Tannases play a role in the clarification of beer and fruit juices, malt polyphenol stabilization, manufacture of coffee flavored soft drinks, etc. (Mehta et al., 2013; Darah et al., 2011).

Pectinases are used in the preparation of clear fruit juices, tea, and coffee fermentation. Several new applications using pectinases have been identified like in the oil extraction processes or pharmaceutical or cosmetics, where they help in higher extraction of products through cellular degradation. Degraded pectins are also used in tablet preparation, nutritive fiber preparation, etc. Though by definition, pectinases are identified as enzymes capable of degrading pectic substances like pectinic acids, propectins, pectins, and pectic acids. Thus, several subclasses of pectinases are recognized like polygalacturonase, pectinesterase, protopectinase, and pectin lyase. Predominant fungal producers include various members of *Aspergillus, Penicillium*, and *Fusarium* (Satapathy et al., 2020).

TABLE 15.1 Diversity of Fungal Enzymes and Their Producers

Fungi	Species Identified	Enzymes
Aspergillus	*A. melleus, A. niger, A. sojae, A. sydowi, A. terreus, A. nidulans HA10, A. funiculosus, A. tamarii*	Aminopeptidase, arabinofuranidase, catalase, cellulose, beta-glucanase, beta-galactosidase, protease, tannase, lipase, pectinase, lysophospholipase, and xylanase
Penicillium	*Penicillium citricum* Thom, *Penicillium funiculosum* Thom, *Penicillium vitale, Penicillium pinophilum*	Cellulase, catalase, dextranase, β-galactosidase
Rhizopus	*Rhizopus oligosporus, Rhizopus oryzae*	Protease, lipase
Mucor	*Mucor miehie*	Rennet
Humicola	*Humicola fuscoatra* MTCC 1409	Cellulase
Cepahalosporium	*Cepahalosporium acremonium* (Corda)	Proteinase
Conidiobolus	*C. coronatus*	Alkaline protease
Ganoderma	*G. neo-japonicum*	Beta-glucosidase, cellulase, avicelase, pectinase, xylanase, protease, amylase, ligninase
Trichoderma	*Trichoderma reesei, Trichoderma viride*	α-Amylase, xylanase, cellulase, β-glucanase
Saccharomyces	*Saccharomyces cerevisiae*	Invertase, raffinase

15.4 MECHANISM OR MODE OF ACTION

Depending upon the enzyme type and the specific substrate the mechanism of action varies. The enzymatic processes have been given a lot of emphasis because of their conformation to the principles of green chemistry. The enzymatic route gives a more concise way of organic synthesis with lesser wastes, it also reduces the need for different preparatory steps like functional-group activation, protection, and deprotection steps. This leads to a more cost-effective manufacturing process. When joined to the ability of functioning at ambient conditions makes them a better candidate for environmentally sustainable processes. Enzymatic processes are also highly selective and capable of distinguishing between regio- and stereoisomers along with various functional groups (chemoselective). These processes traditionally have been commercially used in the hydrolysis and formation of, inter alia, ester, amide, and glycoside bonds, especially where very high selectivities were required (chemo-, regio-, or stereo-selectivity). Here, the overall mechanism of few selected industrially relevant enzymes is presented (Sheldon et al., 2020).

15.4.1 LIPASES

Lipases catalyze esterification reaction, particularly in the polar-nonpolar interface and in different organic solvents. These are monomeric proteins with molecular weight generally between 19 and 60 kDa. Presence of calcium enhances its activity, but other divalent cations can lead to suppression of enzyme action. Lipases can carry out hydrolysis, esterification, interesterification, acidolysis, alcoholysis, and aminolysis. Based on the regiospecificity two types of lipases are identified. Those that liberate fatty acids from all three positions and those which specifically break and liberate the acylglycerols from 1,3 position. Lipases along with other hydrolases like esterases, epoxy hydrolases, peroxidases, dehalogenases, and proteases belong to the αß-hydrolase family which are characterized by the presence of an active site possessing a highly conserved catalytic triad. The nucleophilic residue may be composed of serine, cysteine or aspartic acid where serine is always associated with lipases. The remaining parts of the catalytic triad include one histidine residue and one catalytic acidic residue composed of aspartic or glutamic acid (Anobom et al., 2014).

When lipase comes in contact with the substrate, the active site serine residue performs a nucleophilic attack on the carbonyl carbon of the ester bond, resulting in an acyl-enzyme intermediate which releases an alcohol. Thus, a diacylglycerol is obtained from a trialcyglycerol, and the process is stabilized due to the presence of histidine and aspartic acid on the active site. This is followed by another nucleophilic attack. As many compounds can act as acyl donors or nucleophilic compounds which can break the acyl enzyme intermediate, broad substrate specificity is noted. This enables the lipases to carry out other reactions like transesterification, esterification, interesterification, and acidolysis (Jaeger & Reetz, 1998; Nardini & Dijkstra, 1999; Gupta et al., 2004).

15.4.2 PROTEASES

These enzymes also known as peptidases cleave peptide bonds in proteins leading to their breakdown into smaller peptide units or amino acids. Their mode of action can be general or specific where selective protein cleavage is seen like milk clotting in cheese or post-translational modification. Proteases can show preference for substrate chain length and residues on the active site along with the nucleophilic agent. These exhibit variation in the pH range of catalysis along with their requirement for divalent cations to augment their activity. da Silva (2017) stated proteases can be classified according to the source of harvest being plant or animal or fungal or bacterial. Another can be on the preferred site of cleavage, as endo or exopeptidases. While endopeptidases characteristically cleave the peptide bonds within the polypeptide chain, exopeptidases act only towards the end of the polypeptides either at the N or C terminus. Proteases can further be classified on the basis of active site main residue and nucleophile agent involved in the catalytic process along with the homology and molecular structure (Rao et al., 1998; Silva et al., 2014; Halabi et al., 2009; Krem & Di Cera, 2001; de Souza et al., 2015). These differences divide proteases into seven categories (Rao et al., 1998; Klein et al., 2019):

1. **Serine Proteases:** In the hydrolysis reaction catalyzed by these proteases, serine residue plays the role of the nucleating agent. A two-step hydrolysis is observed where loss of the amino acid/peptide fragment takes place with the formation of a covalently linked enzyme peptide intermediate. The acylation process is followed by a diacylation process where the serine residue acting as a nucleophile enables the hydrolysis of peptide in presence of water.
2. **Aspartic Proteases:** They possess an Asp-Thr-Gly motif in the active site. Studies have shown that their mechanism involves an acid-base catalysis with lytic water molecule involved in the reaction. The nucleophilic agent is a water molecule activated by two aspartates.
3. **Cysteine Proteases:** These proteases follow a double-displacement mechanism, wherein a nucleophilic cysteine residue first attacks the carbon of the peptide bond of the substrate, producing the first tetrahedral thioester intermediate with simultaneous release of an amine or amino terminus fragment of the substrate. This is followed by the hydrolysis of the thioester bond which produces a carboxylic acid moiety from the hydrolyzed substrate (Coulombe et al., 1996).
4. **Metalloproteases:** The catalytic activity of these proteases depends on a divalent cation (such as zinc), which is held in place by amino acid ligands (e.g., histidine, aspartate, glutamate, or lysine). The divalent metal ion activates the water molecule that plays the role of a nucleophile. The metal ligand (i.e., the amino acid residue) may play the role of an electrophile and is also required for catalysis.

5. **Threonine Peptidase:** These peptidases employ a Thr residue as a central catalytic residue and nucleophile. They utilize a catalytic charge relay system to activate their secondary hydroxyl nucleophile for catalysis. These residues are found in multiple catalytic subunits of the proteasome. The N-terminal amino group (i.e., serine or cysteine) or a water molecule present between the α-amine and Thr side chain acts as proton donor which is necessary for the hydrolytic reaction (Brannigan et al., 1995; Klein et al., 2018).
6. **Glutamic Peptidase:** This endopeptidase group was identified in fungi in 2004 as eqolysins. Since then, renamed as glutamate peptidases, they possess glutamic acid and glutamine as the catalytic dyad. Their three-dimensional structure is different from other peptidases.
7. **Asparagine Peptide Lyase:** The mechanism of this category of protease is different from the other six groups. These enzymes show self-cleavage which does not include hydrolysis. An asparagine residue dependent cleavage is noted upon the breakage of peptide bonds amidine lyase. The nucleophile is Asp residue which attacks its own main chain carbonyl to form via intramolecular cyclization a five membered succinimide intermediate. Ultimately a peptide bond will get cleaved in an elimination reaction in the absence of water (Klein et al., 2018).

15.4.3 AMYLASES

Amylases are generally categorized as alpha, beta or gamma amylase. All three exhibit different specificities. Alpha amylases are metalloenzymes dependent on calcium ions causing random cleavage in long chain carbohydrates resulting in the formation of maltotriose and maltose from amylose, or maltose, glucose, and "limit dextrin" from amylopectin. On the other hand, beta amylases are highly specific cleaving off a maltose unit at a time as it acts on the second α-1,4 glycosidic bond. γ-amylase cleaves the terminal α-1,4 glycosidic linkage at the non-reducing ends of amylose and amylopectin to produce glucose. It also cleaves α-1,6 glycosidic linkages (Saranraj & Stella., 2013).

In the active site of amylases, glutamic acid and aspartic acid form the catalytic diad. Here the glutamic acid works as the acid/base catalyst and helps in transition state stabilization during hydrolysis. Whereas the aspartic acid residue acts as a nucleophile during intermediate formation. Besides these, amino acid residues such as histidine, arginine, and tyrosine help in orienting the substrate into the active site. They also help in proper orientation of the nucleophile and polarization of substrate molecules (Nakamura et al., 1993; Uitdehaag et al., 1999; Strokopytov et al., 1996; Muralikrishna & Nirmala, 2005; Kumari et al., 2012).

15.5 FACTORS AFFECTING FUNGAL ACTIVITY/ENVIRONMENTAL CONDITIONS

Most of the filamentous fungi used for enzyme production belong either to genera *Aspergillus*, *Penicillium*, *Rhizopus*, *Mucor* or *Trichoderma*. The advantage that these fungi have over other microbial producers is the higher enzyme yield using a great variety of cheap substrates with ease in scale up and down streaming. During cultivation, fungi grow in different forms like free mycelia and pellet in liquid and pellet and mycelia to conidia at solid-air interface (Musoni et al., 2015). The production strategies for fungal cultivation can be:

- Submerged fermentation with two evident fungal morphologies of free mycelia of various lengths and pellets.
- Semi solid fermentation with mycelial growth on the surface and within the solid substrate.
- Solid state fermentation (SSF) with mycelial growth on substrates.
- Biofilm formation where mycelial development occurs on an inert support (Barrios-Gonzalez, 2012; Musoni et al., 2015).

During production, certain factors play a significant role in the production of the enzymes. With fungi, the SSF strategy has been reported to be a better method. There are several reports (Sandhya et al., 2005; Sun & Xu, 2009) which attempt to summarize these methodologies for enzyme production. The advantages conferred by SSF over SMF are low production cost, use of raw materials particularly agro wastes as substrate, less energy and space requirements along with ease in harvesting and stable end product in higher concentration per unit volume (Das & Mukherjee, 2007; Sun & Xu, 2009). The disadvantage, however, is the water availability and heat dispersal. These problems are overcome in the SMF route which also supports easy recovery but high dilution of product. Still the process has abundant water and buffering with respect to temperature, oxygen, and nutrients (Souza et al., 2015).

15.5.1 PROTEASE

Large scale production of extracellular protease is heavily influenced by pH, temperature, media composition, C/N ratio, presence or absence of easily metabolizable sugar/nitrogen, aeration, inoculum density and incubation period (Abidi et al., 2011; Gupta et al., 2002). Species of *Penicillium*, *Aspergillus*, thermophilic fungi like *Thermoascus aurantiacus* and *Thermomyces lanuginosus* have been cultivated by the SSF system (Merheb et al., 2007; Germano et al., 2003; Vishwanatha et al., 2010). Cultivation pH can alter from acidic to alaline range and the temperature also shows variation depending upon the characteristic of the fungi. Generally, the temperature can be as low as 25°C (Mucor circinelloides) or as high as 60°C (*Thermoascus auranticus*) (Andrade et al., 2002; Merheb et al., 2007).

15.5.2 LIPASES

The various physicochemical parameters like temperature, pH, and agitation play a significant role in lipase production. However, the substrates used along with the inorganic salts and nitrogen play a crucial role in determining the extracellular lipase yield. Since the inducible nature of fungal lipases is well established, use of different fatty acids/triglycerides (TGs), hydrolysable bile salts along with tweens and glycerol have been evaluated by various researchers (Fickers et al., 2004; Singh & Mukhopadhyay, 2012). Simple and complex carbon sources such as monosaccharides, sugars, sugar alcohol, polysaccharides, whey, amino acids and other complex sources also play a pivotal role (Gilbert et al., 1991; Rashid et al., 2001). Corn steep liquor, yeast extract, and peptone have been reported as the best nitrogen sources for lipase production from *Penicillium verrucosumin* (Fickers et al., 2004; Mehta et al., 2017). SMF and SSF strategies have been used for lipase production with *Rhizopus, Penicillium*, and *Aspergillus* grown predominantly via SSF methods using agro wastes (Mehta et al., 2017). In a review, Singh & Mukhopadhyay (2012), different agroresidues like soybean meal, wheat bran (WB), soybean bran, oil cakes (e.g., Gingelly oil cake), cassava bagasse, apple pulp and beans residue have been successfully used for lipase production. Furthermore, use of low-cost supplements like soybean oil, slaughterhouse wastewater, corn steep liquor, yeast hydrolysate or several vegetable oils including castor, soybean, olive, corn, and palm oil have inducible effect on lipase production.

15.5.3 AMYLASE

SSF has again been cited as a better strategy for amylase production by fungi (Saranraj & Stella, 2013). There are reports of variation in fungal morphology in *Aspergillus oryzae* upon alteration of pH where at lower pH freely dispersed hyphae was noted, but at pH 6.0 and above only pellets were seen. Oil cakes have been identified as good substrates along with wheat straw (WS), rye straw, corncob leaf, and WB for fungal cultivation via SSF (Ramachandran et al., 2004; Bilal & Figen, 2006). Other suggested substrates are maltose, sucrose, glucose, starch, and glycogen. These substrates are generally used in SMF amylase production by fungi like *Aspergillus tamarii* (Kundu & Das, 1970). SSF has been successfully carried out with *Aspergillus oryzae*, *Penicillium janthinellum* at elevated temperatures in the range of 45–50C. Increasing temperature beyond this range resulted in loss of enzyme production probably due to the negative effects on metabolism caused by water loss (Prakasham et al., 2007). Variations in incubation period up to 96 h have been reported (Erdal & Taskin, 2010). Both organic and inorganic nitrogen has been used for amylase production, but better yields are reported with organic sources like peptone, casein hydrolysate, meat extract, etc., when *Penicillium fellutanum, P. expansum, Aspergillus fumigatus, A. niger*, and *Thermomyces lanuginosus* were used (Sundarram et al., 2014).

15.6 EXAMPLES OF COMMERCIAL TECHNOLOGIES/FUNGAL PRODUCTS

The global industrial enzyme market has developed from 6 billion USD (1917) to USD 9.9 billion (2019) with a future projection of compound annual growth rate (CAGR) of 7.1% from 2020 to 2027 (Figure 15.1). Their market potential stems from the fact that currently more than 700 commercial products are dependent on enzymes in over 40 different industrial sectors ranging from household care to bioenergy to agriculture, animal health and food including baking, brewing, beverage, juice, wine, dairy, etc. Even in India, the food enzyme market is expected to register a CAGR of 3.8% during 2020–2025 (Patel et al., 2016; www.grandviewresearch.com; www.mordorintelligence.com). Though enzymes can be sourced from animals, plants, and microbes, at least 243 commercial enzymes are sourced from microbes. Around 47% of these enzymes are sourced from recombinant strains. In Figure 15.1, the global share of enzymes sourced from microorganisms is depicted. The major players in the field of enzyme production (www.marketsandmarkets.com)

are DuPont (US), BASF (Germany), DSM (Netherlands), Associated British Foods (UK), Novozymes (Denmark), Dyadic International (US), Kerry Group (Ireland), Advanced Enzymes (India), Chr. Hansen (Denmark), Amano Enzymes (Japan), Augene Biosciences (India), Roche Holding (Switzerland), Codexis (US), Sanofi (France), Enzyme Supplies (UK), Merck (Germany), Creative Enzymes (US), Enzymatic Deinking Technologies (US). Enzyme Solutions (US) and Biocatalysts (UK). In Table 15.2, some commercially available fungal enzymes are cited along with their source and function. It is very interesting to note that though significant literature is available on filamentous fungi and single-celled yeasts as enzyme producers, the maximum commercial strains belong to that of *Aspergillus* and in particular *A. niger*.

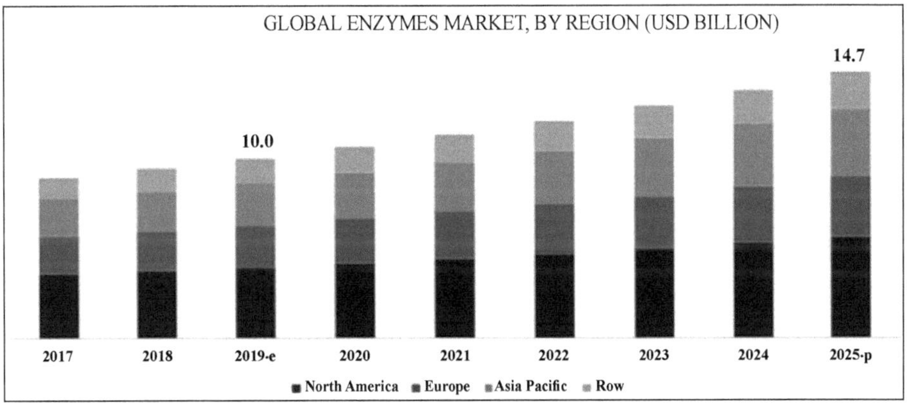

FIGURE 15.1 Global enzyme market: Past, present, and future.

Source: https://www.marketsandmarkets.com/Market-Reports/enzyme-market-46202020.html

15.7 MODERN TOOLS AND TECHNIQUES

Metagenomics has been proven to be a mighty tool for identifying new enzymes with better industrial potential. This method is based on searching for the genes of the desired enzymes in the DNA extracted directly from soil, compost or other environmental samples. Most of the enzymes used for industrial and other applications are sourced from microorganisms and which would normally be obtained through qualitative screening methods using enzyme specific agar media. However, microbiologists realize that the diversity of culturable microbes obtained from any natural environmental sample is extremely small as compared to the actual variety identified by microbial community analysis based on DNA extracted from the same

TABLE 15.2 List of Commercially Available Fungal Enzymes

Enzyme	Commercial Source	Company	Function	Organism
Lipase	Lipolase TM	Novo Nordisk	Detergent additive	Humicola lanuginose
	ChiroCLEC-CR	Atlus Biologics	Food processing	Candida cylindraceae
	Lipase AY	Amano		*C. cylindraceae*
	Chirazyme® L-3	Boehringer Mannheim		*C. cylindraceae*
	Lipomod™ 34P-L034P	Biocatalysts		*C. cylindraceae*
	Palatase®	Novozymes		*Rhizopus meihei*
	Lipase A "Amano" 6	Amano		*Aspergillus niger*
	Lipomod™ 338P-L338P	Biocatalysts		*Penicillium roquefortii*
	Lipase AY "Amano" 30	Amano	Organic synthesis	*Candida rugosa*
	Resinase®	Novozymes		*C. rugosa*
	Lipomod™ 621P-L621	Biocatalysts		*Candida antarctica* A/B
	Chirazyme®L-5	Boehringer Mannheim		*C. antarctica* A/B
Protease	Acid Protease DS	Amano Enzyme U.S.A. Co. Ltd.	Food processing	*Aspergillus niger*
	ALBAMAX Liquid	Clerici-Sacco Group	Milk clotting enzyme	*Aspergillus niger var. awamori.*
	Acid stable protease (SAPU/g)	American Laboratories Inc.	Low pH applications in the pharmaceutical, food, and feed industries	*Aspergillus niger*
	Microclerici	Clerici-Sacco Group	Food industry for cheese production	*Rhizomucor miehei*
	Milase ® Premium	CSK Food Enrichment		*Rhizomucor miehei*

TABLE 15.2 *(Continued)*

Enzyme	Commercial Source	Company	Function	Organism
Carbohydrases	Amylase DS	Amano Enzyme U.S.A. Co. Ltd.	Starch saccharification	*Aspergillus oryzae*
	Amyloglucosidase	American Laboratories Inc.	Brewing, baking, and distilling industries	*A. niger*
	SQzyme FA L	Suntaq International Ltd.	Beer, high maltose syrup and many other food industries	*A. oryzae*
	Beta Glucanase	American Laboratories Inc.		*Trichoderma longibrachiatum*
	Cellulase DS	Amano Enzyme U.S.A. Co. Ltd.	Food processing applications	*A. niger*
	Glucoamylase	Winovazyme Biological Science and Technology Co. Ltd.	Production of dry beer	*A. niger*
	SQzyme XS L	Suntaq International Ltd.	Brewery, distillery, and animal feed.	*T. reesei*
	Cellulase	American Laboratories Inc.	Releases bound water molecules from the absorbent gums, improve filtration, reduces viscosity, degrade and modify soluble gums and fiber, and releases substances from surrounding cellulosic matrices.	*T. longibrachiatum.*
Other hydrolases	Catalase	American Laboratories Inc.	Used as antimicrobial and bleaching agent.	*A. niger*
	HYDERASE HC	Amano Enzyme U.S.A. Co. Ltd.		*A. niger*

TABLE 15.2 *(Continued)*

Enzyme	Commercial Source	Company	Function	Organism
	Phytase	American Laboratories Inc.	Used in soy protein isolate production	*A. niger*
	DEAMIZYME 50000G	Amano Enzyme U.S.A. Co., Ltd.	Flavor generation	*A. melleus*
	ENZYME RP-1 G	Amano Enzyme U.S.A. Co. Ltd.	Flavor enhancer	*Penicillium citrinum*
	LACTASE 14-DS	Amano Enzyme U.S.A. Co. Ltd.	Dietary supplemental grade enzyme preparation	A. oryzae

Source: UL prospector material database; https://www.ulprospector.com/.

source sample. These DNA based identification showed that the variety of microorganisms was over 1,000 times more than that could be isolated using agar plate techniques (https://gtr.ukri.org/projects). It has been known from the environmental DNA (eDNA) analysis methods that only 0.1% to 1% of the prokaryotes can be cultured through traditional microbiological methods (Culligan et al., 2014a; Berini et al., 2017). Comparatively, the eukaryotic microbes are more culturable with 10% to 30% of fungal isolates obtained from different soils (Magnuson & Lasure, 2002; Hirst, 2011; Berini et al., 2017).

Lately, culture independent methods particularly the metagenomics (also referred to as environmental genomics or community genomics) have been developed for assessing the microbial diversity by directly analyzing the genomes collectively recovered from the environment (Cuadros-Orellana et al., 2013). Recently, metatranscriptomics and metaproteomics have been added to the culture independent methods (Berini et al., 2017). Considered as a very promising tool, metagenomics is used for identifying novel enzymes from prokaryotic eDNA (Lorenz & Eck, 2005; Wilmes, Heintz-Buschart, & Bond, 2015) leading to the discovery of novel lipases (Sukul et al., 2017) and cellulases (Speda et al., 2017).

The genes of prokaryotic microbes can be directly transcribed and translated in a suitable heterologous host, but the presence of large introns in genes of eukaryotic organisms limits their metagenomic analyzes. A solution to this problem was given by Grant et al. (2006), who used metatranscriptomics (i.e., he carried out reverse transcription of selectively enriched eukaryotic 3-polyadenylated mRNAs) to discover novel eukaryotic enzymes. Later this approach was used to characterize industrial enzymes like phosphatase (Kellner et al., 2011) and cellulase (Takasaki et al., 2013).

15.7.1 ANALYSIS OF EDNA FOR SEARCHING NOVEL ENZYMES

Metagenomics refers to the analysis of the genetic material present in a habitat by direct extraction and subsequent cloning of DNA from an assemblage of microorganisms (Handelsman et al., 1998). Metagenomics, the study of the metagenome uses the whole genetic material of the microbial community existing in certain environments (Sleator et al., 2008). It has provided to be a strong tool for exploring the unexplored microbial diversity from unusual environments for novel products (Pereira, 2019). Developments in structural and functional metagenomics have overcome the traditional barriers to discovery of novel microbial products including enzymes with better properties. Reduction of product (semi-processed and processed forms of enzymes) cost is a major drive-in enzyme manufacturing. This has been achieved through technologies for large-scale production of stable recombinant enzymes from fungal (as well as bacterial) hosts.

Efforts were also made to make the enzymes-based processes sustainable so as to replace the chemical processes which have been energy intensive, required adverse reaction conditions like very high temperatures, extreme pH, pressure much above the atmospheric pressure and produce are hazardous by products or wastes, which caused environmental pollution. These techniques were essential to meet the huge demand for carrying out the industrial catalysis, which required huge quantity of the catalysts.

The core technologies for enzyme production are now complemented by a broad set of newer technologies such as omics, i.e., genomics, transcriptomics, proteomics, and metabolomics (Arnau et al., 2020). Progress in omics has been of great significance as it has facilitated in affordable and high throughput genome sequencing, ready synthesis of codon-optimized synthetic genes, synthesis of robust systems through gene editing for efficient production of recombinant enzyme and, optimized production strains such as *Saccharomyces cerevisiae, Pichia pastoris, Yarrowia lipolytica*, etc., for use in Research and Development sectors. Advancements in the area of metabolic engineering resulted in optimal production of value added products in, *Saccharomyces cerevisiae* (Bao et al., 2017). New gene editing technologies such as CRISP-Cas 9 will add momentum to this field. In industrial biotechnology, the use of CRISPR (clustered regularly interspaced short palindromic repeats) technique has provided a shorter and faster tool for construction of production strain (Ferreira et al., 2018). Synthetic biology, an emerging and disruptive technology is claimed to be a further advancement to the current techniques which offers innovative approaches for engineering new biological systems or re-designing existing ones for useful purposes (Karoui et al., 2019).

Bioinformatics based approaches to analyze genome sequences and genome annotation have helped in discovery of novel enzyme activities and which are continuously being explored using various approaches, such as, SFLD (Akiva et al., 2014), Hotpep (Busk et al., 2017), SACCHARIS (Jones et al., 2018), dbCAN2 (Zhang et al., 2018), ECPred (Dalkiran et al., 2018), MSA, MEGA-X, ML method, JTT model and NCBI BLASTp (Verma et al., 2019) lastly CUPP (conserved unique peptide patterns), an automated, peptide-based functional annotation method (Barrett & Lange, 2019). For the last few decades, there has been a focus on genome sequencing projects for various organisms. In this context, 1,000 fungal genome started by Joint

Genome Institute of the US Department of Energy in collaboration with an international research team has sequenced around 2,030 fungal genomes (https://mycocosm.jgi.doe.gov/fungi/fungi.info.html). The ultimate goal of this project is to provide information for further research on plant-microbe interactions, microbial emission, capture of greenhouse gas and environmental metagenomic sequencing (https://mycocosm.jgi.doe.gov/mycocosm/home/1000-fungal-genomes) (Arnau et al., 2020).

Metagenomics (along with transcriptomics and proteomics) have provided important platform for insights in the discovery of mechanisms used by the fungi for biodegradation activities, e.g., for biodegradation of agricultural wastes (Kuuskeri et al., 2016; Borin et al., 2017; Miyauchi et al., 2018; Umezawa et al., 2020) or degradation of non-natural compounds (Blasi et al., 2017), understanding and controlling plant diseases (e.g., Collins et al., 2017); and for automation in biotechnological processes have brought out great progress resulting in better productivity for enzymes and metabolites (https://www.rdworldonline.com/taking-biotechnology-to-the-next-level-with-automation/). Online databases for supporting enzyme research have been developed (BRENDA, ExPASy–ENZYME, KEGG ENZYME database, explorenz, enzyme structures database – EMBL-EBI, etc.), which are of great use to researchers and industrialists as well. Using the modern molecular biology tools by different culture collection centers, like ATCC (USA), AHU (Japan), and BCRC (Chinese Taipei), have maintained them as invaluable support system for both basic and applied research. With the formation of new disciplines of study like biophysics more advancement in structural biology have been made resulting in the application of techniques like NMR, MS, and bioimaging in better understanding of the structure and function of biomolecules and their functions (Lange et al., 2020).

Use of modern molecular biology techniques have been of great use in the study of microbial ecology, wherein genetic fingerprinting techniques providing genetic diversity profile have been applied in several environmental studies of microbial communities. Another technique for studying the composition and mechanism of microbial communities in situ is whole-cell hybridization which can also be used to identify expression of key enzymes. Further developments in these techniques will result in greater insights into functional interactions, temporal, and spatial relationships between different microorganisms (Duan et al., 2006). Currently, researchers are focused on studying comparative genomics rather than evaluating a genome in isolation. The study of interactions between organisms and in gene regulation along with prediction of gene function will help in generating new information leading to better biological understanding (Arnau et al., 2020). Rapid growth in high throughput data processing, analyzes, and classification performed through bioinformatics has allowed the comparative studies of fungal genomes for identification of novel enzyme gene on a Personal Computer or laptop. Besides, enzymes' efficacies in terms of stability, specificity, substrate range, etc., have been enhanced through the combinatorial studies like of, crystallographic studies, structural analysis, protein engineering, *in vitro* evolution, and bioinformatics (Arnau et al., 2020). This fast increase in the number of substantive techniques and tools and methods has allowed a great progress in mycology, giving relevant information that would translate into conceptually interesting mycobiological knowledge. The tools, techniques, and methods developed in the last decade are now being used to revolutionize the fungal research. However, harnessing the full potential of the fungal bioeconomy will require many more biotechnological interventions. Further developments in the field of fungal enzymes would depend on the collaborative approaches, wherein the developments achieved in basic biology, molecular biology, systems biology, biophysics, bioinformatics, robotics, bioengineering, and fermentation technology have to be integrated to obtain superior production and application technologies for fungal enzymes (Saxena et al., 2005; Meghwanshi & Vashishtha, 2018; Verma et al., 2019, 2021).

15.8 PRESENT CHALLENGES AND FUTURE PROSPECTS

15.8.1 FUNGI ENZYMES IN MANAGEMENT OF ENVIRONMENT

Production of recombinant fungal enzymes for industrial applications requires suitable hosts. Although a number of fungal strains such as *Candida rugosa, Aspergillus oryzae, A. melleus, Pichia pastoris, Penicillium funiculosum, Chaetomium erraticum, Talaromyces emersonii*, etc., are all used commercially, but industrial level production of recombinant enzymes has been carried out mostly using *A. niger, A. oryzae, A. nidulans* and *Trichoderma reesei* (Christensen, 1994; Ntana et al., 2020; Arnau et al., 2020). This possible is due to their extensive safe use spanning decades. However, these fungal hosts are generally used as mono enzyme expression systems while several enzyme applications require simultaneous and synergistic action using multiple enzymes such as the degradation of complex and/or recalcitrant biomasses, e.g., keratin or lignocellulose requires an entire set of enzymes (Merino & Cherry, 2007).

Production of one enzyme at a time using a recombinant host strain would no longer be an economically sustainable process. So, there is a need for co-expression of multiple genes of the desired products (or enzymes) in a single host. This would require the identification of the enzymes most suitable for the conversion of waste biomasses to the desired products or substrates for other application, optimizing their blend composition and then expressing them in a multigene recombinant production host like *Trichoderma* spp. (Lange et al., 2020), a challenge for the fungal biotechnology (Meyer et al., 2009; Bunterngsook et al., 2018). Growth of industrial biotechnology would further require sustainable production of more recombinant fungal enzymes at large scale for meeting the desired industrial scale biotransformation of wastes to commodities.

15.8.2 REDUCING THE EMISSION OF GREEN HOUSE GASES (CO_2 AND METHANE)

Agro-industrial wastes which make up a huge part of the biodegradable wastes on the planet are often disposed in the environment or left unattended. These waste materials become the part of biogeochemical cycle and through microbial degradation broken into simple forms, which are recycled to be utilized by the primary producers, i.e., the photosynthetic organisms (plants, algae, microalgae, and cyanobacteria). Microbial degradation processes of these biodegradable wastes result in a significant rise in CO_2 levels in the atmosphere, which may add to the greenhouse effect. A study on plastic degradation by *Rhodococcus rhodochrous*, showed that this bacterial sp. released 32% of available carbon as CO_2 at the stationary phase, with the majority of carbon being assimilated into biomass (Flickinger et al., 2010; Rose et al., 2020). Agricultural wastes generated in the form of leftover plant parts after harvesting of the produce can be hydrolyzed using fungal enzymes like cellulases, hemicelluases, lignin peroxidases, tannases, amylases, and related enzymes into fermentable substrates which can be converted into useful valuable products like organic solvents, biofuels, organic acids, paper, textile, biopolymers, bulk enzymes, etc. (Meghwanshi, 2018). Similarly industrial effluents can be treated using fungal lipases, proteases, and amylases to convert them to single cell protein, SCO, acetone, butanol, ethanol, methanol, and other organic solvents (Meghwanshi & Vashishtha, 2014, 2018; Meghwanshi et al., 2020). Currently, around 35–50% of total global harvest is being wasted. However, fungal enzymes can be used to improve the recycling and reuse of such agro-industrial resources (Lange et al., 2020).

Methane is mainly released into environment by the ruminants or a by-product of animal husbandry. Cattles are known to harbor methane producing bacteria in their rumen, which produce methane from the food consumed by these animals. The methane so produced is released by the animals into the environment, increasing the levels of methane into the atmosphere and adding to the greenhouse effect. A substantial decrease in methane emission can be achieved by modulating the microbial diversity of the rumen microbiome and also by changing the composition of animal feed. Fungi can help in achieving both these goals: firstly, the rumen fungi are non-methane producing and their composition in rumen microbiome can be altered to achieve less methane production; secondly fungal enzymes can be used in formulating new and better (prebiotic) and as anti-methane feed additives for ruminants (Dupont et al., 2018; Chuang et al., 2020). Most of the studies related to the diversity of rumen microflora or gut microbiome have remained confined to bacteria and archaea. Only a few studies of rumen fungi, have been performed which showed the role of these fungi in plant cell wall opening through their rhizoids and thus exposing the plant polymers like starch to bacterial decomposition (Nagpal et al., 2009). However, the complete spectrum of the possible applications of rumen fungi is yet to be discovered, which is hampering its use to the industrial biotechnology.

15.8.3 FUNGI AS SOURCE OF FOOD AND FEED

Mushrooms, the sexual structures of basidiomycetes and ascomycetes (e.g., truffles) have been incorporated in the human diet as low-fat high protein source. Some varieties like boletus, truffles, morels are considered as delicacies and highly priced. Another interesting aspect is that apart from a few exceptions like *Morchella rufobrunnea*, other varieties exhibit unsuccessful cultivation (Dupont et al., 2017). Only a few species of mushrooms can be produced industrially, such as *Agaricus bisporus*, *Lentinus edodes* (shiitake), and *Pleurotus ostreatus*, and their production mainly confined to Asia for thousands of years (Morin et al., 2012; Dupont et al., 2017). Some other fungi like shiitake and oyster mushrooms are cultivated on trees in appropriate natural habitats to increase the production of fruiting bodies with variable outcome (De Miguel et al., 2014).

Fungi have been used as feed additive to enhance animal health. Common fungi used in feed are *Saccharomyces cerevisiae, Aspergillus oryzae, Pleurotus* spp., *Antrodia cinnamomea*, and *Cordyceps militaris* (Chuang et al., 2020).

These fungi are rich in glucans (Awaad et al., 2011), polysaccharides (Song et al., 2019), polyphenols (Mohanta, 2020), triterpenes (Ma et al., 2018), ergosterol (Baur et al., 2019), adenosine (Caiazzo et al., 2016), and laccases (Chuang et al., 2020). These fungal metabolites act as natural and safe anti-inflammatory, antioxidant, anti-obesity agents, and immune regulators (Chuang et al., 2020). Fungi as feed have been shown to increase body mass as well as enhanced egg laying in poultry (Wang et al., 2015). In addition, constituents of fungal cell wall and their enzymes have been claimed to reduce the toxicity of mycotoxins (Haque et al., 2020). On the whole, fungi as feed or probiotics can be of huge worth; however, their safety and long-term effects on health should also be studied (Chuang et al., 2020).

15.8.4 EFFICIENT REUSE OF AGRICULTURAL AND INDUSTRIAL WASTES

The easiest approach to feed the rapidly growing human population in the world is to reuse biowastes originating from agriculture and industries. This may be done through the proper use of fungal enzymes to transform agricultural residues and industrial by-products and effluents for producing more food and feed. This could be achieved through applying the biorefinery approach, wherein nothing goes waste out of a production technology. The organic residues from agriculture, industries, and households can be reused as soil improvers such as biochar, biogas slurry and sewage (treated) sludge (enriched in phosphorous and other minerals). Similarly, enzymatically hydrolyzed lignocellulic biomases (after sugar extraction) can be used as soil improver. This approach could hold in increasing the agricultural coverage of non-arable lands as well for the production of more crops both for food and feed.

Some examples of waste to wealth from industrial perspective are cited here. Whey (the watery liquid leftover after cheese manufacturing) has been used for the production of various new products such as food formulations for premature babies, pregnant women, infants or athletes. Whey-based products are also used in yogurt, ice cream and in low-fat cheese (https://www.healthline.com, https://www.healthline.com/health/pregnancy/protein-powder-for-pregnancy#how-much-you-need). Similarly, soy protein has been in use as a major ingredient of many health supplements rich in proteins (https://www.healthline.com/health/pregnancy/protein-powder-for-pregnancy#how-much-you-need). Other noteworthy commercial process is the upgradation of potato starch production to a modern biorefinery process (http://www.kmc.dk/).

15.8.5 FUNGAL PROBIOTICS/ENZYMES FOR HUMAN HEALTH

Recently several fungal strains have been identified as probiotics. Research findings have shown that many fungi can be used to inhibit and resist the colonization of the intestine by enteric bacterial pathogens; and are used for maintenance of epithelial barrier integrity, anti-inflammatory effects, immunomodulation activity, eradication of *Helicobacter pylori* infection, reducing blood cholesterol levels. The exact mechanisms of these actions may not be known, but some involvement of the fungal enzymes may be there in mediating these actions. However, some of the actions such as production of all B-complex vitamins, enhanced bioavailability of significant ions, e.g., iron, zinc, calcium, and magnesium, biodegradation of phytate, release of oligosaccharides, are certainly the function of some hydrolytic or metabolic enzymes of the fungal origin. Introducing novel fungal strains for therapeutics is tricky until a proper understanding of the mechanism of action, efficacy, and dosage level is developed. Among several fungal candidates, *Saccharomyces boulardii* var. *cerevisiae* is the most promising commercialized probiotic yeast due to the several health benefits to the host under both normal and adverse conditions. Evaluating the efficacy of newly isolated probiotic fungi for better bioavailability and food safety is an ongoing research theme globally (Banik et al., 2019).

15.8.6 FUNGAL FERMENTED PRODUCTS FOR GOOD HEALTH

Fungi (both yeasts and molds) have been used from the ancient time for the production of fermented food products worldwide. Among major fermented food products from which fungi have been isolated include cheese (*Debaryomyces hansenii*, *Geotrichum candidum*, *Penicillium camemberti*, and *P. roqueforti*), Chhu (*Candida* sp. and *Saccharomycopsis* sp.), Airag (*Kluyveromyces marxianus*, Dahi (*Saccharomycopsis* sp. and *Candida* sp.), kefir (*Torulaspora delbruechii* and *Tor. holmii*), Boza (*Saccharomyces cerevisiae*), Dosa (*Tor. candida* and *Trichosporon pullulans*), Gowé (*Candida tropicalis*, *C. krusei*, *Kluyveromyces marxianus*), Kimchi (*Pichia, Saccharomyces, Sporisorium*, and *Trichosporon*), Olives (*Candida* cf. *apicola*, *Pichia* sp., *Pic. manshurica*/*Pic. galeiformis*, *Sacch. cerevisiae*), Bhallae (*Sacch. cerevisiae, Trichosporon beigelii, Trichosporon pullulans* and *Wingea robertsii*) (Tamang et al., 2016). The crucial function of filamentous fungi in fermented foods and alcoholic beverages is the production of enzymes and the breakdown of anti-nutritive factors (Aidoo & Nout, 2010;

Tamang et al., 2016). Similar role may be played by various spp. of yeast as well in the fermented foods. The characteristic flavor and taste developed in the fermented foods are mediated by the fungal/yeast metabolites which they produce during the process of fermentation (Maicas, 2020). Currently, the new challenge in the field of fermented food products is to use the normally discarded parts of vegetables and plants, which are rich in sugars and can be made palatable, nutritive, and digestible using the microbial fermentation. A notable example is the production of pickles from the petioles and studs of the cauliflower (Nai & Meghwanshi, 2014).

15.8.7 UPGRADING KERATINACEOUS BIOMASS USING FUNGAL ENZYMES

Chickens' feathers and animal hairs and hooves are very abundant protein-rich waste streams, worldwide (Baid et al., 2016). Preparation of animal feed using these keratinaceous wastes using the conventional methods, which includes grinding and chemical hydrolysis results in a very poor-quality feed, which is non-assimilable with destroyed essential, branched amino acids. Fungal species which specifically grow on keratinaceous materials in nature have been identified and their specific keratinolytic proteases acting on the recalcitrant keratin wastes have also been studied (Lange et al., 2020). Furthermore, overproduction of special AA11 type of the lytic polysaccharide monooxygenases (LPMO) enzymes have been observed in both keratin degrading fungi and the skin-invading dermatophytic fungi (Busk & Lange, 2015). It therefore seems possible to perform microbial conversion of the recalcitrant protein rich wastes into usable forms such as animal feed and biofertilizers (Baid et al., 2016). However, still there remain numerous biotechnology challenges which need to be tackled to make this process commercially viable.

15.9 FUTURE PERSPECTIVES

Transforming the big biological data (gathered through genomics, meta-genomics, proteomics, transcriptomics, transcriptomics, transcriptomics, omics, and metabolomics) into understandable information which can be used and applied for the progression of fungal biotechnology for the welfare and progression of the human race. Dedicated and aggressive mycological research is need of the hour to ensure that the fungi are included in the future omic studies to study the unexplored microbiome the plants (rhizosphere, phyllosphere, and spermosphere), animals' gut, rumen, skin, mucous surfaces, etc.). Studies need to be divergent which may include, phylogeny, taxonomy, genetics, nutrition, biochemistry, molecular biology, and how they are correlated with fungal behavior like their beneficial or harmful manifestations in the nature, such as their utility as industrial work horses for the production of numerous foods, pharmaceuticals, and environmental management solutions; otherwise, their role as plant, animal, and human pathogens. Inclusion of investigation approaches like the use of new digitalization, bioinformatics, machine learning and big data options will result in improved understanding of various aspects of mycology that will help us in better management of fungi and their optimal utilization for getting sustainable solutions to many current and future challenges whether related to energy crisis, agriculture, food production, and security, human health and disease and many more unexplored avenues.

KEYWORDS

- **amylase**
- **biofuels**
- **compound annual growth rate**
- **fungal enzymes**
- **industrial application**
- **lignocelluloses**
- **lytic polysaccharide monooxygenases**
- ***Saccharomyces cerevisiae***

REFERENCES

Abidi, F., Chobert, J. M., Haertlé, T., & Marzouki, M. N., (2011). Purification and biochemical characterization of stable alkaline protease Prot-2 from *Botrytis cinerea*. *Process Biochemistry, 46*, 2301–2310.

Abu, E. A., Ado, S. A., & James, D. B., (2005). Raw starch degrading amylase production by mixed culture of *Aspergillus Niger* and *Saccharomyces cerevisiae*. *African Journal of Biotechnology, 4*, 785–790.

Agrawal, S., & Kango, N., (2019). Development and catalytic characterization of L-asparaginase nanobioconjugates. *International Journal of Biological Macromolecules, 145*, 1145–1150.

Ahmed, A., & Bibi, A., (2018). Fungal cellulase; production and applications: Minireview. Life: *International Journal of Health and Life-Sciences, 4*(1), 19–36.

Ahmed, A., Ul-Hassan, N. F., Batool, K., & Bibi, A., (2017). Microbial β-glucosidase: Sources, production and applications. *Journal of Applied & Environmental Microbiology, 5*(1), 31–46.

Akiva, E., Brown, S., Almonacid, D. E., Barber, A. E. 2nd., Custer, A. F., Hicks, M. A., Huang, C. C., et al., (2014). The structure-function linkage database. *Nucleic Acids Research, 42(Database issue)*, D521–D530. doi: 10.1093/nar/gkt1130.

Andrade, V. S., Sarubbo, L. A., Fukushima, K., Miyaji, M., Nishimura, K., & De Campos-Takaki, G. M., (2002). Production of extracellular proteases by *Mucor circinelloides* using D-glucose as carbon source / substrate. *Brazilian Journal of Microbiology, 33*, 106–110.

Anobom, C. D., Pinheiro, A. S., De-Andrade, R. A., Aguieiras, E. C. G., Andrade, G. C., Moura, M. V., Almeida, R. V., & Freire, D. M., (2014). From structure to catalysis: Recent developments in the biotechnological applications of lipases. *BioMed Research International*, 11. Article ID 684506. http://dx.doi.org/10.1155/2014/684506.

Arnau, J., Yaver, D., & Hjort, C. M., (2020). Strategies and challenges for the development of industrial enzymes using fungal cell factories. In: Nevalainen, H., (ed.), *Grand Challenges in Fungal Biotechnology, Grand Challenges in Biology and Biotechnology*. Springer, Cham. https://doi.org/10.1007/978-3-03. 0-29541-7_7.

Awaad, M., Atta, A. M., El-Ghany, W. A., Elmenawey, M., Ahmed, A., Hassan, A. A., Nada, A., et al., (2011). Effect of a specific combination of mannan-oligosaccharides and β-glucans extracted from yeast cell wall on the health status and growth performance of ochratoxicated broiler chickens. *Journal of Animal Science, 7*, 82–96.

Baid, S., Vashishtha, A., Ahmed, A., Verma, S., & Meghwanshi, G. K., (2016). Bacterial keratinase catalyzed bioremediation of keratin rich wastes for potential agricultural and other applications. *Journal of Phytological Research, 29*(1, 2). 17–30. ISSN 0970-5767.

Balkan, B., & Ertan, F., (2006). Production of amylase from *Penicillium chrysogenum* under SSF by using some agricultural by-product. *Food Technology and Biotechnology, 45*, 439–442.

Banerjee, G., & Ray, A. K., (2017). Impact of microbial proteases on biotechnological industries. *Biotechnology and Genetic Engineering Reviews, 33*(2), 119–143. doi: 10.1080/02648725.2017.1408256.

Banik, A., Halder, S. K., Ghosh, C., & Mondal, K. C., (2019). Fungal probiotics: Opportunity, challenge, and prospects. In: Yadav, A., Singh, S., Mishra, S., & Gupta, A., (eds.), *Recent Advancement in White Biotechnology Through Fungi: Fungal Biology*. Springer, Cham. https://doi.org/10.1007/978-3-030-14846-1_3.

Bao, J., Huang, M., Petranovic, D., & Nielsen, J., (2017). Moderate expression of SEC16 increases protein secretion by *Saccharomyces cerevisiae*. *Applied and Environmental Microbiology, 83*, e03400–e03416. https:// doi.org/10.1128/AEM.03400-16.

Barrios-González, J., (2012). Solid-state fermentation: Physiology of solid medium, its molecular basis and applications. *Process Biochemistry, 47*, 175–185.

Basheer, S. M., Chellappan, S., Beena, P. S., Sukumaran, R. K., Elyas, K. K., & Chandrasekaran, M., (2011). Lipase from marine *Aspergillus awamori* BTMFW032: Production, partial purification and application in oil effluent treatment. *New Biotechnology, 28*(6), 627–638.

Baur, A. C., Kühn, J., Brandsch, C., Hirche, F., & Stangl, G. I., (2019). Intake of ergosterol increases the vitamin D concentrations in serum and liver of mice. *Journal of Steroid Biochemistry, 194*, 105435. doi: 10.1016/j.jsbmb.2019.105435.

Binod, P., Sindhu, R., Madhavan, A., Abraham, A., Mathew, A. K., Beevi, U. S., Sukumaran, R. K., et al., (2017). Recent developments in L-glutaminase production and applications – An overview. *Bioresource Technology, 245*(B), 1766–1774.

Blasi, B., Tafer, H., Kustor, C., Poyntner, C., Lopandic, K., & Sterflinger, K., (2017). Genomic and transcriptomic analysis of the toluene degrading black yeast *Cladophialophora immunda. Scientific Reports, 7*, 11436. https://doi.org/10.1038/s41598-017-11807-8.

Borin, G. P., Sanchez, C. C., De Santana, E. S., Zanini, G. K., Dos Santos, R. A. C., De Oliveira, P. A., De Souza, A. T., et al., (2017). Comparative transcriptome analysis reveals different strategies for degradation of steam-exploded sugarcane bagasse by *Aspergillus Niger* and *Trichoderma reesei. BMC Genomics, 18*, 501. https://doi.org/10.1186/s12864-017-3857-5.

Brannigan, J. A., Dodson, G., Duggleby, H. J., Mood, P. C. E., Smith, J. L., Tomchick, D. R., & Murzin, A. G. A., (1995). Protein catalytic framework with an N-Terminal nucleophile is capable of self-activation. *Nature, 378*, 416–419.

Caiazzo, E., Maione, F., Morello, S., Lapucci, A., Paccosi, S., Steckel, B., Lavecchia, A., et al., (2016). Adenosine signaling mediates the anti-inflammatory effects of the COX-2 inhibitor nimesulide. *Biochemical Pharmacology, 112*, 72–81. doi: 10.1016/j.bcp.2016.05.006.

Chandra, P., & Enespa, E., (2019a). Fungal enzymes for bioremediation of contaminated soil. In: *Recent Advancement in White Biotechnology Through Fungi* (pp. 189–215). Springer, Cham.

Chandra, P., & Enespa, E., (2019b). Mycoremediation of environmental pollutants from contaminated soil. In: *Mycorrhizosphere and Pedogenesis* (pp. 239–274). Springer, Singapore.

Chandra, P., Enespa, E., Singh, R., & Arora, P. K., (2020). *Microbial Cell Factories, 19*, 169. https://doi.org/10.1186/s12934-020-01428-8.

Chang, S. T., (1977). The origin and early development of straw mushroom cultivation. *Economic Botany, 31*, 374–376.

Christensen, T., (1994). Application: *Aspergillus Oryzae* as a host for the production of industrial enzymes. In: Powell, K. A., Renwick, A., & Peberdy, J. F., (eds.), *The Genus Aspergillus. Federation of European Microbiological Societies Symposium Series* (Vol. 69). Springer, Boston, MA. https://doi.org/10.1007/978-1-4899-0981-7_19.

Chuang, W. Y., Hsieh, Y. C., & Lee, T. T., (2020). The effects of fungal feed additives in animals: A review. *Animals: An Open Access Journal from MDPI, 10*(5), 805. https://doi.org/10.3390/ani10050805.

Costa, T. M., Hermann, K. L., Garcia-Roman, M., Valle, R. D. C. S. C., & Tavares, L. B. B., (2012). Lipase production by *Aspergillus Niger* grown in different agro-industrial wastes by solid-state fermentation. *Brazilian Journal of Chemical Engineering, 34*(2), 419–427.

Cowan, D., Meyer, Q., Stafford, W., Muyanga, S., Cameron, R., & Wittwer, P., (2005). Metagenomic gene discovery: Past, present and future. *Trends in Biotechnology, 23*(6), 321–329.

Crueger, W., & Crueger, A., (1984). Enzymes. In: *Biotechnology. A Textbook of Industrial Microbiology*, 161–186.

Da Silva, R. R., (2017). Bacterial and fungal proteolytic enzymes: Production, catalysis and potential applications. *Applied Biochemistry and Biotechnology, 183*(1), 1–19. doi: 10.1007/s12010-017-2427-2.

Da Silva, R. R., Caetano, R. C., Okamoto, D. N., De Oliveira, L. C. G., Bertolin, T. C., Juliano, M. A., Juliano, L., et al., (2014). The identification and biochemical properties of the catalytic specificity of a serine peptidase secreted by *Aspergillus fumigatus* Fresenius. *Protein and Peptide Letters, 21*, 663–671.

Dalkiran, A., Rifaioglu, A. S., Martin, M. J., Cetin-Atalay, R., Atalay, V., & Doğan, T., (2018). ECPred: A tool for the prediction of the enzymatic functions of protein sequences based on the EC nomenclature. *BMC Bioinformatics, 19*, 334. https://doi.org/10.1186/s12859-018-2368-y.

Dange, V. U., & Harke, S., (2018). Production and purification of pectinase by fungal strain in solid-state fermentation using agro-industrial bioproduct. *International Journal of Life Sciences Research, 6*(4), 85–93.

Darah, I., Sumathi, G., Jain, K., & Hong, L. S., (2011). Involvement of physical parameters in medium improvement for tannase production by *Aspergillus Niger* FETL FT3 in submerged fermentation. *Biotechnology Research International*, 1–7. doi: 10.4061/2011/897931.

Das, K., & Mukherjee, A. K., (2007). Comparison of lipopeptide biosurfactants production by *Bacillus subtilis* strains in submerged and solid-state fermentation systems using a cheap carbon source: Some industrial applications of biosurfactants. *Process Biochemistry, 42*, 1191–1199.

Dawood, A., & Ma, K., (2020). Applications of microbial β-mannanases. *Frontiers in Bioengineering and Biotechnology, 8*, 598630. doi: 10.3389/fbioe.2020.598630.

De Almeida, A. F., Tauk-Tornisielo, S. M., & Carmona, E. C., (2013). Acid lipase from *Candida viswanathii*: Production, biochemical properties, and potential application. *BioMed Research International, 435818*, 1–10.

De Miguel, A. M., A′ gueda, B., Sa′nchez, S., & Parlade, J., (2014). Ectomycorrhizal fungus diversity and community structure with natural and cultivated truffle hosts: Applying lessons learned to future truffle culture. *Mycorrhiza, 24*(1), 5–18.

De Souza, P. M., De Assis, B. M. L., Caprara, C. C., De Freitas, M., De Almeida, R. P. C., Silveira, D., Fonseca, Y. M., et al., (2015). A biotechnology perspective of fungal proteases. *Brazilian Journal of Microbiology, 46*(2), 337–346. doi: http://dx.doi.org/10.1590/S1517-838246220140359.

Demain, A. L., & Adrio, J. L., (2008). Contributions of microorganisms to industrial biology. *Molecular Biotechnology, 38*, 41–55.

Derwing, T. M., Rossiter, M. J., & Munro, M. J., (2002). Teaching native speakers to listen to foreign-accented speech. *Journal of Multilingual and Multicultural Development, 23*(4), 245–259.

Desai, S. S., Gachhi, D. B., & Hungund, B. S., (2016). Glucose isomerizing enzymes. In: *Microbial Enzyme Technology in Food Applications* (pp. 69–84). CRC Press. doi: 10.1201/9781315368405-6, eBook ISBN 9781315368405.

Duan, Z. Y., Guo, Y. Q., & Liu, J. X., (2006). Application of modern molecular biology techniques to study micro-ecosystem in the rumen. *Wei Sheng Wu Xue Bao.* (*Acta Microbiologica Sinica*), *46*(1), 166–169. Chinese. PMID: 16579489.

Dupont, J., Dequin, S., Giraud, T., Le Tacon, F., Marsit, S., Ropars, J., Richard, F., & Selosse, M. A., (2017). Fungi as a source of food. *Microbiology Spectrum, 5*(3). doi: 10.1128/microbiolspec.FUNK-0030-2016.

Dupont, J., Dequin, S., Giraud, T., Le Tacon, F., Selosse, S. M., Ropars, J., Richard, F., & Selosse, M. A., (2018). Fungi as a source of food. In: Heitman, J., Howlett, B. J., Crous, P. W., Stukenbrock, E. H., James, T. Y., & Gow, N. A. R., (eds.), *The Fungal Kingdom*. American Society for Microbiology. Washington, DC. doi: 10.1128/microbiolspec.FUNK-0030-2016.

El Enshasy, H. A., Elsayed, E. A., Suhaimi, N., Malek, R. A., & Esawy, M., (2018). Bioprocess optimization for pectinase production using *Aspergillus Niger* in a submerged cultivation system. *BMC Biotechnology, 18*, 71. https://doi.org/10.1186/s12896-018-0481-7.

El Karoui, M., Hoyos-Flight, M., & Fletcher, L., (2019). Future trends in synthetic biology—A report. *Frontiers in Bioengineering and Biotechnology, 7*, 175.

Erdal, S. E. R. K. A. N., & Taskin, M. E. S. U. T., (2010). Production of alpha-amylase by *Penicillium expansum* MT-1 in solid-state fermentation using waste loquat (*Eriobotrya japonica* Lindley) kernels as substrate. *Romanian Biotechnological Letters, 15*(3), 5342–5350.

Erdal, S., & Taskin, M., (2010). Production of α-amylase by *Penicillium expansum* MT-1 in solid-state fermentation using waste loquat (*Eriobotrya japonica* Lindley) kernels as substrate. *Romanian Biotechnological Letters, 15*(3), 5342–5350.

Erjavec, J., (2012). Proteins of higher fungi-from forest to application. *Trends in Biotechnology, 30*(5), 259–273.

Fickers, P., Nicaud, J. M., Gaillardin, C., Destain, J., & Thonart, P., (2004). Carbon and nitrogen sources modulate lipase production in the yeast *Yarrowia lipolytica*. *Journal of Applied Microbiology, 96*(4), 742–749. https://doi.org/10.1111/j.1365–2672.2004.02190.x.

Flickinger, M. C., & Panikov, N. S., (2010). Kinetics, microbial growth. In: *Encyclopedia of Industrial Biotechnology*. John Wiley & Sons, Inc., Hoboken, NJ, USA.

Germano, S., Pandey, A., Osaku, C., Rocha, S., & Soccol, C., (2003). Characterization and stability of proteases from *Penicillium* sp. produced by solid-state fermentation. *Enzyme and Microbial Technology, 32*, 246–251.

Gilbert, E. J., Drozd, J. W., & Jones, C. W., (1991). Physiological regulation and optimization of lipase activity in *Pseudomonas aeruginosa* EF2. *Journal of General Microbiology, 137*, 2215–2221.

Godfredson, S. E., Fogarty, W. M., & Kelly, E. T., (1990). Microbial enzymes and biotechnology (Vol. 10, pp. 255–273). Elsevier, Amsterdam.

Gupta, R., Beg, Q. K., & Lorenz, P., (2002). Bacterial alkaline proteases: Molecular approaches and industrial applications. *Applied Microbiology and Biotechnology, 59*, 15–32.

Gupta, R., Gupta, N., & Rathi, P., (2004). Bacterial lipases: An overview of production, purification and biochemical properties. *Applied Microbiology and Biotechnology, 64*(6), 763–781.

Hajji, M., Hmidet, N., Jellouli, K., Vallaeys, T., Nasri, M., & Sellami-Kamoun, A., (2010). Gene cloning and expression of a detergent stable alkaline protease from *Aspergillus clavatus* ES1. *Process Biochemistry, 45*, 1746–1752.

Halabi, N., Rivoire, O., Leibler, S., & Ranganathan, R., (2009). Protein sectors: Evolutionary units of three-dimensional structure. *Cell, 138*, 774–786.

Haque, M. A., Wang, Y., Shen, Z., Li, X., Saleemi, M. K., & He, C., (2020). Mycotoxin contamination and control strategy in human, domestic animal and poultry: A review. *Microbial Pathogenesis, 142*, 104095. doi: 10.1016/j.micpath.2020.104095.

Heerd, D., Yegin, S., Tari, C., & Fernandez-Lahore, M., (2012). Pectinase enzyme-complex production by *Aspergillus* spp. in solid-state fermentation: A comparative study. *Food and Bioproducts Processing, 90*(2), 102–110.

https://www.marketsandmarkets.com/Market-Reports/enzyme-market-46202020.html (accessed on 13 January 2023).

https://www.ulprospector.com/en/na/Food/Products/15593/Food-Enzymes?st=1 (accessed on 13 January 2023).

Jaeger, K., & Reetz, M. T., (1998). Microbial lipases form versatile tools for biotechnology. *Trends in Biotechnology, 16*(9), 396–403.

Johnvesly, B., & Naik, G. R., (2001). Studies on production of thermostable alkaline protease from thermophilic and alkaliphilic *Bacillus* sp. JB-99 in a chemically defined medium. *Process Biochemistry, 37*, 139–144.

Kavanaugh, K., (2005). *Fungal Fermentation Systems and Products in Fungi: Biology and Applications* (pp. 89–111). Wiley.

Keil, B., (1992). *Specificity of Proteolysis*. Springer-Verlag KG, Berlin, Germany.

Klein, T., Eckhard, U., Dufour, A., Solis, N., & Overall, C. M., (2018). Proteolytic cleavage-mechanisms, function, and "omic" approaches for a near-ubiquitous posttranslational modification. *Chemical Reviews, 118*, 1137–1168.

Knox, A. M., Du Preez, J. C., & Kilian, S. G., (2004). Starch fermentation characteristics of *Saccharomyces cerevisiae* strains transformed with amylase genes from *Lipomyces kononenkoae* and *Saccharomyces fibuligera. Enzyme and Microbial Technology, 34*, 453–460.

Krem, M. M., & Di Cera, E., (2001). Molecular markers of serine protease evolution. *The EMBO Journal, 20*, 3036–3045.

Kubra, K. T., Ali, K., Walait, M., & Sundus, H., (2018). Potential applications of pectinases in food, agricultural and environmental sectors. *Journal of Pharmaceutical, Chemical and Biological Sciences, 6*(2), 23–34.

Kumari, A., Singh, K., & Kayastha, A. M., (2012). α-amylase: General properties, mechanism and biotechnological applications: A review. *Current Biotechnology, 1*, 98–107.

Kumari, M., Sharma, A., & Jagannadham, M. V., (2012). Religiosin B, a milk-clotting serine protease from *Ficus religiosa. Food Chemistry, 131*, 1295–1303.

Kundu, A. K., & Das, S., (1970). Production of amylase in liquid culture by a strain of *Aspergillus oryzae. Applied Microbiology, 19*(4), 598–603.

Lange, L., Agger, J. W., & Meyer, A. S., (2020). Fungal biotechnology: Unlocking the full potential of fungi for a more sustainable world. In: Nevalainen, H., (ed.), *Grand Challenges in Fungal Biotechnology, Grand Challenges in Biology and Biotechnology*. https://doi.org/10.1007/978-3-030-29541-7_1.

Lee, I. H., Huang, R. L., Chen, C. T., Chen, H. C., Hsu, W. C., & Lu, M. K., (2002). Antrodia camphorate polysaccharides exhibit anti-hepatitis B virus effects. *FEMS Microbiology Letters, 209*, 63–67. doi: 10.1111/j.1574-6968.2002.tb11110.x.

Ma, G., Yang, W., Zhao, L., Pei, F., Fang, D., & Hu, Q., (2018). A critical review on the health promoting effects of mushrooms nutraceuticals. *Food Science and Human Wellness, 7*, 125–133. doi: 10.1016/j.fshw.2018.05.002.

Maicas, S., (2020). The role of yeasts in fermentation processes. *Microorganisms, 8*(8), 1142. https://doi.org/10.3390/microorganisms8081142.

Meghwanshi, G. K., & Vashishtha, A., (2014). Microalgae as a potential source of biofuels. *Journal of Phytological Research, 27*(1, 2), 41–56. ISSN 0970-5767.

Meghwanshi, G. K., & Vashishtha, A., (2018). Biotechnology of fungal lipases. In: *Fungi and Their Role in Sustainable Development: Current Perspectives* (pp. 383–411). Springer Nature Singapore Pte Ltd. https://doi.org/10.1007/978-981-13-0393-7_22. Print ISBN: 9789811303920, eText ISBN: 9789811303937.

Meghwanshi, G. K., Kaur, N., Verma, S., Dabi, N. K., Vashishtha, A., Charan, P. D., Purohit, P., et al., (2020). Enzymes for pharmaceutical and therapeutic applications. *Biotechnology and Applied Biochemistry*. 10.1002/bab.1919. doi: 10.1002/bab.1919.

Mehta, A., Bodh, U., & Gupta, R., (2017). Fungal lipases: A review. *Journal of Biotech Research, 6*, 58–77.

Mehta, M., Muddapur, U. M., & Shanmuga Priya, V. G., (2013). Fungal production of tannase: A review. *International Journal of Scientific Engineering and Technology, 2*(8), 752–755.

Mensink, R. P., Sanders, T. A., Baer, D. J., Hayes, K. C., Howles, P. N., & Alejandro, M., (2016). The increasing use of interesterified lipids in the food supply and their effects on health parameters. *Advances in Nutrition, 7*(4), 719–729. doi:10.3945/an.115.009662. ISSN 2161-8313. PMC 4942855. PMID 27422506.

Merheb, C. W., Cabral, H., Gomes, E., & Da-Silva, R., (2007). Partial characterization of protease from a thermophilic fungus, *Thermoascus aurantiacus*, and its hydrolytic activity on bovine casein. *Food Chemistry, 104*, 127–131.

Metagenomics for New Enzyme Discovery and Industrial Biocatalysis. gtr.ukri.org/projects?ref=BB%2FL007444%2F1 (accessed on 13 January 2023).

Mohanta, T. K., (2020). Fungi contain genes associated with flavonoid biosynthesis pathway. *Journal of Functional Foods, 68*, 103910. doi: 10.1016/j.jff.2020.103910.

Morin, E., Kohler, A., Baker, A. R., Foulongne-Oriol, M., Lombard, V., Nagy, L. G., Ohm, R. A., et al., (2012). Genome sequence of the button mushroom *Agaricus bisporus* reveals mechanisms governing adaptation to a humic-rich ecological niche. *Proceedings of the National Academy of Sciences of the United States, 109*, 17501–17506.

Muralikrishna, G., & Nirmala, M., (2005). Cereal α-amylases: An overview. *Carbohydrate Polymers, 60*, 163–173.

Murphy, D. J. Jr., (2014). Coarctation of the aorta. In: Moodie, D. S., (ed.), *Clinical Management of Congenital Heart Disease: From Infancy to Adulthood* (pp. 142–170). Minneapolis, Cardiotext Publishing, Minnesota.

Musoni, M., Destain, J., Thonart, P., Bahama, J. P., & Delvigne, F., (2015). Bioreactor design and implementation strategies for the cultivation of filamentous fungi and the production of fungal metabolites: From traditional methods to engineered systems. *Biotechnology, Agronomy and Society and Environment, 19*(4), 430–442.

Nai, S., & Meghwanshi, G. K., (2015). Production of pickle from cauliflower petiole studs. *Journal of Phytological Research, 28*(1, 2), 1–6. ISSN 0970-5767.

Nakamura, A., Haga, K., & Yamane, K., (1993). Three histidine residues in the active center of cyclodextrin gluconotransferase from alkalophilic *Bacillus* sp. 1011 effects replacement on pH dependence and transition-state stabilization. *Biochemistry, 32*, 6624–6631.

Nardini, M., & Dijkstra, B. W., (1999). α/β hydrolase fold enzymes: The family keeps growing. *Current Opinion in Structural Biology, 9*(6), 732–737.

Neugnot, V., Moulin, G., Dubreucq, E., & Bigey, F., (2002). The lipase/acyltransferase from *Candida parapsilosis*: Molecular cloning and characterization of purified recombinant enzymes. *European Journal of Biochemistry, 269*(6), 1734–1745.

Panchapakesan, A., & Shankar, N., (2016). Chapter 2 – fungal cellulases: An overview. In: Gupta, V. K., (ed.), *New and Future Developments in Microbial Biotechnology and Bioengineering* (pp. 9–18). Elsevier.

Patel, A. K., Singhania, R. R., & Pandey, A., (2016). Novel enzymatic processes applied to the food industry. *Current Opinion in Food Science, 7*, 64–72. https://doi.org/10.1016/j.cofs.2015.12.002.

Pereira, F., (2019). Metagenomics: A gateway to drug discovery, In: Meena, S. N., & Naik, M. M., (eds.), *Advances in Biological Science Research* (pp. 453–468). Academic Press, ISBN 9780128174975.

Polizeli, M. L. T. M., Rizzatti, A. C. S., Monti, R., Terenzi, H. F., Jorge, J. A., & Amorim, D. S., (2005). Xylanase from fungi: Properties and industrial applications. *Applied Microbiology and Biotechnology, 67*(5), 577–591.

Prakasham, R. S., Subba, R. Ch., Sreenivas, R. R., & Sarma, P. N., (2007). Enhancement of acid amylase production by an isolated *Aspergillus awamori*. *Journal of Applied Microbiology, 102*(1), 204–211. https://doi.org/10.1111/j.1365-2672.2006.03058.x.

Qing, L., Yi, L., Marek, P., & Iverson, B. L., (2013). Commercial proteases: Present and future. *FEBS Letters, 587*, 1155–1163.
Ramachandran, S., Patel, A. K., Nampoothiri, K. M., Francis, F., Nagy, V., Szakacs, G., & Pandey, A., (2004). Coconut oil cake--a potential raw material for the production of alpha-amylase. *Bioresource Technology, 93*(2), 169–174. doi: 10.1016/j.biortech.2003.10.021. PMID: 15051078.
Rao, M. B., Tanksale, A. M., Ghatge, M. S., & Deshpande, V. V., (1998). Molecular and biotechnological aspects of microbial proteases. *Microbiology and Molecular Biology Reviews, 62*, 597–635.
Rashid, N., Shimada, Y., Ezaki, S., Atomi, H., & Imanaka, T., (2001). Low temperature lipase from psychrotrophic *Pseudomonas* sp. Strain KB700A. *Applied and Environmental Microbiology, 67*, 4064–4069.
Rawat, H. K., Soni, H., Treichel, H., & Kango, N., (2017). Biotechnological potential of microbial inulinases: Recent perspective. *Critical Reviews in Food Science and Nutrition, 57*(18), 3818–3829.
Rodionova, M. V., Poudyal, R. S., Tiwari, I., Voloshin, R. A., Zharmukhamedov, S. K., Nam, H. G., Zayadan, B. K., et al., (2017). Biofuel production: Challenges and opportunities. *International Journal of Hydrogen Energy, 42*(12), 8450–8461, https://doi.org/10.1016/j.ijhydene.2016.11.125.
Ruiz, H. A., Rodriguez-Jasso, R., Rodríguez, R., Contreras-Esquivel, J. C., & Aguilar, C. N., (2012). Pectinase production from lemon peel pomace as support and carbon source in solid-state fermentation column-tray bioreactor. *Biochemical Engineering Journal, 65*, 90–95.
Salam, L. B., (2019). Metagenomic insights into the diversity and functions of microbial assemblages in lakes, In: Bandh, S. A., Shafi, S., & Shameem, N., (eds.), *Freshwater Microbiology* (pp. 175–223). Academic Press, ISBN 9780128174951.
Sandhya, C., Sumantha, A., Szakacs, G., & Pandey, A., (2005). Comparative evaluation of neutral protease production by *Aspergillus oryzae* in submerged and solid-state fermentation. *Process Biochemistry, 40*, 2689–2694.
Saranraj, P., & Stella, D., (2013). Fungal amylase: A review. *International Journal of Microbiological Research, 4*(2), 203–211.
Satapathy, S., Rout, J. R., Kerry, R. G., Thatoi, H., & Sahoo, S. L., (2020). Biochemical prospects of various microbial pectinase and pectin: An approachable concept in pharmaceutical bioprocessing. *Frontiers in Nutrition, 7*, 117. doi: 10.3389/fnut.2020.00117.
Savitha, S., Sadhasivam, S., Swaminathan, K., & Lin, F. H., (2011). Fungal protease: Production, purification and compatibility with laundry detergents and their wash performance. *Journal of the Taiwan Institute of Chemical Engineers, 42*, 298–304.
Saxena, A., Upadhyay, R., & Kango, N., (2015). Isolation and identification of actinomycetes for production of novel extracellular glutaminase free L-asparaginase. *Indian Journal of Experimental Biology, 53*, 786–793.
Seon, G., Kim, H. S., Cho, J. M., Kim, M., Park, W. K., & Chang, Y. K., (2020). Effect of post-treatment process of microalgal hydrolysate on bioethanol production. *Scientific Reports, 10*, 16698. https://doi.org/10.1038/s41598-020-73816-4.
Sharma, N., (2019). A review on fungal alkaline protease. *Journal of Emerging Technologies and Innovative Research, 6*(6), 261–273.
Sheldon, R. A., Brady, D., & Bode, M. L., (2020). The hitchhiker's guide to biocatalysis: Recent advances in the use of enzymes in organic synthesis. *Chemical Science, 11*, 2587–2605. doi: 10.1039/c9sc05746c.

Singh, A. K., & Mukhopadhyay, M., (2012). Overview of fungal lipase. *Applied Biochemistry and Biotechnology, 166*, 486–520.

Song, X., Ren, Z., Wang, X., Jia, L., & Zhang, C., (2019). Antioxidant, anti-inflammatory and renoprotective effects of acidic hydrolytic polysaccharides by spent mushroom compost (*Lentinula edodes*) on LPS-induced kidney injury. *International Journal of Biological Macromolecules*. doi: 10.1016/j.ijbiomac.2019.10.173.

Srilakshmi, J., Madhavi, J., Lavanya, S., & Ammani, K., (2015). Commercial potential of fungal protease: Past, present and future prospects. *Journal of Pharmaceutical, Chemical and Biological Sciences, 2*(4), 218–234.

Strokopytov, B., Knegtel, R. M. A., Penninga, D., Roozeboom, H. J., Kalk, K. H., Dijkhuizen, L., & Dijkstra, B. W., (1996). Structure of cyclodextrin glycosyltransferase complexed with a maltononaose inhibitor at 2.6 Å resolution implications for product specificity. *Biochemistry, 35*, 4241–4249.

Sudeep, K. C., Upadhyaya, J., Joshi, D. R., Lekhak, B., Chaudhary, D. K., Pant, B. R., Bajgai, T. R., et al., (2020). Production, characterization, and industrial application of pectinase enzyme isolated from fungal strains. *Fermentation, 6*, 59. doi: 10.3390/fermentation 6020059.

Sun, S. Y., & Xu, Y., (2009). Membrane-bound 'synthetic lipase' specifically cultured under solid-state fermentation and submerged fermentation by *Rhizopus chinensis*: A comparative investigation. *Bioresource Technology, 100*, 1336–1342.

Sundarram, A., & Krishna, M. T. P., (2014). *α*-amylase production and applications: A review. *Journal of Applied & Environmental Microbiology, 2*(4), 166–175.

Sunitha, V. H., Ramesha, A., Savitha, J., & Srinivas, C., (2012). Amylase production by endophytic fungi *Cylindrocephalum* sp. isolated from medicinal plant *Alpinia calcarata* (Haw.) Roscoe. *Brazilian Journal of Microbiology, 43*(3), 1213–1221.

Tamang, J. P., Watanabe, K., & Holzapfel, W. H., (2016). Review: Diversity of microorganisms in global fermented foods and beverages. *Frontiers in Microbiology, 7*, 377. https://doi.org/10.3389/fmicb.2016.00377.

Tavano, O. L., (2013). Protein hydrolysis using proteases: An important tool for food biotechnology. *Journal of Molecular Catalysis B: Enzymatic, 90*, 1–11.

Teixeira, J. A., Gonçalves, D. B., De Queiroz, M. V., & De Araújo, E. F., (2011). Improved pectinase production in *Penicillium griseoroseum* recombinant strains. *Journal of Applied Microbiology, 111*(4), 818–825.

Thomas, H. K., (2004). *Training Strategies for Improving Listeners' Comprehension of Foreign-Accented Speech*. Doctoral dissertation. University of Colorado, Boulder.

Tomar, R., Kumar, R., & Jagannadham, M. V., (2008). A stable serine protease, wrightin, from the latex of the plant *Wrightia tinctoria* (Roxb.) R. Br.: Purification and biochemical properties. *Journal of Agricultural and Food Chemistry, 56*, 1479–1487.

Uitdehaag, J. C. M., Mosi, R., Kalk, K. H., Van, D. V. B. A., Dijkhuizen, L., Withers, S. G., & Dijkstra, B. W., (1999). X-ray structures along the reaction pathway of cyclodextrin glycosyltransferase elucidate catalysis in the α-amylase family. *Nature Structural Biology, 6*, 432–436.

Umezawa, K., Niikura, M., Kojima, Y., Goodell, B., & Yoshida, M., (2020). Transcriptome analysis of the brown rot fungus *Gloeophyllum trabeum* during lignocellulose degradation. *PLoS One, 15*(12), e0243984. https://doi.org/10.1371/journal.pone.0243984.

Verma, S., Meghwanshi, G. K., & Kumar, R., (2021). Current perspectives for microbial lipases from extremophiles and metagenomics. *Biochimie, 182*, 23–36.

Vishwanatha, K. S., Appu, R. A. G., & Singh, S. A., (2010). Production and characterization of a milk-clotting enzyme from *Aspergillus oryzae* MTCC 5341. *Applied Microbiology and Biotechnology, 85*, 1849–1859.

Walsh, M. C., Geraert, P. A., Maillard, R., Kluess, J., & Lawlor, P. G., (2012). The effect of a non-starch polysaccharide-hydrolyzing enzyme (Rovabio® Excel) on feed intake and body condition of sows during lactation and on progeny growth performance. *Animal, 6*(10), 1627–1633.

Wang, C. L., Chiang, C. J., Chao, Y. P., Yu, B., & Lee, T. T., (2015). Effect of *Cordyceps Militaris* waster medium on production performance, egg traits and egg yolk cholesterol of laying hens. *Journal of Poultry Science, 52*, 188–196. doi: 10.2141/jpsa.0140191.

Ward, O. P., (2011). Proteases. In: Moo-Young, M., (ed.), *Comprehensive Biotechnology* (2nd edn., pp. 571–582). Waterloo, Canada.

CHAPTER 16

Current Status of Bioactive Compounds Derived from the Endophytic Mycoflora of *Azadirachta indica*

DEEPAK KUMAR[1], SANTOSH K. ARYA[2], SANJAY K. SINGH[3], MANJUSHA TYAGI[2], and S. PRANANK[4]

[1]*R&D Division, Nextnode Bioscience Pvt. Ltd., Kadi, Gujarat, India*

[2]*Department of Microbiology, Shri Guru Ram Rai University, Patel Nagar, Dehradun, Uttarakhand, India*

[3]*Biodiversity and Paleobiology Group, Agharkar Research Institute, Pune, Maharashtra, India*

[4]*Department of Environmental and Applied Sciences, Algonquin College, Perth, Ontario, Canada*

ABSTRACT

Neem (*Azadirachta indica* A. Juss) is an Indian lilac that is utilized medicinally in a variety of ways. More than 400 chemicals have been identified from various portions of neem, including key bioactive secondary metabolites, like azadirachtin, nimbin, nimbidin, salanin, gedunin, and many others. It also has various bioactive qualities, which are used in agriculture (as pesticides, molluscicides, antifungals, acaricides, antimalarials, and antibiotics), as well as those used in medicine (as anti-inflammatory, analgetic, hypoglycemic, and antioxidant). Endophytic mycoflora associated with neem have been identified in a variety of forms, and they produce a wide range of bioactive

Applied Mycology for Agriculture and Foods: Industrial Applications. Sanjay K. Singh, Deepak Kumar, Md. Shamim, & Rohit Sharma (Eds.)

compounds having potential value in the agricultural, pharmaceutical, and industrial sectors.

16.1 INTRODUCTION

An 'endophyte' is any type of microbe that lives entirely or partially inside of a plant's tissues (Hardoim et al., 2015). They infect plants invisibly and asymptomatically, and they only exist within the tissues of plants. They don't show any signs of sickness when they're infected by the host plant. They infect plants invisibly and asymptomatically, and they only exist within the tissues of plants. They don't show any signs of sickness when they're infected by the host plant (Chutulo & Chalannavar, 2018; Wilson, 1995). Endocytic microbes, or endophytes, are the biochemical synthesizers in plants and have been intensively studied (Ramesha & Srinivas, 2014). Endophytes are a category of microorganisms that have received little attention, despite the fact that they constitute a rich and renewable supply of bioactive and chemically novel substances that have the potential to be used in a wide range of pharmaceuticals, industrial, and agricultural contexts (Strobel et al., 2004). Their taxonomic and ecological heterogeneity distinguishes them from other groups of organisms, with the majority of them related to the hyphomycetes, Ascomycota, coelomycetes, and among others (Chutulo & Chalannavar, 2018; Mahesh et al., 2005). The various microbial communities that live inside plants include a wide range of organisms, such as bacteria, mycoflora, and protista (Hardoim et al., 2015). An endophyte's ability to coordinate its growth with the host plant's is linked to a specific adaption that allows the endophyte to survive. Such endophytic microorganisms are responsible for either partial or entire biosynthesis of secondary metabolites produced by the host plant (Ludwig-Müller, 2015; Rajagopal et al., 2012).

Endophytic microbial biotechnology can be utilized to produce industrially, agriculturally, and economically relevant plant-based bioactive molecules and products efficiently. The use of endophytes for microbiota manipulation can help improve agricultural output, improve the volume of essential metabolites in medicinal and aromatic plants, and adapt to new bio-geographic areas through adaptation to different biotic and abiotic environments (Wani et al., 2015). Recently, there has been significant interest in endophytic mycoflora among those working in the field of microbial biochemistry because of their immense potential for contributing to new findings of bioactive chemicals. In contrast to epiphytes or plant microbiota, it has been proposed that the intimate biological interaction among endophytes and their plant host leads to the production of a significant quantity and diversity of bioactive molecules as compared to epiphytes or soil-associated microbes (Chutulo & Chalannavar, 2018; Strobel, 2003). Endophytic bioactive molecules are less hazardous to cells since they don't damage the eukaryotic host system, according to the symbiotic nature of the interaction. This is crucial for the health sector, as new medications may not harm human cells if they are properly tested (Alvin et al., 2014). Endophytes may produce new bioactive substances depending on a variety of parameters, including culture conditions, tissue type, and host age, as well as the degree to which endophytes interact with their photobionts and the genetic makeup of photobionts and mycobionts. Leaving most of these aspects aside, several researchers have reported new endophytic isolates and novel bioactive substances in recent investigations by altering the culture conditions (Ashkezari & Fotouhifar, 2017). Endophytes survive in a symbiotic connection with the plants that they parasitize. It is presumed that in many contexts, endophytic microbes serve as the plant's physiological defense system against phytopathogens. This is partly related to the concept of antimicrobial activity of the majority of secondary metabolites produced by endophytics, and these bioactive molecules have been involved in safeguarding the host cells against phytopathogens (Figure 16.1).

Azadirachta indica is prominently employed for the exploration of endophytes and their bioactive metabolites, but it is critical to continue to expand the scope of the research because it is one of the most important therapeutic plants used by approximately 80% of developing countries, according to the World Health Organization (WHO). This is due to the identification of novel bioactive chemicals as well as the finding of new fungal endophytes from the plant *Azadirachta indica*, which has only barely been published. The presence of endophytes, which exist on neem plants in a variety of ecological niches, has been linked to the production of more than 30 chemicals, including *Azadirachta indica* "mimetic" metabolites. The latest research advancements on endophytic mycoflora from the neem tree are discussed in this chapter, which includes the kinds of fungal endophytes defined and characterized, as well as biologically active compounds and the bioactivities of endophytic fungi to detect the existing level of knowledge about endophytes. Recent research on the antimicrobial activity of endophytic fungi found in neem, as well as recent breakthroughs in understanding their efficacy as prospective antimicrobials to combat the current health concerns caused by emerging multidrug-resistant infections. This chapter also focuses on the well-studied endophytic origins of prospective antimicrobials, with a particular emphasis on the economic future of endophytes on human diseases, as evidenced by contemporary research findings (Kharwar et al., 2020; Chutulo & Chalannavar, 2018).

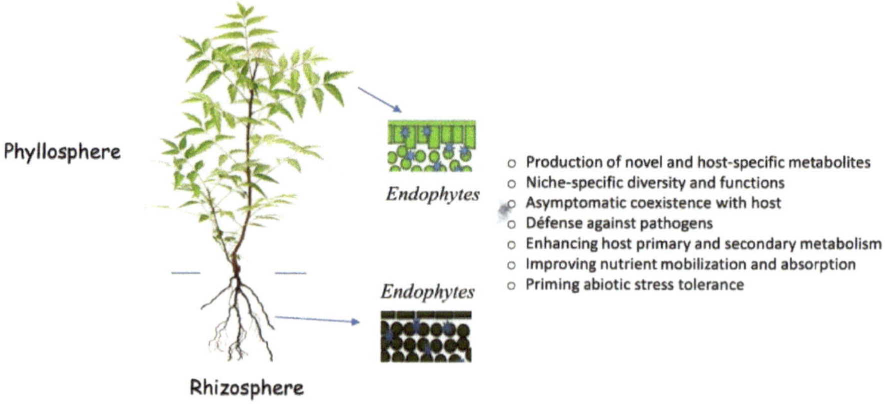

FIGURE 16.1 Coexistence and interaction between neem plants and their associated endophytic mycoflora.

16.2 DIVERSITY IN THE MYCOFLORA OF AZADIRACHTA INDICA

The highest species diversity of endophytic mycoflora is found in various tissue samples of a single plant taken from various sites. There have been reports of endophytes growing on neem plants, including *Periconia*, *Stenella*, and *Drechslera* (Verma et al., 2007). Moreover, they identified *Cladosporium cladosporioides*, *Phomopsis oblonga*, *Pestalotiopsis* sp., *Trichoderma* sp., and *Aspergillus* as the most prevalent endophytic fungal isolates in their study. The maturity of the plant parts is most likely a factor in determining the diversity of endophytic fungus. Some endophytes are separated solely from certain specific places, regardless of whether or not the endophytes are specific to different tissues (Qi et al., 2012; Tejesvi et al., 2006). As a result, endophytic fungi from various tissues of *Azadirachta indica* have been assessed and are shown in Table 16.1.

The leaf is an excellent tissue for studying endophytic mycoflora and their bioactive metabolites since it contains a high concentration of these compounds. The leaf has received the highest rating for species richness and endophyte colonization, according to the reports (Chutulo & Chalannavar, 2018; Verma, 2007). *Fusarium avenaceum*, an asymptomatic endophyte, was isolated from the leaves of *Azadirachta indica* A. Juss. by Rajagopal

TABLE 16.1 Diversity in Endophytic Mycoflora from Different Tissues of Neem Plant

Endophytic Fungi	Division	Class/Subclass	Family	Associated Plant Tissue	Identified by	Literature Cited
Acremonium sp., *Gliomastix* sp., *Trichoderma* sp.	Ascomycota (Pezizomycotina)	Sordariomycetes/ Hypocreomycetidae	Hypocreaceae	Root, bark, leaf, stem, and fruit	Phylogeny and multigene sequence identification (TEF, LSU, SSU, RPB2)	Chutulo & Chalannavar (2018); Maharachchikumbura et al. (2015); Verma et al. (2011); Summerbell et al. (2011)
Fusarium sp.	Ascomycota	Sordariomycetes/ Hypocreomycetidae	Nectriaceae	Root, bark, leaf, stem, and fruit	Multigene phylogeny, ITS-5.8S rDNA analysis	Hongsanan et al. (2017); Maharachchikumbura et al. (2015); Verma et al. (2011)
Verticillium sp.	Ascomycota	Sordariomycetes/ Hypocreomycetidae	Plectosphaerellaceae	Root	Multigene sequence identification (TEF, LSU, SSU, RPB2) data	Chutulo & Chalannavar (2018); Maharachchikumbura et al. (2015); Verma et al. (2011)
Colletotrichum sp.	Ascomycota	Sordariomycetes/ Hypocreomycetidae	Glomerellaceae	Root, leaf	ITS-5.8S rDNA analysis	Kumaresan et al. (2015); Verma et al. (2011)
Chaetomium sp.	Ascomycota	Sordariomycetes/ Sordariomycetidae	Chaetomiaceae	Root, bark, and leaf	ITS-5.8S rDNA analysis	Chutulo & Chalannavar (2018); Tenguria & Khan (2011); Verma et al. (2011)

TABLE 16.1 *(Continued)*

Endophytic Fungi	Division	Class/Subclass	Family	Associated Plant Tissue	Identified by	Literature Cited
Humicola sp.	Ascomycota	Sordariomycetes/ Sordariomycetidae	Chaetomiaceae	Fruit	ITS-5.8S rDNA analysis	Chutulo & Chalannavar (2018); Maharachchikumbura et al. (2015)
Chloridium sp.	Ascomycota	Sordariomycetes/ Sordariomycetidae	Chaetosphaeriaceae	Root	ITS analysis	Verma et al. (2011)
Xylaria sp.	Ascomycota	Sordariomycetes/ Xylariomycetidae	Xylariaceae	Stem	Phylogenetic analysis	Hongsanan et al. (2017); Wu et al. (2011)
Pestalotiopsis sp.	Ascomycota	Sordariomycetes/ Xylariomycetidae	Amphisphaeriaceae	Bark, stem, leaf	Phylogenetic analysis	Chutulo & Chalannavar (2018); Maharachchikumbura et al. (2015); Tejesvi et al. (2008)
Bartalinia sp.	Ascomycota	Sordariomycetes/ Xylariomycetidae	Bartaliniaceae	Bark	ITS rDNA analysis	Wijayawardene et al. (2016); Senanayake et al. (2015)
Nigrospora sp.	Ascomycota	Sordariomycetes/ Diaporthomycetidae	Trichosphaeriaceae	Root, stem, bark, and leaf	ITS-5.8S rDNA analysis	Chutulo & Chalannavar (2018); Maharachchikumbura et al. (2015); Verma et al. (2011)
Diaporthe (*Phomopsis*) sp.	Ascomycota	Sordariomycetes/ Diaporthomycetidae	Diaporthaceae	Stem, bark, and leaf	Multigene phylogeny	Maharachchikumbura et al. (2015); Wu et al. (2008)

TABLE 16.1 *(Continued)*

Endophytic Fungi	Division	Class/Subclass	Family	Associated Plant Tissue	Identified by	Literature Cited
Aspergillus sp.	Ascomycota	Eurotiomycetes	Diaporthaceae	Root, bark, leaf, stem, and fruit	Multigene phylogeny analysis	Chutulo & Chalannavar (2018); James et al. (2006)
Penicillium sp.	Ascomycota	Eurotiomycetes	Diaporthaceae	Root, fruit	ITS-5.8S rDNA analysis	Chutulo & Chalannavar (2018); Verma et al. (2011)
Eupenicillium sp.	Ascomycota	Eurotiomycetes	Diaporthaceae	Root, stem, bark, leaf	Multigene phylogeny analysis	Kusari et al. (2012); Spatafora et al. (2006)
Curvularia sp.	Ascomycota	Dothideomycetes	Diaporthaceae	Root, bark, and fruit	ITS-5.8S rDNA analysis	Chutulo & Chalannavar (2018); Verma et al. (2011); Tejesvi et al. (2008)
Alternaria sp.	Ascomycota	Dothideomycetes	Diaporthaceae	Root, leaf, and fruit	ITS-5.8S rDNA analysis	Chutulo & Chalannavar (2018); Verma et al. (2007, 2011)
Drechslera sp.	Ascomycota	Dothideomycetes	Diaporthaceae	Bark, fruit	ITS-5.8S rDNA analysis	Chutulo & Chalannavar (2018); Verma et al. (2007, 2011)
Phoma sp.	Ascomycota	Dothideomycetes	Diaporthaceae	Leaf	Phylogenetic analysis	Chutulo & Chalannavar (2018); Maharachchikumbura et al. (2015)

TABLE 16.1 *(Continued)*

Endophytic Fungi	Division	Class/Subclass	Family	Associated Plant Tissue	Identified by	Literature Cited
Cladosporium sp.	Ascomycota	Dothideomycetes	Davidiellaceae	Root, leaf, bark, and fruit	ITS-5.8S rDNA analysis	Verma et al. (2011); Crous et al. (2007)
Stenella sp.	Ascomycota	Dothideomycetes	Davidiellaceae	Bark	Multigene sequences	Chutulo & Chalannavar (2018); Verma et al. (2007)
Phyllosticta sp.	Ascomycota	Dothideomycetes	Botryosphaeriaceae	Root, fruit	ITS-5.8S rDNA analysis	Senanayake et al. (2015); Wikee et al. (2013)
Cochlonema sp.	Zygomycota (Zoopago-mycotina)	Tremellomycetes/ Tremellomycetidae	Cochlonemataceae	Bark	Phylogeny analysis	Chutulo & Chalannavar (2018); Hibbett et al. (2007)
Cercinella sp.	Zygomycota (Zoopago-mycotina)	Tremellomycetes/ Tremellomycetidae	Mucoraceae	Root	ITS-5.8S rDNA analysis	Verma et al. (2011)
Cryptococcus sp.	Basidiomycota (Agarico-mycotina)	Tremellomycetes/ Tremellomycetidae	Tremellaceae	Stem	Multigene phylogeny	Chutulo & Chalannavar (2018); Hibbett et al. (2007); James et al. (2006)

& Suryanarayanan (2000). Endophytes were found more frequently in the lowest leaflets than in the lamina tissue. The environment, as well as the type and chemistry of the host tissues, all influence the diversification of endophytes in tropical environments. When it comes to tropical trees, the environment, the types of host tissue and their chemical composition are all important factors to consider. The nature of the connection between a mycobiotic and a phycobiotic organism is also important (Rajagopal & Suryanarayanan, 2000). With the help of sterile filamentous fungi, they were able to isolate and document *Fusarium avenaceum*. Kumaresan et al. (2015) did research in which they recovered four endophytic fungus species (*Trichoderma* sp., *Curvularia* sp., *Colletotrichum* sp., and *Chaetomium* sp.). Similarly, Kaur et al. (2013) isolated *Alternaria alternata* from the leaves of *Azadirachta indica* A. Juss.

Tenguria & Khan (2011) identified 85 endophytic mycoflora from 200 portions of freshly picked *Azadirachta indica* leaves, representing a total of 85 species from 10 genera. The diversity of mycoflora occurred at the species level of *Penicillium* sp., *Trichoderma* sp., *Fusarium* sp., *Chaetomium globosum*, *Aspergillus* sp., *Alternaria alternata*, *Phoma* sp., *Pestalotiopsis* sp., and sterile mycelia in this category. *Azadirachta indica* leaves were used to isolate two endophytic fungi (AIn-L4 and AIn-L5). The fungi were grown in solid form on rice medium, and the fungal bioactive compounds were extracted with ethyl acetate. These endophytic fungi were shown to be interesting sources of potent bioactive molecules with medicinal or industrial applications, according to the study (Nnanna et al., 2018). Lwin & Soe (2018) investigated the isolation, characterization, and antibacterial properties of endophytic fungi isolated from leaves, inner barks, and stems of *Azadirachta indica* A. Juss. *Colletotrichum* sp., *Curvularia* sp., *Pestalotiopsis* sp., *Nigrospora* sp., *Aspergillus* sp., *Trichoderma* sp., *Fusarium* sp. and *Penicillium* species were identified among the fungal isolates. *Cladosporium* sp. predominated the leaves with a colonization frequency of 11.3% in *Azadirachta indica* leaf tissue, as reported by Abubakar et al. (2017).

Huang et al. (2015) isolated the endophytic mycoflora strain YM 311647 from the healthy stems of *Azadirachta indica* in China. Similarly, an endophytic fungus, *Phomopsis* sp. YM 311483, was isolated from a fresh stem of an *Azadirachta indica* specimen (Wu et al., 2008).

Moreover, Wu et al. (2011) identified the fungus *Xylaria* sp. YC-10 from *Azadirachta indica* A. Juss stems. One of the most common isolates, *Cryptococcus* sp., was found to be the most prevalent endophytic fungi on *A. indica* stems (twigs), accounting for 13.9% of the total number of isolates by

Abubakar & Ndana (2016). Treasure et al. (2020) isolated four endophytic fungi from fresh leaves and stems of *A. indica,* collected from different provinces of Nigeria. *Sodaria* sp., *Penicillium* sp., *Aureobasidium* sp., and *Aspergillus* sp. were identified as the endophytic mycoflora of *Azadirachta indica.* It's possible that secondary metabolites with antibacterial characteristics found in *Azadirachta indica* could be used in drug research.

From fresh leaves and twigs of *Azadirachta indica,* Chaturvedi & Shrivastava (2020) isolated the endophytic fungi *Rosellinia sanctae-cruciana.* Different secondary metabolites like tannins, alkaloids, flavonoids, and other organic acids were found to have potential activity in this endophytic mycoflora. In Oman, Al-Daghari et al. (2020) studied *Azadirachta indica's* leaf and stem tissues for the presence of fungal endophytes. A total of 144 neem plant sections (stem and leaf tissues) were examined for fungal isolates. The ascomycetes' classes, Sordariomycetes, Dothideomycetes, and Eurotiomyetes, each contained 23 taxa belonging to 15 genera. The most predominant endophytic mycoflora were *Aspergillus niger, Alternaria* sp., and *Cladosporium sphaerospermum* among isolated endophytic mycoflora. Several endophytic fungi have been found in *Azadirachta indica* leaves (midrib, no midrib, petiole, and stem). Overall, 16 different endophytic fungi were identified, with five belonging to the genus *Mycelia sterilla,* most of which are hypomycetes and a few were coleomycetes. *Alternaria* sp. was found to be the most prevalent endophytic fungus in the area (Taware & Rajurkar, 2015).

All plant's tissues contain endophytic fungi, which have long been thought to be a source of potential natural compounds with promising therapeutic uses. Endophytic fungi were isolated from the *Azadirachta indica* for pharmacological activities and studied for variety and biological activity of endophytic mycoflora by Naveen & Sathivelu (2017). In their findings, a total of five endophytic fungi have been identified, with one fungus being chosen for biochemical analysis based on phytochemical research.

Fruits of *Azadirachta indica* are rarely colonized by endophytic fungi because they are more commonly found on the bark, leaves, and stems of the plant. Verma et al. (2011) used unripe fruits and roots to isolate endophytic fungi. They found a total of 105 endophytic fungi in these immature fruits, representing 29 different taxa, or 68.0% of the total. Their isolates contained 81.69% hyphomycetes, 11.06% mycelia sterilia, and 7.25% coelomycetes, and whereas *Colletotrichum* sp., *Humicola, Drechslera* sp., were only found in fruit samples.

The neem (*Azadirachta indica* A. Juss) stem, leaf, and root samples were used by Qureshi et al. (2019) to isolate 11 endophytic species. In terms of colonization, *Aspergillus flavus* was the most prevalent, followed by *Penicillium* sp., *Aspergillus niger*, *Aspergillus fumigatus*, *Alternaria alternata*, *Phoma* sp., *Fusarium* sp., *Cladosporium cladsporioides*, *Colletotrichum* sp., and *Chaetomium globosum*, respectively. Tenguria & Khan (2011) isolated 85 endophytic fungi spanning 10 genera from fresh *Azadirachta indica* leaves collected from the Panchmarhi biosphere reserve. They found the fungi on 200 segments of leaves. Various types of endophytic fungi, including hyphomycetes, coelomycetes, ascomycetes, and sterile mycelium, were found to be present in the samples. Three endophytes were found to be particularly common, namely *Penicillium* sp., *Trichoderma* sp., and *Pestalotiopsis* sp. Roots of *A. indica* A. Juss. were found in Varanasi, India, to contain the endophytic mycoflora *Chloridium* sp. The fungus secretes javanicin, a potent antibacterial compound that may help protect its host from pathogenic microorganisms during infection (Kharwar et al., 2009). Moreover, in their study, Kusari et al. (2012) led to the isolation and characterization of a newly endophytic mycoflora, *Eupenicillium parvum*, which was previously unknown.

Despite the fact that endophytic fungi have been identified from various plant tissues in multiple studies, there are very limited findings of endophytic mycoflora from neem plant stems and twigs. Neem plant seeds and twigs have been found to harbor the virulent endophyte *Phomopsis azadirachtae*, which is responsible for the severe die-back disease (Vedashree et al., 2013). Endophytes can be studied on *Azadirachta indica* twigs, which have received little attention in the past. As a result, more research into endophytic mycoflora and their bioactive properties is needed to find new uses for them.

16.3 BIOACTIVE METABOLITES FROM ENDOPHYTIC MYCOFLORA OF *AZADIRACHTA INDICA*

Antimicrobial bioactive chemical synthesis by endophytes has recently gained attention due to the advent of chemotherapy-resistant infections that are resistant to many drugs. Contrary to chemical synthetic materials, endophytic sources of antimicrobials are required for both biological and chemical toxicity effects because of health and environmental concerns. Endophytic fungi can be exploited for the production of a large number of new, physiologically active secondary metabolites. Microorganisms' secondary metabolites have been shown to be enormously valuable and long-lasting resources (Chutulo & Chalannavar, 2018; Chowdhary et al., 2012). These bioactive compounds can be found in different parts of medicinal plant endophytes that have been identified (Prakash et al., 2011). Endophytic fungal mycoflora from the Neem tree produce a variety of bioactive chemicals, which are discussed in subsections with their biological activity (Table 16.2).

16.3.1 NATURAL BIOPESTICIDES, NEMATICIDES, AND ANTIPARASITES

Kusari et al. (2012) identified that the endophytic mycoflora *Eupenicillium parvum* can produce the natural insecticides azadirachtin A and B. Neem trees (*Azadirachta indica* A. Juss) are the only place these chemicals may be found. This chemical is produced by the fungus in both the mycelium and the culture media. Finding this azadirachtin-producing endophytic mycoflora has important ecological and biochemical ramifications for future research. Kaur et al. (2013) found that *Alternaria alternata* showed antifeedant and toxic activity against the tobacco caterpillar *Spodoptera litura* in their findings. The larvae of *Spodoptera litura* fed an artificial meal supplemented with a fungal extract had much lower survival rates and had their development and reproduction hampered because of the extreme toxicity of *A. alternata. Spodoptera litura*, a parasite, was found to be immunosuppressed by this endophytic mycoflora (Kaur et al., 2015). Moreover, Wu et al. (2011) isolated 11 bioactive metabolites (5-Methylmellein, Halorosellinic acid, Hymatoxin C&D, and Cerebroside C, etc.) from endophytic mycoflora *Xylaria* sp. of *Azadirachta indica* A. Juss that showed insecticidal activity against *Plutella xylostella*.

Chen et al. (2018) isolated a total of 85 actinomycetes isolates from neem trees, and these were tested for their insecticidal effectiveness against *Myzus persicae*. An interaction effect on *M. persicae* was seen with crude extracts from 24 distinct strains, with varying insecticidal efficacies. Among eight different actinomycete strains, crude extracts had the strongest insecticidal effects (above 60%). *Streptomyces albidoflavus*, an actinomycete with insecticidal action, was identified through molecular characterization. Bio-insecticide development could be aided by *S. albidoflavus* and its crude extract.

There were four compounds found in the endophytic *Geotrichum* sp. AL4 fungus grown from *Azadirachta indica* leaves by Li et al. (2007). They found two new chlorinated oxazinane derivatives that have nematicides activities. The nematodes *Panagrellus redivivus* and *Bursaphelenchus xylophilus* were

TABLE 16.2 Prominent Bioactive Compounds Produced from Endophytic Mycoflora of *Azadirachta indica*

Endophytic Mycoflora	Bioactive Compound	Derivative	Activity	References
Diaporthe sp./ *Phomopsis* sp.	3,4-dihydroxy-phenylalanine (DOPA)	Also known as levodopa and derivative of amino acid L-tyrosine	Precursor of biological pigment Melanin, and neurotransmitters dopamine (adrenaline and noradrenaline)	Chutulo et al. (2018); Rajagopal et al. (2012); Suryanarayanan et al. (2004)
Fusarium sp., *Trichoderma reesei*	Tyrosinase enzyme	A copper containing enzyme that controlling the melanin synthesis	Extracellular tyrosinase activity catalyzes tyrosine to melanin and other pigments by oxidation	Zaidi et al. (2013); Selinheimo et al. (2006)
Penicillium sp.	Saponins	Triterpene glycosides	Saponins modulate the immune system to prevent cancer and decrease cholesterol	Abubakar et al. (2017)
Fusarium solani	Cajaninstilbene acid	Methyl 6-formyl-2-hydroxy-4-methoxy-3-(3-methylbut-2-en-1-yl)benzoate is precursor building blocks of Cajaninstilbene acid	Neuroprotective property, anti-inflammatory effects	Zhao et al. (2012)
Nigrospora sp.	Solanapyrones N and O	Solanapyrone analogs	Antimicrobial secondary metabolites	Kharwar et al. (2020); Wu et al. (2009)
Phomopsis sp.	8R-acetoxy-5R-hydroxy-7-oxodecan-9-olide, and 7R-acetomy-multiplolide A	Lactones	Antimicrobial secondary metabolites	Kharwar et al. (2020); Wu et al. (2008)
Phomopsis sp.	7R,8R-dihydroxy-3,5-decadien-10-olide, and 8R-acetoxymultiploide A	Lactones	Antifungal activity	Tan et al. (2007)

TABLE 16.2 *(Continued)*

Endophytic Mycoflora	Bioactive Compound	Derivative	Activity	References
Xylaria sp.	Guaiane sesquiterpenes	Sesquiterpenes	Weak antifungal activities	Huang et al. (2015)
Pestalotiopsis microspora	Oxygenated guaiane sesquiterpenes and Isopimarane	Sesquiterpenes	Antioxidant, antihypertensive, and antibacterial activities	Tejesvi et al. (2008)
Chloridium sp.	Javanicin	Naphthazarin derivatives	Antimicrobial potential activities against plant and human pathogens	Kharwar et al. (2020, 2009)
Eupenicillium parvum	Azadirachtin A&B	limonoid group derivative secondary metabolite	Natural insecticides	Kusari et al. (2012)
Alternaria alternata	Unknown	–	Toxic, antifeedant, and immunomodulatory activities, natural insecticide against *Spodoptera litura*	Kaur et al. (2015, 2013)
Xylaria sp.	5-Methylmellein, 5-Carboxylmellein, Hymatoxin C, Hymatoxin D, Cerebroside C, Halorosellinic acid, Cerevisterol, (2S,3S, 4R, 2′R)-2-(2′Hydroxytetracosanoy lamino)-octadecane-1,3,4-triol)	Catechol derivatives	Weak insecticidal activity against *Plutella xylostella*	Wu et al. (2011)
Geotrichum sp.	1-[(2R*, 4S*, 5S*)-2-chloro-4-methyl-1,3-oxazinan-5-yl] ethanone and, its epimer	Chlorinated oxazinane derivatives	Nematicidal activities against *Bursaphelenchus xylophilus* and *Panagrellus radivivus* nematodes	Li et al. (2007)
Pestalotiopsis sp.	Unknown	–	Anticestodal activity against cysts *Echinococcus granulosus*	Verma et al. (2013)

TABLE 16.2 *(Continued)*

Endophytic Mycoflora	Bioactive Compound	Derivative	Activity	References
Endophytic fungi (AIn-L4 and AIn-L5)	Scytalone, cladosporin, and citreoisocoumarinol	Derivatives of norephedrine	Antimicrobial and cytotoxic activities	Nnanna et al. (2018)
Alternaria alternata	Sulfurous acid, butyl heptadecyl ester, oxalic acid, allyl hexadecyl ester, longifolenaldehyde, 2-Hexyl-1-decanol, 4,8,13-cyclotetradecatriene-1,3-diol,1,5,9-trimethyl-12-(1-methylethyl), longifolenaldehyde, d-norandrostane (5a,14a), 1H-3a,7-Methanoazulen-5-ol, octahydro-3,8,8-trimethyl-6-methylene, 2,2,6-Trimethyl-1-(2-methyl-cyclobut-2-enyl)-hepta-4,6-dien-3-one, 2,4,7,14-tetramethyl-4-vinyl-tricyclo[5.4.3.0(1,8)]tetradecan-6-ol, dichloronitromethane, L-5-propylthiomethylhydantoin, 4-O-methyl-d-arabinose, 6,7-dimethyl-triazolo(4,3-b)(1,2,4)-triazine	Chain hydrocarbons, chain branched hydrocarbons, acids, alcohols, *esters*-based derivatives	Antibacterial and antioxidant potential activities	Chatterjee et al. (2019)
Aspergillus aculeatus	Chrysophanol	Anthraquinone	Antibacterial	Kharwar et al. (2020); Wu et al. (2009); Garcia-Sosa et al. (2006)

TABLE 16.2 *(Continued)*

Endophytic Mycoflora	Bioactive Compound	Derivative	Activity	References
Aspergillus aculeatus	Emodin	Napthodianthrone derivative	Anticancerous, antibacterial	Kharwar et al. (2020); Wu et al. (2009); Kusari et al. (2008)
Aureobasidium sp.	Glucobrassicin	Glucosinolate	Antimicrobial activity	Treasure et al. (2020)
Sodaria sp.	Ferulic acid	4-hydroxy-3-methoxycinnamic acid	Antioxidant activity	Treasure et al. (2020)
Aspergillus sp.	4-methoxybenzaldehyde, 12-Hydroxy-16-scalaren, 12-O-deacetyl-12-epi-scalarin, Ixoside	Organic compounds of benzaldehydes class, sesterterpenes analogs, and barbituric and thiobarbituric acid *derivatives*	Antimicrobial activity	Treasure et al. (2020)
Penicillium sp.	Citreodrimene F, Cytosporin D	Sesquiterpene lactones secondary metabolites	Antimicrobial activity, immunosuppressant, anticancerous	Treasure et al. (2020)
Colletotrichum sp.	Unknown	Secondary metabolites	Antimicrobial activities against pathogenic *Candida albicans*, *Staphylococcus aureus*, *Escherichia coli*, *Bacillus cereus*	Lwin & Soe (2018)
Curvularia sp.	Unknown	Secondary metabolites	Antimicrobial activities	Lwin & Soe (2018)
Pestalotiopsis sp.	Unknown	Secondary metabolites	Antimicrobial activities	Lwin & Soe (2018)
Nigrospora sp.	Unknown	Secondary metabolites	Antimicrobial activities	Lwin & Soe (2018)
Colletotrichum acutatum, *Colletotrichum gloeosporioides*, *Colletotrichum johnstonii*	Unknown	Extracellular secondary metabolites	Antimicrobial activities against growth of multidrug-resistant (MDR) bacterial phytopathogenic fungi	Raihan et al. (2021)

TABLE 16.2 *(Continued)*

Endophytic Mycoflora	Bioactive Compound	Derivative	Activity	References
Cladosporium perangustum	Unknown	Extracellular secondary metabolites	Antimicrobial activities	Raihan et al. (2021)
Phyllostica capitalensis	Unknown	Extracellular secondary metabolites	Antimicrobial activities	Raihan et al. (2021)
Aspergillus oryzae	Unknown	Extracellular secondary metabolites	Antimicrobial activities	Raihan et al. (2021)
Macrophomina phascolina	Unknown	Extracellular secondary metabolites	Antimicrobial activities	Raihan et al. (2021)
Neofusicoccum kwambonambiense	Unknown	Extracellular secondary metabolites	Antimicrobial activities	Raihan et al. (2021)

tested for nematicidal activity against chlorinated oxazinane compounds and both showed significant bioactivity. Therefore, an effective antinematicidal formulation can be made from *Geotrichum* sp. AL4 endophyte.

Endophytic mycoflra isolate *Pestalotiopsis* sp. of *Azadirachta indica* A. Juss was reported to have strong anticestodal ability against *Echinococcus granulosus* by Verma et al. (2013). Within 30 minutes of culturing, *Pestalotiopsis* sp. demonstrated up to 97% mortality against *S. colemani*. Since this was the first study to cope with the anticestodal significant effects of the fungal endophyte *Pestalotiopsis* sp., additional findings are required to solidify its anticestodal activities by optimizing culture conditions.

16.3.2 ANTIMICROBIAL ACTIVITIES

Six *Azadirachta indica* A. Juss fungal endophytic mycoflora were tested and assessed for their efficacy against the fungi *Microsporum* and *Trichophyton* by Verma et al. (2008). These bioactive compounds produced by endophytic mycoflora have antimicrobial activities, but the bioactive compounds have yet to be identified by the researchers. Similarly, Abubakar et al. (2017) isolated bioactive compounds from *Penicillium* sp. from the leaves of *Azadirachta indica* A. Juss and that have antibacterial and antifungal properties. An endophytic fungus, *Nigrospora* sp. YB-141, was isolated from *Azadirachta indica* A. Juss. of tropical Southwest China and produced two solanapyrone analogs (solanapyrones N and O) molecules under optimized fermentation conditions (Wu et al., 2009).

Tejesvi et al. (2008) found that *Pestalotiopsis microspora* extracts had antimicrobial activity against a variety of pathogens. Moreover, antihypertensive and antioxidant properties were also found in the same species across different isolates. The research concludes that extracts of *Pestalotiopsis microspora* can be used commercially as an antihypertensive or an antioxidant for the diagnosis of the aforementioned diseases. Endophytic mycoflora *Xylaria* sp. YM 311647 produced guaiane sesquiterpenes in broth culture and sowed moderate antifungal activity against five pathogens of *Fusarium avenaceum* (Huang et al., 2015). Kharwar et al. (2009) identified javanicin after studying the secondary metabolites found in *Chloridium* sp. root isolates. The new compound was found to be similar to javanicin, a commonly known naphthaquinone. This makes it potentially appropriate for a variety of chemicals, including useful targeted antimicrobial formulations for plant and human infections.

16.3.3 ANTIOXIDANT METABOLITES

Kumaresan et al. (2015) investigated the biologically active antioxidants produced by endophytic fungi that colonized the leaf of a plant. They have looked at the ingredients of antioxidants such as phenols, tannins, flavonoids, ascorbic acid, and β-carotene, among other substances. The scientists found that all of the fungal endophytes that were analyzed had antioxidant activity *in vitro*.

The investigation of bioactive molecules from *Penicillium* species confirmed the production of saponins in addition to the other metabolites previously described by Abubakar et al. (2017). This is one among the very few findings on the antioxidant properties of neem endophytes that have been published. Another work by Zhao et al. (2012) identified the bioactive antioxidant Cajaninstilbene acid in endophytic fungi *Fusarium solani* that was isolated from pigeon pea and reported it as a natural antioxidant.

16.3.4 MELANIN PIGMENT

The enzyme tyrosinase, which is produced by endophytic mycoflora *Fusarium* sp. from the *Azadirachta indica*, is responsible for the formation of melanin pigment. Extracellular tyrosinase activity was produced by the fungus at a concentration of 2.8 U/mol (L-DOPA). This enzyme is required for pigmentation, and it is also a critical element in wound healing and the development of the initial immune response. In the environmental, pharmaceutical, cosmetics, and food industries, it has a wide range of applications (Zaidi et al., 2013). Despite the fact that endophytes are well-known for their antibacterial, anticancer, immunomodulatory, anti-inflammatory, and antiviral properties, there have only been a few publications on the use of endophytes in the manufacture of tyrosinase from Neem plants (Zaidi et al., 2013). Extracellular synthesis of fungal tyrosinase has been documented, initially from *Trichoderma reesei*, and afterwards from other fungi (Selinheimo et al., 2006).

Diaporthe (formerly known as *Phomopsis*) is a phellophytic mold that dwell on the outermost surface of the bark of trees and other plants. It was discovered in the plant *Azadirachta indica*, where it produced and deposited DOPA (3,4-dihydroxyphenylalanine) type melanin on the hyphae of the plant. The endophyte, *Phomopsis*, produces melanin, which provides adaptations such as mechanisms for overcoming host barriers, successful competition with other phylloplane fungi, and surviving harsh environmental conditions. This is reflected in the widespread occurrence of the endophyte (Manawasinghe et al., 2016; Rajagopal et al., 2012; Suryanarayanan et al., 2004). Consequently, future study should concentrate on exploring tyrosinase from *Azadirachta indica* A. Juss due to the greater number of industrial applications for this enzyme. The detailed description of the bioactive metabolites produced by neem fungal endophytes is listed in Table 16.2.

16.4 CONCLUSION

The plant *Azadirachta indica* and its endophytic mycoflora comprise an extensive repertory of various natural compounds with a wide range of biochemical and biological activity. For example, azadirachtins (A&B) and similar tetranortriterpenoids, which are found in abundance in neem (the host plant), are a valuable source of insecticides and antioxidants. Endophytes linked with neem, on the other hand, offer a tremendous potential for the production of bioactive and chemically new chemicals, which are now unavailable. Among the most notable findings from *Azadirachta indica* is the vast number of different fungal endophytes and actinomycetes that have been identified in a relatively short period of time. The result has been an explosion in the discovery and isolation of natural products, many of which are from neem and its endophytes, despite the fact that their biochemistry and overall environmental implications are still largely unknown. To date, only a small number of bioactive chemicals derived from endophytic fungi of neem have been described and identified, as well as the majority of these studies have been constrained to the level of fungal identification and bioactivity assay. The many groups of endophytic fungi (Basidiomycota, Ascomycota, and Zygomycota) that have been identified in the neem tree are genotypically assembled using contemporary taxonomy derived from several systematic reviews and molecular annotations. With the exception of a few research studies, endophytic microorganisms of neem have received little attention. Therefore, there is a need for more attention to exploring the bioactive molecules of endophytic mycoflora of *Azadirachta indica.* These bioactive molecules may play a prominent role in the fields of agriculture, pharmaceuticals, and industrial chemicals in the future.

KEYWORDS

- **antibiotics**
- ***Azadirachta indica***
- **bioactive molecules**
- **endocytic microbes**
- **endophytes**
- **endophytic mycoflora**
- **microorganisms**
- **multidrug-resistant**

REFERENCES

Abubakar, S. & Ndana, R.W. (2016). Preliminary study of Endomycodiversity among three ethnomedicinal plants from family Meliaceae in Nigeria. *J. BioSci. Biotechnol. 5*, 195–201.

Abubakar, S., Ndana, R. W., & Afolabi, A. S., (2017). Bioprospective potentials of endophytic fungi *Penicillium* spp. isolated from leaves of *Azadirachta indica* (A. JUSS). *Int. J. Biol. Res., 5*, 15–21.

Al-Daghari, N. R., Maharachchikumbura, S. S. N., Al-Moqbali, D., Al-Saady, N., & Al-Sadi, A. M., (2020). Fungal diversity in leaves and stems of neem (*Azadirachta indica*). *International Journal of Scientific & Technology, 9*(6), 793–798.

Alvin, A., Miller, K. I., & Neilan, B. A., (2014). Exploring the potential of endophytes from medicinal plants as sources of antimycobacterial compounds. *Microbiol. Res., 169*, 483–495.

Ashkezari, S. J., & Fotouhifar, K. B., (2017). Diversity of endophytic fungi of common yew (*Taxus baccata* L.) in Iran. *Mycol. Prog., 16*, 247–256.

Chatterjee, S., Ghosh, R., & Mandal, N. C., (2019). Production of bioactive compounds with bactericidal and antioxidant potential by endophytic fungus *Alternaria alternata* AE1 isolated from *Azadirachta indica* A. Juss. *PLoS One, 14*(4), e0214744. https://doi.org/10.1371/journal.pone.0214744.

Chaturvedi, A., & Shrivastava, D., (2020). Antioxidant potential of Rosellina sanctae *Crucian sp.*-endophytic fungi isolated from *Azadirachta indica. International Journal of Recent Scientific Research, 11*(2), 37616–37625.

Chaturvedi, U. N., Christiana, A. C., Maduabuchi, E. P., Nnamdi, O. A., Ebiye, E. C., Chigozie, U. M., Basden, O. F., et al., (2020). *GSC Biological and Pharmaceutical Sciences, 11*(03), 115–124.

Chen, Y., Shafi, J., Li, M., Fu, D., & Ji, M., (2018). Insecticidal activity of endophytic actinomycetes isolated from *Azadirachta indica* against *Myzus persicae*. *Archives of Biological Sciences, 70*(2), 349–357.

Chowdhary, K., Kaushik, N., Coloma, A. G., & Raimundo, C. M., (2012). Endophytic fungi and their metabolites isolated from Indian medicinal plant. *Phytochem. Rev., 11*, 467–485.

Chutulo, E. C., & Chalannavar, R. K., (2018). Endophytic mycoflora and their bioactive compounds from *Azadirachta indica*: A comprehensive review. *J. Fungi, 4*, 42. doi: 10.3390/jof4020042.

Clark, R. & Soil, Appalachian. (1997). Arbuscular mycorrhizal adaptation, spore germination, root colonization, and host plant growth and mineral acquisition at low pH. Plant and Soil - PLANT SOIL. 192. 15-22. 10.1023/A:1004218915413

Crous, P. W., Braun, U., Schubert, K., & Groenewald, J. Z., (2007). Delimiting *Cladosporium* from morphologically similar genera. *Stud. Mycol., 58*, 33–56.

Garcia-Sosa, K., Villarreal-Alvarez, N., Lubben, P., & Peña-Rodriguez, L. M., (2006). Chrysophanol, an antimicrobial anthraquinone from the root extract of *Colubrina greggii. J Mex Chem Soc., 50*, 76–78.

Hardoim, P. R., Van, O. L. S., Berg, G., Pirttilä, A. M., Compant, S., Campisano, A., & Döring, M., (2015). The hidden world within plants: Ecological and evolutionary considerations for defining functioning of microbial endophytes. *Microbiol. Mol. Biol. Rev., 79*, 293–320.

Hibbett, D. S., Binder, M., Bischoff, J. F., Blackwell, M., Cannon, P. F., Eriksson, O. E., Huhndorf, S., James, T., Kirk, P. M., Lücking, R., Lumbsch, H. T., et al., (2007). A higher-level phylogenetic classification of the fungi. *Mycol. Res., 111*, 509–547.

Hongsanan, S., Maharachchikumbura, S. S. N., Hyde, K. D., Samarakoon, M. C., Jeewon, R., Zhao, Q., Al-Sadi, A. M., & Bahkali, A. H., (2017). An updated phylogeny of *Sordariomycetes* based on phylogenetic and molecular clock evidence. *Fungal Divers., 84*, 25–41.

Huang, R., Xie, X. S., Fang, X. W., Ma, K. X., & Wu, S. H., (2015). Five new guaiane sesquiterpenes from the endophytic fungus *Xylaria* sp. YM 311647 of *Azadirachta indica. Chem. Biodivers., 12*, 1281–1286.

James, T. Y., Kauff, F., Schoch, C. L., Matheny, P. B., Hofstetter, V., Cox, C. J., Celio, G., Gueidan, C., Fraker, E., Miadlikowska, J., et al., (2006). Reconstructing the early evolution of fungi using a six-gene phylogeny. *Nature, 443*, 818–822.

Kaur, H. P., Singh, B., Kaur, A., & Kaur, S., (2013). Anti-feedent and toxic activity of endophytic *Alternaria alternate* against tobacco caterpillar *Spodoptera litura. J. Pest Sci., 86*, 543–550.

Kaur, H. P., Singh, B., Thakur, A., Kaur, A., & Kaur, S., (2015). Studies on immunomodulatory effect of endophytic fungus *Alternaria alternata* on *Spodoptera litura. J. Asia. Pac. Entomol., 18*, 67–75.

Kharwar, N. R., Sharma, V. K., Mishra, A., Kumar, J., Singh, D. K., Verma, S. K., Gond, S. K., et al., (2020). Harnessing the phytotherapeutic treasure troves of the ancient medicinal plant *Azadirachta indica* (Neem) and associated endophytic microorganisms. *Planta Med., 86*, 906–940.

Kharwar, R. N., Verma, V. C., Kumar, A., Gond, S. K., Harper, J. K., Hess, W. M., Lobkovosky, E., et al., (2009). Javanicin, an antibacterial naphthaquinone from an endophytic fungus of neem, *Chloridium* sp. *Curr. Microbiol., 58*, 233–238.

Kumaresan, S., Karthi, V., Senthilkumar, V., Balakumar, B. S., & Stephen, A., (2015). Biochemical constituents and antioxidant potential of endophytic fungi isolated from the leaves of *Azadirachta indica* A. Juss (Neem) from Chennai. *J. Acad. Ind. Res., 3*, 355–361.

Kusari, S., Lamshöft, M., Zühlke, S., & Spiteller, M., (2008). An endophytic fungus from *Hypericum perforatum* that produces hypericin. *J. Nat. Prod., 71*, 159–162.

Kusari, S., Verma, V. C., Lamshoeft, M., & Spiteller, M., (2012). An endophytic fungus from *Azadirachta indica* A. Juss. that produces azadirachtin. *World J. Microbiol. Biotechnol., 28*, 1287–1294.

Li, G. H., Yu, Z. F., Li, X., Wang, X. B., Zheng, L. J., & Zhang, K. Q., (2007). Nematicidal metabolites produced by the endophytic fungus *Geotrichum* sp. AL4. *Chem. Biodivers., 4*, 1520–1524.

Ludwig-Müller, J., (2015). Plants and endophytes: Equal partners in secondary metabolite production? *Biotechnol. Lett., 37*, 1325–1334.

Lwin, A. T., & Soe, K. K., (2018). Isolation, characterization and antimicrobial activities of endophytic fungi from leaves, stem and inner barks of *Azadirachta indica* A. Juss. *Mandalay University Journal, 9*(2), 1–12.

Maharachchikumbura, S. S., Hyde, K. D., Jones, E. G., McKenzie, E. H., Huang, S. K., Abdel-Wahab, M. A., Daranagama, D. A., Dayarathne, M., D'souza, M. J., Goonasekara, I. D., et al., (2015). Towards a natural classification and backbone tree for *Sordariomycetes*. *Fungal Divers., 72*, 199–301.

Mahesh, B., Tejesvi, M. V., Nalini, M. S., Prakash, H. S., Kini, K. R., Subbiah, V., & Shetty, H. S., (2005). Endophytic mycoflora of inner bark of *Azadirachata indica*. *Curr. Sci., 88*, 218–220.

Manawasinghe, I. S., Phillips, A. J. L., Hyde, K. D., Chethana, K. W. T., Zhang, W., Zhao, W. S., Yan, J. Y., & Li, X., (2016). Mycosphere essays 14: Assessing the aggressiveness of plant pathogenic *Botryosphaeriaceae*. *Mycosphere, 7*, 883–892.

Naveen, S., & Sathiavelu, A., (2017). Diversity and biological activities of endophytic fungi isolated from *Azadirachta indica*. *Research Journal of Pharmacy and Technology, 10*(12), 4198–4202.

Nnanna, J. C., Eze, P. M., Anyanwu, O. O., Ujam, T. N., Ikegbunam, M. N., Okoye, F. B. C., Charles, O., & Esimone, C. O., (2018). Screening of metabolites of endophytic fungi isolated from leaves of *Azadirachta indica* for antimicrobial and cytotoxic activities. *The Pharmaceutical and Chemical Journal, 5*(3), 20–27.

Prakash, H. S., Ruma, K., Shailasree, S., Kumara, K., Sampath, S., Ramachandrappa, N., & Prakash, H. S., (2011). Diversity of fungal endophytes from two endemic tree species *Artocarpus hirsutus* Lam. and *Vateria indica* Linn. of Western Ghats, India. *World J. Agric. Sci., 7*, 577–582.

Qi F, Jing T, Zhan Y (2012) Characterization of Endophytic Fungi from Acer *ginnala* Maxim. in an Artificial Plantation: Media Effect and Tissue-Dependent Variation. PLoS ONE 7(10): e46785. https://doi.org/10.1371/journal.pone.0046785.

Qureshi, A. H., Shoket, A. S., & Pandey, A. K., (2019). Isolation and identification of endophytic fungi inhabiting *Azadirachta indica* A. Juss from different regions of Jabalpur (M.P.) India. *Research Journal of Life Sciences, Bioinformatics, Pharmaceuticals and Chemical Sciences, 5*(3), 274–281.

Raihan, T., Azad, A. K., Ahmed, J., Shepon, M. R., Dey, P., Chowdhury, N., Aunkor, T. H., Ali, H., & Suhani, S., (2021). Extracellular metabolites of endophytic fungi from *Azadirachta indica* inhibit multidrug-resistant bacteria and phytopathogens. *Future Microbiology, 16*, 8. https://doi.org/10.2217/fmb-2020-0259.

Rajagopal, K., & Suryanarayanan, T. S., (2000). Isolation of endophytic fungi from leaves of neem (*Azadirachta indica* A. Juss). *Current Science, 78*(11), 1375–1378.

Rajagopal, K., Maheswari, S., & Kathiravan, G., (2012). Diversity of endophytic fungi in some tropical medicinal plants—A report. *Afr. J. Microbiol. Res., 6*, 2822–2827.

Ramesha, A., & Srinivas, C., (2014). Antimicrobial activity and phytochemical analysis of crude extracts of endophytic fungi isolated from *Plumeria acuminata* L. and *Plumeria obtusifolia* L. *Eur. J. Exp. Biol., 4*, 35–43.

Selinheimo, E., Saloheimo, M., Ahola, E., Westerholm-Parvinen, A., Kalkkinen, N., Buchert, J., & Kruus, K., (2006). Production and characterization of a secreted, C-terminally processed tyrosinase from the filamentous fungus *Trichoderma reesei*. *FEBS J., 273*, 4322–4335.

Senanayake, I. C., Maharachchikumbura, S. S. N., Hyde, K. D., Bhat, J. D., Jones, E. B. G., McKenzie, E. H. C., Dai, D. Q., Daranagama, D. A., Dayarathne, M. C., Goonasekara, I. D., et al., (2015). Towards unraveling relationships in *Xylariomycetidae* (*Sordariomycetes*). *Fungal Divers., I*, 73–144.

Spatafora, J. W., Sung, G. H., Johnson, D., Hesse, C., O'Rourke, B., Serdani, M., Spotts, R., Lutzoni, F., Hofstetter, V., Miadlikowska, J., et al., (2006). A five-gene phylogeny of pezizomycotina. *Mycologia, 98*, 1018–1028.

Strobel, G. A., (2003). Endophytes as sources of bioactive products. *Microbes Infect., 5*, 535–544.

Strobel, G., Daisy, B., Castillo, U., & Harper, J., (2004). Natural products from endophytic microorganisms. *J. Nat. Prod., 67*, 257–268.

Summerbell, R. C., Gueidan, C., Schroers, H. J., De Hoog, G. S., Starink, M., Rosete, A. Y., Guarro, J., & Scott, J. A., (2011). *Acremonium* phylogenetic overview and revision of *Gliomastix, Sarocladium, and Trichothecium. Stud. Mycol., 68*, 139–162.

Suryanarayanan, T. S., Ravishankar, J. P., Venkatesan, G., & Murali, T. S., (2004). Characterization of the melanin pigment of a cosmopolitan fungal endophyte. *Mycol. Res., 108*, 974–978.

Tan, Q., Yan, X., Lin, X., Huang, Y., Zheng, Z., Song, S., Lu, C., & Shen, Y., (2007). Chemical constituents of the endophytic fungal strain *Phomopsis* sp. NXZ-05 of *Camptotheca acuminata. Helv. Chim. Acta, 90*, 1811–1817.

Taware, A. S., & Rajurkar, S. K., (2015). Diversity assessment of endophytic fungi from *Azadirachta indica* A. Juss. from various regions of Aurangabad, Maharashtra (India). *International Journal of Innovative Science, Engineering & Technology, 2*(7), 96–107.

Tejesvi, M. V., Kini, K. R., Prakash, H. S., Subbiah, V., & Shetty, H. S., (2008). Antioxidant, antihypertensive, and antibacterial properties of endophytic *Pestalotiopsis* species from medicinal plants. *Can. J. Microbiol., 54*, 769–780.

Tejesvi, M. V., Mahesh, B., Nalini, M. S., Prakash, H. S., Kini, K. R., Subbiah, V., & Shetty, H. S., (2006). Fungal endophyte assemblages from ethnopharmaceutically important medicinal trees. *Can. J. Microbiol., 52*, 427–435.

Tenguria, R. K., & Khan, F. N., (2011). Distribution of endophytic fungi in leaves of *Azadirachta indica* A. JUSS. (Neem) of Panchmarhi biosphere reserve. *Curr. Bot., 2*, 27–29.

Vedashree, S., Sateesh, M. K., Lakshmeesha, T. R., & Mohammed, S. S., (2013). Screening and assay of extracellular enzymes in *Phomopsis azadirachtae* causing die-back disease of neem. *J. Agric. Technol., 9*, 915–927.

Verma, V. C., Gangwar, M., Yashpal, M., & Nath, G., (2013). Anti-cestodal activity of endophytic *Pestalotiopsis* sp. on protoscoleces of hydatid cyst *Echinococcus granulosus. Biomed Res. Int., 2013*, 308515.

Verma, V. C., Gond, S. K., Kumar, A., Kharwar, R. N., & Strobel, G., (2007). The endophytic mycoflora of bark, leaf, and stem tissues of *Azadirachta indica* A. Juss (Neem) from Varanasi (India). *Microb. Ecol., 54*, 119–125.

Verma, V. C., Gond, S. K., Kumar, A., Kharwar, R. N., Boulanger, L. A., & Strobel, G. A., (2011). Endophytic fungal flora from roots and fruits of an Indian neem plant *Azadirachta indica* A. Juss., and impact of culture media on their isolation. *Indian J. Microbiol., 51*, 469–476.

Wani, Z. A., Ashraf, N., Mohiuddin, T., & Riyaz-Ul-Hassan, S., (2015). Plant-endophyte symbiosis, an ecological perspective. *Appl. Microbiol. Biotechnol., 99*, 2955–2965.

Wijayawardene, N. N., Hyde, K. D., Wanasinghe, D. N., Papizadeh, M., Goonasekara, I. D., Camporesi, E., Bhat, D. J., McKenzie, E. H. C., Phillips, A. J. L., Diederich, P., et al., (2016). Taxonomy and phylogeny of dematiaceous coelomycetes. *Fungal Divers, 77*, 1–316.

Wikee, S., Lombard, L., Nakashima, C., Motohashi, K., Chukeatirote, E., Cheewangkoon, R., McKenzie, E. H. C., Hyde, K. D., & Crous, P. W., (2013). A phylogenetic re-evaluation of *Phyllosticta* (*Botryosphaeriales*). *Stud. Mycol., 76*, 1–29.

Wilson, D., (1995). Endophyte: The evolution of a term, and clarification of its use and definition. *Oikos, 73*, 274–276.

Wu, S. H., Chen, Y. W., & Miao, C. P., (2011). Secondary metabolites of endophytic fungus *Xylaria* sp. YC-10 of *Azadirachta indica. Chem. Nat. Compd., 47*, 858–861.

Wu, S. H., Chen, Y. W., Shao, S. C., Li, Z. Z., Yang, L. Y., & Li, S. L., (2009). Studies on the metabolites of *Aspergillus aculeatus* YM 311498, an endophytic fungus from *Azadirachta indica* A. Juss. *Nat Prod Res Dev., 21*, 731–732.

Wu, S. H., Chen, Y. W., Shao, S. C., Wang, L. D., Li, Z. Y., Yang, L. Y., Li, S. L., & Huang, R., (2008). Ten-membered lactones from *Phomopsis* sp. an endophytic fungus of *Azadirachta indica. J. Nat. Prod., 71*, 731–734.

Wu, S. H., Chen, Y. W., Shao, S. C., Wang, L. D., Yu, Y., Li, Z. Y., Yang, L. Y., Li, S. L., & Huang, R., (2009). Two new solanapyrone analogues from the endophytic fungus *Nigrospora* sp. YB-141 of *Azadirachta indica. Chem. Biodivers., 6*, 79–85.

Zaidi, K. U., Mani, A., Ali, A. S., & Ali, S. A., (2013). Evaluation of tyrosinase producing endophytic fungi from *Calotropis gigantea, Azadirachta indica, Ocimum tenuiflorum* and *Lantana camara. Annu. Rev. Res. Biol., 3*, 389–396.

Zhao, J., Fu, Y., Luo, M., Zu, Y., Wang, W., Zhao, C., & Gu, C., (2012). Endophytic fungi from pigeon pea [*Cajanus cajan* (L.) Millsp.] produce antioxidant cajaninstilbene acid. *J. Agric. Food Chem., 60*, 4314–4319.

CHAPTER 17

Bio-Prospecting Fungi for Hydrocarbons with a Potential for Biofuels Production

SANJAI SAXENA[1,2]

[1]Department of Biotechnology, Thapar Institute of Engineering and Technology (Deemed to be University), Patiala, Punjab, India

[2]Agpharm Bioinnovations LLP, Incubatee: Science and Technology Entrepreneurs Park (STEP), Thapar Institute of Engineering and Technology, Patiala, Punjab, India

ABSTRACT

Fungi are a diverse assemblage of organisms exhibiting tremendous metabolic plasticity in terms of utilizing complex substrates and therefore are ubiquitous in their existence. Fungi, as well as their endophytic forms, have been implicated to be producers of volatile as well as non-volatile secondary metabolites, which find applications in different sectors of industry. As alternative fuels are in great demand with the depletion of carboniferous reserves of fuels in nature, the present chapter, the prospects of fungi and their endophytic counterparts as potent producers of hydrocarbons which may hold a promise to replace the current fuels for the transport industry as well as other applications for the survival of mankind.

17.1 INTRODUCTION

Global demand for fuels is continuously increasing with the pace of industrialization happening in the developing countries as well as in developed

Applied Mycology for Agriculture and Foods: Industrial Applications. Sanjay K. Singh, Deepak Kumar, Md. Shamim, & Rohit Sharma (Eds.)

countries. This demand is also increasing with the transportation industry gaining a strength hold with enhanced continental as well as intercontinental travel (Strobel, 2015).

Fossil fuel is one of the primary sources of industrial, economical, transport, and agricultural activities. Hence enhanced consumption of these have let to depletion of these natural resources. They have also been responsible for global warming apart from particulate pollutants in the atmosphere. Hence, the need of cleaner alternatives which could be perpetually produced by environment friendly methods is become imperative (Luque et al., 2008; UN-GAP Report, 2014).

One of the prominent alternatives which can reduce or mitigate the effects of fossil fuels is the use of biofuels. The practical definition of biofuel is one which is derived from the biomass, i.e., produced from plants, algae, microorganisms, or animal wastes. Generally, the term biofuel is reserved for liquid or gaseous fuels used for transportation purpose. The most common types of biofuels are bioethanol and biodiesel (Demirbas, 2007; Luque et al., 2008).

Bioethanol is generally referred to as ethyl alcohol produced via microbial fermentation. It generally involves conversion of carbohydrates present in biomass (cellulosic and non-cellulosic) is used as feedstock for the ethanol production. During the beginning of the automobile industry, the primary fuel developed was ethanol (Giebelhaus, 1980). Henry Ford was fully convinced that the internal combustion engines which are the heart of the automobile industry could be effectively driven by ethanol. He designed various models developed from 1908 to be driven on ethanol (Kovarik, 1998). Today ethanol is being used as pure fuel (E100), but more often it is being used as an additive to petroleum to increase the octane number as well as improve the vehicular emissions. It is extensively being used in the USA and Brazil. Blending of petrol with ethanol was a common practice which was being followed during World War II since at that time there were very less crude oil producing countries in the world (Lapuerta et al., 2007). India has also started blending petroleum with 20% ethanol to improve emissions from the vehicles as well as extend the current petroleum reserves.

Biodiesel on the other hand is a biofuel which is derived from oils produced by plants, generally by transesterification process. Vegetable oils were initially used to demonstrate the functioning of compression ignition engine by Rudolph Diesel in 1898, until the 1920's when alteration in engine was brought in to use the residue of petroleum (No. 2 Diesel fuel) (Ma & Hanna, 1999). Biodiesel (B100) can be used as an alternative fuel in its pure form, however currently it is being used as an additive in diesel for the reduction of particulates, hydrocarbons, and carbon dioxide from diesel powered vehicles. Chemically biodiesel is largely composed of fatty acid methyl/ ethyl esters produced as a result of transesterification process (Sharma & Singh, 2009).

The feedstock for biodiesel production is largely comprising of animal fats, vegetable oils (virgin) produced from various oleaginous species such as soy, rapeseed, jatropha, mustard, flax, sunflower, hemp, and *Pongamia pinnata* (Kulkarni & Dalai, 2006; Sarin et al., 2007). Various species of algae and fungi also accumulate oils and could be mass produced for the production of biodiesel/biofuel (Ma, 2006; Ma et al., 2018).

There is a vast literature which highlights various methodologies of using feedstocks (biomass) for the production of bioethanol, bio-methanol, and biobutanol. The role of various microorganisms has also been highlighted in optimizing these processes to industrial scale. However, in this chapter, we would largely discuss the role of fungi in the production of hydrocarbons exhibiting the potential for the development into biodiesel/biofuel.

17.2 FUNGI AS SOURCES OF BIOFUELS

Fungi possess excellent biosynthetic capabilities due to their intricate metabolic pathways and capabilities to utilize a diversity of substrates to derive nutrition for their survival and growth. In these processes, they produce a variety of primary and secondary metabolites which could be exploited in different industrial sectors. Majority of the secondary metabolites have been exploited in the agrochemical (Saxena & Pandey, 2001; Saxena, 2014) and pharmaceutical (Saxena et al., 2019; Dufosse et al., 2021) sectors towards the development of biorational agrochemicals and in development of new drugs, respectively. These biosynthetic capabilities of fungi can be advantageously diverted to producing hydrocarbons-based fuels and their precursors. One of the prominent pathways existing in the fungal system is the mevalonate pathway which can be modulated towards the biosynthesis of terpenes.

17.3 OLEAGINOUS FUNGI

One of the major advantages of biodiesel over conventional diesel is it does not contribute to the production of carbon dioxide or sulfur into the environment apart from emission of less gaseous pollutants (Lund et al., 2014). Cutting down the cost of biodiesel production is one of the most prominent hurdles in realizing the true commercial potential. This is largely attributed to the cost of oil raw materials; for instance, when plant oil is used for the production of biodiesel, it covers up 75–80% of the total biodiesel production cost. Hence there is a demand for alternative sources of mass production of raw oil cost effectively (Ling et al., 2014). Microorganisms (fungi and algae) exhibit potential to be used as alternatives to oil being recovered from agricultural and animal sources (Subramaniam et al., 2010; Ma et al., 2018).

When a microorganism accumulates or produces lipid (microbial) in excess to 20% on a dry weight basis, then it is referred as oleaginous (Carvalho et al., 2015). Thus, single cell oil (SCO), has garnered great attention across the globe as a bioresource for biodiesel production. Among the yeast and fungi, some of the prominent oil-bearing species were *Cryptococcus albidus* (65%), *Rhodotorula glutinis* (72%) and *Mortierella vinacea* (66%), *Humicola lanuginosa* (75%) and *Mortierella isabellina* (86%), respectively (Adrio, 2017; Yang et al., 2019). In the recent past, there has been an upsurge in screening of microorganisms which have accumulated large amounts of hydrophobic lipids for possible use in SCO production as well as biodiesel production (Dey et al., 2011; Tanimura et al., 2014).

Largely oleaginous yeasts and fungi (especially molds) accumulate triacylglycerols rich in polyunsaturated fatty acids (PUFAs) or somewhat limited: Oleic acid (18:1) and Linoleic acid (18:2) together with palmitic (16:0) and palmitoleic acid (16:1) are the most frequently found free fatty acids (FFA) (Zhu et al., 2002; Shafiqe & Ali, 2017). On such study was conducted in Mucoromycota wherein FAME (fatty acid methyl esters) were analyzed to determine the saturated fatty acids, monoene, and polyenes. High value fatty acids viz. C18:1, C18:2, C18:3, C20:3, C20:4, and C20:5 was reported from isolates belonging to all three orders viz., *Mortierellales, Mucorales, and Umbelopsidales* (Zhao et al., 2020). In the recent past, several oleaginous fungi have been explored as they have major advantages in lipid production over microalgae in terms of space, venue, and climatic conditions. Oleaginous yeasts and fungi possess the capability of transforming inexpensive carbon sources, such as lignocellulosic biomass and waste materials into lipids (Dey et al., 2011). Thus, they are preferred for biomass cultivation for commercial biofuel/biodiesel production.

The lipid accumulation in fungi and yeasts is largely driven by the nature of the isolate, culture conditions (such as temperature, pH, nutrients, culture time, etc.) and the substrate. Inexpensive organic materials or wastes are also being used as substrate for the production of lipids by fungi and yeasts. Yet another advantage unlike the oleaginous plants is they do not lead to fuel versus food crisis as well as they exhibit inherent capacity of upscaling the process with practical and commercial viability. Several oleaginous yeasts and fungi have been explored and optimized for the production of SCO (Table 17.1).

An added dimension is endophytic microorganisms which also offer themselves to be producers of SCO. Endophytes are microorganism which reside within the plants without any obvious signatures of their existence (Bacon & White, 2015). It has been established that these evolve with their host and acquire biosynthetic capabilities to produce their host compounds after the seminal discovery of *Taxomyces andreane*, the taxol producing fungi (Stierle et al., 1993). Thus, endophytic fungi existing in the oleaginous plants could be of immense potential in production of SCO. Endophytic fungi like *Alternaria* sp. and *Colletotrichum* sp. isolated from *Brassica juncea* (oil yielding plant) and *Ocimum sanctum* were found to possess significant oleaginous properties with lipid content ranging between 30% and 58% of their biomass (Dey et al., 2011). Recently over 170 endophytic isolates were obtained from different plant parts of oleaginous plants viz. *Jatropha curcas*, *Pongamia pinnata*, *Sapindus mukorossi*, *Terminalia bellerica* and *Ricinus communis*. These were bioprospected for oil production, of which four isolates exhibited oleaginous characteristics, viz. *Lasiodiplodia exigua* SPSRJ27, *Phomopsis* sp. SPSRJ28, *Pestalotiopsis microspora* SPSRJL35, and *Phomopsis* sp. SPSRJL36, were found to have more than 20% of the total lipid content of their dry biomass. Further studies on these isolates indicated that *Phomopsis* sp. SPSRJ28 was the most promising isolate for biodiesel production (Paul et al., 2020). Another endophytic fungus which accumulated ~58% of lipid was found to be *Fusarium* sp. ML-GEN.1, which was isolated from *Strobilanthus cusia*. The oleic acid accumulation by *Fusarium* sp. ML-GEN.1 was to the tune of 41.66%, palmitic acid 23.26% and linoleic acid 19.18% (Xie et al., 2013). Another oleaginous fungus, namely *Pesudofusicoccum* sp. has been existing as an endophyte in *Annona muricata* growing in Southeast Nigeria (Abba & Eze, 2020). *Colletotrichum truncatum* was identified as an endophytic oleaginous fungus residing in *Jatropha curcas* from New Delhi and exhibited promising potential for production of biodiesel (Kumar & Kaushik, 2013).

Endophytic fungi and yeasts, especially from oleaginous plants hold promise in the production of new generation of biofuels as their biosynthetic gene clusters have evolved along with their host during their course of association or interaction.

TABLE 17.1 Oleaginous Yeasts and Fungi with Potential for Biodiesel Production

SL. No.	Name of the Organism	Type	Lipid Content >30% (w/w)	References
1.	*Cryptococcus aerius*	Yeast	63.3	Sitepu et al. (2014)
2.	*Cryptococcus albidus*	Yeast	40.1	Li et al. (2015)
3.	*Cryptococcus curvatus*	Yeast	69.0	Iassonova et al. (2008)
4.	*Cryptococcus terreus*	Yeast	51.0	Sitepu et al. (2014)
5.	*Cryptococcus terricola*	Yeast	62.0	Tanimura et al. (2014)
6.	*Candida diddensiae*	Yeast	37.0	Malkhas'ian et al. (1982)
7.	*Candida freyschussii*	Yeast	32.0	Malkhas'ian et al. (1982)
8.	*Kodamaea ohmeri*	Yeast	53.3	Kitcha & Cheirsilp (2011)
9.	*Lipomyces doorenjongii*	Yeast	72.3	Oguri et al. (2012)
10.	*Lipomyces kockii*	Yeast	77.8	Oguri et al. (2012)
11.	*Lipomyces lipofer*	Yeast	51.5	Evans & Ratledge (1983)
12.	*Lipomyces starkeyi*	Yeast	68.4	Juanssilfero et al. (2018)
13.	*Lipomyces tetrasporus*	Yeast	66.5	Sitepu et al. (2014)
14.	*Leucosporidiella creatinivora*	Yeast	62.0	Rossi et al. (2009)
15.	*Metschnikowia gruessi*	Yeast	34.0	Sitepu et al. (2014)
16.	*Rhodosporidium babjevae*	Yeast	58.0	Sitepu et al. (2013)
17.	*Rhodosporidium sphaerocarpum*	Yeast	43.0	Franklin et al. (2011)
18.	*Rhodotorula glutinis*	Yeast	66.0	Ageitos et al. (2011)
19.	*Rhodotorula graminis*	Yeast	52.3	Galafassi et al. (2012)
20.	*Rhodotorula glacialis*	Yeast	68.0	Amaretti et al. (2010)
21.	*Torulaspora delbrueckii*	Yeast	40.0	Franklin et al. (2011)
22.	*Trichosporon coremiiforme*	Yeast	37.8	Huang et al. (2012)
23.	*Trichosporon dermatis*	Yeast	35.2	Huang et al. (2012)
24.	*Aspergillus terreus*	Fungi	37.4	Zheng et al. (2012)
25.	*Aspergillus niger*	Fungi	57.4	Andre et al. (2012)
26.	*Cunninghamella elegans*	Fungi	33.6	Zheng et al. (2012)
27.	*Cunninghamella echinulata*	Fungi	40–57	Fakas et al. (2009)
28.	*Humicola lanuginosa*	Fungi	75	Meng et al. (2009)
29.	*Mortierella isabellina*	Fungi	60–86	Zheng et al. (2012)
30.	*Mortierella vinacea*	Fungi	66	Zheng et al. (2012)
31.	*Mucor circinelloides*	Fungi	46	Mitra et al. (2012)
32.	*Zygorhynchus moelleri*	Fungi	43.2	Chatzifragkou et al. (2011)

17.4 FUNGAL SECONDARY METABOLITES AS SOURCES OF HYDROCARBONS

A wide array of secondary metabolites produced by filamentous fungi can be classified as terpenes, terpenoids, polyketides, alkaloids, and non-ribosomal polyketides (Keller et al., 2005). From the viewpoint of an organic chemist, these secondary metabolites can be grouped as simple linear, branched or aromatic (having ring structure). They can also be grouped as those analogous to paraffins, isoparaffins, and cyclo-paraffins or else could be a unique chemical scaffold comprising of both rings and branches in its structure. Polyketides, terpenes, and terpenoids have drawn greater interest when liquid hydrocarbon fuels are being explored since they have very few or no oxygen atoms in their structure and hence could be upgraded or converted into a fuel via known conversion methods (Butcher et al., 2018).

17.5 FUNGAL POLYKETIDES AS PROMISING BIOFUELS

Polyketides are natural products which contain multiple β-hydroxyketone or β-hydroxyaldehyde functional groups. The precursor of the polyketides are acyl-CoA thioesters. Fungal polyketides have immensely contributed to the pharmaceuticals and agrochemical industry. Some promising fungal polyketides have been Lovastatin from *Aspergillus terreus* which acts as a cholesterol reducing agent (Alberts, 1988), Strobilurin A from *Strobilurus tenacellus* which was used for the development of the most sold fungicide Azoxystrobin (Anke et al., 1977). The polyketides comprise of several groups of compounds belonging to xanthones, anthraquinones, naphthoquinones, depsidones, pyrenocines, isocoumarins, cyclohexanone, naphthalene, chromone, preussomerins. Polyketide biosynthetic pathways are generally considered as the most versatile pathways for the production of hydrocarbons with diverse structures. Reduced polyketides serve as a good starting point for the production of hydrocarbons which can be used to produce biofuels as derivatives. The enzyme polyketide synthase type-1 (PKS) plays an important role in the biosynthesis of polyketides and cis closely related to eukaryotic fatty acid synthases. Bioengineering of PKS is one of the most plausible ways for designing novel versions of polyketides which could be used as biofuels. Though the importance of polyketide-based biofuels has been greatly acknowledged, however, the supporting literature on polyketide-based fuels is grossly lacking.

17.6 POTENTIAL OF TERPENES AS BIOFUELS

Terpenes are largely natural products, predominantly unsaturated hydrocarbons which are polymers of isoprene units (C_5H_8). Terpenes which largely are suitable for use as liquid biofuels belong to monoterpenes (C_{10}), diterpenes (C_{15}) and sesquiterpenes (C_{20}). The enzyme responsible for terpene biosynthesis is terpene synthase. Further functionalized derivatives of terpenes with aldehydes, ketones, and hydroxyl groups are known as terpenoids. Fungi have been a prolific source of terpenes and terpenoids (Peralta-Yahya et al., 2012).

Monoterpenes are derived from geranyl phosphate while sesquiterpenes are generated from farnesyl phosphate whereas diterpenes are synthesized from geranylgeranyl pyrophosphate. Terpenes are already known as constituents of flavors and fragrances, as pharmaceutical agents and as alternative fuels (Ignea et al., 2011; Peralta-Yahya et al., 2012). There is a great deal of interest in fungi producing monoterpenes and sesquiterpenes as possible alternative fuels or additives since they possess low freezing point and high ignition stability (Lee et al., 2015). Farnesane is the hydrogenated form of β-farnesene which has been approved as a 10% blend component in commercial jet fuel as per ASTM Designation D7566 (ASTM, 2017).

Endophytic fungi have the inherent potential of producing the plant host metabolites as a result of their co-evolution during the course of time. Hence these appear as lucrative sources of production of plant-based monoterpenes. Until now 1,8-cineole was being produced from plants, so having a fungal source of 1,8-cineole could lead to significant industrial applications if the production could be increased to economically viable levels (Shaw et al., 2015). An endophytic isolate of *Hypoxylon* and two isolates of *Nodulisporium* have been found to produce 1,8-cineole in volatile state. *Phomposis* species isolated from *Odontoglossum* sp. (Orchidaceae) growth in forest in Northern Ecuador was found to produce a monoterpene, Sabinene which was previously reported only from higher plants (Singh et al., 2011). Sabinene is a C_{10} bicyclic hydrocarbon which has a boiling point of 163°C and 92.4 research octane number (RON) and has a potential to be used as a blend in jet fuel (Zhang et al., 2014). Hence, endophytic fungi hold promise to produce a variety of terpenes which have the potential to be used as components of diesel fuel or jet fuel (Table 17.2).

TABLE 17.2 Hydrocarbons with Potential as Biofuels Produced by Fungi and Yeasts

SL. No.	Compound (Formula)	Compound Class/Type	Organism (Y/F)	References
1.	Pinene ($C_{10}H_{16}$)	Bicyclic monoterpene	*Saccharomyces cerevisiae* (Y)	Peralta-Yahya et al. (2012)
2.	Limonene ($C_{10}H_{16}$)	Cyclic monoterpene	*S. cerevisiae* (Y)	Peralta-Yahya et al. (2012)
3.	Sabinene ($C_{10}H_{16}$)	Bicyclic monoterpene	*Phomopsis* sp. (F)	Morath et al. (2012)
4.	1,8-Cineole ($C_{10}H_{18}O$)	Monoterpenoid	*Hypoxylon* sp. (F)	Shaw et al. (2015)
5.	β-Farnesene ($C_{15}H_{24}$)	Sesquiterpenoid	*Aspergillus fumigatus* (F)	Kramer & Abraham (2012)
6.	Longifolene ($C_{15}H_{24}$)	Tricyclic sesquiterpenes	*Fomitopsis pinicola* (F), *Fusarium graminearum* (F)	Kramer & Abraham (2012)
7.	β-Cubebene ($C_{15}H_{24}$)	Tricyclic sesquiterpenes	*Monascus purpureus* (F)	Patakova-Juzlova et al. (1998)
8.	β-Ylangene ($C_{15}H_{24}$)	Sesquiterpenoid	*Fomitopsis pinicola* (F)	Kramer & Abraham (2012)
9.	β-Patchoulene ($C_{15}H_{24}$)	Tricyclic sesquiterpenes	*Penicillium roquefortii* (F)	Kramer & Abraham (2012)
10.	A-Santalene ($C_{15}H_{24}$)	Sesquiterpene	*Inonotus obliquus* (F)	Kramer & Abraham (2012)
11.	Aristolochene ($C_{15}H_{24}$)	Bicyclic sesquiterpene	*Penicillium roquefortii* (F)	Hanson (2008)
12.	β-Caryophyllene ($C_{15}H_{24}$)	Bicyclic sesquiterpene	*Phialophora fastigiate* (F)	Sunesson et al. (1995)
13.	Bisabolene ($C_{15}H_{24}$)	Sesquiterpene	*S. cerevisiae* (Y)	Kramer & Abraham (2012)
14.	Farnesol ($C_{15}H_{26}O$)	Sesquiterpene alcohol	*Candida albicans* (Y), *Ceratocystis coerulescens* (F)	Kramer & Abraham (2012)
15.	Geraniol ($C_{10}H_{18}O$)	Monoterpenoid	*Ceratocystis species* (F)	Hanson (2008)
16.	Linalool ($C_{10}H_{18}O$)	Terpene alcohol	*Ceratocystis species* (F)	Hanson (2008)
17.	Elemene ($C_{15}H_{24}$)	Sesquiterpene	*Lentinus lepideus* (F)	Kramer & Abraham (2012)
18.	Cerapicol ($C_{15}H_{28}O$)	Sesquiterpenoid	*Ceratocystis piceae* (F)	Hanssen & Abraham (1988)
19.	α-Bisabolol ($C_{15}H_{26}O$)	Monocyclic sesquiterpene alcohol	*Phlebia radiata Fr.* (F)	Ebel (2010)
20.	Isoculmorin ($C_{15}H_{21}O_2$)	Tricyclic sesquiterpene	*Kallichroma tethys* (F)	Ebel (2010)

TABLE 17.2 *(Continued)*

SL. No.	Compound (Formula)	Compound Class/Type	Organism (Y/F)	References
21.	3,4-dimethylhexane (C_8H_{18})	Alkane	*Gliocladium species* (F)	Ahamed & Ahring (2011)
22.	Cyclopropane, 1-butyl-2-(2-methyl propyl)-($C_{11}H_{22}$)	Alkane	*Aspergillus carbonarius* (F)	Sinha et al. (2015)
23.	Octane, 2,2-dimethyl-($C_{10}H_{22}$)	Alkane	*Aspergillus carbonarius* (F)	Sinha et al. (2015)
24.	2,4-dimethyl-2-heptane (C_9H_{20})	Alkane	*Aspergillus carbonarius* (F)	Sinha et al. (2015)
25.	2-Ethyl-5,5-dimethyl-1,3-cyclopentadiene (C_9H_{14})	Alkene	*Gliocladiun roseum* (F) (NRRL 50072)	Strobel et al. (2008)
26.	Isoamyl acetate ($C_7H_{14}O_2$)	Alcohol ester	*Muscodor crispans* (F)	Mitchell et al. (2010)
27.	3,3,5-Trimethyldecane ($C_{13}H_{28}$)	Alkane	*Gliocladiun roseum* (F) (NRRL 50072)	Strobel et al. (2008)
28.	3-Octanone ($C_8H_{16}O$)	Dialkyl ketone	*Tricholoma matsutake* (F)	Cho et al. (2008)
29.	(2Z, 4Z)-4,6-Dimethyl-octa-2,4-dienoic acid ($C_{10}H_{15}O_2$)	Monocarboxylic acid	*A. nidulans* (F)	Yaegashi et al. (2014)
30.	Isobutyric acid ($C_4H_8O_2$)	Carboxylic acid	*S. cerevisiae* (Y)	Suomalaine & Lehtonen (1979)

17.7 ENDOPHYTIC FUNGI AS SOURCES OF HC (HYDROCARBONS) AND BIOFUELS

The concept the endophytic fungi could be the source of fuel related hydrocarbons is traced back to the discovery of a novel endophytic fungus, *Muscodor albus*. *Muscodor albus* and its related species have proven to be a treasure trove of biological activities relevant to pharmaceutical, agrochemical, environment, and energy sectors. It is largely attributed to the volatile as well as non-volatile secondary metabolites which are being bioprospected as well as exploited (Saxena & Strobel, 2020). The discovery of *Muscodor* species and its spectrum of activities, evoked the interest of mycologists to explore endophytic fungi for their antimicrobial spectrum as well as fuel (long chain hydrocarbon) synthesizing properties (Strobel et al., 2001). Another promising endophytic fungus which was isolated from Ulmo tress (*Eucryphia cordifolia*) was *Gliocladium roseum* NRRL 50072 now known as *Ascocoryne sarcoides* is found to produce series of volatile hydrocarbons and hydrocarbon derivatives such as heptane, octane, benzene, and some branched hydrocarbon on cellulose based and oatmeal-based agar medium under microaerophilic conditions as analyzed through SPME-GC-MS technique (Strobel, 2008; Gainoulis et al., 2012). The fungus also produced alkanes, i.e., straight chain and branched alkanes such as pentane, hexane, heptane, octane, etc., and branched hydrocarbons like 3-methyl nonane and many others which constitute the class one compounds of diesel. Other hydrocarbons present in the headspace of the fungus culture comprised of cycloalkanes primarily comprising of 1-methyl cyclohexene, cyclodecane, and cyclodecene, which essentially belong to class 2 compounds of diesel fuel (Strobel, 2010). Another promising outcome of studies on *Ascocoryne sarcoides* was the yield of hydrocarbons was much higher on cellulose-based medium, indicating the potential valorization of lignocellulosic biomass and agriculture wastes for commercial production of these hydrocarbons (Strobel, 2014)

Yet another endophytic fungus which was found to produce hydrocarbon *Hypoxylon* sp. (CI-4A). It was isolated from *Persea indica*, an evergreen tree native to the Canary Islands in an imperfect stage *Nodulisporium* sp. from the small stem of the tree. The volatile organic compounds (VOCs) emanated from *Hypoxylon* sp. (CI-4A) comprised of 1,8-cineole, 1-methyl-1,4-cyclohexadiene, and (+)-α-methylene-α-fenchocamphorone (Tomsheck et al., 2010). 1,8-cineole has already been reported to be used as an additive in ethanol-gasoline blends. Similarly, another fungus, *Daldinia* sp. (EC-12) was isolated as an endophyte from *Myroxylon balsamum* found in the upper Napo region of the Ecuadorian Amazon. This organisms under microaerophilic growth conditions emanated, 1-methyl-, 1-4 pentadiene; 1,4-cyclohexadiene and cyclohexene, 1-methyl-4-(1-methylethenyl)—along with production of some terpenoids and alcohols of interest as potential fuels (Mends et al., 2012). Another endophytic isolate of *Nodulisporium* sp. isolated from *Thelypteris angustifolia* (Broadleaf Maiden Fern) growing in central America was found to produce 1,8-cineole in abundance apart from 2-methyl-1-butanol and phenylethyl alcohol (Hassan et al., 2013). *Phomopsis* sp. isolated as an endophyte from *Odontoglossum* sp. apart from producing Sabinene also produced 3-methyl-1-butanol, benezenethanol, and 2-methyl-1-propanol as other volatile compounds holding promise to be used as a biofuel (Singh et al., 2011).

Annulohypoxylon sp. was another promising endophyte isolated from *Citrus aurantifolia*, in the swamps of Florida which was found to produce 1,8-cineole, 2-butanone, methyl ester, (1-methylethyl)-benzene, benzophenone, and azulene, 1,2,3,5,6,7,8,8a-octahydro-1,4-dimethyl-7-(1-methylethenyl), [1S-(1α,7α,8αβ)]- and 2-methyl-propanoic acid in solid and liquid potato dextrose media. The concept of Paleobiosphere has been put forward for the presumptive role of these endophytic fungi and other microorganisms along with plant materials in production of accumulation of these VOCs and fuel like compounds during the course of degradation in ancient time, i.e., Cretaceous era which led to possible origin of crude oil and hydrocarbons in the bentonite shale (Strobel et al., 2013). This has further propelled the exploration of endophytic fungi for their secondary metabolites with fuel like properties. *Nigrograna mackinnonii* E5202H, an endophyte of *Guazuma ulmifolia* tree collected from the Cerro Blanco Protected Forest near Guayaquil, Ecuador was found to produce volatile natural products, including terpenes and odd chain polyenes. The main polyene identified was (3E,5E,7E)-nona-1,3,5,7-tetraene (NTE), a novel natural product. Non-oxygenated hydrocarbons of this chain length are considered as desirable gasoline surrogate biofuels (Shaw et al., 2015).

From India, *Myrothecium innundatum* isolated from *Acalypha indica* as an endophyte produced 1,4-cyclohexadiene, 1-methyl- and cyclohexane (1-ethyl propyl) and octane when grown under microaerophilic conditions (Banerjee et al., 2010). *Muscodor* species have also been a prolific producer of volatile terpenes and terpenoids which could be used as biofuels (Table 17.3). Apart from terpenes and terpenoids *Muscodor* has been found to be producers of a variety of volatile compounds which constituted the gas, such as alkanes, alkenes, alkyl esters, benzene derivatives and polyaromatic hydrocarbons.

TABLE 17.3 Volatile Terpenes and Terpenoids from *Muscodor* Species with Biofuel Potential

SL. No.	*Muscodor* Species	Host	Location	Compounds	References
1.	*M. albus*	*C. zeylanicum*	Honduras	α-bulnesene; α-cedrene; caryophyllene; valencene	Strobel et al. (2001)
2.	*M. yucatanensis*	*Bursera simaruba*	Mexico	Caryophyllene; aromadendrene	Gonzalez et al. (2009)
3.	*M. fengyangensis*	*Actinidia chinensis, Pseudotaxus chienii*	China	β-phellandrene; caryophyllene	Zhang et al. (2010)
4.	*M. cinnamomi*	*C. bejolghota*	Thailand	β-humulene; Valencene; γ-selinene	Suwannarach et al. (2010)
5.	*M. sutura*	*Prestonia trifida*	Colombia	Thujopsene; β-chamigrene isocaryophyllene	Kudalkar et al. (2012)
6.	*M. suthepensis*	*C. bejolghota*	Thailand	α-bulnesene	Suwannarach et al. (2013)
7.	*M. oryzae*	*Oryza rufipogon*	Thailand	Caryophyllene; α-Bulnesene	Suwannarach et al. (2013)
8.	*M. musae*	*Musa acuminata*	Thailand	α-Bulnesene	Suwannarach et al. (2013)
9.	*M. kashayum*	*Aegle marmelos*	India	β-Bisabolol; Farnesene epoxide, E-; Terpinolene	Meshram et al. (2013)
10.	*M. strobelii*	*C. zeylanicum*	India	Terpinolene; Viridiflorol	Meshram et al. (2014)
11.	*M. darjeelingensis*	*C. camphora*	India	Seudenone	Saxena et al. (2014)
12.	*M. ghoomensis*	*C. camphora*	India	Pogostol	Meshram et al. (2015)
13.	*M. indica*	*C. camphora*	India	Pogostol	Meshram et al. (2015)
14.	*M. brasiliensis*	*Schinus terebinthifolius*	Brazil	B-elemene; α-bulnesene; β-selinene; Pogostol	Pena et al. (2019)

17.8 FUTURE PERSPECTIVES

It will not be out of scope to say that fungi and their interactions play a tremendous role in different ecosystems to produce a variety of bioactive compounds which apart from agrochemicals and pharmaceuticals could be exploited for use as biofuels directly or indirectly.

Both endophytic as well as non-endophytic fungi possess the potential to convert the lignocellulosic biomass into bioethanol, accumulate SCO which could be used for conversion into biodiesel-based components as well as produce VOCs which could be exploited as gasoline or jet fuel after development of proper mass production technologies.

Genomic studies of these fungi could provide an insight to the biosynthetic machinery of these compounds and further help in their optimization for development into commercial strains. Biosynthetic machinery of *Ascocoryne sarcoides* has now been completed using transcriptomic and metabolomic data, comprising over 80 biosynthetic gene clusters. These studies will also help in understanding the role of non-coding RNA in gene regulation for biofuel production.

Finally, the potential of taming any of the Biofuel/VOC producing fungi depends upon the type of fermentation system that is used as well as developing robust and steady mutants of the organism that are facilitated for the production of one or more products of importance.

Majority of the biofuel and VOC producing fungi reported till date have been excavated out of unique habitats across the globe. The geographic variability and climatic condition of India, added with immense biodiversity makes it a goldmine of unique fungal inhabitants which could be bioprospected for its biofuel potential.

KEYWORDS

- **fossil fuel**
- **polyketide synthase type 1**
- **polyunsaturated fatty acids**
- **research octane number**
- **single cell oil**
- **volatile organic compounds**

REFERENCES

Abba, C. C., & Eze, P. M., (2020). Fatty acid profile of an oleaginous endophytic *Pseudofusicoccum* sp. isolated from *Annona muricata. American Journal of Essential Oils and Natural Products, 8*(1), 01–05.

Adrio, J. L., (2017). Oleaginous yeasts, promising platforms for the production of oleochemicals and biofuels. *Biotechnol. Bioeng., 114*, 1915–1920.

Ageitos, J. M., Vallejo, J. A., Veiga-Crespo, P., & Villa, T. G., (2011). Oily yeasts as oleaginous cell factories. *Appl. Microbiol. Biotechnol. 90,* 1219–1227.

Ahamed, A., & Ahring, B. K., (2011). Production of hydrocarbon compounds by endophytic fungi *Gliocladium* species grown on cellulose. *Bioresource Technology, 102,* 9718–9722.

Alberts, A. W., (1988). Discovery, biochemistry and biology of lovastatin. *American J. Cardiology, 62*(15), J10–J15.

Amaretti, A., Raimondi, S., Sala, M., Roncaglia, L., Lucia, M. D., Leonardi, A., & Rossi, M., (2010). Single cell oils of the cold-adapted oleaginous yeast *Rhodotorula glacialis* DBVPG 4785. *Microb. Cell Fact., 9,* 73. doi:10.1186/1475-2859-9-73.

André, P., Diamantopoulou, A., Philippoussis, A., Sarris, D., Komaitis, M., & Papanikolaou, S., (2010). Biotechnological conversions of bio-diesel derived waste glycerol into added-value compounds by higher fungi, production of biomass, single cell oil and oxalic acid. *Ind. Crops Prod., 31,* 407–416.

Anke, T., Oberwinkler, F., Steglich, W., & Schramm, G., (1977). The strobilurins – New antifungal antibiotics from the basidiomycete *Strobilurus tenacellus* (Pers. ex. Fr.) sing. *Journal of Antibiotics, 30*, 806–810.

ASTM D7566-17, (2017). *Standard Specification for Aviation Turbine Fuel Containing Synthesized Hydrocarbons.* ASTM International, 100 Barr Harbor Drive, PO Box C700, West Conshohocken, Pennsylvania. http,//dx.doi.org/10.1520/D7566-17.

Bacon, C. W., & White, J. F., (2015). Functions, mechanisms, and regulation of endophytic and epiphytic microbial communities of plants. *Symbiosis, 67*, 1–12.

Banerjee, D., Strobel, G. A., Booth, E., Geary, B., Sears, J., Spakowicz, D., & Busse, S., (2010). An endophytic *Myrothecium inundatum* producing volatile organic compounds. *Mycosphere, 1*(3), 229–240.

Butcher, M. G., Meyer, P. A., Hallen, R. T., Albrecht, K. O., Clayton, C. K., Polikarpov, E., Rappe, K. G., et al., (2018). Fungal metabolites as precursors to renewable transportation fuels. *Fuels, 215*, 123–141.

Carvalho, A. K., Rivaldi, J. D., Barbosa, J. C., & deCastro, H. C., (2015). Biosynthesis, characterization and enzymatic transesterification of single cell oil from *Mucor circinelloides* – A sustainable pathway for biofuel production. *Bioresource Technology, 181*, 47–53.

Chatzifragkou, Makri, A., Belka, A., Bellou, S., Mavrou, M., Mastoridou, M., Mystrioti, P., Onjaro, G., Aggelis, G., & Papanikolaou, S., (2011). Biotechnological conversions of biodiesel derived waste glycerol by yeast and fungal species. *Energy, 36*, 1097–1108.

Cho, I., Namgung, H. J., Choi, H. K., & Kim, Y. S., (2008). Volatiles and key odorants in the pileus and stipe of pine-mushroom (*Tricholoma matsutake* Sing.). *Food Chem., 106*, 71–6.

Demirbas, A., (2007). Progress and recent trends in biofuels. *Progr. Energy Combust Sci., 33*, 1–18.

Dey, P., Banerjee, J., & Maiti, M., (2011). Comparative lipid profiling of two endophytic fungal isolates – *Colletotrichum* sp. and *Alternaria* sp. having potential utilities as biodiesel feedstock. *Bioresour. Technol., 102*, 5815–5823.

Dufosse, L., Fouillaud, M., & Caro, Y., (2021). Fungi and fungal metabolites for the improvement of human and animal nutrition and health. *J. Fungi, 7*(4), 274. https,//doi.org/10.3390/jof7040274.
Ebel, R., (2010). Terpenes from marine-derived fungi. *Mar Drugs, 8,* 2340–2368.
Evans, C. T., & Ratledge, C., (1983). A comparison of the oleaginous yeast, *Candida curvata*, grown on different carbon sources in continuous and batch culture. *Lipids, 18,* 623–629.
Fakas, S., Papanikolaou, S., Batsos, A., Galiotou-Panayotou, M., Mallouchos, A., & Aggelis, G., (2009). Evaluating renewable carbon sources as substrates for single cell oil production by *Cunninghamella echinulata* and *Mortierella isabelline*. *Biomass Bioenergy, 33,* 573–580.
Franklin, S., Piechocki, J., Norris, L. M., Decker, S. M., Wee, J., & Zdanis, D., (2011). *Oleaginous Yeast Food Compositions.* WO2011130576.
Gianoulis, T. A., Griffin, M. A., Spakowicz, D. J., Dunican, B. F., Alpha, C. J., Sboner, A., Sismour, A. M., et al., (2012). Genomic analysis of the hydrocarbon-producing, cellulolytic, endophytic fungus *Ascocoryne sarcoides*. *PLoS Genet, 8*(3), e1002558. https,//doi.org/10.1371/journal.pgen. 1002558.
Giebelhaus, A. W., (1980). Farming for fuel, the alcohol motor fuel movement in the 1930s. *Agric Hist, 54,* 173–184.
Gonzalez, M. C., Anaya, A. L., Glenn, A. E., Macias-Rubalcava, M., Hernandez-Bautista, B. E., & Hamlin, R. T., (2009). *Muscodor yucatanensis*, a new endophytic ascomycete from Mexican chakah, *Bursera simaruba*. *Mycotaxon, 110,* 363–372.
Hanson, J., (2008). *The Chemistry of Fungi* (pp. 47–50). Royal Society of Chemistry Publishing, Cambridge, England. ISBN, 978-0-854045-136-7.
Hanssen, H. P., & Abraham, W., (1988). Sesquiterpene alcohols with novel skeletons from the fungus *Ceratocystis piceae* (*Ascomycotina*). *Tetrahedron, 44,* 2175–2180.
Huang, C., Chen, X. F., Xiong, L., Chen, F. D., & Ma, L. L., (2012). Oil production by the yeast *Trichosporon dermatis* cultured in enzymatic hydrolysates of corncobs. *Bioresource Technology, 110,* 711–714.
Iassonova, D. R., Hammond, E. G., & Beattie, S. E., (2008). Oxidative stability of polyunsaturated triacylglycerols encapsulated in oleaginous yeast. *J. Am. Oil Chem. Soc, 85,* 711–716.
Ignea, C., Cvetkovic, I., Loupassaki, S., Kefalas, P., Johnson, C., Kampranis, S. C., & Makris, A. M., (2011). Improving yeast strains using recyclable integration cassettes for the production of plant terpenoids. *Microb. Cell Fact, 10*(4), 1–18. http,//dx.doi.org/10.1186/1475-2859-10-4.
Juanssilfero, A. B., Kahara, P., Amza, R. L., Miyamoto, N., Otsuka, H., Matsumoto, H., Kihira, C., et al., (2018). Selection of oleaginous yeasts capable of high lipid accumulation during challenges from inhibitory chemical compounds. *Biochemical Engg. Journal, 137,* 182–191.
Keller, N., Turner, G., & Bennett, J., (2005). Fungal secondary metabolism—From biochemistry to genomics. *Nat. Rev. Microbiol., 3,* 937–947. https,//doi.org/10.1038/nrmicro 1286.
Kitcha, S., & Cheirsilp, B., (2011). Screening of oleaginous yeasts and optimization for lipid production using crude glycerol as a carbon source. *Energy Procedia, 9,* 274–282.
Kovarik, B., (1998). Henry Ford, Charles Kettering, and the "fuel of the future." *Automot. Hist. Rev., 32,* 7–27.
Kramer, R., & Abraham, W., (2012). Volatile sesquiterpenes from fungi, what are they good for? *Phytochem. Rev., 11,* 15–37.
Kudalkar, P., Strobel, G., Riyaz-Ul-Hassan, S., Geary, B., & Sears, J., (2012). *Muscodor sutura*, a novel endophytic fungus with volatile antibiotic activities. *Mycoscience, 53,* 319–325.
Kulkarni, M. G., & Dalai, A. K., (2006). Waste cooking oil an economical source for biodiesel, a review. *Ind. Eng. Chem. Res., 45,* 2901–2913.

Kumar, S., & Kaushik, N., (2013). Endophytic fungi isolated from oilseed crop *Jatropha curcas* produces oil and exhibit antifungal activity. *PLoS One, 8*(2), e56202. doi, 10.1371/ journal.pone. 0056202.

Lapuerta, M., Armas, O., & Garcia-Contreras, R., (2007). Stability of diesel–bioethanol blends for use in diesel engines. *Fuel, 86,* 1351–1357.

Lee, S., Kim, H., & Cheon, S., (2015). Metabolic engineering for the production of hydrocarbon fuels. *Curr. Opin. Biotechnol., 33,* 15–22. http,//dx.doi.org/10.1016/ j.copbio. 2014.09.008.

Li, X., Xiong, L., Chen, X., Huang, C., Chen, X., & Ma, L., (2015). Effects of acetic acid on growth and lipid production by *Cryptococcus albidus. Journal of the American Oil Chemists Society, 92*(8), 1113–1118.

Ling, J., Nip, S., Cheok, W. L., De Toledo, R. A., & Shim, H., (2014). Lipid production by a mixed culture of oleaginous yeast and microalga from distillery and domestic mixed wastewater. *Bioresource Technology, 173*, 132–139.

Lund, M. T., Berntsen, T. K., & Fuglestvedt, J. S., (2014). Climate impacts of short-lived climate forcers versus CO_2 from biodiesel, a case of the EU on-road sector. *Environ. Sci. Technol., 48,* 14445–14454.

Luque, R., Herrero-Davila, L., Campelo, J. M., Clark, J. H., Hidalgo, J. M., Luna, D., Marinasa, J. M., & Romero, A. R., (2008). Biofuels, a technological perspective. *Energy Environ. Sci., 1,* 542–564.

Ma, F., & Hanna, M. A., (1999). Biodiesel production: A review. *Bioresource Technol., 70,* 1–15.

Ma, Y. L., (2006). Microbial oils and its research advance. *Chin. J. Bioprocess Eng., 4,* 7–11.

Ma, Y., Gao, Z., Wang, Q., & Liu, Y., (2018). Biodiesels from microbial oils, Opportunity and challenges. *Bioresource Technology, 263,* 631–641.

Malkhas'ian, A., Takashima, M., Sugita, T., Endoh, R., Kikukawa, M., Yamaguchi, S., Sakuradani, E., et al., (2014). *Cryptococcus terricola* is a promising oleaginous yeast for biodiesel production from starch through consolidated bioprocessing. *Scientific Reports, 4,* 4776. doi: 10.1038/srep04776.

Malkhas'ian, S. S., Nechaev, A. P., Gavrilova, N. N., Zotova, E. E., & Doronina, O. D., (1982). Group and fatty acid composition of the lipids in yeasts of the genus *Candida. Prikladnaia Biokhimiia i Mikrobiologiia, 18*(5), 621–629.

Mends, M. T., Yu, E., Strobel, G. A., Hassan, S. R. U., Booth, E., Geary, B., Sears, J., et al., (2012). An endophytic *Nodulisporium* sp. producing volatile organic compounds having bioactivity and fuel potential. *J. Petrol. Environ. Biotechnol.* doi,10.4172/2157-7463.1000117.

Meshram, V., Gupta, M., & Saxena, S., (2015). *Muscodor ghoomensis* and *Muscodor indica*, new endophytic species based on morphological features, molecular and volatile organic analysis from Northeast India. *Sydowia, 67*, 133–146.

Meshram, V., Kapoor, N., & Saxena, S., (2013). *Muscodor kashayum* sp. Nov. – a new volatile anti-microbial producing endophytic fungus. *Mycology, 4,* 196–204.

Meshram, V., Saxena, S., & Kapoor, N., (2014). *Muscodor strobelii*, a new endophytic species from South India. *Mycotaxon, 128,* 93–104.

Mitchell, A., Strobel, G., Moore, E., Robison, R., & Sears, J., (2010). Volatile antimicrobials from *Muscodor crispans*, a novel endophytic fungus. *Microbiology, 156,* 270–277.

Mitra, D., Rasmussen, M. L., Chand, P., Chintareddy, V. R., Yao, L., Grewell, V., Verkade, J. G., et al., (2012). Value-added oil and animal feed production from corn-ethanol stillage using the oleaginous fungus *Mucor circinelloides*. *Bioresource Technology, 107,* 368–375.

Morath, S., Hung, R., & Bennett, J., (2012). Fungal volatile organic compounds, a review with emphasis on their biotechnological potential. *Fungal Biotechnol. Re.v, 26,* 73–83.

Oguri, E., Masaki, K., Naganuma, T., & Iefuji, H., (2012). Phylogenetic and biochemical characterization of the oil-producing yeast *Lipomyces starkeyi*. *Antonie Van Leeuwenhoek, 101*, 359–368.

Patakova-Juzlova, P., Rezanka, T., & Viden, I., (1998). Identification of volatile metabolites from rice fermented by the fungus *Monascus purpureus* (Ang-kak). *Folia Microbiol., 43*(4), 407–410.

Pena, L. C., Jungklaus, G. H., Savi, D. C., Ferriera-Maba, L., Servienski, A., Maia, B. H. L. N. S., Annies, V., et al., (2019). *Muscodor brasiliensis* sp.nov. produced volatile organic compounds with activity against *Penicillium digitatum*. *Microbiological Research, 221*, 28–35.

Peralta-Yahya, P., Ouellet, M., Chan, R., Mukhopadhyay, A., Keasling, J., & Lee, T., (2011). Identification and microbial production of a terpene-based advanced biofuel. *Nat. Commun., 2,* 1–8. http,//dx.doi.org/10.1038/ncomms 1494.

Peralta-Yahya, P., Zhang, F., Cardayre, S., & Keasling, J., (2012). Microbial engineering for the production of advanced biofuels. *Nature, 488,* 320–328. http,//dx.doi.org/10.1038/nature11478.

Riyaz-Ul-Hassan, S., Strobel, G., Geary, B., & Sears, J., (2013). An endophytic *Nodulisporium* sp. from central America producing volatile organic compounds with both biological and fuel potential. *J. Microbiol. Biotechnol, 23*(1), 29–35. doi, 10.4014/jmb.1208.04062.

Rossi, M., Buzzini, P., Cordisco, L., Amaretti, A., Sala, M., Raimondi, S., Ponzoni, C., Pagnoni, U. M., & Matteuzzi, D., (2009). Growth, lipid accumulation, and fatty acid composition in obligate psychrophilic, facultative psychrophilic, and mesophilic yeasts. *FEMS Microbiol. Ecol., 69*, 363–372.

Sarin, R., Sharma, M., Sinharay, S., & Malhotra, R. K., (2007). Jatropha–Palm biodiesel blends, An optimum mix for Asia. *Fuel, 86*(10, 11), 1365–1371.

Saxena, S., & Pandey, A. K., (2001). Microbial metabolites as eco-friendly agrochemicals for the next millennium. *Applied Microbiology and Biotechnology, 55*(4), 395–403.

Saxena, S., & Strobel, G. A., (2020). Marvellous *Muscodor* spp., update on their biology and applications. *Microb. Ecol.* https,//doi.org/10.1007/s00248-020-01644-0.

Saxena, S., (2014). Microbial metabolites for the development of eco-friendly agrochemicals. *Allelopathy Journal, 33*(1), 1–24.

Saxena, S., Chhibber, M., & Singh, I. P., (2019). Fungal bioactive compounds in pharmaceutical research and development. *Current Bioactive Compounds, 15*(2), 211–231.

Shafiqe, S. A., & Ali, R. H., (2017). Myco-diesel production by oleaginous fungi. *Research Journal of Pharmaceutical, Biological and Chemical Sciences, 8*(2), 1252.

Sharma, Y., & Singh, B., (2009). Development of biodiesel, Current scenario. *Renewable and Sustainable Energy Reviews, 13*(6, 7), 1646–1651.

Shaw, J. J., Berbasova, T., Sasaki, T., Jefferson-George, K., Spakowicz, D., Dunican, B., Portero, C. E., Narvaez-Trujillo, A., & Strobel, S. A., (2015). Identification of a fungal 1,8-cineole synthase from *Hypoxylon* sp. with specificity determinants in common with the plant synthases. *J. Biol. Chem., 290,* 8511–8526. http,//dx.doi.org/10.1074/ jbc.M114.636159.

Singh, S., Strobel, G., Knighton, B., Geary, B., Sears, J., & Ezra, D., (2011). An endophytic *Phomopsis* sp. possessing bioactivity and fuel potential with its volatile organic compounds. *Microb. Ecol., 61*(4), 729–739. http,//dx.doi.org/ 10.1007/ s00248-011-9818-7.

Sinha, M., Sorensen, A., Ahamed, A., & Ahring, B. K., (2015). Production of hydrocarbon by *Aspergillus carbonarious* ITEM 5010. *Fungal Biology, 119,* 274–282.

Sitepu, I., Selby, T., Lin, T., Zhu, S., & Boundy-Mills, K., (2014). Carbon source utilization and inhibitor tolerance of 45 oleaginous yeast species. *Journal of Industrial Microbiology and Biotechnology, 41*(7), 1061–1070.

Stierle, A., Strobel, G., & Stierle, D., (1993). Taxol and taxane production by *Taxomyces andreanae*, an endophytic fungus of Pacific yew. *Science, 260*(5105), 214–216.

Strobel, G. A., (2014). Methods of discovery and techniques to study endophytic fungi producing fuel-related hydrocarbons. *Nat. Prod. Rep., 31,* 259–272. https,//doi.org/10.1039/C3NP70129H.

Strobel, G. A., Booth, E., Schaible, G., Mends, M. T., Sears, J., & Geary, B., (2013). The Paleobiosphere, a novel device for the *in vivo* testing of hydrocarbon producing-utilizing microorganisms. *Biotechnol. Lett., 35,* 539–552.

Strobel, G. A., Dirkse, E., Sears, J., & Markworth, C., (2001). Volatile antimicrobials from *Muscodor albus*, a novel endophytic fungus. *Microbiology, 147,* 2943–2950. doi, 10.1099/00221 287-147-11-2943. PMID, 11700345.

Strobel, G. A., Knighton, B., Kluck, K., Ren, Y., Livinghouse, T., Griffin, M., Spakowicz, D., & Sears, J., (2008). The production of myco-diesel hydrocarbons and their derivatives by the endophytic fungus *Gliocladium roseum* (NRRL 50072). *Microbiology, 154,* 3319–3328.

Strobel, G. A., Tomsheck, A., Geary, B., Spakowicz, D., Strobel, S., Mattner, S., & Mann, R., (2010). Endophytic strain NRRL 50072 producing volatile organics is a species of *Ascocoryne*. *Mycology, 1,* 187–194.

Strobel, G., (2015). Bioprospecting—Fuels from fungi. *Biotechnol. Lett., 37,* 973–982.

Strobel, G., Dirksie, E., Sears, J., & Markworth, C., (2001). Volatile antimicrobials from a novel endophytic fungus. *Microbiol., 147,* 2943–2950.

Subramaniam, R., Dufreche, S., Zappi, M., & Bajpai, R., (2010). Microbial lipids from renewable resources, production and characterization. *J. Ind. Microbiol. Biotechnol., 37,* 1271–1287.

Sunesson, A. L., Vaes, W., Nilsson, C. A., Blomquist, G., Andersson, B., & Carlson, R., (1995). Identification of volatile metabolites from five fungal species cultivated on two media. *Appl Environ. Microbiol., 61*(8), 2911–2918.

Suomalainen, H., & Lehtonen, M., (1979). The production of aroma compounds by yeast. *J. Inst. Brew. 85,* 149–156.

Suwannarach, N., Kumla, J., Bussaban, B., Hyde, K. D., Matsui, K., & Lumyong, S., (2013). Molecular and morphological evidence support four new species in the genus *Muscodor* from northern Thailand. *Ann. Microbiol., 63,* 1341–1351.

Tanimura, A., Takashima, M., Sugita, T., Endoh, R., Kikukawa, M., Yamaguchi, S., Sakuradani, E., Ogawa, J., & Shima, J., (2014). Selection of oleaginous yeasts with high lipid productivity for practical biodiesel production. *Bioresour. Technol., 153,* 230–235.

Tomsheck, A. R., Strobel, G. A., Booth, E., Geary, B., Spakowicz, D., Knighton, B., Floerchinger, C., et al., (2010). *Hypoxylon* sp., an endophyte of *Persea indica*, producing 1,8-Cineole and other bioactive volatiles with fuel potential. *Microbial. Ecology, 60*(4), 903–914.

UN-GAP, (2014). *The Emissions GAP Report (2014).* A United Nations Environment Program. Online.

Xie, J., Song, L., Li, X., Wu, Q., Wang, C., Xu, H., Cao, Y., & Qiao, D., (2013). Isolation and identification of oleaginous endophytic fungi. *African Journal of Microbiology Research, 7*(19), 2014–2019.

Xin, M., Jianming, Y., Xu, X., Zhang, L., Nie, Q., & Xian, M., (2009). Biodiesel production from oleaginous microorganisms. *Renewable Energy, 34*(1), 1–5.

Yaegashi, J., Oakley, B., & Wang, C., (2014). Recent advances in genome mining of secondary metabolite biosynthetic gene clusters and the development of heterologous expression systems in *Aspergillus nidulans*. *J. Ind. Microbiol. Biotechnol., 41*(2), 433–442.

Yang, Y., Heidari, F., & Hu, B., (2019). Fungi (mold)-based lipid production. In: Balan, V., (ed.), *Microbial Lipid Production. Methods in Molecular Biology* (Vol. 1995). Humana, New York, NY. https,//doi.org/10.1007/978-1-4939-9484-7_3.

Zhang, C. L., Wang, G. P., Mao, L. J., Komon-Zelazowska, M., Yuan, Z. L., Lin, F. C., Druzhinina, I. S., & Kubicek, C. P., (2010). *Muscodor fengyangensis* sp. Nov. from southeast China, morphology, physiology and production of volatile compounds. *Fungal Biol., 114*(10), 797–808.

Zhang, H., Liu, Q., Feng, X., Zheng, Y., Zou, H., Liu, H., Yang, J., & Xian, M., (2014). Microbial production of sabinene—A new terpene-based precursor of advanced biofuel. *Microb Cell Fact, 13*(20). https,//doi.org/10.1186/1475-2859-13-20.

Zhao, H., Lv, M., Liu, Z., Zhang, M., Wang, Y., Ju, X., Song, Z., et al., (2020). High-yield oleaginous fungi and high-value microbial lipid resources from *Mucoromycota*. *Bioenerg. Res*. https,//doi.org/10.1007/s12155-020-10219-3.

Zheng, Y., Yu, X., Zeng, J., & Chen, S., (2012). Feasibility of filamentous fungi for biofuel production using hydrolysate from dilute sulphuric acid pre-treatment of wheat straw. *Biotechnol. Biofuel., 5*, 50.

Zhu, M., Zhou, P. P., & Yu, L. J., (2002). Extraction of lipids from *Mortierella alpina* and enrichment of arachidonic acid from the fungal lipids. *Bioresource Technology, 84*(1), 93–95.

CHAPTER 18

Role of Fungi in Converting Agro-Residues into Cellulase and Bioethanol

GURLEEN KAUR SODHI and SANJAI SAXENA

Department of Biotechnology, Thapar Institute of Engineering and Technology, Patiala, Punjab, India

ABSTRACT

Globally ~998 million tons of agricultural waste, including rice husk, wheat straw (WS), corn cobs (CCs), sugarcane bagasse, etc., is accumulated as a result of various agricultural operations. The accelerated rate of production of greens, staples, etc., by agro-industries have led to the dumping of waste in huge quantities. Owing to the environmental risks associated with direct disposal and loss of economically important carbohydrates, a quest for sustainable waste management techniques has become a priority these days. Conventional method of waste management includes burning of field or leaving the agricultural waste in the field, but over the years these methods have contributed immensely to air and soil pollution. Bioconversion of agricultural waste is an environment friendly and promising technique of keen research interest which is being optimized in order to replace the conventional methods. Micro-organisms such as fungi play various ecological roles in our ecosystem. Exploration of the diverse ecological niche of fungi (both endophytic and epiphytic) and their use in sustainable waste management techniques is the most eco-friendly and cost-effective method. Fungi play a key role in the production of cellulase via solid-state fermentation (SSF) and submerged fermentation (SmF). Fungi from Dipodascaceae family have been reported for their use in cellulase production by SSF using multiple substrates such as rice straw, wheat bran (WB), etc. A vast array of *Aspergillus*

Applied Mycology for Agriculture and Foods: Industrial Applications. Sanjay K. Singh, Deepak Kumar, Md. Shamim, & Rohit Sharma (Eds.)

sp. has also been reported for the production of cellulase which is secreted in fermentation medium during SmF. They have also been employed in various other applications such as Mucoralean which is a typically fast-growing fungus has been widely used in sustainable production of bioethanol using the abundant lignocellulosic biomass. Other than bioethanol they have also been known to produce other commercially valuable by-products such as chitosan and polyunsaturated fatty acids (PUFAs). The search for sustainable agro-waste disposal methods has led biotechnologists to explore micro-organisms such as fungi and produce beneficial products. Thus, this chapter comprises of the recent advancements in the utilization of fungi for their role in agricultural waste disposal.

18.1 INTRODUCTION

The waste material generated as a result of agricultural activities is known as agro waste and contributes 9% to the global energy consumption. Developing countries have reported 35% of the total energy consumption from agro-wastes. More than 80% of the total agro-waste of any farm comprises of organic matter (Duque-Acevedo et al., 2020). Owing to their different composition many agro wastes such as rice husk, wheat straw (WS), corn cob, sugarcane bagasse, pine sawdust (SD), coconut shell, sunflower stalk, etc., holds immense potential for the production of value-added products. Specifically rice husk is one of the major contributors of agro waste globally. As a disposal method burning and gasification of agro-waste has been an eternal practice but the loss of economically important materials as well as its immense contribution to the environmental pollution is one of the greatest concerns (Iqbal et al., 2020). Although the lure for the high content of pozzolanic elements such as silica produced by burning rice husk is pleasing but the amount of pollution produced by this practice has been taking a great toll on our environment (Motlagh et al., 2020). With the extensive amount of agro waste produced each year and its amount increasing every year, extensive research is being carried out in order to find a sustainable substitute for the agro waste disposal/utilization.

Lignocellulosic biomass contains three major components namely, cellulose, hemi-cellulose, and lignin. Among this cellulose is a key raw material also known as 'biological currency,' used as a precursor for the production of industrially important products via micro-organism systems (Saldarriaga-Hernández et al., 2020). Although there are many sustainable and economically important uses of agro waste, but the conventional methods used for the production of bio-chemicals involve the use of hazardous chemicals for pre-treatment and costly biocatalysts (enzymes) for further conversion into industrially valuable products. Owing to these drawbacks, an emerging and eco-friendly system involving the use of micro-organisms such as fungi is critically necessary to deal with aforementioned issues (Tsegaye et al., 2019). Although the use of microbial system for hydrolysis and fermentation of lignocellulosic biomass is still at lab scale but due to its eco-friendly attribute it is garnering much attention. Enzymes are one of the prominent by-products which can be effectively produced by the utilization of the agro-waste. Enzyme production industry has witnessed exponential boom in the recent time, owing to their environment friendly nature and superior product specificity, however their production costs make them unavailable for a large-scale process in the industry. The employment of agro waste for the production of enzyme using fungi could pave the path for economically feasible alternative instead of current enzyme production approaches (Bezerra et al., 2021).

Researchers have reported production of many industrially important enzymes such as cellulase (which degrades cellulose) from agro waste. Many species of *Trichoderma* have been reported to produce cellulose. Their low β-glycosidase activity lead to a new quest and many other fungi such as *Chaetomium thermophile*, *Sporotrichum thermophile* and *Thermoascus aurantiacus* were reported having two to three times higher cellulolytic rates (Jampala et al., 2017; Hua et al., 2018; Bajaj et al., 2014; Schuerg et al., 2017). Many species of *Aspergillus,* such as *Aspergillus niger*, *Aspergillus terreus* are known to produce cellulase using corn cobs (CCs) and rice straw as carbon source, respectively (Dhillon et al., 2012; Sohail et al., 2016). The industrial demand of cellulase is met by processes such as SmF, but the high cost and low productivity has led to a shift towards solid state fermentation (SSF). The various advantages and disadvantages associated with using fungal species for different fermentation process have been discussed later in the chapter.

With the ever so depleting renewable energy supply of the world and soaring oil prices, utilization of individual valuable components of agro waste especially carbohydrates as a raw material for energy production is an emerging area. The use of inedible lignocellulosic biomass produced as the end product of agro waste for the production of bioethanol is one of the key applications (Atitallah et al., 2020). Bioethanol is a renewable energy source produced through fermentation as compared to customary fossil fuels. Owing to its high oxygen content there is complete combustion as well as low emission of harmful gases. In terms of toxicity, ethanol is less toxic to humans as compared to gasoline. Moreover, its low volatility and photochemical reactivity reduces the chance of smog formation (Sharma et al., 2020). Many lignin degrading fungi such as *Phanerochaete chrysosporium, Trametes versicolor*, etc., have been reported for effective delignification (Kuppuraj et al., 2021; Montoya et al., 2021). The mechanism by which the fungi are able to degrade lignin can be broadly classified into hydrolytic and oxidative types. In the hydrolytic mechanism, the hydrolytic enzymes produced by fungi break down the glycosidic bonds whereas in oxidative mechanism free radicals of reactive oxygen species are produced which break down lignin into low molecular weight products (Bugg et al., 2011). After delignification, fermentation of the hydrolysate product is the key step for its conversion into bioethanol. Many fungi, especially *Saccharomyces cerevisiae* have been widely exploited for this purpose. *Saccharomyces cerevisiae* has been reported to yield 90% ethanol however its inability to utilize pentose sugar limits its performance, for this pentose consuming fungi such as *Candida shehatae, Pachysolen tannophilus*, etc., were investigated, and found to improve the overall ethanol yield (Yesmin et al., 2020; Tanimura et al., 2012; El-Harchi et al., 2018). Moreover, different methods of hydrolysis and fermentation such as simultaneous saccharification and fermentation process improve the bio-ethanol yield by another 13% (Sethi et al., 2017). Due to the high market value of these products, researchers have been extensively trying to screen multiple fungi possessing promising properties which can be implied in the process. Other than the eco-friendly nature of using the biological system, the specificity, low energy demand and mild operating conditions makes it a convenient method. Thus, this chapter focuses on the direct conversion of lignocellulosic biomass into value added products, i.e., cellulase and bioethanol without any chemical or commercial enzymatic pre-treatment step.

18.2 GLOBAL AGRO-WASTE PRODUCTION

The expansion of soils for agriculture, advancement in technology, increasing population and green revolution are some of the key contributors which have led to an increase in agricultural production by more than three times in the last 50 years (FAO, 2017). This increased production henceforth contributes to the agro waste produced as a result of these agricultural activities. Owing to the different sources of agro waste, it consists of both usable and non-usable materials. It is estimated that ~998 million ton of agricultural waste/agro waste is produced each year out of which 80% can be converted to organic manure, industrially important value-added products, etc., in a sustainable manner (Duque-Acevedo et al., 2020). This makes the lignocellulosic biomass one of the most abundant raw materials (Figure 18.1).

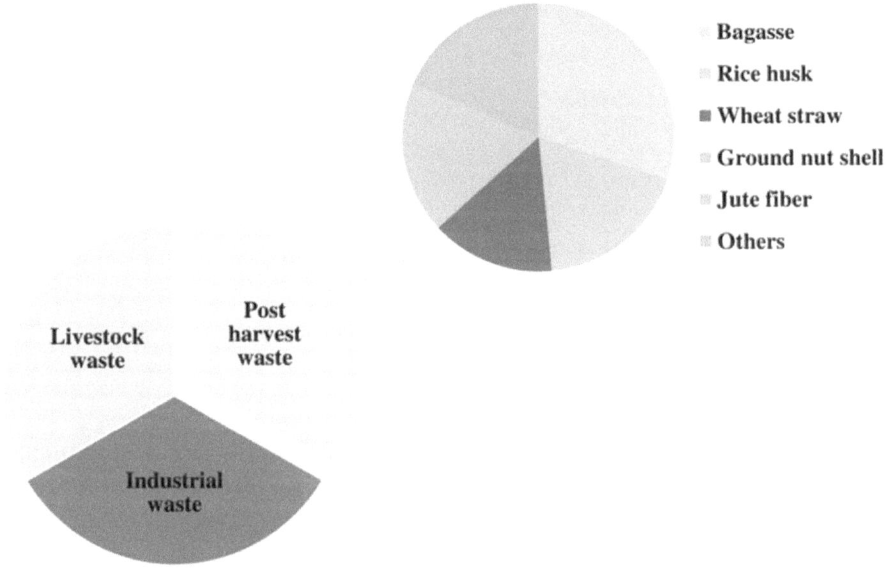

FIGURE 18.1 Types of agriculture waste.

Agro-waste can be broadly classified into three categories namely livestock manure, industrial processing waste and post-harvest agro waste (Iqbal et al., 2020). The post-harvest agro waste further consists of many chief components of industrial and economic importance. Out of the various types of post-harvest agro waste, bagasse, rice husk, WS, sweet sorghum, CCs, fruit waste, etc., are some of the major components. The effective and sustainable utilization rather than disposal of this agro waste cannot only decrease the waste generation but produce many economical, green, value-added products. The lignocellulosic biomass is chiefly composed of three main components, i.e., cellulose, hemicelluloses, and lignin. The cellulose and hemicellulose are composed of carbohydrate polymers whereas the lignin is composed of aromatic polymer. The lignin binds the different sugar monomers present in the carbohydrate polymers (Sindhu et al., 2016). The cellulose, hemicelluloses, lignin content of different agro-residues is listed in Table 18.1.

TABLE 18.1 Composition of Different Agro-Waste

Agro Residue	Cellulose	Hemicellulose	Lignin
Banana peel	20–24	16–19	7–10
Barley straw	28–34	27–30	18–21
Casava peel	38–40	25–28	8–10
Cocoa pod husk	26–30	12–14	21–24
Corn cobs	45–50	35–40	15–18
Cotton stalk	36–38	21–25	28–30
Ground nut husk	35–38	18–22	23–26
Nutshells	25–30	25–30	30–40
Rice straw	30–40	18–20	5–7
Soybean hulls	21–51	10–25	1–4
Sugarcane bagasse	32–40	19–24	25–32
Sweet sorghum	12–15	10–13	4–6
Wheat straw	33–40	20–25	15–20

In order to efficiently utilize the carbohydrate polymers, pre-treatment is done which removes the lignin. This delignification allows the cellulose and hemicellulose to be released. As discussed in the previous section, there are many lignin degrading fungi which can replace the chemical or enzymatic pre-treatment methods. This not only makes the process cost efficient but environment friendly as well. Further, the conversion of lignocellulosic biomass to industrially important products involves the hydrolysis of carbohydrate polymers into simpler sugars which are then fermented to produce ethanol. The use of microbial system, as well as the different techniques available for the fermentation processes is discussed in the succeeding sections.

18.3 ROLE OF FUNGI IN PRODUCTION OF VALUE-ADDED PRODUCTS FROM AGRO-WASTE

18.3.1 CELLULASE PRODUCTION

As previously mentioned, the agro-waste comprises three major components, i.e., cellulose, hemicelluloses, and lignin. The hydrolysis of cellulose into fermentable sugars is a must in order to completely utilize the agro-waste. Chemical processes exist for carrying to the hydrolysis, however owing to the environmentally hazardous nature, these are not preferred. Enzymatic hydrolysis of cellulose on the other hand using cellulase is an expensive process. Thus, to reduce the operational cost and process cost effective, use of microbial systems possessing capability to hydrolyze cellulose and also produce other hydrolytic enzymes is generally bioprospected as an eco-friendly alternative. Three enzymes namely endo-glucanase, cellobiohydrolase, and β-glucosidase (BGL) are required in a specific proportion for the catalytic conversion of cellulose (Andlar et al., 2018) (Figure 18.2).

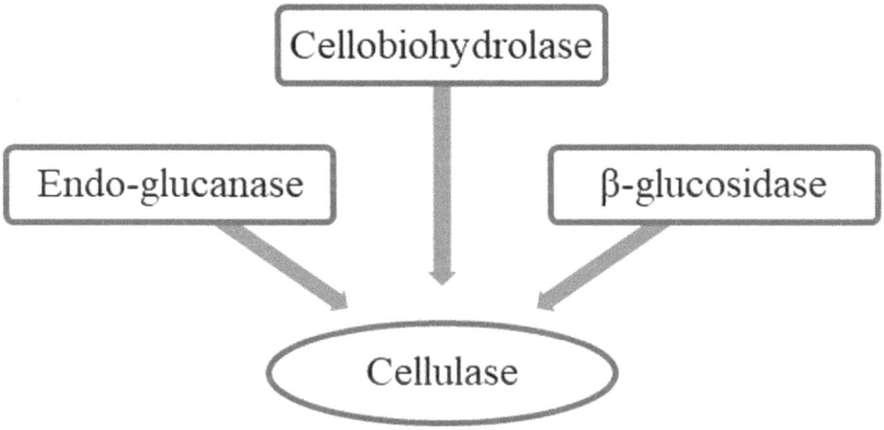

FIGURE 18.2 Cellulase enzyme components.

Of the many microbes available for these enzymatic activities, fungi exhibit the ability to produce copious amounts of enzymes. For instance, *Aspergillus niger* NRRL-567, a potent strain for citric acid production was evaluated for its ability to produce different cellulolytic enzymes such as BGL, carboxymethyl cellulase (CMCase) and filter paper cellulase (FPase). Apple pomace, a polyphenol rich residue left after milling and processing of apples mixed with 1% w/w rice husk was used as a substrate in SSF system which resembles the natural growing habitat of the fungi. Different inducers such as copper sulfate ($CuSO_4$), lactose, etc., were evaluated for their effect on enzyme production. Highest FPase activity of 133.68±5.44 IU/gds followed by 108.76 ± 5.13 IU/gds was observed on adding $CuSO_4$ and lactose, respectively. Similarly, highest BGL activity of 60.09 ± 3.43 IU/gds followed by 57.04 ± 3.40 IU/gds was observed on addition of veratryl alcohol (VA) and lactose, respectively. Activity of other enzymes such as CMCase (172.31 ± 6.8 IU/gds) and xylanase (1412.58 ± 27.9 IU/gds) was also seen on addition of lactose and VA, respectively (Dhillon et al., 2012). *Aspergillus fumigatus* SK1 was utilized to induce enzyme production by using untreated oil palm trunk as a substrate in SSF system. The optimized conditions of 80% moisture level, 5.5 pH produced the highest cellulolytic and xylanase activity of 54.27 (CMCase), 3.36 (FPase), 4.54 (BGL) and 418.70 U/g, respectively (Ang et al., 2013). The same researchers later screened the gene encoding for xylanase activity. Multiple sequence alignment confirmed that the gene belonged to xylanase glycoside hydrolase family 11 (GH11). The gene encoded for a protein of 24.49 kD. *A. fumigatus* was further tested for saccharification of lemon grass, as a result highest reducing sugar production of 6.84 g/L was achieved by day 6 (Ang et al., 2015) Since efficient conversion of lignocellulosic biomass requires both cellulase and xylanase, this major key step achieved by utilizing a microbial system implies that the vast unexplored mycoflora holds immense potential which can aid the process to produce commercial, value added products in an eco-friendly manner. Many other studies using different *Apergillus* sp. have also reported the production of lignocellulolytic enzymes from various substrates such as untreated SD, rice straw, barley straw, pomelo, banana peel, sugarcane bagasse, ragi husk, etc., are listed in Table 18.2.

Another extensively studied filamentous fungus known for secreting large amounts of cellulases, hemicellulases, chitinases, etc., is *Trichoderma sp.* Various hypercellulolytic strains of *Trichoderma* have been developed with the sole purpose of obtaining a carbon catabolite depressed strain which would allow it to produce cellulase even in the presence of glucose. The employment of advanced genetic engineering techniques has led to the production of strains known as 'hyperproducers,' which can produce up to 100 g/L cellulase. The utilization of such fungal strains in the conversion of agro waste holds promising results. In a study, *Trichoderma reesei* was utilized for cellulase production using straw substrate. On optimizing the fermentation temperature and time, maximum cellulase production (FPase) of 0.041-unit gm/L was reported (Rosyida et al., 2015).

Sweet pods of *Prosopis juliflora* which is an invasive weed in countries like Asia, Africa, and Australia, etc., were used as a potential substrate for cellulase production. *Trichoderma reesei* NCIM 1186 was used, and the fermentation parameters were optimized by one factor at a time (OFAT) approach. This approach is being used immensely in biological systems where the experimental design requires monothetic analysis. In single response optimization at the corresponding optimized process parameters cellulase production of 3055.65 U/L was observed. The sweet pods used as a substrate have high cellulosic content which further aids the growth

TABLE 18.2 Fungi Used for Cellulase Production from Different Lignocellulosic Biomass

Lignocellulosic Biomass	Fungi	Reaction Condition	Cellulase Production	References
Wheat bran	*Trichoderma reesei*	Solid-state fermentation	8.2 g^{-1} DM in 96 hrs.	Ortiz et al. (2015)
Grass, corncob, sugarcane-bagasse	*Aspergillus terreus MS105*	Submerged and solid-state fermentation	8.5, 2.275 IU/g grass 12.591, 29.0 IU/g corn cob 13.83, 12.96 IU/g sugarcane bagasse	Sohail et al. (2016)
Untreated sawdust	*Aspergillus* sp.	Solid-state fermentation	β-glucosidase (22.97 U/g)	Kamsani et al. (2016)
Rotten rice straw	*Aspergillus tubingenesis*	Submerged and solid-state fermentation	0.23 U/g	El-Nahrawy et al. (2017)
Parthenium hysterophorus	*Trichoderma reesei*	Solid-state fermentation	CMCase 20.49 U/gds and FPase 2.42 U/gds	Saini et al. (2017)
Barley straw	*Aspergillus fumigatus N2*	Submerged fermentation	CMCase 5.61 U/mL	Lin et al. (2017)
Rice straw	*Trichoderma reesei* and *Humicola insolens*	Subsequent saccharification	CMCase activity ~30 unit/mL	Kogo et al. (2017)
Raw rice straw, raw wheat straw	*Inonotus obliquus*	Solid-state fermentation	CMCase (19.83 ± 0.18 IU/g), FPase (2.62 ± 0.08 IU/g) and β-glucosidase (2.53 ± 0.01 IU/g)	Xu et al. (2018)
Rice straw (RS), rice husk (RH), wheat straw (WS), wheat bran (WB), and sugarcane bagasse (SB).	*Schizophyllum commune NAIMCC-F-03379*	Submerged fermentation	CMCase 7.6 IU/mL, 10.9 IU/mL, 7.6 IU/mL and 12.5 IU/mL, Fpase 8.2 IU/mL, 10.9 IU/mL, 10.3 IU/mL and 13.1 IU/mL	Kumar et al. (2018)
Soybean hull	*Penicillium* sp.	Solid-state fermentation	FPase 1.02 U/g CMCase 8.77 U/g	Salazar et al. (2019)
Food manufacturing wastes	*Trichoderma* sp.	–	–	Gordillo-Fuenzalida et al. (2019)

TABLE 18.2 *(Continued)*

Lignocellulosic Biomass	Fungi	Reaction Condition	Cellulase Production	References
Corn cob	*Fusarium oxysporum* and *Sporothrix carnis*	Saccharification and fermentation	586.93 U/mL from *F. oxysporum* 590.7 U/mL from *S. carnis*	Olajuyigbe et al. (2019)
Pomelo and banana peels	*Aspergillus niger* and *Saccharomyces cerevisiae*	Simultaneous saccharification and co-fermentation	2.79 g/L pomelo, 1.49 g/L banana peel	Chamchoi (2019)
Banana pseudostem and jatropha	*Pleurotus ostreatus PLO 9*	Solid-state fermentation	19.5 FPU g^{-1} of banana pseudostem	Da Silva et al. (2019)
Coffee pulp and green tea residues	*Aspergillus niger* and *Trichoderma asperellum*	Solid-state fermentation	2.5 U/g	Saldaña-Mendoza et al. (2020)
Sweet lemon peels	Co-culturing of *Bacillus megaterium* and *Saccharomyces cerevisiae*	Submerged fermentation	0.435± 0.02 IU/mL after 24 h	Irfan et al. (2020)
Rice straw, sugarcane bagasse, rice bran and sawdust	*Aspergillus flavus KUB2*	Submerged fermentation	1.27 U/mL after 14 days	Namnuch et al. (2020)
Raw oil palm frond leaves	*Rhizopus oryzae UC2*	Solid-state fermentation	94.68 U/g	Ezeilo et al. (2020)
Wheat bran	*Rhizopus oryzae SN5*	Solid-state fermentation	452.16 ± 2.38 U/gds	Pandey & Negi (2020)
Sugarcane bagasse	*Cladosporium cladosporioides NS2*	Solid-state fermentation	FPase (16.9 IU/gds), β-glucosidase (200 IU/gds) and EG (150 IU/gds) activity	Srivastava et al. (2020)
Ragi husk	*Aspergillus fumigates*	Solid-state fermentation	95.2 IU/mL	Saroj et al. (2021)
Peach-palm waste	*Trichoderma stromaticum AM7*	Solid-state fermentation	120 U/g cellulase activity of dry initial substrate (gds)	Bezerra et al. (2021)

and metabolism of *T. reesei* NCIM 1186 (Jampala et al., 2017). Similarly, soybean hulls are one of the major by-product/waste materials after the extraction of oils. A *Trichoderma* strain was engineered which could produce cellulase even in the presence of repressing sugars. Invertase gene isolated from *Aspergillus niger* was introduced into the *Trichoderma* strain to enable it to utilize sucrose from sugarcane molasses, a viscous by-product produced as a result of sugarcane refining. There was simultaneous saccharification of the biomass and fermentation process. At the optimized conditions of 83% soybean hulls and 17% sugarcane molasses as the carbon source, endoglucanase production of 14.5 U/mg was recorded. Other than production of cellulase, the quality of the enzyme was also improved by adding a heterologous BGL from *T. emersonii* (Ellila et al., 2017).

A novel cellulolytic strain *Trichoderma asperellum* UC1, identified using 18s rRNA and phylogenetic analysis was studied for its cellulase production capability. Raw oil palm frond leaves, a by-product of oil palm industry was used as a substrate. More than 50% of the biomass produced as a result is left to decompose thus making it readily available, cheap agro waste. Under optimum conditions of 30°C, 60–80% moisture content, and 2.5×10^6 spores/g inoculum size, *T. asperellum* UC1 produced 59.64 U/g (CMCase), 9.58 U/g (FPase), and 118.1 U/g (BGL), which were 5.0, 5.8, 6.0, and 18.8-fold higher, respectively, than the *T. asperellum* RCK2011 strain cultivated on wheat bran (WB) under similar SSF conditions (Ezeilo et al., 2019). The same researchers later discovered another novel cellulolytic fungi *Rhizopus oryzae* UC2 which produced the same enzymes at 41.62 U/g (CMCase), 7.65 U/g (FPase) and 113.07 U/g (BGL) under similar fermentation conditions (Ezeilo et al., 2020). This was the first study which made the use of raw oil palm frond leaves as a substrate for cellulase production. Further optimization of fermentation conditions and large-scale production of these organisms could help in yielding higher titers of the enzymes.

Among all the traits, thermostability is one of the most desired traits of any industrial enzyme. Thermostability not only allows a faster rate of reaction at elevated temperatures but also decreases the usage of enzyme. Using thermostable cellulase, the process of hydrolysis could be operated at elevated temperatures, thereby increasing the efficiency of biomass hydrolysis and making the technology economically feasible. Many thermophilic fungi such as *Myceliophthora thermophila* have been employed for the process. Its ability to grow at higher temperatures elevates the enzyme yield and decreases the chance of contaminations as well. In a study *Myceliophthora thermophila* was utilized for thermophilic cellulase production from

WB and sugarcane bagasse. They are the by-products of wheat milling and crushing of sugarcane, respectively. It is estimated that for every five tons of sugarcane crushed, almost a ton and a half-wet bagasse is produced. Using 9:1 ratio of WB and sugarcane bagasse for cultivation of *Myceliophthora heterothallica* under SSF conditions, endoglucanase production of 1,170.6 ± 0.8 U/g after 192 hours was observed. Whereas on using 1:1 ratio of WB and sugarcane bagasse under similar cultivation conditions, BGLs activity of 10.9 ± 0.3 U/g was observed after 240 h (Teixeira da Silva et al., 2016). Another thermophilic fungal strain *Talalromyces thermophilus* AX4, isolated from Tunisian soil sample was studied for the production of extracellular BGL. *T. thermophilus* AX4 under SmF condition with a mixture of 2% WB supplemented with 0.8% lactose yielded highest BGL activity of 8.5 ± 0.28 U/mL, proving that the substrate composition and *Talaromyces* strain holds promising production of the thermostable enzyme at cost effective rates. The β-glucosidae rich enzymatic juice of *T. thermophilus* was further utilized for saccharification of wastepaper, it enhanced the waste paper saccharification by 423% (Mallek-Fakhfakh et al., 2017). Saccharification is one of the key steps in the production of various industrially important products such as bioethanol. The employment of fungal strains which can produce cellulolytic enzymes from agro waste biomass, as well as ferment the simpler carbohydrates via simultaneous saccharification and fermentation can provide an economical solution for utilization of agro waste.

There are many studies where different agro waste have been compared for screening the biomass exhibiting highest enzyme production potential using various fungi. Bharti et al. (2018) screened six different substrates including corn cob, rice straw, rice husk, WB, WS, *Parthenium hysterophorous* and sugarcane tops. *Talaromyces stipitatus* MTCC 12687 was used under optimized cultural conditions of SSF. Out of different substrates tested, *P. hysterophorous*, also known as white top weed or famine weed exhibited promising results. Highest endoglucanase activity of 30 IU/gds, 26.4 IU/gds and 16.7 IU/gds was observed on using *P. hysterophorous*, WB, and rice straw, respectively. Maximum FPase activity of 2.86 FPU/gds and 2.22 FPU/gds was observed on using *P. hysterophorous* and WB, respectively. Similarly, *P. hysterophorous* also exhibited highest BGL activity of 33 IU/gds followed by 25.2 IU/gds on using WB. Initially, *T. stipitatus* MTCC 12687 produced 53.3 IU/gds (endoglucanase), 4.51 FPU/gds (FPase) and 62.6 IU/gds (β-glucodisase) utilizing *P. hysterophorous* supplemented with peptone, after 5th day at temperature of 30 C, pH 6.0 and moisture content of 80%. Endoglucanase, FPase, and BGL enzyme activities were increased by 2.0 and 2.7-fold, respectively, after the fermentation conditions were optimised. Although many fungal species such as *Trichoderma reesei* have shown promising results in the production of cellulolytic enzymes from agro waste biomass but its inability to efficiently produce BGL limits its performance. Thus, the use of consortium provides a promising solution to the problem by pooling their intrinsic properties for a common goal.

The use of fungal consortia, i.e., the use of two or more fungal species has been employed for various lab scale and industrial processes. It provides an easy solution by combining the capabilities of two or more different fungal species and thereby increasing the efficiency of the process. A specialized consortium of *Cyberlindnera jardinii* and *Barnettozyma californica*, both lignocellulolytic yeast species along with two bacterial species, was used to access the production of enzymes. Coffee husk, a fibrous product obtained as a result of processing of coffee beans was used as a substrate in a pilot scale SSF system. Under fermentation condition of 3.1 ± 0.5 FPU g^{-1} DM FPase activity was obtained. However, on upscaling the process to a 50 L bioreactor, high temperatures caused approximately 50% reduction in production of cellulase (Cerda et al., 2017). Similarly, various other studies using fungi for cellulase production have been listed in Table 18.1. It can be inferred from the literature that cellulases enzyme production from lignocellulosic biomass could be an excellent opportunity to utilize the agro-waste as well as reduce the enzyme production cost. The hydrolysate products formed from the lignocellulosic biomass can further be converted into various industrially important products such as bioethanol. The following section discusses the direct production of bioethanol from agro-waste using different fungal species.

18.3.2 BIO-ETHANOL PRODUCTION

Ethanol-based fuel industry began in the 1970's as an alternative to the high-priced petroleum-based fuel. The low carbon monoxide emissions of ethanol and its use as oxygenate encouraged its production through the 1990s. The abundance and ease of conversion to ethanol has made edible feedstock the predominant precursor for bio-ethanol production. Easily cultured fungal species have been commonly used for conversion of food crops into bioethanol and other fermentation products (Qiao & Lu, 2021). In a study, different food crops such as corn, carrot, and pumpkin were evaluated for their ability of bio-ethanol production using *Saccharomyces cerevisiae*. Although high ethanol concentration limits the growth of *Saccharomyces cerevisiae* but its ability to grow anaerobically makes it ideal for SmF process. Approximately 63 mL of ethanol with 13.33% v/v purity was produced using 100 g corn as a substrate. Similarly, between red and black pumpkin variety, red pumpkin exhibited a high yield of 53 mL ethanol of 6% v/v purity. Altogether highest ethanol production of 73.67 mL was observed by using carrot as a substrate with the purity of 12.66% v/v (Yesmin et al., 2020). Though food crops are an attractive source for bio-ethanol production, but the amount of arable land required for the production of adequate food crops to feed the population as well as fuel production has always been a concern. The quest for a sustainable source for ethanol production unearthed the potential of agro waste which was previously disregarded. Over the years, many researchers have employed different methods to produce bio ethanol from agro waste, but the use of harsh chemicals and uneconomical enzymes involved in the process has always been a topic of scrutiny (Figure 18.3).

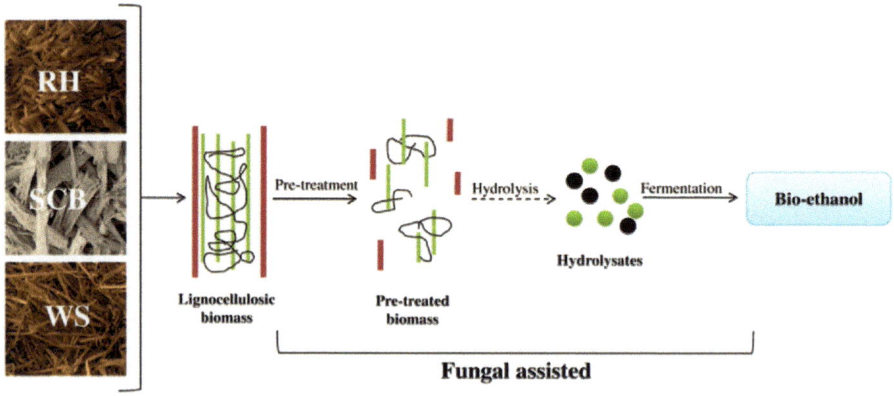

FIGURE 18.3 Bio-ethanol production from agro-waste.

The friend or foe discussion about micro-organisms has always led researchers towards groundbreaking discoveries. The use of micro-organisms has not only provided sustainable solutions to many problems but is also environment friendly and economical. The ability of many microbes to ferment lignocellulosic substrate into valuable products has led to the exploration of various fungi for their ability to produce bioethanol in various fermentation systems. For example, *Candida tropicalis* CICC1779 which is known for fermentation of various carbohydrates such as galactose, sucrose, maltose, trehalose, etc., was screened by UV mutagenesis in order to find a strain that could efficiently utilize xylose. The mutant strain *Candida tropicalis* CICC1779-Dyd showed significant improvement in the production of key enzymes xylose reductase (XR) and xylose dehydrogenase (XD) which led to effective utilization of xylose as compared to the wild type of strain. Thus, the mutant strain was subjected to fermentation by using cassava residue as the substrate for bio ethanol production. It was seen that utilization of xylose from the lignocellulosic biomass increases the bio ethanol production by 25%. Different initial liquid phase oxygen transfer coefficients (k_La) were studied, and the highest ethanol yield of 24.85 g/L was achieved after 72 hours with k_La of 120. Further optimization of aeration rate, agitation speed and dissolved oxygen led to increase in ethanol yield by 1.068-fold taking the final yield to 26.56 g/L (Li et al., 2017).

The ease of maintaining the fungi in the exponential phase for longer periods of time by the addition of fresh media and high turnover rate are some of the key points that call for the use of continuous fermentation system. For example, two *Trichoderma sp.* namely, *Trichoderma reesei* Aq-5 and *Trichoderma viride* NSW-XM were co cultured at different concentrations and investigated for the performance of their lignocellulosic enzymes. They are filamentous fungi known for secreting large amounts of cellulases, chitinases, etc. Air dried rice straw powder containing 35% cellulose, 21% lignin and 28% hemicellulose was used as a substrate. Under optimized conditions the two fungi produced 22.74 g/L reducing sugar. For the fermentation process mixed culture system by co immobilizing *Saccharomyces cerevisiae* and *Candida tropicalis* NSW-NW on sodium alginate, polyvinyl alcohol, and silicon dioxide beads was used. It stabilized the yeast and increased the ethanol production efficiency as a result of which 2.1 g/(L h) ethanol was produced (Wu et al., 2016). Large scale production using continuous fermentation system is a labor extensive process and is more prone to contamination thus the use of batch fermentation is preferred over the continuous system. Salih & Paranjothi (2019) explored the use of rinds syrup of *Cucumis melo* for production of bioethanol. The *Cucumis melo* rind waste produced as a result of nutraceutical manufacturing makes it an economically favorable substrate for the process. *Aspergillus niger 2081* was used in the batch fermentation process at different pH values ranging from 4–7. It produces α-galactosidase (GH27) enzyme which can break down complex sugars by hydrolyzing their terminal α-galactosyl moieties. Highest ethanol production of 384.91 g/L was seen at pH 6 after 120 hours closely followed by 369.43 g/L at pH 5. Even though batch fermentation is more economical and less labor extensive than continuous fermentation, but the meticulous quality control limits its productivity.

Maintaining aerobic conditions in liquid culture demands a constant supply of oxygen, which further needs to be distributed homogenously throughout the fermentation media. This is generally achieved by stirring the fermentation media but applying this technique to filamentous fungi leads to formation of hyphae which then increase the viscosity in the growing medium. The stirring also leads to disruption of the cell network of fungi leading to high cell mortality. Thus, the use of techniques such as solid-state fermentation provides a ground for the fungi to grow on while using the vegetal by-products as a nutrient source. Other than providing a natural growing environment to the fungi in *in-vitro* conditions SSF is more cost effective, has lower water and energy consumption. For instance, on site multi-enzyme production was carried out using *Fusarium oxysporum* F3. Food waste including WS, WB, and CCs were used as a substrate in SSF conditions. The vast array of enzymes produced during the SSF included 17.8 U/g amylase, 0.1 U/g glucoamylase, 65.2 U/g endoglucanase, 27.4 U/g BGL, 3.5 U/g cellobiohydrolase, 221.5 U/g xylanase, 0.7 U/g b-xylosidase and cellulase 1.5 U/g WB. Ethanol production using monoculture of *F. oxysporum* resulted in 16.3 g/L yield after 94 hours. On using mixed culture of *S. cerevisiae* and *F. oxysporum* ethanol yield of 18.3 g/L was achieved after 18 hours. The highest ethanol yield of 29.9 and 30.8 g/L was recorded on supplementing the mixed culture with 20 U/g and 40 U/g glucoamylase, respectively. The use of non-commercial enzymes not only increases the yield of ethanol but also provides an economical alternative to the commercial preparations (Prasoulas et al., 2020). In a similar study different types of fermentation such as liquid state surface fermentation (LSSF), liquid shaking fermentation (LShF) and solid-state fermentation (SSF) was evaluated for highest ethanol production. Spores of *Aspergillus terreus* NCFT 4269.10 were immobilized on sodium alginate beads and inoculated in sterile fermentation broth. The ability of *A. terreus* to produce cellulolytic enzymes enables it to be used in saccharification of the different agro waste. Different substrates namely, banana peels (BPs), SD, and WB were used for the process. Highest cellulase production of 344.88±3.35 U/mL was reported using BP in LSSF. In one flask *S. cerevisiae* and *A. terreus* were co cultivated using BP as substrate whereas in the other partially purified cellulase and *S. cerevisiae* was used. Out of the two, highest ethanol production of 19.8 g/L with 1.9% w/w purity was seen in the latter (Sethi et al., 2017).

The various types of fermentation techniques discussed for the production of bioethanol from agro waste involve three basic unit operations. First is the production of saccharolytic enzymes which is followed by hydrolysis of the complex sugars into simpler forms and lastly the fermentation of monosaccharide to yield bioethanol. Newer techniques such as consolidated bioprocessing (CBP) combine these steps into a single operation. The low-cost biomass processing makes CBP an attractive next generation biofuel production technique. Although fewer fungi possess all the desired features required for CBP but naturally occurring cellulolytic fungi (or consortium of fungi) can be selected and their affinity to ferment sugars to ethanol can be improved by recombinant strategies. Wood-decaying fungi of phlebioid polyporales were used in a study to convert non-pre-treated lignocelluloses into ethanol. Core board which is a waste wood fiber-based material was used as a substrate. Out of the five fungi evaluated, only four (Pace 0004, Prad 0043, Prad 0125, and Poch 0360) were able to produce ethanol from core board. On optimizing the pH conditions, highest ethanol production of 5.9 g/L at pH 3 was exhibited by isolate Prad 0043 identified as *Phlebia radiata*. Other than ethanol, production of fumarate, oxalates, etc., were also reported (Mattila et al., 2017). The same researchers also evaluated the efficiency of *Polyporales phlebioid* white rot fungus *Phlebia radiata* using *Fomitopsis pinicola* as collaborative fungi. Though higher ethanol yield was hypothesized using collaborative fungi including *S. cerevisiae* but no evident increase in ethanol yield was observed, whereas a slight increase in decomposition rate of the lignocellulosic substrate was observed. On single cultivation of *P. radiata* 32.4 ± 4.5 g/L, 7.0 ± 0.01 g/L of ethanol yield was achieved using core board and barley straw as substrate, respectively. Such comparative experiments allow process flexibility and further research can provide insight on increasing the productivity of ethanol fermentation (Mattila et al., 2018). In a similar study, *Fusarium oxysporum* strain 11C (IMI501118) and 7E (IMI501116) were used as microbial agents for their capability to degrade the lignocellulosic biomass into value added chemicals. Further, the effect of peptide transporter (FoPTR) on the microbe's ability to produce ethanol was studied. WB and bran mix in the ratio 10:1 was used as a substrate. Ethanol yield of ~80 mg/g of substrate was observed in the wild type of strain of *F. oxysporum*, decrease in yield was seen on silencing of the peptide transporter (FoPTR). On overexpression of the peptide transporter 17% increase in ethanol yield was observed. It was also observed that the expression of FoPTR is governed by the amount of nitrogen present in the culture medium. This positive contribution towards the ethanol yield implies that the FoPTR can be used as a target gene for improving ethanol yield (Nugent et al., 2019).

Following the zero-waste policy focused on waste prevention Carrillo-Nieves et al. (2020) designed a study for ethanol production using white-rot fungus *Trametes hirsuta* CS5. Different raw material such as Brewer's spent grain (BSG), orange peel (OP) and WB were selected. *T. hirsute* was also investigated for its ability to produce an array of lignocellulolytic enzymes such as cellulase and xylanase. On using OP as a substrate 500 U/L of cellulase, 45 U/mL of xylanase activity and 1.0 g/L ethanol production was recorded in 11 days. BSG as a substrate yielded 0.3 g/L ethanol in four days. Highest ethanol production of 1.4 g/L in seven days was observed on using WB as a substrate. The diauxic behavior was observed during the fermentation process. On comparing the different raw material with glucose and xylose as a carbon source, high ethanol production was seen using OP and WB, implying that *T. hirsute* has the ability to utilize other hexose and pentose sugars present in the lignocellulosic biomass. Further optimization of fermentation conditions using *T. hirsute* can help increasing ethanol production, thereby partially or entirely avoiding the pre-treatment steps which are directly associated with polluting water streams and hence the environment. Similarly, various researchers have employed the use of various fungi for bio-ethanol production using lignocellulosic biomass as listed in Table 18.3. It is evident from these studies discussed that fungi hold great potential in converting the agro waste to value added products. Thus, the screening and utilization of such fungi will not only economical but also be environment friendly. Optimization of fermentation conditions as well as identification and manipulations of certain target genes can further enhance the production of bioethanol.

18.4 USE OF ENDOPHYTIC FUNGI FOR CELLULASE AND BIOETHANOL PRODUCTION FROM AGRO-WASTE

As Louis Pasteur said, "Gentlemen, it is the microbes who will have the last word." The world has seen the versatility of microbe's from flavors of cheese to the discovery of Penicillin. Thus, the unexplored, diverse ecological niche of endophytic fungi has always been of keen interest. In 1809, German botanist – Heinrich Friedrich link first described endophytes. Researchers have indicated that plants cannot function as souvenir individuals and have one or more type of endophytes present in them. They are ubiquitous in nature and reside within the body of the host plant without causing any apparent harm to the host plant (Strobel & Daisy, 2003). Their co-existence and symbiotic relationship with the plants since the ancient times has led them to show a variety of biological activities (Strobel, 2003; Saikkonen et al., 2004; Saxena et al., 2019).

TABLE 18.3 Fungi Used for Bioethanol Production from Different Lignocellulosic Biomass

Lignocellulosic Biomass	Fungi	Reaction Condition	Ethanol Production	References
Hardwood kraft pulp	*Phlebia sp. MG-60*	Consolidated bioprocessing	25.9 g/L	Kamei et al. (2014)
Rye straw	*Trametes hirsuta* MT-24.24 and *Trametes versicolor* IT-1	–	33 g/L after 9 days	Kozhevnikova et al. (2016)
Sugarcane baggage, cotton stalks, banana stems, wheat straws, pigeon pea stalk	*Saccharomyces cerevisiae*	Aerobic + anaerobic fermentation	–	Kulkarni et al. (2017)
Brewers spent grains	*A. oryzae and S. cerevisiae ncyc479*	Consolidated bioprocessing	37 g/L of within 10 days	Wilkinson et al. (2017)
Rotten potato, sawdust, compost, and date palm juice	*Candida wangnamkhiaoensis, Hyphopichia pseudoburtonii (2 isolates), Wickerhamia sp.*	Consolidated bioprocessing	33.92 g/L at 7 days	Hossain et al. (2018)
Watermelon waste	*Saccharomyces cerevisiae*	Anaerobic fermentation	35.5 g/kg watermelon waste	Jahanbakhshi & Saleh (2019)
Corn dry-grind thin stillage	*Aspergillus oryzae and Rhizopus oligosporus, Candida utilis*	Co-culture	4.5 g/L	Pietrzak et al. (2019)
Japanese cedar (*Cryptomeria japonica*) wood	*Schizophyllum commune, Bjerkandera adusta, and Fomitopsis palustris*	Consolidated bioprocessing	–	Horisawa et al. (2019)
Raw flour of *Brosimum alicastrum* Seeds	*Trametes hirsuta Bm-2*	Consolidated bioprocessing	13 g/L after 288 hrs.	Olguin et al. (2019)
Potato peels waste	*Wickerhamomyces anomalus*	Saccharification and fermentation	0.49 g/g consumed sugars	Atitallah et al. (2019)
Fruit waste	*Saccharomyces cerevisiae*	Batch fermentation	20.1 g/L in 48 h	Bisht & Tewari (2019)
Rice bran, rice husk	*Aspergillus niger*	–	12.16% (rice bran) and 11.61% (rice husk)	Sinha et al. (2019)

TABLE 18.3 *(Continued)*

Lignocellulosic Biomass	Fungi	Reaction Condition	Ethanol Production	References
Raw corn starch	*Saccharomyces cerevisiae*	Consolidated bioprocessing	98.13 g/L	Cripwell et al. (2019)
Pea hull	*Saccharomyces cerevisiae*	–	1.65% at 71 hs	Saxena et al. (2019)
Sweet sorghum residues	*Trichoderma citrinoviride C1*	Solid state fermentation	120 mL kg d.m.$^{-1}$	Kancelista et al. (2020)
	Saccharomyces cerevisiae			
Paper wastes	*Aspergillus niger MK543209*	Submerged fermentation	0.9 g/L/h	Darwesh et al. (2020)
Olive mill waste	*Fusarium oxysporum*	Consolidated bioprocessing	2.47 g/L	M'Barek et al. (2020)
Date palm sap	*Wickerhamomyces anomalus*	Fed-batch mode	73 g/L	Atitallah et al. (2020)

In recent years the interest in endophytic fungi for production of enzymes has increased. Though there are only a handful of studies which have reported the use of endophytic fungi for enzyme production. For instance, a novel endophytic fungi *Pestalotiopsis microspora* TKBRR was isolated from Thalakona forest was evaluated for cellulase production. It was cultivated under SSF conditions with WB as the substrate. Under normal conditions it exhibited activity of 13, 12, and 11 U/g DS for filter paperase, CMCase, and BGL, respectively. After optimizing the conditions using OFAT approach activity of 15, 16, and 13 U/g DS for FPase, CMCase, and BGL were reported. The isolated exhibited high ratio of BGL to FPase as compared to the other cellulolytic micro-organisms (Goukanapalle et al., 2020).

Similarly, diverse endophytic fungi were isolated from mangrove tropical forest and subjected to screening for production of carbohydrolases. Avicel plate assay was done to screen a total of 41 endophytic fungi. The cultures showing distinguished growth were selected for further study. For enzyme production, six strains were selected after a plate assay: *Aspergillus niger* 56 (3), *Hypocrealixii* 1.14, *Penicillium* sp. 60 (4), *Diaporthe* sp. 37 (4), *Xylaria enteroleuca* 39.3 (1), and *Fusarium* sp. 21.5. WB was used as a substrate under SSF system. *Aspergillus niger* 56 (3) exhibited the highest endoglucanase and BGL activity of 57 U/g and 47 U/g, respectively. This was followed by *Hypocrealixii* 1.14 which exhibited endoglucanase activity of ~25 U/g and BGL activity of ~15 U/g. The *Penicillium sp.* 60 (4) exhibited high BGL activity of 65 U/g as compared to the *Hypocrealixii* 1.14, but very low endoglucanase activity was recorded. For comparison of the activities shown by the endophytic fungi two previously known reference strains *Aspergillus niger* F12 and *Trichoderma reesei* Rut-C30 were also used. The endoglucanase produced by the endophytic strain was higher than the *T. reesei* Rut-C30 reference strain but comparable to the *A. niger* F12 reference strain. Whereas in BGL production, *Penicillium* sp. 60 (4) showed superior results than the reference strain *A. niger* F12. The enzymes produced by the endophytic fungi were further tested for saccharification of cellulose. Overall high production of reducing sugar, glucose, and high conversion rate was recorded by the endophytic strains as compared to the reference strains (Maroldi et al., 2018). This implies that the endophytic fungi hold great potential in the production of industrially important enzymes. Further optimization of SSF parameters could also increase the yield of these enzymes. In another study, 14 endophytic fungi were screened for production of cellulase. The cultures were cultivated in SSF system with a mixture of sugarcane bagasse and WB in the ratio 1:1 w/w. Out of the total four endophytic fungi exhibiting high endoglucanase activity of 88.51 ± 1.0 U/g (*Cladosporium cladosporioides* PAJ 03) 83.44 ± 7.7 U/g (*Phomopsis stipata* SC 04), 64.56 ± 4.0 U/g (*Trichoderma viridae* PAJ 01) and 42.79 ± 1.6 U/g (*Botryosphaeria* sp. AM 01) were selected. Whereas four endophytic fungi exhibiting high BGL activity of 51.56 ± 2.7 U/g (*Saccharicola* sp. EJC 04), 33.19 ± 9.2 U/g (*Paecilomyces* sp. SF 021), 29.75 ± 0.8 U/g (*Ustilaginoidea* sp. CV 04) and 21.72 ± 3.05 U/g (*Ustilaginoidea sp.* XYA04) were selected. The selected endophytic fungi were cultivated on cotton seed meal and WB (1:1 w/w). Under optimum conditions *Botryosphaeria* sp. AM01 and *Saccharicola* sp. EJC04 exhibited highest endoglucanase and BGL activity of 184.74 ± 6.0 and 92.28 ± 9.57 U/g, respectively. They were further used in testing the saccharification of pre-treated sugarcane bagasse. The promising results indicate their potential for enzyme production as well as using the hydrolysates for bio-ethanol production (Marques et al., 2018).

Several studies have reported the production of mycofuel in the form of volatile organic compounds (VOCs) from endophytic fungi. The studies implore the exploration of endophytic fungi which not only produce lignocellulytic enzymes but also covert the degrading lignocelluloses into bioethanol. Using CBP four endophytic fungi (*Hypoxylon* sp. CI4A, *Hypoxylon* sp. EC38, *Hypoxylon* sp. CO27, and *Daldinia eschscholzii* EC12) were evaluated for bio-ethanol production. Avicel, corn stover, switchgrass, and eucalyptus mill powder were used as a substrate. *Daldinia eschscholzii* EC12 exhibited highest endoglucanase activity of 450 U/g mycelia as well as highest BGL activity of 753 U/g mycelia on cultivation with Avicel. The endophytes also produced other carbohydrate active enzymes such as cellobiohydrolase which was also confirmed by genomic analysis. Further the VOCs were collected using a SPME (solid phase microextraction) needle and analyzed by GC-MS. Highest cyclic or phenolic alcohol of 30 and 20% were exhibited by *Hypoxylon* sp. EC38 and *Hypoxylon* sp. CO27, respectively when cultivated with switch grass (Wu et al., 2017). The ability to degrade the lignocellulosic biomass as well as convert it into bio-ethanol makes these endophytes a promising new niche to be explored. Exploration of more genera as well as screening them for bio-ethanol production can provide economical solutions to aforementioned problems.

18.5 CONCLUSION AND FUTURE ASPECT

The rate at which fossil fuels are depleting, it will not be long before the world runs out of oil, gas, and coal. The alarming situation implores the need to find a sustainable and renewable energy source. The large-scale availability and low cost of lignocellulosic biomass makes it an attractive raw material for various industrial applications. The key step involved in the conversion of this lignocellulosic biomass to ethanol is the hydrolysis of biomass into simpler sugars. Many chemical and enzymatic methods are being utilized currently for the same, but their hazardous nature and cost intensiveness calls for an alternative approach. If the key converter of this biomass, i.e., cellulase is produced using the microbial system such as fungi, it can significantly reduce the cost of the process. As discussed in the chapter, there are many known fungi such as *Trametes hirsute,* which not only produce the lignocellulolytic enzymes but simultaneously ferment the hydrolysates into bioethanol. With advancement in technology, newer techniques such as CBP combine the multistep process of ethanol production into one single low-cost conversion method. The employment of various statistical optimizations tools such as OFAT, RSM, etc., makes it convenient for researchers to work on the optimum conditions which can further yield the highest results on large scale. The advent of technology has also led to the discovery of many new genera of fungi. Exploration of endophytic fungi which can help in bioethanol production could be a futuristic approach, considering the promising potential demonstrated by them in various studies.

KEYWORDS

- **agro-waste disposal**
- **bioethanol**
- **carboxymethyl cellulase**
- **cellulase**
- **fungi**
- **solid state fermentation**
- **submerged fermentation**
- **veratryl alcohol**

REFERENCES

Andlar, M., Rezić, T., Marđetko, N., Kracher, D., Ludwig, R., & Šantek, B., (2018). Lignocellulose degradation: An overview of fungi and fungal enzymes involved in lignocellulose degradation. *Engineering in Life Sciences, 18*(11), 768–778.

Ang, S. K., Shaza, E. M., Adibah, Y., Suraini, A. A., & Madihah, M. S., (2013). Production of cellulases and xylanase by *Aspergillus fumigatus* SK1 using untreated oil palm trunk through solid state fermentation. *Process Biochemistry, 48*(9), 1293–1302.

Ang, S. K., Yahya, A., Abd Aziz, S., & Md Salleh, M., (2015). Isolation, screening, and identification of potential cellulolytic and xylanolytic producers for biodegradation of untreated oil palm trunk and its application in saccharification of lemongrass leaves. *Preparative Biochemistry and Biotechnology, 45*(3), 279–305.

Atitallah, I. B., Antonopoulou, G., Ntaikou, I., Alexandropoulou, M., Nasri, M., Mechichi, T., & Lyberatos, G., (2019). On the evaluation of different saccharification schemes for enhanced bioethanol production from potato peels waste via a newly isolated yeast strain of *Wickerhamomyces anomalus*. *Bioresource Technology, 289*, 121614.

Atitallah, I. B., Ntaikou, I., Antonopoulou, G., Alexandropoulou, M., Brysch-Herzberg, M., Nasri, M., & Mechichi, T., (2020). Evaluation of the non-conventional yeast strain *Wickerhamomyces anomalus* (*Pichia anomala*) X19 for enhanced bioethanol production using date palm sap as renewable feedstock. *Renewable Energy, 154*, 71–81.

Bajaj, B. K., Sharma, M., & Rao, R. S., (2014). Agricultural residues for production of cellulase from *Sporotrichum thermophile* LAR5 and its application for saccharification of rice straw. *J Mater. Environ. Sci., 5*(5), 1454–1460.

Bezerra, C. O., Carneiro, L. L., Carvalho, E. A., Das Chagas, T. P., De Carvalho, L. R., Uetanabaro, A. P. T., & Da Costa, A. M., (2021). Artificial intelligence as a combinatorial optimization strategy for cellulase production by *Trichoderma stromaticum* AM7 using peach-palm waste under solid-state fermentation. *BioEnergy Research,* 1–10.

Bharti, A. K., Kumar, A., Kumar, A., & Dutt, D., (2018). Exploitation of *Parthenium hysterophorous* biomass as low-cost substrate for cellulase and xylanase production under solid-state fermentation using *Talaromyces stipitatus* MTCC 12687. *Journal of Radiation Research and Applied Sciences, 11*(4), 271–280.

Bisht, N., & Tewari, L., (2019). Identification and characterization of high efficacy thermotolerant yeast for bioethanol production from fruit waste. *Biotech Today: An International Journal of Biological Sciences, 9*(1), 81–86.

Bugg, T. D., Ahmad, M., Hardiman, E. M., & Rahmanpour, R., (2011). Pathways for degradation of lignin in bacteria and fungi. *Natural Product Reports, 28*(12), 1883–1896.

Carrillo-Nieves, D., Saldarriaga-Hernandez, S., Gutiérrez-Soto, G., Rostro-Alanis, M., Hernández-Luna, C., Alvarez, A. J., & Parra-Saldívar, R., (2020). Biotransformation of agro-industrial waste to produce lignocellulolytic enzymes and bioethanol with a zero waste. *Biomass Conversion and Biorefinery*, 1–12.

Cerda, A., Mejías, L., Gea, T., & Sánchez, A., (2017). Cellulase and xylanase production at pilot scale by solid-state fermentation from coffee husk using specialized consortia: The consistency of the process and the microbial communities involved. *Bioresource Technology, 243,* 1059–1068.

Chamchoi, N., (2019). Utilization of fruit waste for bioethanol production by co-cultures of *Aspergillus Niger* and *Saccharomyces cerevisiae*. *Applied Environmental Research, 41*(2), 63–72.

Cheng, Y., Addy, M., & Ruan, R., (2016). Microbial hydrolysis and fermentation of rice straw for ethanol production. *Fuel, 180,* 679–686.

Cripwell, R. A., Rose, S. H., Favaro, L., & Van, Z. W. H., (2019). Construction of industrial *Saccharomyces cerevisiae* strains for the efficient consolidated bioprocessing of raw starch. *Biotechnology for Biofuels, 12*(1), 1–16.

Da Silva, I. F., Da Luz, J. M. R., Oliveira, S. F., De Queiroz, J. H., & Kasuya, M. C. M., (2019). High yield cellulase and LiP production after SSF of agricultural wastes by *Pleurotus ostreatus* using different surfactants. *Biocatalysis and Agricultural Biotechnology, 22*, 101428.

Darwesh, O. M., El-Maraghy, S. H., Abdel-Rahman, H. M., & Zaghloul, R. A., (2020). Improvement of paper wastes conversion to bioethanol using novel cellulose degrading fungal isolate. *Fuel, 262*, 116518.

Dhillon, G. S., Kaur, S., Brar, S. K., & Verma, M., (2012). Potential of apple pomace as a solid substrate for fungal cellulase and hemicellulase bioproduction through solid-state fermentation. *Industrial Crops and Products, 38*, 6–13.

Duque-Acevedo, M., Belmonte-Urena, L. J., Cortés-García, F. J., & Camacho-Ferre, F., (2020). Agricultural waste: Review of the evolution, approaches and perspectives on alternative uses. *Global Ecology and Conservation, 22*, e00902.

El Harchi, M., Kachkach, F. F., & El Mtili, N., (2018). Optimization of thermal acid hydrolysis for bioethanol production from Ulva rigida with yeast *Pachysolen tannophilus*. *South African Journal of Botany, 115*, 161–169.

Ellilä, S., Fonseca, L., Uchima, C., Cota, J., Goldman, G. H., Saloheimo, M., & Siika-Aho, M., (2017). Development of a low-cost cellulase production process using *Trichoderma reesei* for Brazilian biorefineries. *Biotechnology for Biofuels, 10*(1), 1–17.

El-Nahrawy, S., Metwally, M., El-Kodoos, A., Rizk, Y., Belal, E. S. B., Shabana, S. A., & El-Refai, I. M., (2017). Optimization of culture conditions for production of cellulase by *Aspergillus tubingensis* KY615746 using rice straw waste. *Environment, Biodiversity and Soil Security, 1*(2017), 177–189.

Ezeilo, U. R., Lee, C. T., Huyop, F., Zakaria, I. I., & Wahab, R. A., (2019). Raw oil palm frond leaves as cost-effective substrate for cellulase and xylanase productions by *Trichoderma asperellum* UC1 under solid-state fermentation. *Journal of Environmental Management, 243*, 206–217.

Ezeilo, U. R., Wahab, R. A., & Mahat, N. A., (2020). Optimization studies on cellulase and xylanase production by *Rhizopus oryzae* UC2 using raw oil palm frond leaves as substrate under solid state fermentation. *Renewable Energy, 156*, 1301–1312.

Gordillo-Fuenzalida, F., Echeverria-Vega, A., Cuadros-Orellana, S., Faundez, C., Kähne, T., & Morales-Vera, R., (2019). Cellulases production by a *Trichoderma sp.* using food manufacturing wastes. *Applied Sciences, 9*(20), 4419.

Goukanapalle, P. K. R., Kanderi, D. K., Rajoji, G., Shanthi, K. B. S., & Bontha, R. R., (2020). Optimization of cellulase production by a novel endophytic fungus *Pestalotiopsis microspora* TKBRR isolated from Thalakona forest. *Cellulose, 27*, 6299–6316.

Gur, C., Mandelboim, O., & Bachrach, G., (2015). "Messieurs, c'est les microbes qui auront le dernier mot": Gentlemen, it is the microbes who have the last word (Louis Pasteur)—*Fusobacterium nucleatum* protect tumors from killing by immune cells. *Oncoimmunology, 4*(9), e1038690.

Horisawa, S., Inoue, A., & Yamanaka, Y., (2019). Direct ethanol production from lignocellulosic materials by mixed culture of wood rot fungi *Schizophyllum commune*, *Bjerkandera adusta*, and *Fomitopsis palustris*. *Fermentation, 5*(1), 21.

Hossain, T., Miah, A. B., & Mahmud, S. A., (2018). Enhanced bioethanol production from potato peel waste via consolidated bioprocessing with statistically optimized medium. *Applied Biochemistry and Biotechnology, 186*(2), 425–442.

Hua, C., Li, W., Han, W., Wang, Q., Bi, P., Han, C., & Zhu, L., (2018). Characterization of a novel thermostable GH7 endoglucanase from *Chaetomium thermophilum* capable of xylan hydrolysis. *International Journal of Biological Macromolecules, 117*, 342–349.

Iqbal, N., Agrawal, A., Dubey, S., & Kumar, J., (2020). Role of decomposers in agricultural waste management. In: *Biomass*. Intech Open.

Irfan, M., Bakhtawar, J., Sadia, S., Shakir, H. A., Khan, M., & Ali, S., (2020). Utilization of fruit wastes for enzyme production in submerged fermentation. *International Journal of Biology and Chemistry, 13*(2), 88–95.

Jahanbakhshi, A., & Salehi, R., (2019). Processing watermelon waste using *Saccharomyces cerevisiae* yeast and the fermentation method for bioethanol production. *Journal of Food Process Engineering, 42*(7), e13283.

Jampala, P., Tadikamalla, S., Preethi, M., Ramanujam, S., & Uppuluri, K. B., (2017). Concurrent production of cellulase and xylanase from *Trichoderma reesei* NCIM 1186: Enhancement of production by desirability-based multi-objective method. *3 Biotech, 7*(1), 14.

Kamei, I., Hirota, Y., & Meguro, S., (2014). Direct fungal production of ethanol from high-solids pulps by the ethanol-fermenting white-rot fungus *Phlebia sp.* MG-60. *BioResources, 9*(3), 5114–5124.

Kamsani, N., Salleh, M. M., Yahya, A., & Chong, C. S., (2016). Production of lignocellulolytic enzymes by microorganisms isolated from *Bulbitermes sp.* termite gut in solid-state fermentation. *Waste and Biomass Valorization, 7*(2), 357–371.

Kancelista, A., Chmielewska, J., Korzeniowski, P., & Łaba, W., (2020). Bioconversion of sweet sorghum residues by *Trichoderma citrinoviride* C1 enzymes cocktail for effective bioethanol production. *Catalysts, 10*(11), 1292.

Kogo, T., Yoshida, Y., Koganei, K., Matsumoto, H., Watanabe, T., Ogihara, J., & Kasumi, T., (2017). Production of rice straw hydrolysis enzymes by the fungi *Trichoderma reesei* and *Humicola insolens* using rice straw as a carbon source. *Bioresource Technology, 233,* 67–73.

Kozhevnikova, E. Y., Petrova, D. A., Kopitsyn, D. S., Novikov, A. A., Shnyreva, A. V., Barkov, A. V., & Vinokurov, V. A., (2016). New strains of basidiomycetes that produce bioethanol from lignocellulose biomass. *Applied Biochemistry and Microbiology, 52*(6), 638–642.

Kulkarni, N., Begadwad, A., Pachori, R., & Sadar, P., (2017). Production of ethanol from agro waste by dual fermentation system. *Studies, 3*(8).

Kumar, B., Bhardwaj, N., Alam, A., Agrawal, K., Prasad, H., & Verma, P., (2018). Production, purification and characterization of an acid/alkali and thermo tolerant cellulase from *Schizophyllum commune* NAIMCC-F-03379 and its application in hydrolysis of lignocellulosic wastes. *AMB Express, 8*(1), 1–16.

Kuppuraj, S. P., Venkidasamy, B., Selvaraj, D., & Ramalingam, S., (2021). Comprehensive in silico and gene expression profiles of MnP family genes in *Phanerochaete chrysosporium* towards lignin biodegradation. *International Biodeterioration & Biodegradation, 157,* 105143.

Li, X., Deng, Y., Yang, Y., Wei, Z., Cheng, J., Cao, L., & Wu, X., (2017). Fermentation process and metabolic flux of ethanol production from the detoxified hydrolyzate of cassava residue. *Frontiers in Microbiology, 8,* 1603.

Lin, C., Shen, Z., & Qin, W., (2017). Characterization of xylanase and cellulase produced by a newly isolated *Aspergillus fumigatus* N2 and its efficient saccharification of barley straw. *Applied Biochemistry and Biotechnology, 182*(2), 559–569.

M'barek, H. N., Arif, S., Taidi, B., & Hajjaj, H., (2020). Consolidated bioethanol production from olive mill waste: Wood-decay fungi from central Morocco as promising decomposition and fermentation biocatalysts. *Biotechnology Reports, 28,* e00541.

Mallek-Fakhfakh, H., Fakhfakh, J., Masmoudi, N., Rezgui, F., Gargouri, A., & Belghith, H., (2017). Agricultural wastes as substrates for β-glucosidase production by *Talaromyces*

thermophilus: Role of these enzymes in enhancing wastepaper saccharification. *Preparative Biochemistry and Biotechnology, 47*(4), 414–423.
Maroldi, M. M. C., Vasconcellos, V. M., Lacava, P. T., & Farinas, C. S., (2018). Potential of mangrove-associated endophytic fungi for production of carbohydrolases with high saccharification efficiency. *Applied Biochemistry and Biotechnology, 184*(3), 806–820.
Marques, N. P., De Cassia, P. J., Gomes, E., Da Silva, R., Araújo, A. R., Ferreira, H., & Bocchini, D. A., (2018). Cellulases and xylanases production by endophytic fungi by solid state fermentation using lignocellulosic substrates and enzymatic saccharification of pretreated sugarcane bagasse. *Industrial Crops and Products, 122,* 66–75.
Montoya, S., Patiño, A., & Sánchez, Ó. J., (2021). Production of lignocellulolytic enzymes and biomass of *Trametes versicolor* from agro-industrial residues in a novel fixed-bed bioreactor with natural convection and forced aeration at pilot scale. *Processes, 9*(2), 397.
Motlagh, E. K., Asasian-Kolur, N., & Sharifian, S. (2022). A comparative study on rice husk and rice straw as bioresources for production of carbonaceous adsorbent and silica. *Biomass Conversion and Biorefinery, 12,* 5729–5738. https://doi.org/10.1007/s13399-020-01145-7.
Namnuch, N., Thammasittirong, A., & Thammasittirong, S. N. R. (2021). Lignocellulose hydrolytic enzymes production by *Aspergillus flavus* KUB2 using submerged fermentation of sugarcane bagasse waste. *Mycology, 12*(2), 119–127. https://www.tandfonline.com/doi/full/10.1080/21501203.2020.1806938 (accessed on 22 February 2023).
Nugent, B., Ali, S. S., Mullins, E., & Doohan, F. M., (2019). A major facilitator superfamily peptide transporter from *Fusarium oxysporum* influences bioethanol production from lignocellulosic material. *Frontiers in Microbiology, 10,* 295.
Olajuyigbe, F. M., Fatokun, C. O., & Oni, O. I., (2019). Effective substrate loading for saccharification of corn cob and concurrent production of lignocellulolytic enzymes by *Fusarium oxysporum* and *Sporothrix carnis*. *Current Biotechnology, 8*(2), 109–115.
Olguin-Maciel, E., Larqué-Saavedra, A., Lappe-Oliveras, P. E., Barahona-Pérez, L. F., Alzate-Gaviria, L., Chablé-Villacis, R., & Tapia-Tussell, R., (2019). Consolidated bioprocess for bioethanol production from raw flour of *Brosimum alicastrum* seeds using the native strain of *Trametes hirsuta* bm-2. *Microorganisms, 7*(11), 483.
Ortiz, G. E., Guitart, M. E., Cavalitto, S. F., Albertó, E. O., Fernández-Lahore, M., & Blasco, M., (2015). Characterization, optimization, and scale-up of cellulases production by *Trichoderma reesei* cbs 836.91 in solid-state fermentation using agro-industrial products. *Bioprocess and Biosystems Engineering, 38*(11), 2117–2128.
Pandey, A. K., & Negi, S., (2020). Enhanced cellulase recovery in SSF from *Rhizopus oryzae* SN5 and immobilization for multi-batch saccharification of carboxymethylcellulose. *Biocatalysis and Agricultural Biotechnology, 26,* 101656.
Pietrzak, W., & Kawa-Rygielska, J., (2019). Backset valorization in dry-grind ethanol process by co-culture of edible filamentous fungi and fodder yeast. *Journal of Cleaner Production, 220,* 376–385.
Prasoulas, G., Gentikis, A., Konti, A., Kalantzi, S., Kekos, D., & Mamma, D., (2020). Bioethanol production from food waste applying the multienzyme system produced on-site by *Fusarium oxysporum* F3 and mixed microbial cultures. *Fermentation, 6*(2), 39.
Qiao, Z., & Lü, X., (2021). Industrial bioethanol production: Status and bottlenecks. In: *Advances in 2nd Generation of Bioethanol Production* (pp. 213–227).
Saikkonen, K., Wa¨li, P., Helander, M., & Faeth, S. H., (2004). Evolution of endophyte–plant symbioses. *Trends in Plant Science, 9*(6), 275–280.

Saini, A., Aggarwal, N. K., & Yadav, A., (2017). Cost-effective cellulase production using *Parthenium hysterophorus* biomass as an unconventional lignocellulosic substrate. *3 Biotech, 7*(1), 12.

Salazar, L. N., Dal Maso, S. S., Ogimbosvski, T. A., Daronch, N. A., Zeni, J., Valduga, E., & Cansian, R. L., (2019). Production, partial characterization and application of cellulases by newly isolated *Penicillium sp.* using agro-industrial substrate solid-state fermentation. *Industrial Biotechnology, 15*(2), 79–88.

Saldaña-Mendoza, S. A., Ascacio-Valdés, J. A., Palacios-Ponce, A. S., Contreras-Esquivel, J. C., Rodríguez-Herrera, R., Ruiz, H. A., & Aguilar, C. N. (2021). Use of wastes from the tea and coffee industries for the production of cellulases using fungi isolated from the Western Ghats of India. *Systems Microbiology and Biomanufacturing*. 1, 33–41. https://doi.org/10.1007/s43393-020-00001-z (accessed on 22 February 2023).

Saldarriaga-Hernández, S., Velasco-Ayala, C., Flores, P. L. I., De Rostro-Alanis, M. J., Parra-Saldivar, R., Iqbal, H. M., & Carrillo-Nieves, D., (2020). Biotransformation of lignocellulosic biomass into industrially relevant products with the aid of fungi-derived lignocellulolytic enzymes. *International Journal of Biological Macromolecules.*

Salih, N. K., & Paranjothi, N., (2019). Bioethanol production from *Aspergillus Niger* 2081 by submerged batch fermentation using honeydew melon (*Cucumis melo*) rinds as a substrate. *International Journal of Advanced Research in Engineering Innovation, 1*(1), 1–11.

Saroj, P., Manasa, P., & Narasimhulu, K., (2021). Assessment and evaluation of cellulase production using ragi (*Eleusine coracana*) husk as a substrate from thermo-acidophilic *Aspergillus fumigatus* JCM 10253. *Bioprocess and Biosystems Engineering, 44*(1), 113–126.

Saxena, D., Singh, A., & Sirohi, R., (2019). Bioethanol production from pea hull aqueous extract as a novel substrate using *Saccharomyces cerevisiae*. *Indian Journal of Biotechnology*, 253–259.

Saxena, S., Chhibber, M., & Singh, I. P., (2019). Fungal bioactive compounds in pharmaceutical research and development. *Current Bioactive Compounds, 15*(2), 211–231.

Schuerg, T., Prahl, J. P., Gabriel, R., Harth, S., Tachea, F., Chen, C. S., & Singer, S. W., (2017). Xylose induces cellulase production in *Thermoascus aurantiacus*. *Biotechnology for Biofuels, 10*(1), 1–11.

Sethi, B. K., Das, A. S., Satpathy, A., & Behera, B. C., (2017). Ethanol production by a cellulolytic fungus *Aspergillus terreus* NCFT 4269.10 using agro-waste as a substrate. *Biofuels, 8*(2), 207–213.

Sharma, B., Larroche, C., & Dussap, C. G., (2020). Comprehensive assessment of 2G bioethanol production. *Bioresource Technology,* 123630.

Sindhu, R., Binod, P., & Pandey, A., (2016). Biological pretreatment of lignocellulosic biomass–An overview. *Bioresource Technology, 199,* 76–82.

Sinha, G., Tiwari, S., & Jadhav, S. K. (2019). Simultaneous saccharification and fermentation of rice residues and its comparative analysis for bioethanol production. *Defence Life Science Journal, 4*(3), 158–162. https://doi.org/10.14429/dlsj.4.14188 (accessed on 22 February 2023).

Sohail, M., Ahmad, A., & Khan, S. A., (2016). Production of cellulase from *Aspergillus terreus* MS105 on crude and commercially purified substrates. *3 Biotech, 6*(1), 103.

Srivastava, N., Elgorban, A. M., Mishra, P. K., Marraiki, N., Alharbi, A. M., Ahmad, I., & Gupta, V. K., (2020). Enhance production of fungal cellulase cocktail using cellulosic waste. *Environmental Technology & Innovation, 19,* 100949.

Strobel, G. A., (2003). Endophytes as sources of bioactive products. *Microbes and Infection, 5*(6), 535–544.

Strobel, G., & Daisy, B., (2003). Bioprospecting for microbial endophytes and their natural products. *Microbiology and Molecular Biology Reviews, 67*(4), 491–502.

Tanimura, A., Nakamura, T., Watanabe, I., Ogawa, J., & Shima, J., (2012). Isolation of a novel strain of *Candida shehatae* for ethanol production at elevated temperature. *SpringerPlus, 1*(1), 1–7.

Teixeira Da, S. V. D. C., De Souza, C. A. L., De Carvalho, S. R., Bertoldi, S. N. M., Gomes, E., & Bonilla-Rodriguez, G. O., (2016). Effect of pH, temperature, and chemicals on the endoglucanases and β-glucosidases from the thermophilic fungus *Myceliophthora heterothallica* F. 2.1. 4. Obtained by solid-state and submerged cultivation. *Biochemistry Research International*, 2016.

Tsegaye, B., Balomajumder, C., & Roy, P., (2019). Microbial delignification and hydrolysis of lignocellulosic biomass to enhance biofuel production: An overview and future prospect. *Bulletin of the National Research Centre, 43*(1), 1–16.

Wilkinson, S., Smart, K. A., James, S., & Cook, D. J., (2017). Bioethanol production from brewers spent grains using a fungal consolidated bioprocessing (CBP) approach. *Bioenergy Research, 10*(1), 146–157.

Wu, W., Davis, R. W., Tran-Gyamfi, M. B., Kuo, A., LaButti, K., Mihaltcheva, S., & Gladden, J. M., (2017). Characterization of four endophytic fungi as potential consolidated bioprocessing hosts for conversion of lignocellulose into advanced biofuels. *Applied Microbiology and Biotechnology, 101*(6), 2603–2618.

Xu, X., Lin, M., Zang, Q., & Shi, S., (2018). Solid state bioconversion of lignocellulosic residues by *Inonotus obliquus* for production of cellulolytic enzymes and saccharification. *Bioresource Technology, 247*, 88–95.

Yesmin, M. N., Azad, M. A. K., Kamruzzaman, M., & Uddin, M. N., (2020). Bioethanol production from corn, pumpkin and carrot of Bangladesh as renewable source using yeast *Saccharomyces cerevisiae*. *Acta Chemica Malaysia, 4*(2), 45–54.

Index

D

E

N

O

P

T

Y

Z